Probability and Mathematical Statistics (Continued)

LARSON • Introduction to Probability Theory and Statistical Inference, *Third Edition*
MATTHES, KERSTAN, and MECKE • Infinitely Divisible Point Processes
MUIRHEAD • Aspects of Multivariate Statistical Theory
OLIVER and SMITH • Influence Diagrams, Belief Nets and Decision Analysis
PILZ • Bayesian Estimation and Experimental Design in Linear Regression Models
PRESS • Bayesian Statistics: Principles, Models, and Applications
PURI and SEN • Nonparametric Methods in General Linear Models
PURI and SEN • Nonparametric Methods in Multivariate Analysis
PURI, VILAPLANA, and WERTZ • New Perspectives in Theoretical and Applied Statistics
RANDLES and WOLFE • Introduction to the Theory of Nonparametric Statistics
RAO • Asymptotic Theory of Statistical Inference
RAO • Linear Statistical Inference and Its Applications, *Second Edition*
ROBERTSON, WRIGHT and DYKSTRA • Order Restricted Statistical Inference
ROGERS and WILLIAMS • Diffusions, Markov Processes, and Martingales, Volume II: Îto Calculus
ROHATGI • Statistical Inference
ROSS • Stochastic Processes
RUBINSTEIN • Simulation and The Monte Carlo Method
RUZSA and SZEKELY • Algebraic Probability Theory
SCHEFFE • The Analysis of Variance
SEBER • Linear Regression Analysis
SEBER • Multivariate Observations
SEBER and WILD • Nonlinear Regression
SEN • Sequential Nonparametrics: Invariance Principles and Statistical Inference
SERFLING • Approximation Theorems of Mathematical Statistics
SHORACK and WELLNER • Empirical Processes with Applications to Statistics
STOYANOV • Counterexamples in Probability
STYAN • The Collected Papers of T. W. Anderson 1943—1985, Volumes 1 and 2
WHITTAKER • Graphical Models in Applied Multivariate Statistics
YANG • The Construction Theory of Denumerable Markov Processes

Applied Probability and Statistics

ABRAHAM and LEDOLTER • Statistical Methods for Forecasting
AGRESTI • Analysis of Ordinal Categorical Data
AGRESTI • Categorical Data Analysis
AICKIN • Linear Statistical Analysis of Discrete Data
ANDERSON and LOYNES • The Teaching of Practical Statistics
ANDERSON, AUQUIER, HAUCK, OAKES, VANDAELE, and WEISBERG • Statistical Methods for Comparative Studies
ARTHANARI and DODGE • Mathematical Programming in Statistics
ASMUSSEN • Applied Probability and Queues
*BAILEY - The Elements of Stochastic Processes with Applications to the Natural Sciences
BARNETT • Interpreting Multivariate Data
BARNETT and LEWIS • Outliers in Statistical Data, *Second Edition*
BARTHOLOMEW • Stochastic Models for Social Processes, *Third Edition*
BARTHOLOMEW and FORBES • Statistical Techniques for Manpower Planning
BATES and WATTS • Nonlinear Regression Analysis and Its Applications

*Now available in a lower priced paperback edition in the Wiley Classics Library

Applied Probability and Statistics (Continued)

BECK and ARNOLD • Parameter Estimation in Engineering and Science
BELSLEY • Conditioning Diagnostics: Collinearity and Weak Data in Regression
BELSLEY, KUH, and WELSCH • Regression Diagnostics: Identifying Influential Data and Sources of Collinearity
BHAT • Elements of Applied Stochastic Processes, *Second Edition*
BHATTACHARYA and WAYMIRE • Stochastic Processes with Applications
BLOOMFIELD • Fourier Analysis of Time Series: An Introduction
BOLLEN • Structural Equations with Latent Variables
BOX • R. A. Fisher, The Life of a Scientist
BOX and DRAPER • Empirical Model-Building and Response Surfaces
BOX and DRAPER • Evolutionary Operation: A Statistical Method for Process Improvement
BOX, HUNTER, and HUNTER • Statistics for Experimenters: An Introduction to Design, Data Analysis, and Model Building
BROWN and HOLLANDER • Statistics: A Biomedical Introduction
BUCKLEW • Large Deviation Techniques in Decision, Simulation, and Estimation
BUNKE and BUNKE • Nonlinear Regression, Functional Relations and Robust Methods: Statistical Methods of Model Building
BUNKE and BUNKE • Statistical Inference in Linear Models, Volume I
CHAMBERS • Computational Methods for Data Analysis
CHATTERJEE and HADI • Sensitivity Analysis in Linear Regression
CHATTERJEE and PRICE • Regression Analysis by Example., *Second Edition*
CHOW • Econometric Analysis by Control Methods
CLARKE and DISNEY • Probability and Random Processes: A First Course with Applications, *Second Edition*
COCHRAN • Sampling Techniques, *Third Edition*
COCHRAN and COX • Experimental Designs, *Second Edition*
CONOVER • Practical Nonparametric Statistics, *Second Edition*
CONOVER and IMAN • Introduction to Modern Business Statistics
CORNELL • Experiments with Mixtures: Designs, Models and The Analysis of Mixture Data, Second Edition
COVER and THOMAS · Elements of Information Theory
COX • A Handbook of Introductory Statistical Methods
COX • Planning of Experiments
CRESSIE • Statistics of Spatial Data
DANIEL • Applications of Statistics to Industrial Experimentation
DANIEL • Biostatistics: A Foundation for Analysis in the Health Sciences, *Fourth Edition*
DANIEL and WOOD • Fitting Equations to Data: Computer Analysis of Multifactor Data, *Second Edition*
DAVID • Order Statistics, *Second Edition*
DAVISON • Multidimensional Scaling
DEGROOT, FIENBERG and KADANE • Statistics and the Law
DEMING • Sample Design in Business Research
DILLON and GOLDSTEIN • Multivariate Analysis: Methods and Applications
DODGE • Analysis of Experiments with Missing Data
DODGE and ROMIG • Sampling Inspection Tables, *Second Edition*
DOWDY and WEARDEN • Statistics for Research
DRAPER and SMITH • Applied Regression Analysis, *Second Edition*
DUNN • Basic Statistics: A Primer for the Biomedical Sciences, *Second Edition*
DUNN and CLARK • Applied Statistics: Analysis of Variance and Regression, *Second Edition*
ELANDT-JOHNSON and JOHNSON • Survival Models and Data Analysis
FLEISS • The Design and Analysis of Clinical Experiments

(*cardplate continues in back of book*)

BAYESIAN ESTIMATION AND EXPERIMENTAL DESIGN IN LINEAR REGRESSION MODELS

Bayesian Estimation and Experimental Design in Linear Regression Models

Jürgen Pilz

Bergakademie Freiberg, GDR

JOHN WILEY & SONS

Chichester · New York · Brisbane · Toronto · Singapore

Licensed edition for John Wiley & Sons Limited, 1991
Printed in Germany

John Wiley & Sons Limited
Baffins Lane, Chichester
West Sussex, PQ19 1UD, England

Other Wiley Editorial Offices

John Wiley & Sons, Inc., 605 Third Avenue,
New York, NY 10158-0012, USA

Jacaranda Wiley Ltd, G.P.O. Box 859, Brisbane,
Queensland 4001, Australia

John Wiley & Sons (Canada) Ltd, 22 Worcester Road,
Rexdale, Ontario M9W 1L1, Canada

John Wiley & Sons (SEA) Pte Ltd, 37 Jalan Pemimpin 05-04,
Block B, Union Industrial Building, Singapore 2057

British Library Cataloguing in Publication Data:

Pilz, Jürgen
Bayesian estimation and experimental design in linear regression models. –
(Wiley series in probability and mathematical statistics).
1. Regression analysis
I. Title
519.5′36 QA 278.2
ISBN 0 471 91732 X

Library of Congress Cataloging-in-Publication Data:

Pilz, Jürgen, 1951 –
Bayesian estimation and experimental design in linear regression models.
(Wiley series in probability and mathematical statistics)
Bibliography: p.
1. Experimental design. 2. Estimation theory. 3. Bayesian statistical decision theory. 4. Regression analysis. I. Title. II. Series.
QA279.P55 1991 519.5 87-27915

ISBN 0 471 91732 X

Preface to the first edition

Experimental investigations are used in all fields of technics and scientific research to gain new knowledge and to support results obtained previously. This book deals with the design and analysis of linear regression experiments in presence of prior knowledge about the model parameters. Usual estimation and design methods for linear regression models, which work on the basis of the method of least squares or some modification of it (see Kiefer (1959), (1974), Kiefer/Wolfowitz (1959), Fedorov (1972)), do not make use of prior knowledge which will but be available in almost every practical situation; for example it may come from previous experiments, technological conditions and theoretical reasoning or from subjective imaginations. Here we adopt the Bayesian approach which allows for an efficient and exhaustive use of such prior knowledge and its combination with actual observations. This leads to an improved estimation and a possibly considerable reduction of experimental efforts.

We start our investigations with a given linear regression model assumed to be true, the subproblem of model choice will be excluded since there is a wide variety of literature on this topic (see e.g. Lempers (1971), Zellner (1971)). Though there are also various books and numerous articles concerning the subproblem of regression parameter estimation in presence of prior information, both within a Bayesian framework (e.g. Zellner (1971), Lindley/Smith (1972), Box/Tiao (1973)) and a non-Bayesian framework (e.g. Bibby/Toutenburg (1977), Toutenburg (1982)), yet there appears to be available only a very limited amount of literature on experimental design for estimators using prior knowledge.

The aim of the present monograph is to develop a unified approach to estimation and design from the Bayesian viewpoint, to give a Bayesian alternative to the least squares estimator which is suitable for experimental design and to indicate methods for the construction of optimal designs for the Bayes estimator. The experimental design methods developed here do not only apply to Bayes estimators but also to some well-known estimators using prior knowledge which does not come available in form of a prior distribution for the model parameters, for example mixed linear, minimax linear and also ridge-type estimators.

The book consists of three parts: In Part I, after a decision theoretic formulation of the compound estimation and design problem, we deal with the specification of the prior distribution to summarize our prior knowledge, which is one of the major problems of any Bayesian analysis. We consider some typical kind of prior knowledge and give some advice how it can be used to form a prior distribution. Then, in Part II, our main concern is with the well-known linear Bayes estimator and the investigation of the robustness of optimality of this estimator. Particularly, we investigate the extent to which this estimator is robust under a change of error and prior distributions and the loss function. This is of great importance since in practical situations our prior knowledge will be more or less incomplete and Bayesian procedures are often criticized because of the assumption of a fully specified probabilistic model. In the third part we give an approach to experimental design for the linear Bayes estimator which shows some analogies to the classical experimental design approach developed in the pioneering works of J. Kiefer and then we deal extensively with the construction of optimal designs for this estimator. For the important special case of multiple linear regression we give optimal designs in an explicit form. Furthermore, we set up iterative procedures for the construction of optimal designs which, in general, converge very rapidly since there is no problem with the coice of the starting design, which is a crucial point in the application of iterative procedures in experimental design for the least squares estimator. Finally, in a self-contained appendix we recall, in a rough form, some basic facts and notions from probability theory, statistical decision theory and matrix algebra which are used throughout the book.

It is the author's pleasant duty to express his thanks to all who helped him in preparing this text: I am indebted to Prof. H. Bandemer for his stimulating interest and steady encouragement in my writing the text and, moreover, for his generous support of my research work. Also, I would like to thank him and Dr. W. Näther for providing useful comments and suggestions that helped to improve the manuscript. Further, I am grateful to Dr. J. Gladitz (Academy of Sciences, Berlin) for the fruitful co-operation over the last years; much of the material presented in Chapters 12 and 13 emerged in the course of this co-operation. Finally, I would like to thank Dr. R. Thiele and Mr. J. Weiss from the Teubner Publishing House for the considerate collaboration and obligingness.

Freiberg, March 1983 J. Pilz

Preface to the second edition

This edition represents an enlarged version of the first edition which appeared in 1983 in the Teubner Publishing House in Leipzig (GDR). The volume includes new material on integral geometric prior distributions (Section 3.2), nonlinear modifications of the conjugate Bayes estimator (Section 5.5), Bayes and admissible linear estimation (Sections 5.3, 7, 16), Bayesian robustness measures and restricted minimax optimality (Sections 6.2, 6.3, 15.6), and also on Bayes and minimax linear estimation in inadequate models (Sections 9.3, 17). Finally, we largely extended the list of references by a lot of actual papers which appeared over the last five years and we included a completely new part (Part IV) on minimax linear regression estimation in case of prior knowledge representable by some restricted compact parameter set. The resulting minimax estimators, which effect a uniform improvement over the least squares estimator, can be obtained as Bayes linear estimators with respect to least favourable prior distributions on the restricted parameter set. We establish a duality between the problems of minimax linear estimation and Bayesian experimental design and demonstrate how these results can be applied to the construction of minimax estimators both in case of true models and of approximately true models.

Since the appearance of the first edition we can record a further rapid progress in research and application of Bayesian methods to aid in model building and parameter estimation, in particular I mention the rich material contained in the proceedings of the Valencia meeting of Bayesian Statisticians (edited by Bernardo et al. (1980), (1985), (1989)), of the Innsbruck meeting in honour of B. de Finetti (edited by Viertl (1987)) and the recent monographs of Hartigan (1983) and Broemeling (1985).

I would like to express my gratitude to the John Wiley Publishing Company for giving me the opportunity to present the material included in this monograph to a broader community of readers interested in the state of art and in current activities in the modern and rapidly expanding area of Bayesian regression estimation and experimental design.

Freiberg, Spring 1989 J. Pilz

Contents

I. Bayesian regression and prior distributions

1. Introduction

A very frequent and important task of experimental research is the investigation of the functional dependence of the experimental results on a predetermined number of controlled factors $x_1, \ldots, x_k$ which can be adjusted at certain levels. In general, the experimental outcomes have random characters since they are influenced by a large number of disturbances which are not under control of the investigator. The experimental results are then interpreted as realizations of some random variable $\tilde{y} = \tilde{y}(\boldsymbol{x})$ at the level $\boldsymbol{x} = (x_1, \ldots, x_k)^\top$ taking values in the k-dimensional Euclidean space $\mathbb{R}^k$. Their functional dependence on the factors can be thought to be representable by the relationship

$$\forall \boldsymbol{x} \in X \subseteqq \mathbb{R}^k : \mathbf{E}\,\tilde{y}(\boldsymbol{x}) = \eta(\boldsymbol{x}) \tag{1.1}$$

combining the expectation $\mathbf{E}\,\tilde{y}(\boldsymbol{x})$ of the random variable $\tilde{y}(\boldsymbol{x})$ with the values taken by the factors. Here $\eta | X \to \mathbb{R}^1$ is a nonstochastic function and X is the closed and bounded subset of $\mathbb{R}^k$ for which the representation (1.1) makes sense. It is the aim of regression analysis to make statements about the unknown function η.

In the context of regression analysis the factor components of $\boldsymbol{x} = (x_1, \ldots, x_k)^\top$ are called *regressors* (or independent variables), $\tilde{y}(\boldsymbol{x})$ is called the *response variable* (or dependent variable). The function η is called the *response surface* of the regressors.

Practically it is impossible to incorporate all the factors which could be of importance for the response. Therefore, in (1.1) it is assumed that the process of screening the essential factors is finished and the k regressors are determined, i.e. nonstochastic. The remaining factors which are not included in the functional relationship are assumed to be stochastic. At this stage of model building the detailed knowledge of the subject matter and the experience of the investigator must be taken into account carefully. For the problem of the choice of regressor variables out of a large number of given quantities we refer to LINDLEY (1968), BROOKS (1972), (1974), HOCKING (1976) and TRADER (1983).

We assume the family $\{\tilde{y}(x)\}_{x \in X}$ of response variables to be defined over some probability space $[\Omega, \mathcal{B}_\Omega, P]$. From the relationship (1.1) it then follows that there exists a family $\{\tilde{e}(x)\}_{x \in X}$ of random variables such that

$$\forall x \in X: \tilde{y}(x) = \eta(x) + \tilde{e}(x), \qquad \mathbf{E}\,\tilde{e}(x) = 0. \tag{1.2}$$

The random variables $\tilde{e}(x)$ can be interpreted as random disturbances representing the total influence of all those factors which cannot (or shall not) be controlled.

Let the response variable be observable at the points of a subset $X_E \subseteqq X$, i.e., for any $x \in X_E$ we can obtain realizations $y(x)$ of the random variable $\tilde{y}(x)$. This subset X_E we call the *experimental region.* Any n-tuple

$$v_n := (x_1, \ldots, x_n) \in (X_E)^n$$

of experimental points (not necessarily distinct) is called an *exact design* of size n. The random vector

$$\tilde{\boldsymbol{y}}(v_n) := (\tilde{y}(x_1), \ldots, \tilde{y}(x_n))^\top$$

is called the *observation vector* for the design v_n and

$$\tilde{\boldsymbol{e}}(v_n) := (\tilde{e}(x_1), \ldots, \tilde{e}(x_n))^\top$$

is called the *error vector* for v_n.

Throughout this book we will be concerned exclusively with the estimation problem for the response surface: On the basis of experimental data, observations and our prior knowledge about the functional relationship we have to find an estimation function approximating $\eta(\cdot)$ over some specified region $X_P \subseteqq X$ as good as possible. The subset X_P of points for which an estimation of the mean response $\eta(x)$ is of interest is called the *prediction region.*

In order to find an estimation function for η it is suggested to choose at first a set of functions approximately describing the unknown response surface. Here we assume that η can be approximated by a *linear regression setup,* i.e., η is contained in a set of functions $\bar{\eta}(\cdot, \vartheta)$ on X depending on some parameter $\boldsymbol{\vartheta} = (\vartheta_1, \ldots, \vartheta_r)^\top$ and having the linear structure

$$\bar{\eta}(x, \boldsymbol{\vartheta}) = \sum_{i=1}^{r} \vartheta_i f_i(x) = \boldsymbol{f}(x)^\top \boldsymbol{\vartheta}, \qquad x \in X, \qquad \boldsymbol{\vartheta} \in \Theta \subseteqq \mathbb{R}^r.$$

Here $\Theta \subseteqq \mathbb{R}^r$ is the set of possible parameters and $\boldsymbol{f} = (f_1, \ldots, f_r)^\top | X \to \mathbb{R}^r$ is a vector of functions $f_1, \ldots, f_r$ which are linear independent over X. In many cases this can be motivated by the fact that η can be expanded in a Taylor-series converging sufficiently fast so that a partial series of finite order serves as a good approximation for η.

Assumption 1.1. $\exists \boldsymbol{\vartheta} \in \Theta \quad \forall x \in X: \eta(x) = \bar{\eta}(x, \boldsymbol{\vartheta}) = \boldsymbol{f}(x)^\top \boldsymbol{\vartheta}$.

This assumption means that $\bar{\eta}$ is a true linear regression setup for the response surface η. In whatever way one has obtained a linear regression setup for η, it must be proved to be a true setup. If this does not immediately result from knowledge of the structure of η then this assumption can be verified by a corresponding test procedure (see, e.g. BANDEMER/NÄTHER (1980)).

If there is not only one, but a finite number $s > 1$ of different setups, which are thought to serve as possible approximations for η, then a "best" setup can be chosen according to the criterion of maximum posterior probability (see, e.g. BOX/HILL (1967), LEMPERS (1971), ZELLNER (1971), ATKINSON (1978)).

For further discussions of Bayesian criteria for choosing among alternative linear models we refer to SMITH/SPIEGELHALTER (1980), SAN MARTINI/SPEZZAFERRI (1984) and the literature cited there; for the special case of polynomial settings see, e.g. HALPERN (1973).

By Assumption 1.1 the response surface is characterized through a regression parameter (true parameter) $\boldsymbol{\vartheta} \in \Theta$. Introducing the *design matrix*

$$\boldsymbol{F}(v_n) := (\boldsymbol{f}(\boldsymbol{x}_1), \ldots, \boldsymbol{f}(\boldsymbol{x}_n))^\top = (f_j(\boldsymbol{x}_i))_{\substack{i=1,\ldots,n \\ j=1,\ldots,r}}$$

we obtain with (1.2) the model

$$\tilde{\boldsymbol{y}}(v_n) = \boldsymbol{F}(v_n)\boldsymbol{\vartheta} + \tilde{\boldsymbol{e}}(v_n), \qquad \mathbf{E}\,\tilde{\boldsymbol{e}}(v_n) = \boldsymbol{0}. \tag{1.3}$$

We call it the *linear regression model* based on the design v_n. Thus, the estimation problem for η changes into an estimation problem for the true parameter $\boldsymbol{\vartheta}$. The task of optimal experimental design is then to find experimental arrangements, i.e. designs v_n, improving the goodness of estimation in a sense yet to be defined (see Section 2).

Usually, the compound problem of estimation and design for response surfaces is treated in the literature in such a way that the method of least squares is chosen as a fixed estimation procedure and an optimal design is to be found by minimizing some functional of the mean squared error which comes out for this estimation. In this way, taking the least squares estimator (LSE)

$$\hat{\eta}_{\mathrm{LS}}(\boldsymbol{x}, \tilde{\boldsymbol{y}}(v_n)) = \boldsymbol{f}(\boldsymbol{x})^\top[\boldsymbol{F}(v_n)^\top \boldsymbol{F}(v_n)]^{-1}\boldsymbol{F}(v_n)^\top \tilde{\boldsymbol{y}}(v_n) \tag{1.4}$$

for $\eta(\boldsymbol{x})$ as a basis for experimental design (in case of singularity of $\boldsymbol{F}(v_n)^\top \boldsymbol{F}(v_n)$ a generalized inverse is to be taken), the classical approach leads to the minimization of some functional of the so-called *information matrix*

$$\boldsymbol{M}(v_n) = \frac{1}{n}\,\boldsymbol{F}(v_n)^\top \boldsymbol{F}(v_n) \tag{1.5}$$

(see KIEFER/WOLFOWITZ (1959), FEDOROV (1972), BANDEMER et al. (1977)).

Thus, for example, the minimization of the trace (A-optimality), determinant (D-optimality), maximum eigenvalue (E-optimality) or some quadratic form (C-optimality) of $\boldsymbol{M}(v_n)^{-1}$ are design criteria which have been studied extensively.

The classical approach can be justified by the following well-known Gauss-Markov theorem.

Theorem 1.1. *Let the observations be uncorrelated and homoscedastic with variance $\sigma^2 > 0$, i.e.,* $\mathbf{Cov}\,\tilde{y}(v_n) = \sigma^2 I_n$. *Then* $\tilde{\eta}$ *from* (1.4) *is the best linear unbiased estimator for* $\eta(x)$, *i.e., there is no other linear unbiased estimator for* $\eta(x)$ *which has smaller variance than* $\hat{\eta}_{\mathrm{LS}}$. *(Here "linear" means linearity in* $\tilde{y}(v_n)$*).*

With $\mathbf{Cov}\,\tilde{y}(v_n) = \sigma^2 I_n$, the estimator $\hat{\eta}_{\mathrm{LS}}$ has variance (identical with mean square error due to unbiasedness)

$$\operatorname{Var}\hat{\eta}_{\mathrm{LS}}(x, \tilde{y}(v_n)) = \frac{\sigma^2}{n} f(x)^\top M(v_n)^{-1} f(x) \tag{1.6}$$

so that by minimization of some functional of $M(v_n)^{-1}$ the variance ratio of the LSE can be optimized in a predetermined sense. If we wish to minimize the maximum variance of the LSE over the whole prediction region X_{P} or the integrated variance with respect to a given nonnegative weight function over X_{P}, then we are led to the design criteria of G-optimality,

$$\frac{1}{n} \sup_{x \in X_{\mathrm{P}}} f(x)^\top M(v_n)^{-1} f(x) \overset{!}{=} \inf_{v_n} \tag{1.7}$$

and of I-optimality,

$$\frac{1}{n} \int_{X_{\mathrm{P}}} f(x)^\top M(v_n)^{-1} f(x)\, p(x)\, \mathrm{d}x \overset{!}{=} \inf_{v_n} \tag{1.8}$$

which are also widely used in the design literature.

The classical approach, however, must be criticized for the fact that the LSE and the above mentioned design criteria in no manner take account of prior knowledge about the regression parameter, the error distribution, requirements concerning the frequency and accuracy of prediction within X_{P} etc. Such additional information will but be available in almost every practical application and its use allows for a more precise estimation and a possibly considerable reduction of experimental efforts, which is of special importance in small sample situations.

Within the classical framework there have also been developed several approaches to incorporate prior knowledge via equality and inequality constraints on the regression parameters. These approaches can be criticized, however, for the unrealistic way a priori information is expressed. In practice, prior information will be more or less imprecise and rarely can be expressed in an exact equation. Inequality constraints may ignore the fine detail of our knowledge and, in addition, are inconvenient to use and little is known about the sampling properties of the resulting estimators.

As an alternative to the classical approach described above, we will choose a *Bayesian approach* which requires the reformulation of the prior knowledge by means of a probability distribution for the model parameters.

In this way the LSE is replaced by the Bayesian version

$$\hat{\eta}_{\mathrm{B}}(\boldsymbol{x}, \tilde{\boldsymbol{y}}(v_n)) = \boldsymbol{f}(\boldsymbol{x})^{\top}(\boldsymbol{F}(v_n)^{\top}\boldsymbol{F}(v_n) + \boldsymbol{\Phi}^{-1})^{-1}(\boldsymbol{F}(v_n)^{\top}\tilde{\boldsymbol{y}}(v_n) + \boldsymbol{\Phi}^{-1}\boldsymbol{\mu}),$$

where $\boldsymbol{\mu}$ and $\boldsymbol{\Phi}$ stand for the mean and the covariance matrix (apart from a constant factor) of the prior distribution for the regression coefficients. In a sense, $\boldsymbol{\mu}$ summarizes the prior knowledge about the location of the coefficients and the elements of $\boldsymbol{\Phi}^{-1}$ measure the precision of this knowledge and the possible interrelations between the coefficients. The problem of the specification of such a prior distribution is discussed in Sections 3 and 4 whereas the derivation of the above Bayes estimator, its properties and its interrelations with other alternatives to the LSE such as mixed linear (Theil-) estimators, minimax linear estimators of the Kuks-Ol'man type, ridge and shrunken estimators, which are all Bayesian in spirit, are studied in Part II.

An obvious disadvantage associated with the use of the LSE in case of unknown $\boldsymbol{\mu}$ and $\boldsymbol{\Phi}$ is that the design matrix must have full rank. If $\boldsymbol{F}(v_n)$ has less than full rank, which happens often in practical applications, then the regression parameter $\boldsymbol{\vartheta}$ cannot be identified without additional information. With a Bayesian estimation procedure however, no identifiability problems will arise. Moreover, even in case of a full rank $\boldsymbol{F}(v_n)$, the information matrix will often be ill-conditioned which may imply a numerical instability of the LSE, whereas the inversion of $\boldsymbol{F}(v_n)^{\top}\boldsymbol{F}(v_n) + \boldsymbol{\Phi}^{-1}$ needed for the computation of $\hat{\eta}_{\mathrm{B}}$ will be numerically more stable.

The most striking argument for the use of the Bayes estimator is but the fact that it may lead to a substantial reduction of the mean square error (risk) over the LSE (see Section 5). This, in turn, may imply a large reduction of the experimental efforts.

If the regression parameter region can be restricted to a compact subset of some Euclidean space then the class of (proper) Bayes linear estimators coincides with the class of all admissible linear estimators, i.e., any non-Bayesian linear estimator such as, for example, the LSE can be uniformly improved (with respect to quadratic risk) by some Bayes linear estimator.

One of the main criticisms raised by opponents of the Bayesian approach is that the investigator will rarely have sufficient prior knowledge to specify a prior probability distribution in full detail. But often it suffices to know some qualitative features and summaric values (e.g. first and second order moments) of the prior distribution, i.e., it can be chosen within a specified class of priors and the resulting Bayes strategy will be robust against the choice of prior distributions from this class. Moreover, even if the assumption of independence and homoscedasticity of errors, which is usually made in linear models, is dropped the Bayes estimator has a minimax property (see Section 6).

For the Bayes estimator $\hat{\eta}_{\mathrm{B}}$ the expected mean square error (with respect to the prior distribution) is proportional to

$$\boldsymbol{f}(\boldsymbol{x})^{\top}(\boldsymbol{F}(v_n)^{\top}\boldsymbol{F}(v_n) + \boldsymbol{\Phi}^{-1})^{-1}\boldsymbol{f}(\boldsymbol{x}).$$

Thus, the design problem for $\hat{\eta}_{\mathrm{B}}$ leads to the minimization of appropriate functionals of the inverse of

$$M_B(v_n) = \frac{1}{n}(F(v_n)^\top F(v_n) + \Phi^{-1})$$

which represents the Bayesian analogue to the classical information matrix. The experimental design problem for the Bayes estimator $\hat{\eta}_B$ is the subject matter of Part III.

For large numbers of observations n, or alternatively, for matrices Φ with large eigenvalues corresponding to a high degree of prior uncertainty, the optimal choice of a design in a Bayesian setting may be very close to that in a classical setting. For moderate sizes of n and informative prior knowledge, however, the designs will be very different.

The Bayesian design criterion of minimizing some functional of the inverse of $M_B(v_n)$ has also meaning in some non-Bayesian situations. Actually, the experimental design considerations and construction methods set forth in Part III apply to any estimator whose mean square error or some corresponding risk is in the form of some functional of $(F(v_n)^\top F(v_n) + A)^{-1}$ with some positive definite matrix A, including, for example, the above mentioned mixed and minimax linear estimators, and also apply to the important problems of experimental design to augment previous observations and for fitting biased response surfaces.

Ideas and methods of Bayesian experimental design can be successfully applied to optimal linear regression estimation when prior knowledge comes available in form of a restricted compact parameter set. In this case the LSE as well as some other well-known linear estimators such as principal component estimators, Marquardt-type estimators and certain types of ridge and shrunken estimators are inadmissible, even within the class of linear estimators. The class of admissible linear estimators consists of exactly those linear estimators which are proper Bayes with respect to some prior distribution over the restricted parameter set. If we are not able or willing to restrict the set of possible priors then the minimax criterion would be appropriate. The minimax linear estimators effect a uniform improvement over the LSE and can be characterized as Bayes linear estimators with respect to least favourable prior distributions which, in turn, can be obtained as solutions of appropriate convex Bayesian experimental design problems. The theory of minimax and admissible linear estimation and the duality between the problems of linear estimation in case of a restricted parameter set and Bayesian experimental design forms the subject of Part IV of the present book, where we also deal with approximate regression estimation for fitting biased response surfaces.

2. Estimation and design as a Bayesian decision problem

2.1. Specification of the decision problem

We are now going to formulate the estimation problem for the response surface η as a statistical decision problem. In this way, the estimation problem is considered as a statistical two-person zero-sum game of the statistician against Nature which is in the role of a ficticious counterpart.

Nature is choosing a parameter value $\boldsymbol{\vartheta} \in \boldsymbol{\Theta}$ such that $\bar{\eta}(\boldsymbol{x}, \boldsymbol{\vartheta}) = \boldsymbol{f}(\boldsymbol{x})^\top \boldsymbol{\vartheta} = \eta(\boldsymbol{x})$ is the expected (true) response. Furthermore, Nature chooses a probability distribution for the random error terms $\tilde{e}(\boldsymbol{x})$ and the covariance structure of the error vector $\tilde{\boldsymbol{e}}(v_n)$ arising in jointly observing the responses at the points of the design $v_n \in (X_{\mathrm{E}})^n$. For any design v_n we denote by

$$\mathcal{P}_{\tilde{e}} = \{P_{\tilde{e}}^{(w)}\}_{w \in W}, \qquad W \subseteqq \mathbb{R}^s \tag{2.1}$$

the family of possible distributions of the error vector $\tilde{\boldsymbol{e}}(v_n)$, i.e., the distribution of $\tilde{\boldsymbol{e}}(v_n)$ is assumed to be uniquely determined by some parameter $\boldsymbol{w}$ from a nonempty index set $W \subseteqq \mathbb{R}^s$, $s \geqq 1$. The distributions in $\mathcal{P}_{\tilde{e}}$ are assumed to be identifiable in the following sense:

Assumption 2.1. $\forall (\boldsymbol{w}_1, \boldsymbol{w}_2) \in W \times W\colon \boldsymbol{w}_1 \neq \boldsymbol{w}_2 \Rightarrow P_{\tilde{e}}^{(\boldsymbol{w}_1)} \neq P_{\tilde{e}}^{(\boldsymbol{w}_2)}$.

For example, in case of normally independently and identically distributed observations we have

$$\mathcal{P}_{\tilde{e}} = \left\{ \mathrm{N}\left(\boldsymbol{0}, \frac{1}{w} \boldsymbol{I}_n\right) \right\}_{w \in \mathbb{R}^+}$$

for any design v_n of size n (see A.2, $\boldsymbol{I}_n$ denotes the n-dimensional unity matrix). Here the index w indicates the precision (reciprocal of the variance) of observation, i.e., $\tilde{\boldsymbol{e}}(v_n)$ has the probability density

$$p_{\tilde{e}}(\boldsymbol{t}) = (w/2\pi)^{n/2} \exp\left(-\frac{w}{2} \boldsymbol{t}^\top \boldsymbol{t}\right), \qquad \boldsymbol{t} \in \mathbb{R}^n$$

with respect to Lebesgue measure.

Finally, the choice of the prediction point $\boldsymbol{x} \in X_{\mathrm{P}}$, for which an estimation of the response $\eta(\boldsymbol{x})$ is required, is interpreted as a task of Nature. This assumption is made purely for technical reasons of representation of the problem in a Bayesian context as it will be developed in the sequel.

Hence, Nature is choosing a triplet $\boldsymbol{z} = (\boldsymbol{\vartheta}, w, \boldsymbol{x})$ from the *state space*

$$Z = [\Theta \times W \times X_{\mathrm{P}}]. \tag{2.2}$$

The statistician, on the contrary, chooses a design and an estimation function for η. Let

$$V \subset \bigcup_{n \in \mathbb{N}} (X_E)^n \tag{2.3}$$

be a set of designs of finite size (set of possible designs) and let $C(v_n)$ denote the sample space of the observation vector $\tilde{y}(v_n)$ with $v_n \in V$. For any $x \in X_P$ and given realization y of $\tilde{y}(v_n)$ the response $\eta(x) = \bar{\eta}(x, \vartheta)$ will be estimated by some value $\hat{\eta}(x, y)$ according to an estimation function

$$\hat{\eta} | X_P \times C(v_n) \to \mathbb{R}^1 .$$

The goodness of estimation is valued by a loss function

$$L_\eta | \Theta \times W \times X_P \times \mathbb{R}^1 \times V \to \mathbb{R}^1 . \tag{2.4}$$

Thereby $L_\eta(\vartheta, w, x; \hat{\eta}(x, y), v_n)$ indicates the numerical value for the loss arising if observation at the points of v_n results in the realization y and the true response $\eta(x)$, which we have to expect at the point $x \in X_P$, is estimated by $\hat{\eta}(x, y)$. For the application it is worth noting that already simple qualitative assumptions on the loss function lead to practically important optimality statements for estimation and design. In certain cases, for example, it suffices to know that the loss is some continuous and nondecreasing (or convex) function of the distance $|\eta(x) - \hat{\eta}(x, y)|$ between estimated and true response.

Without loss of generality, let be satisfied

Assumption 2.2. *L_η is nonnegative and measurable with respect to all arguments.*

According to our model (1.3), the distribution $P_{\tilde{e}}^{(w)} \in \mathcal{P}_{\tilde{e}}$ of the error vector induces a conditional probability distribution $P_{\tilde{y}|\vartheta, w}$ of the observation vector $\tilde{y}(v_n)$ over the measurable sample space. For any design $v_n \in V$ let $D_\eta(v_n)$ contain all the estimation functions $\hat{\eta}$ for which the expectation

$$R_\eta(\vartheta, w, x; \hat{\eta}, v_n) := \mathsf{E}_{\tilde{y}|\vartheta, w} L_\eta(\vartheta, w, x; \hat{\eta}(x, \tilde{y}(v_n)), v_n) \tag{2.5}$$

takes finite values for all states $(\vartheta, w, x) \in \Theta \times W \times X_P$.

Now, the *space of strategies* S available to the statistician can be defined as

$$S = \bigcup_{v_n \in V} (D_\eta(v_n) \times \{v_n\}) , \tag{2.6}$$

he first chooses a design $v_n \in V$ and then an estimation function $\hat{\eta} \in D_\eta(v_n)$. With the above definitions and denotations the triplet

$$G_\eta = [Z, S, R_\eta] \tag{2.7}$$

defines a (nonrandomized) statistical decision problem (see A.10).

Definition 2.1. The statistical decision problem $G_\eta = [Z, S, R_\eta]$ with Z, S and R_η given by (2.2), (2.6) and (2.5), respectively, is called the compound *estimation and design problem* for the response surface η. The function R_η is called the *risk function* of G_η.

For fixed values of ϑ, w and x it is possible to compare the different strategies $(\hat{\eta}, v_n) \in S$ on the basis of the risk function. But, in general there will be no strategy in S which is uniformly best, i.e., which has smaller risk than any other strategy irrespective of the state of Nature. Therefore we look for strategies, i.e., estimators and designs, minimizing the risk function in the sense of a given criterion of optimality for which the comparison no longer depends on the state of Nature. To be specific, let Q be an operator transforming the risk function $R_\eta | Z \times S \to \mathbb{R}^1$ into a function $R_\eta^Q | S \to \mathbb{R}^1$, i.e.,

$$\forall (\hat{\eta}, v_n) \in S : R_\eta^Q(\hat{\eta}, v_n) := QR_\eta(\cdot\,, \cdot\,, \cdot\,; \hat{\eta}, v_n)\,. \tag{2.8}$$

Definition 2.2. The strategy $(\hat{\eta}^*, v_n^*) \in S$ is called *Q-optimal* in G_η if it holds

$$R_\eta^Q(\hat{\eta}^*, v_n^*) = \inf_{(\hat{\eta}, v_n) \in S} R_\eta^Q(\hat{\eta}, v_n)$$

(cp. Bandemer et al. (1977), Laycock (1972)).

The operator Q provides a joint criterion of optimality for the choice of an estimator and a design. For example, the supremum operator over Z and the expectation operator with respect to some probability distribution $P_{\tilde{z}}$ over Z,

$$Q = \sup_{(\vartheta, w, x) \in Z} \quad \text{and} \quad Q = \mathbf{E}_{\tilde{z}} = \mathbf{E}_{(\tilde{\vartheta}, \tilde{w}, \tilde{x})},$$

respectively, would lead us to minimax and Bayes strategies in G_η, the latter criterion will be adopted in the sequel.

Under rather weak conditions, a Q-optimal strategy can be composed of optimal strategies obtained in two subproblems of G_η. In the first step, an optimal estimator $\hat{\eta}^*$ is obtained in an estimation problem G_η^{e} for fixed $x \in X_{\mathrm{P}}$ and $v_n \in V$ and hereafter an optimal design v_n^* is obtained in a design problem G_η^{d} for the optimal estimator $\hat{\eta}^*$. For the criterion of Bayes optimality ($Q = \mathbf{E}_{\tilde{z}}$) this is demonstrated in Section 2.2.

Often it is desirable to incorporate the cost aspect of experimentation, concerning, for example, time, material and devices needed to carry out the experiment, into the decision problem G_η. One possibility would be the decomposition of the loss so that L_η is the sum of a function purely evaluating the loss due to misestimation of the response and a cost function $c | V \to \mathbb{R}^+$ indicating the expense of experimentation according to the chosen design. Another possibility would be the minimization of the risk R_η^Q over all estimation functions $\hat{\eta} \in D_\eta(v_n)$ satisfying a cost constraint, say $c(v_n) \leqq c_0$ with a predetermined constant c_0 (see Section 14.2).

2.2. The classical and the Bayesian approach

Let us return for a moment to the classical approach to estimation and design for the response surface as described in Section 1. It is well-known that, for very general loss functions and wide classes $\mathcal{P}_{\tilde{e}}$ of error distributions, the LSE as given by (1.4) is an unrestricted minimax estimator (see RADNER (1959), NÄTHER (1974)). Furthermore, the design criteria of A-, C-, D-, E-, G- and I-optimality mentioned before can be characterized as special cases of the minimax criterion with respect to particular loss functions (see BANDEMER et al. (1977)). This means that the classical approach to the compound problem G_η can be valued such that in the first step an unrestricted minimax estimator is obtained and then in the second step an optimal design for this estimator is chosen as a minimax strategy in the subproblem G_η^{d}. This leads to a minimax strategy in the compound problem G_η with Q chosen as supremum operator over Z.

There are, however, several serious objections that are to be made to the use of the unrestricted minimax principle as an optimality criterion for the choice of strategies in G_η. As already discussed before, an essential objection results from the fact that it ignores additional information which, in almost every practical situation, is available on the subject matter underlying the decision problem G_η. Another serious objection results from the fact that, in general, the unrestricted minimax principle does not lead to admissible strategies. Thus, for example, the estimators obtained by JAMES/STEIN (1961) and BARANCHIK (1973), in case of quadratic loss and normally distributed observations, are better (i.e., their risk is uniformly smaller, see A.11) than the LSE for the regression parameter which is minimax in this case. For this reason, a minimax strategy in G_η on the basis of the LSE is not necessarily admissible in G_η (see LÄUTER (1974)).

Here, as an alternative, we adopt the Bayesian approach to the decision problem G_η. This approach allows for an efficient and exhaustive use of the prior knowledge available to us before experimentation. By "prior knowledge" we understand information on the state $\boldsymbol{z} = (\boldsymbol{\vartheta}, \boldsymbol{w}, \boldsymbol{x}) \in Z$ of Nature that we have additional to the basic knowledge of the pair $(Z, \mathcal{P}_{\tilde{e}})$ of possible states and error distributions.

Prior information on the states of Nature may arise in various ways. In many cases the results of past samples provide information about the states, in other cases theoretical reasoning or introspection suggests constraints on the state space Z. Thus, for example, prior knowledge about the regression parameter $\boldsymbol{\vartheta}$ may be given by linear or quadratic constraints, prior estimates or relative frequencies of parameter values which had arisen in previous regression problems of the same type. Furthermore, we may have knowledge about certain quantities characterizing the error distribution $P_{\tilde{e}}^{(\boldsymbol{w})}$ (e.g. location and scale parameters, fractiles) or its type or shape (symmetry, sphericity etc.) may be known. Also, prior knowledge about the prediction parameter $\boldsymbol{x} \in X_{\mathrm{P}}$ is possible; for example, we may think of a weight function on X reflecting certain requirements to the accuracy of estimation of $\eta(\boldsymbol{x})$ at different points $\boldsymbol{x} \in X_{\mathrm{P}}$.

The Bayesian approach provides a natural and mathematically convenient way of combining our prior knowledge with the observations obtained in an actual sample. The approach is based on the idea that the prior knowledge about the parameters of the model can be represented by a probability distribution, the so-called *prior distribution.*

On the probability space $[\Omega, \mathcal{B}_\Omega, P]$ let be defined the random vectors $\tilde{\vartheta}|\Omega \to \Theta$, $\tilde{w}|\Omega \to W$ and $\tilde{x}|\Omega \to X_P$. Further, let $\mathcal{B}_\Theta$, $\mathcal{B}_W$ and $\mathcal{B}_X$ be corresponding σ-algebras over Θ, W and X which contain all single-element subsets. By $\tilde{z}$ we denote the random vector $\tilde{z} = (\tilde{\vartheta}, \tilde{w}, \tilde{x})$ and $P_{\tilde{z}}$ denotes the distribution of $\tilde{z}$ over $[Z, \mathcal{B}_\Theta \otimes \mathcal{B}_W \otimes \mathcal{B}_X]$ induced by P.

Assumption 2.3. *For all $v_n \in V$ let $\tilde{e}(v_n)$ and $(\tilde{\vartheta}, \tilde{x})$ be stochastically independent, i.e., let be $P^{(w)}_{\tilde{e}|\vartheta, x} = \Pi^{(w)}_{\tilde{e}}$ for any state $z = (\vartheta, w, x) \in Z$.*

With the above definitions and denotations, the state of Nature is considered a random vector $\tilde{z}$ having a prior probability distribution $P_{\tilde{z}}$ summarizing our prior knowledge. In any way, the assumption of a prior distribution should not be misinterpreted such that the parameters ϑ, w and x are random outcomes of some actual or hypothetical experiment. Rather it should be considered a means to incorporate the knowledge of the investigator which he has available, often unconsciously, before taking actual observations, as a means to express and to formalize the uncertainty of the prior knowledge in a flexible way.

According to Assumption 2.3, the prior distribution $P_{\tilde{z}}$ does not depend on the design used for actual observation. With Assumption 1.1 we thus obtain from (1.2) and (1.3) the Bayesian model

$$\begin{aligned} \forall(\vartheta, w, x) \in Z \,\forall v_n \in V: \mathsf{E}(\tilde{y}(x)\,|\,\tilde{\vartheta} = \vartheta, \tilde{w} = w, \tilde{x} = x) &= f(x)^\top \vartheta \\ \mathsf{E}(\tilde{y}(v_n)\,|\,\tilde{\vartheta} = \vartheta, \tilde{w} = w, \tilde{x} = x) &= F(v_n)\vartheta. \end{aligned} \tag{2.9}$$

The prior knowledge is incorporated into estimation of response by choosing a strategy which minimizes the expectation of the risk function with respect to the prior distribution $P_{\tilde{z}}$.

Definition 2.3. (i) The function ϱ defined on S by

$$\forall(\hat{\eta}, v_n) \in S: \varrho(P_{\tilde{z}}; \hat{\eta}, v_n) = \mathsf{E}_{\tilde{z}} R_\eta(\tilde{\vartheta}, \tilde{w}, \tilde{x}; \hat{\eta}, v_n)$$

is called the *Bayes risk* in G_η with respect to $P_{\tilde{z}}$.

(ii) The strategy $(\hat{\eta}^*, v_n^*) \in S$ is called a *Bayesian strategy* in G_η with respect to $P_{\tilde{z}}$ if it holds

$$\varrho(P_{\tilde{z}}; \hat{\eta}^*, v_n^*) = \inf_{(\hat{\eta}, v_n) \in S} \varrho(P_{\tilde{z}}; \hat{\eta}, v_n).$$

Hence, in the sense of Definition 2.2, a Bayesian strategy in G_η is a Q-optimal strategy in G_η when choosing $Q = \mathsf{E}_{\tilde{z}}$, i.e., the expectation operator with respect to the prior distribution $P_{\tilde{z}}$.

The problem of finding Bayesian strategies can be subdivided into a

Bayesian estimation problem and a design problem for the resulting Bayes estimator. Obviously, we can restrict attention to such estimators and designs for which the Bayes risk is finite since, if $(\hat{\eta}_0, v_n^0) \in S$ is a strategy with $\varrho(P_{\tilde{z}}; \hat{\eta}_0, v_n^0) = \varrho_0 < \infty$, then $(\hat{\eta}^*, v_n^*)$ is Bayesian in G_η if and only if $\varrho(P_{\tilde{z}}; \hat{\eta}^*, v_n^*) \leqq \varrho(P_{\tilde{z}}; \hat{\eta}, v_n)$ for all strategies $(\hat{\eta}, v_n)$ such that $\varrho(P_{\tilde{z}}; \hat{\eta}, v_n) \leqq \varrho_0$. Thus we consider for any $v_n \in V$ the class

$$D_\eta(v_n) := \{\hat{\eta} : \varrho(P_{\tilde{z}}; \hat{\eta}, v_n) < \infty\} \tag{2.10}$$

of estimators for η. An optimal estimator $\hat{\eta}^*$ as a subcomponent of a Bayesian strategy $(\hat{\eta}^*, v_n^*)$ in G_η can be obtained for fixed $\boldsymbol{x} \in X_{\mathrm{P}}$ and $v_n \in V$ in the estimation problem

$$\begin{gathered} G_\eta^{\mathrm{e}} = [\Theta \times W, D_\eta(v_n), R_\eta^{\mathrm{e}}] \\ \forall(\boldsymbol{\vartheta}, \boldsymbol{w}) \in \Theta \times W \ \forall \hat{\eta} \in D_\eta(v_n): R_\eta^{\mathrm{e}}(\boldsymbol{\vartheta}, \boldsymbol{w}; \hat{\eta}) = R_\eta(\boldsymbol{\vartheta}, \boldsymbol{w}, \boldsymbol{x}; \hat{\eta}, v_n). \end{gathered} \tag{2.11}$$

In the following we assume, for simplicity of presentation, $(\tilde{\boldsymbol{\vartheta}}, \tilde{\boldsymbol{w}})$ and $\tilde{\boldsymbol{x}}$ to be independent.

Assumption 2.4. $\forall \boldsymbol{x} \in X_{\mathrm{P}} : P_{\tilde{\boldsymbol{\Theta}}, \tilde{\boldsymbol{w}}|\boldsymbol{x}} = P_{\tilde{\boldsymbol{\vartheta}}, \tilde{\boldsymbol{w}}}$.

Then $\hat{\eta}^*$ is determined as Bayes estimator in G_η^{e} with respect to $P_{\tilde{\boldsymbol{\vartheta}}, \tilde{\boldsymbol{w}}}$:

$$\mathbf{E}_{\tilde{\boldsymbol{\vartheta}}, \tilde{\boldsymbol{w}}} R_\eta^{\mathrm{e}}(\tilde{\boldsymbol{\vartheta}}, \tilde{\boldsymbol{w}}; \hat{\eta}^*) = \inf_{\hat{\eta} \in D_\eta(v_n)} \mathbf{E}_{\tilde{\boldsymbol{\vartheta}}, \tilde{\boldsymbol{w}}} R_\eta^{\mathrm{e}}(\tilde{\boldsymbol{\vartheta}}, \tilde{\boldsymbol{w}}; \hat{\eta}). \tag{2.12}$$

Hereafter an optimal design v_n^* can be determined in the design problem

$$\begin{gathered} G_\eta^{\mathrm{d}} = [X_{\mathrm{P}}, V, R_\eta^{\mathrm{d}}] \\ \forall \boldsymbol{x} \in X_{\mathrm{P}} \ \forall v_n \in V: R_\eta^{\mathrm{d}}(\boldsymbol{x}; v_n) = \mathbf{E}_{\tilde{\boldsymbol{\vartheta}}, \tilde{\boldsymbol{w}}} R_\eta(\tilde{\boldsymbol{\vartheta}}, \tilde{\boldsymbol{w}}, \boldsymbol{x}; \hat{\eta}^*, v_n). \end{gathered} \tag{2.13}$$

The optimal component v_n^* is obtained as Bayesian solution to G_η^{d} with respect to the prior distribution $P_{\tilde{\boldsymbol{x}}}$ over the prediction region

$$\mathbf{E}_{\tilde{\boldsymbol{x}}} R_\eta^{\mathrm{d}}(\tilde{\boldsymbol{x}}; v_n^*) = \inf_{v_n \in V} \mathbf{E}_{\tilde{\boldsymbol{x}}} R_\eta^{\mathrm{d}}(\tilde{\boldsymbol{x}}; v_n). \tag{2.14}$$

Theorem 2.1. *Under the Assumptions* 2.1–2.4, $(\hat{\eta}^*, v_n^*) \in S$ *defined by* (2.12) *and* (2.14) *is a Bayesian strategy in* G_η *with respect to* $P_{\tilde{z}}$ (cp. Bandemer et al. (1977)).

We remark that a decomposition of G_η into a Bayesian estimation and design problem is also possible without Assumption 2.4 being valid. Here the independence of $(\tilde{\boldsymbol{\vartheta}}, \tilde{\boldsymbol{w}})$ and $\tilde{\boldsymbol{x}}$ has been assumed, on the one side, to keep the presentation clear and, on the other side, because it seems natural that the prior knowledge about the regression parameter and the error distribution is not conditioned on prior knowledge about the prediction parameter (in the form of a weighting over X_{P}).

In the sequel we will deal with the above formulated subproblems of G_η and give solutions. The Bayesian estimation problem G_η^{e} will be the subject of Part II and the Bayesian design problem G_η^{d} will be treated in Part III.

3. Choice of a prior distribution

Having specified the elements of the estimation and design problem, all what is required to determine an optimal estimator for the response surface is the prior distribution of $(\tilde{\vartheta}, \tilde{w}, \tilde{x})$ summarizing our prior knowledge. Whereas the prior $P_{\tilde{\vartheta}, \tilde{w}}$ enters both the estimation and design problem, the prior $P_{\tilde{x}}$ will be needed only for the choice of an optimal design for the estimated response but does not play a role in estimation itself.

The specification of the prior distribution is one of the major problems of any Bayesian analysis. Principially, there are two different ways for the choice of a prior distribution:

a) *objective priors* based on historical frequencies;

b) *subjective priors* based on subjective probabilities.

Objective prior distributions may be available if, on the basis of extensive experience, data lists etc., there is a solid basis for the statement of relative frequencies of the parameter values which had occured in the past. Then, if it is clear that the data are generated by a fairly stable random process, the prior distribution should be well fitted to the frequency distribution.

On the other hand, if there is no solid objective basis for the determination of such a distribution, we will always have some prior imaginations about the likely occurence of the different parameter values. Formalizing this knowledge according to a preference structure for different sets of plausible parameters we can obtain a subjective probability distribution (see SAVAGE (1954), JEFFREYS (1961)). Such an analysis will be typical of the specification of $P_{\tilde{x}}$.

In general, the parameter sets Θ, W and X_p will be overcountable so that it is impossible to specify prior probabilities for every single parameter. A practical way for the determination of a prior distribution is to start with a family of probability distributions, the algebraic expression of which depends on a fixed number of free parameters, and then to specify these free parameters according to our objective experience, theoretical reasoning or subjective considerations.

The Bayes estimator for the response surface depends on the characteristics of the so-called *posterior distribution* $P_{\tilde{\vartheta}, \tilde{w}|y}$ representing the probability distribution of $(\tilde{\vartheta}, \tilde{w})$ after the realization of the observation vector in the actual experiment (see Section 5.1). This distribution can be obtained from the prior and error distributions using Bayes' formula (see Section 4.1). It will be seen that the Bayes estimator is relatively robust against the type of prior distribution, often the Bayesian analysis requires only the specification of some moments or fractiles of this distribution. This opens the possibility to fit the prior to a suitable type of distributions which can be treated in a mathematically convenient way. An attractive starting point for this analysis is the family of conjugate priors which will be the subject of Section 4.

In the following we will consider some typical kind of prior knowledge and give some advice how it can be used to form a prior distribution $P_{\tilde{\vartheta}, \tilde{w}}$. The choice of $P_{\tilde{x}}$ will not be considered explicitly, it may be chosen in an analogous way, particularly by one of the methods indicated in the paragraphs (e) and (f) below. For the sake of simplicity let us assume the observations to be uncorrelated and having the same precision w (reciprocal of the variance σ^2), i.e.,

$$\mathbf{Cov}\, \tilde{e}(v_n) = w^{-1} I_n, \qquad w \in W = \mathbb{R}^+ .$$

3.1. Some ad-hoc proposals

(a) Suppose from previous investigations we have estimations $\hat{\vartheta}_1, \ldots, \hat{\vartheta}_m$ and $\hat{\sigma}_1^2, \ldots, \hat{\sigma}_m^2$ for ϑ and $\sigma^2 = w^{-1}$ but without further statements on the data used to obtain these values, i.e., especially we have no knowledge of the corresponding sample sizes and covariance matrices. In this case the data should be viewed as independent realizations of $\tilde{\vartheta}$ and $\tilde{w}$, respectively, all of which stem from the same data generating process. With these data we can construct the histogram of the frequency distributions of $\tilde{\vartheta}$ and $\tilde{w}$ which then can be fitted to suitable priors. Often it suffices to know the following moments of the prior distribution: the expected values

$$\sigma_0^2 := \mathbf{E}\, \tilde{w}^{-1}, \qquad \mu := \mathbf{E}\, \tilde{\vartheta} \tag{3.1}$$

and the covariance matrix

$$\Lambda := \mathbf{Cov}\, \tilde{\vartheta} = \mathbf{E}\, (\tilde{\vartheta} - \mu)(\tilde{\vartheta} - \mu)^\top . \tag{3.2}$$

In some cases we additionally require knowledge of the (conditional or unconditional) prior distribution of $\tilde{\vartheta}$. Using our previous data the above moments can be well approximated by setting

$$\sigma_0^2 = \frac{1}{m} \sum_{s=1}^{m} \hat{\sigma}_s^2, \qquad \mu = \frac{1}{m} \sum_{s=1}^{m} \hat{\vartheta}_s$$

$$\Lambda = \frac{1}{m-1} \sum_{s=1}^{m} (\hat{\vartheta}_s - \mu)(\hat{\vartheta}_s - \mu)^\top .$$

Thus, for example, if the regression coefficients can be assumed to be symmetrically distributed, then the normal distribution $P_{\tilde{\vartheta}} = \mathrm{N}(\mu, \Lambda)$ could serve as a prior for $\tilde{\vartheta}$. But in any case, if we have no further knowledge of the origin of the prior data, they should be taken with care and should be checked critically. In particular, we should guard against outliers and manipulations.

(b) Suppose that we have available estimations $\hat{\vartheta}_1, \ldots, \hat{\vartheta}_m$ and $\hat{\sigma}_1^2, \ldots, \hat{\sigma}_m^2$ from previous investigations and, additionally, have knowledge of the sample sizes $n_1, \ldots, n_m$ used to obtain these values and of the covariance matrices

$$\Lambda_s := \mathbf{Cov}\, \hat{\vartheta}_s, \qquad s = 1, \ldots, m$$

of the prior estimations for $\boldsymbol{\vartheta}$. In this case the quantities σ_0^2, $\boldsymbol{\mu}$ and $\boldsymbol{\Lambda}$ from (3.1) and (3.2) can be approximated by the weighted versions

$$\sigma_0^2 = \sum_{s=1}^{m} n_s \hat{\sigma}_s^2 \Big/ \sum_{s=1}^{m} n_s, \qquad \boldsymbol{\mu} = \sum_{s=1}^{m} n_s \hat{\boldsymbol{\vartheta}}_s \Big/ \sum_{s=1}^{m} n_s$$

$$\boldsymbol{\Lambda} = \sum_{s=1}^{m} n_s^2 \boldsymbol{\Lambda}_s \Big/ \sum_{s=1}^{m} n_s^2 .$$

The above methods of specification of $\boldsymbol{\mu}$ and $\boldsymbol{\Lambda}$ can be generalized to the case that we have one or more prior estimators for a linear combination, say $\boldsymbol{\beta} = \boldsymbol{C}\boldsymbol{\vartheta}$ with some matrix $\boldsymbol{C} \in \mathcal{M}_{q \times r}$, $q < r$. This could be the case if, for example, prior estimations for a subset $\{\vartheta_{i1}, \ldots, \vartheta_{iq}\}$ of the regression coefficients are available from previous investigations in related regression problems. Then, having specified a prior mean $\boldsymbol{\mu}_\beta$ and a prior covariance matrix $\boldsymbol{\Lambda}_\beta$ from the data in the same way as demonstrated above, these quantities can be used for a Bayesian analysis concerning $\boldsymbol{\vartheta}$ by setting formally

$$\begin{aligned} \boldsymbol{\Lambda}^{-1} &= \boldsymbol{C}^\top \boldsymbol{\Lambda}_\beta^{-1} \boldsymbol{C} \\ \boldsymbol{\Lambda}^{-1}\boldsymbol{\mu} &= \boldsymbol{C}\boldsymbol{\Lambda}_\beta^{-1} \boldsymbol{\mu}_\beta . \end{aligned}$$

With these specifications it follows that $\boldsymbol{\Lambda}$ is a singular covariance matrix and the prior distribution of $\tilde{\boldsymbol{\vartheta}}$ ist singular too, which is obviously clear if we have prior knowledge only about $\boldsymbol{\beta} = \boldsymbol{C}\boldsymbol{\vartheta}$ but not about the complete set $\{\vartheta_1, \ldots, \vartheta_r\}$ of actual regression coefficients. The Bayesian estimator with regard to this type of prior knowledge is considered in Section 8.2.

Further approaches to the use of prior estimates for actual estimation of the regression parameter are the mixed model approach (see THEIL/GOLDBERGER (1961) and THEIL (1963)) and the double stage estimation procedures as proposed for example in Al-BAYYATI/ARNOLD (1972) and MAYER/SINGH/WILLKE (1974).

(c) Suppose we have complete knowledge of the data and outcomes of m previous experiments (model replications)

$$\begin{aligned} &\tilde{\boldsymbol{y}}_i = \boldsymbol{F}_i \boldsymbol{\vartheta}_i + \tilde{\boldsymbol{e}}_i \\ &\mathsf{E}\,\tilde{\boldsymbol{e}}_i = \boldsymbol{0}, \qquad \mathsf{Cov}\,\tilde{\boldsymbol{e}}_i = \sigma_i^2 \boldsymbol{V}_i, \qquad i = 1, \ldots, m \end{aligned} \tag{3.3}$$

with $(n_i \times r)$-design matrices $\boldsymbol{F}_i$ of rank $r < n_i$, known positive definite matrices $\boldsymbol{V}_i$, independent error vectors $\boldsymbol{e}_i$ and randomly varying regression parameters assumed to have common expectation and covariance matrix

$$\mathsf{E}\,\tilde{\boldsymbol{\vartheta}}_i = \boldsymbol{\mu}, \qquad \mathsf{Cov}\,\tilde{\boldsymbol{\vartheta}}_i = \boldsymbol{\Lambda}, \qquad i = 1, \ldots, m. \tag{3.4}$$

In this case an empirical Bayes approach can be recommended, the aim of which is to estimate the prior distribution from the previous observations and to construct a Bayes strategy with this prior. Estimating the prior parameters $\boldsymbol{\mu}$ and $\boldsymbol{\Lambda}$ by

$$\hat{\boldsymbol{\mu}} = \frac{1}{m} \sum_{i=1}^{m} \hat{\boldsymbol{\vartheta}}_i = \frac{1}{m} \sum_{i=1}^{m} (\boldsymbol{F}_i^\top \boldsymbol{V}_i^{-1} \boldsymbol{F}_i)^{-1} \boldsymbol{F}_i^\top \boldsymbol{V}_i^{-1} \tilde{\boldsymbol{y}}_i$$

$$\hat{\boldsymbol{\Lambda}} = \frac{1}{m-1} \sum_{i=1}^{m} (\hat{\boldsymbol{\vartheta}}_i - \hat{\boldsymbol{\mu}})(\hat{\boldsymbol{\vartheta}}_i - \hat{\boldsymbol{\mu}})^\top - \frac{1}{m} \sum_{i=1}^{m} \hat{\sigma}_i^2 (\boldsymbol{F}_i^\top \boldsymbol{V}_i^{-1} \boldsymbol{F}_i)^{-1},$$

where $$\hat{\sigma}_i^2 = \frac{1}{n_i - r} (\tilde{\boldsymbol{y}}_i^\top \boldsymbol{V}_i^{-1} \tilde{\boldsymbol{y}}_i - \tilde{\boldsymbol{y}}_i^\top \boldsymbol{V}_i^{-1} \boldsymbol{F}_i \hat{\boldsymbol{\vartheta}}_i)$$

will lead us to an empirical Bayes estimator for $\boldsymbol{\vartheta}$ which is optimal in some sense (see Section 9.2).

Moreover, if the variance components are assumed to have common expectation $\mathbf{E}\,\tilde{\sigma}_i^2 = \sigma_0^2$, then it can be estimated by

$$\hat{\sigma}_0^2 = \frac{1}{m} \sum_{i=1}^{m} \hat{\sigma}_i^2 .$$

(d) Very often it is possible to express the prior knowledge by bounds for the response to be expected. Thus, from objective reasons it may be clear that

$$0 \leqq \eta_1 \leqq \eta(\boldsymbol{x}) \leqq \eta_u \quad \text{for all} \quad \boldsymbol{x} \in X$$

with predetermined numbers η_1 and η_u. This may be the case, for example, if the response considered is the output of a chemical process or the life time of a given item or equipment. Of course, the bounds may also depend on the points $\boldsymbol{x}$ at which observations can be taken.

In a number of cases such prior knowledge can be formulated as prior bounds for the regression coefficients of the setup. This, particularly, will be the case if they can be interpreted as physical or technological parameters of which we know, for example, that they can have only positive sign, i.e.,

$$\vartheta_j \geqq 0 \quad \text{for some} \quad j \in \{1, \ldots, r\},$$

or we know upper and lower bounds

$$\vartheta_{j1} \leqq \vartheta_j \leqq \vartheta_{ju} \quad \text{for some} \quad j \in \{1, \ldots, r\}.$$

If we have, for example, a simple linear regression setup

$$\mathbf{E}\,\tilde{y}(x) = \vartheta_0 + \vartheta_1 x, \qquad x \in [-1, 1] \tag{3.5}$$

and we know that the response at the origin can take values only in some interval $[\eta_1, \eta_u]$, then it follows in turn that $\eta_1 \leqq \vartheta_1 \leqq \eta_u$. More general, let us assume that we know prior bounds for a linear combination of the parameter vector, i.e.,

$$\boldsymbol{C}\boldsymbol{\vartheta} \leqq \boldsymbol{b} \quad \text{or} \quad \boldsymbol{a} \leqq \boldsymbol{C}\boldsymbol{\vartheta} \leqq \boldsymbol{b}, \tag{3.6}$$

where $\boldsymbol{C}$ is some $(m \times r)$-matrix and $\boldsymbol{a}, \boldsymbol{b} \in \mathbb{R}^m$ (the inequalities should be read component-wise). As an example, let be given bounds

$$a_i \leqq \eta(x_i) \leqq b_i, \qquad i = 1, \ldots, m \tag{3.7}$$

for the expected response at some points $\boldsymbol{x}_1, \ldots, \boldsymbol{x}_m \in X$. These inequalities can be written in the form of the right-hand side inequality of (3.6) by setting

$$\boldsymbol{C} = (\boldsymbol{f}(\boldsymbol{x}_1), \ldots, \boldsymbol{f}(\boldsymbol{x}_m))^\top,$$

$$\boldsymbol{a} = (a_1, \ldots, a_m)^\top, \qquad \boldsymbol{b} = (b_1, \ldots, b_m)^\top.$$

Prior knowledge of the type (3.6) means that the regression parameter is subject to linear constraints so that it belongs to some convex polyhedron. But our prior knowledge may also include nonlinear constraints such as

$$(\boldsymbol{\vartheta} - \boldsymbol{\mu})^\top \boldsymbol{A}(\boldsymbol{\vartheta} - \boldsymbol{\mu}) \leqq 1 \tag{3.8}$$

with some positive definite matrix $\boldsymbol{A}$, which means that the regression parameter may vary in an ellipsoid with center point $\boldsymbol{\mu} \in \mathbb{R}^r$. A special case of such a knowledge would be that on the basis of observations at the points of a design $v_m = (\boldsymbol{x}_1, \ldots, \boldsymbol{x}_m)$ an upper bound can be given for the norm of the vector of expected responses:

$$\|\boldsymbol{\eta}(v_m)\|^2 = \boldsymbol{\eta}(v_m)^\top \boldsymbol{\eta}(v_m) \leqq \eta_{\mathrm{u}}, \qquad \boldsymbol{\eta}(v_m) := (\eta(\boldsymbol{x}_1), \ldots, \eta(\boldsymbol{x}_m))^\top.$$

Observing that $\boldsymbol{\eta}(v_m) = \boldsymbol{F}(v_m)\boldsymbol{\vartheta}$, this is equivalent to (3.8) with $\boldsymbol{\mu} = \boldsymbol{0}$ and $\boldsymbol{A} = \eta_{\mathrm{u}}^{-1}\boldsymbol{F}(v_m)^\top \boldsymbol{F}(v_m)$. Another special case of (3.8) would be the formulation of a prior bound to the length of the regression parameter

$$\|\boldsymbol{\vartheta}\|^2 = \boldsymbol{\vartheta}^\top \boldsymbol{\vartheta} \leqq \vartheta_{\mathrm{u}}, \qquad \vartheta_{\mathrm{u}} > 0$$

which coincides with (3.8) by setting $\boldsymbol{\mu} = \boldsymbol{0}$ and $\boldsymbol{A} = \vartheta_{\mathrm{u}}^{-1}\boldsymbol{I}_r$.

(e) In its rough form, the linear and quadratic constraints just considered are inconvenient to work with. Moreover, we are never absolutely sure that our theoretical constraints are valid and in most cases we have further detailed knowledge such that not all parameter values in the specified regions have the same chance actually to occur.

Further, we may have available not only objective prior bounds but also subjective prior bounds which can be formulated on the basis of extensive experience and knowledge of the subject matter. This way we will be able to determine sets or intervals of plausible parameter values. Thus it seems more natural to express our prior information in terms of a subjective probability distribution allowing for a greater flexibility to describe our feeling about the likely values of the parameter. The construction of such a distribution starts with a partitioning of the parameter space into intervals containing the parameter with high probability and such containing the parameter with less or zero probability. Hereafter a detailed evaluation of the different intervals with subjective probabilities leads to a subjective prior. This distribution then simply reflects the chances which we are willing to give to the single parameter values.

Let us shortly reconsider the linear and quadratic constraints from above. The essential "trick" of using them for a Bayesian analysis lies in the change from the exact constraints to probabilistic constraints

$$P(\boldsymbol{a} \leqq \boldsymbol{C}\tilde{\boldsymbol{\vartheta}} \leqq \boldsymbol{b}) = 1 - \delta \quad \text{or} \qquad P((\tilde{\boldsymbol{\vartheta}} - \boldsymbol{\mu})^\top \boldsymbol{A}(\tilde{\boldsymbol{\vartheta}} - \boldsymbol{\mu}) \leqq 1) = 1 - \delta \quad (3.9)$$

$\delta \geqq 0$ sufficiently small or zero, and then assuming a special type of probability distribution satisfying (3.9), thus inducing a subjective prior. For example, if we have the feeling that all values within the polyhedron determined by the bounds $\boldsymbol{a}$ and $\boldsymbol{b}$ have the same chance to occur and values outside are impossible (i.e. $\delta = 0$), then we could specify a uniform prior distribution for $\tilde{\boldsymbol{\beta}} = \boldsymbol{C}\tilde{\boldsymbol{\vartheta}}$ with mean and covariance matrix given by

$$\boldsymbol{\mu}_\beta = \frac{1}{2}(\boldsymbol{a} + \boldsymbol{b}), \qquad \boldsymbol{\Lambda}_\beta = \frac{1}{12}\,\mathrm{diag}\,((b_1 - a_1)^2, \ldots, (b_m - a_m)^2).$$

If, instead, we have the feeling that values in the neighbourhood of the midpoint are most likely to occur and the chances of the parameter values decrease with increasing distance from the midpoint then we could specify a normal distribution

$$P_{\tilde{\beta}} = \mathrm{N}\left(\frac{1}{2}(\boldsymbol{a} + \boldsymbol{b}), \mathrm{diag}\,(\lambda_1^2, \ldots, \lambda_m^2)\right)$$

with variances $\lambda_1^2, \ldots, \lambda_m^2$ chosen according to the "3σ-rule" (which means $1 - \delta = 0.997$) as

$$\lambda_i^2 = \frac{1}{36}(b_i - a_i)^2, \qquad i = 1, \ldots, m.$$

In case of quadratic constraints (3.8) a subjective prior for $\tilde{\vartheta}$ can be induced, for example, by interpreting the constraint as a confidence ellipsoid and assuming a normal distribution $P_{\tilde{\vartheta}} = \mathrm{N}(\boldsymbol{\mu}, \boldsymbol{\Lambda})$ with $\boldsymbol{\Lambda} = \boldsymbol{A}^{-1}$.

The use of the above types of prior bounds for the construction of (Bayes) estimators for the regression parameter is reconsidered in Sections 8.3, 15.4 and 15.5.

(f) To support in the specification of a subjective prior it is recommendable to apply to experts' opinions. Suppose we are given the possibility to consult $m \geqq 1$ experts E_j each of them formulating his own subjective prior $P_{\tilde{\vartheta}}^{(i)}$ concerning the regression problem at hand. Evaluating the possibly different levels of knowledge and experience of the experts by *weights of competence*

$$c_j > 0 \quad (j = 1, \ldots, m), \qquad \sum_{j=1}^{m} c_j = 1 \tag{3.10}$$

we can obtain a final subjective prior distribution according to

$$P_{\tilde{\vartheta}} = \sum_{j=1}^{m} c_j P_{\tilde{\vartheta}}^{(j)}. \tag{3.11}$$

Hereafter we can still try to fit $P_{\tilde{\vartheta}}$ to a suitable distribution which is convenient to work with. If the number m of experts increases then $P_{\tilde{\vartheta}}$ can be well approximated by a normal distribution. The procedure just described is applicable even if the experts are not able to give a distribution because it would be too cumbersome or because of lack of knowledge. It then suffices that the experts E_j indicate a region $\Theta_j \subset \mathbb{R}^r$ which, in their opinion, includes the regression parameter and a subjective prior can be obtained as a mixture (3.11) of uniform distributions $P_{\tilde{\vartheta}}^{(j)}$ over Θ_j having density

$$p_{\tilde{\vartheta}}^{(j)}(\boldsymbol{\vartheta}) = \begin{cases} 1/\lambda_r(\Theta_j) & \text{if } \boldsymbol{\vartheta} \in \Theta_j, \\ 0 & \text{if } \boldsymbol{\vartheta} \notin \Theta_j \end{cases}$$

($\lambda_r(\cdot)$ denotes the r-dimensional Lebesgue measure).

Example 3.1. Let us consider the simple linear regression model (3.5) with independently distributed observations of the same precision w. Further, suppose that $m = 2$ experts, receiving the weights of competence $c_1 = 2/3$ and $c_2 = 1/3$, respectively, indicate us the following regions Θ_1 and Θ_2 for $\boldsymbol{\vartheta} = (\vartheta_0, \vartheta_1)$ and W_1, W_2 for the precision w of observation:

$$\Theta_1 = [-1, 0] \times [1, 2], \qquad W_1 = [1, 5],$$

$$\Theta_2 = \left[-\frac{1}{2}, \frac{1}{2}\right] \times \left[0, \frac{3}{2}\right], \qquad W_2 = [2, 4].$$

Hence, assuming the regression and precision parameter to be distributed independently, we have the following prior densities:

$$p_{\tilde{\vartheta}}^{(1)}(\boldsymbol{\vartheta}) = \begin{cases} 1 & \text{if } \boldsymbol{\vartheta} \in \Theta_1, \\ 0 & \text{else} \end{cases}, \qquad p_{\tilde{w}}^{(1)}(w) = \begin{cases} 1/4 & \text{if } w \in W_1, \\ 0 & \text{else} \end{cases},$$

$$p_{\tilde{\vartheta}}^{(2)}(\boldsymbol{\vartheta}) = \begin{cases} 2/3 & \text{if } \boldsymbol{\vartheta} \in \Theta_2, \\ 0 & \text{else} \end{cases}, \qquad p_{\tilde{w}}^{(2)}(w) = \begin{cases} 1/2 & \text{if } w \in W_2, \\ 0 & \text{else} \end{cases}.$$

Mixing these uniform distributions, we obtain the following densities of the subjective prior distributions for $\tilde{\boldsymbol{\vartheta}}$ and $\tilde{w}$:

$$p_{\tilde{w}}(w) = \begin{cases} \frac{2}{3}\cdot\frac{1}{4} + \frac{1}{3}\cdot 0 = \frac{1}{6} & \text{if } w \in [1, 2] \cup (4, 5], \\ \frac{2}{3}\cdot\frac{1}{4} + \frac{1}{3}\cdot\frac{1}{2} = \frac{1}{3} & \text{if } w \in (2, 4], \\ 0 & \text{else}, \end{cases}$$

$$p_{\tilde{\vartheta}}(\boldsymbol{\vartheta}) = \begin{cases} \frac{2}{3}\cdot 1 + \frac{1}{3}\cdot 0 = \frac{2}{3} & \text{if } \boldsymbol{\vartheta} \in \left[-1, -\frac{1}{2}\right] \times [1,2] \cup \left[-\frac{1}{2}, 0\right] \times \left[\frac{3}{2}, 2\right], \\ \frac{2}{3}\cdot 0 + \frac{1}{3}\cdot\frac{2}{3} = \frac{2}{9} & \text{if } \boldsymbol{\vartheta} \in \left[-\frac{1}{2}, 0\right] \times [0,1] \cup \left[0, \frac{1}{2}\right] \times \left[0, \frac{3}{2}\right], \\ \frac{2}{3}\cdot 1 + \frac{1}{3}\cdot\frac{2}{3} = \frac{8}{9} & \text{if } \boldsymbol{\vartheta} \in \left[-\frac{1}{2}, 0\right] \times \left[1, \frac{3}{2}\right], \\ 0 & \text{else}. \end{cases}$$

3.2. Integral geometric prior distributions

We will now describe a special method of assessing a prior distribution for the regression parameters in case that prior knowledge is available in form of inequality constraints

$$a \leqq \eta(\mathbf{x}) = \vartheta_0 + \vartheta_1 x_1 + \ldots + \vartheta_k x_k \leqq b, \quad \forall \mathbf{x} = (x_1, \ldots, x_k)^\top \in X \subset \mathbb{R}^k \quad (3.12)$$

for the response surface $\eta(\cdot)$ which is assumed to be a hyperplane. We derive so-called *integral geometric prior distributions* which are assessed in such a way that, within the region implicitly defined by our prior knowledge (3.12), any hyperplane has the same chance actually to occur. This way, the prior distribution has a natural interpretation in terms of geometric probability.

We first develop the basic idea for the simple case of straight line regression ($k = 1$) and then we indicate the generalization to the multiple linear regression case (3.12) with $k \geqq 1$. Consider the model $\mathbf{E}\tilde{y}(x) = \vartheta_0 + \vartheta_1 x$, $x \in \mathbb{R}^1$, and assume we are given prior knowledge such that

$$a \leqq \mathbf{E}\tilde{y}(x) = \vartheta_0 + \vartheta_1 x \leqq b, \qquad \forall x \in H \subset \mathbb{R}^1$$

for some interval H and predetermined bounds a, b with $-\infty < a < b < \infty$. Without loss of generality we put $H = [0, h]$ with some $h > 0$. Denote

$$S = \{\eta | \mathbb{R}^1 \to \mathbb{R}^1 : a \leqq \eta(x) = \vartheta_0 + \vartheta_1 x \leqq b, \quad \forall x \in [0, h]\}. \quad (3.13)$$

Then our goal is to specify a prior probability distribution $P_{\tilde{\vartheta}}$ for $\tilde{\vartheta} = (\tilde{\vartheta}_0, \tilde{\vartheta}_1)^\top$ which satisfies the following principles:

P 1: S receives prior probability one.

P2: The straight lines within S have equal chances of occurence.

Because of the $1-1$ correspondence between $\vartheta = (\vartheta_0, \vartheta_1)^\top$ and $\eta(x) = \vartheta_0 + \vartheta_1 x$ the problem of specifying $P_{\tilde{\vartheta}}$ is equivalent to the specification of a distribution of straight lines within S. Principle P 2 means that there is no preference among the straight lines within S. Thus, the implementation of this principle is best accomplished by considering the straight lines (hyperplanes) as random geometric objects and then searching for a measure with respect to which all the lines in S have equal mass.

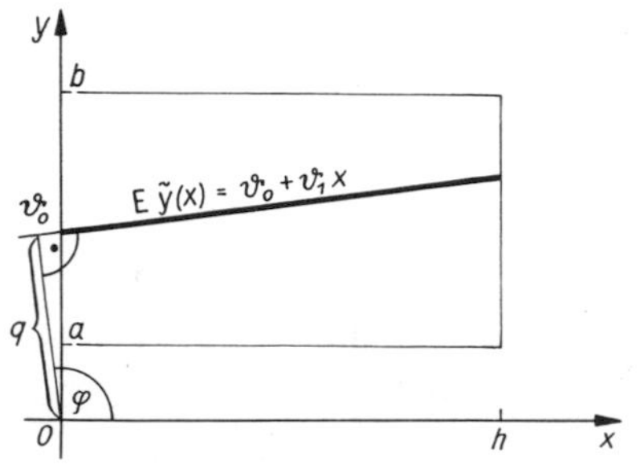

Fig. 1

Proceeding from the Hessian normal coordinates representation

$$x \cos\varphi + y \sin\varphi - q = 0,$$

where (x, y) are the coordinates of the points on the straight line, q is its distance from the origin and φ is the angle which the normal forms with the x-axis (see Fig. 1), the measure can be defined by

$$m(S) = \int_S g(q, \varphi)\, \mathrm{d}q\, \mathrm{d}\varphi, \tag{3.14}$$

where the function g is to be chosen such that the measure becomes invariant under the group of motions in the Euclidean plane. This is the usual invariance criterion chosen in integral geometry and in the theory of geometric probability (see, e.g. SANTALÓ (1976) and the introductory discussion in MILES (1974)).

Lemma 3.1. *The motion invariant measure of a given set S of straight lines is uniquely determined (up to a constant factor) by*

$$m(S) = \int_S \mathrm{d}q\, \mathrm{d}\varphi.$$

(For a proof see SANTALÓ (1976), pp. 27.)

The differential form under the integral sign is called the straight line density $\mathrm{d}s = \mathrm{d}q\, \mathrm{d}\varphi$. This density corresponds to a uniform distribution with respect to the parameters q and φ, the Hessian parametrization is the only one for which the associated invariant prior distribution is uniform. Since the Hessian parameters q, φ and the original parameters ϑ_0, ϑ_1 are related through

$$q = \vartheta_0/(1 + \vartheta_1^2)^{1/2} \quad \text{and} \quad \varphi = \operatorname{arccot}(-\vartheta_1),$$

it follows that

$$m(S) = \int_\Theta (1 + \vartheta_1^2)^{-3/2}\, \mathrm{d}\vartheta_0\, \mathrm{d}\vartheta_1, \tag{3.15}$$

where

$$\Theta = \{(\vartheta_0, \vartheta_1)^\top \in \mathbb{R}^2 : a \leqq \vartheta_0 \leqq b, \qquad (a - \vartheta_0)/h \leqq \vartheta_1 \leqq (b - \vartheta_0)/h\} \tag{3.16}$$

is the parameter set associated with S from (3.13).

Hence, we obtain the integral geometric prior density

$$p_{\tilde{\vartheta}}(\vartheta_0, \vartheta_1) = \frac{1}{m(S)} \begin{cases} (1 + \vartheta_1^2)^{-3/2} & \text{if} \quad \vartheta \in \Theta \\ 0 & \text{otherwise}, \end{cases} \tag{3.17}$$

where the normalizing constant (3.15) comes out as

$$m(S) = 2([h^2 + (b - a)^2]^{1/2} - h).$$

Then, using the abbreviations

$$c^2 = h^2 + (b - a)^2, \qquad c_1 = \frac{1}{m} \ln \frac{h}{c - b + a}, \qquad c_2 = \frac{1}{m} \ln \frac{h}{c + b - a}$$

with $m = m(S)$, straightforward calculations yield the following expressions for the first and second moments of the joint prior distribution of ϑ_0 and ϑ_1:

$$\mathbf{E}\,\tilde{\vartheta}_0 = \frac{1}{m}\,[(b+a)(c-1) - h^2(c_1+c_2)/2]\,, \qquad \mathbf{E}\,\tilde{\vartheta}_1 = h(c_1+c_2)\,,$$

$$\mathbf{E}\,\tilde{\vartheta}_0^2 = \frac{2}{3m}(c^3-h^3) - \frac{h}{m}(a^2+b^2) + \frac{2}{m}abc - h\,\mathbf{E}\,\tilde{\vartheta}_0\tilde{\vartheta}_1\,,$$

$$\mathbf{E}\,\tilde{\vartheta}_1^2 = (b-a)(c_1-c_2) - 2\,,$$

$$\mathbf{E}\,\tilde{\vartheta}_0\tilde{\vartheta}_1 = h(1+ac_1+bc_2)\,.$$

The more general case of prior bounds

$$y_1(x) \leqq \eta(x) = \vartheta_0 + \vartheta_1 x \leqq y_2(x)\,,$$

with predetermined straight lines $y_1(x)$ and $y_2(x)$ is discussed in BANDEMER/PILZ/FELLENBERG (1986), where also the marginal densities for $\tilde{\vartheta}_0$ and $\tilde{\vartheta}_1$, as well as a numerical example are given.

Let us now consider the multiple linear regression case with $k \geqq 1$ regressor variables and prior knowledge as indicated in (3.12), where X is some bounded region and a, b are predetermined real numbers. Again, we look for a prior distribution $P_{\tilde{\vartheta}}$ for $\tilde{\boldsymbol{\vartheta}} = (\tilde{\vartheta}_0, \ldots, \tilde{\vartheta}_k)^\top$ such that any hyperplane satisfying (3.12) actually has the same chance to occur. The regression equation $\mathbf{E}\,\tilde{y}(\boldsymbol{x}) = \vartheta_0 + \vartheta_1 x_1 \ldots + \vartheta_k x_k$, $\boldsymbol{x} \in \mathbb{R}^k$, defines a hyperplane in the $r = (k+1)$-dimensional Euclidean space $\mathbb{R}^r$. Then, an integral geometric prior distribution $P_{\tilde{\vartheta}}$ is defined as a probability distribution which leaves the measure for these regression hyperplanes invariant under the group of motions in $\mathbb{R}^r$. Generally, a motion in $\mathbb{R}^r$ is an affine transformation which preserves distance.

Theorem 3.1. *The integral geometric prior distribution $P_{\tilde{\vartheta}}$ for the multiple linear regression case with prior knowledge given by* (3.12) *takes the form*

$$\mathrm{d}P_{\tilde{\vartheta}} \propto \begin{cases} \left(1 + \sum_{i=1}^{k} \vartheta_i^2\right)^{-(k+2)/2} \mathrm{d}\vartheta_0\,\mathrm{d}\vartheta_1 \ldots \mathrm{d}\vartheta_k & \text{if } \boldsymbol{\vartheta} \in \Theta \\ 0 & \text{otherwise}, \end{cases}$$

where Θ is the parameter set corresponding to the constraints imposed by (3.12).

(For a proof see BANDEMER/PILZ/FELLENBERG (1986), Theorem 3.)

For the special case of straight line regression ($k = 1$) we have $\mathrm{d}P_{\tilde{\vartheta}} \propto (1+\vartheta_1^2)^{-3/2}\,\mathrm{d}\vartheta_0\,\mathrm{d}\vartheta_1$, which coincides with the result (3.17). The crucial point in the application of the above prior distribution $P_{\tilde{\vartheta}}$ is to identify the restricted parameter region Θ implicitly defined by the constraints (3.12), which is needed for the computation of the normalizing constant of the measure $P_{\tilde{\vartheta}}$.

For the special case where X describes an ellipsoidal region

$$X = \{x \in \mathbb{R}^k : x^\top H x \leqq 1\} \tag{3.18}$$

with some positive definite matrix H, the explicit form of Θ has been given by OMAN (1983), who also derives a restricted maximum likelihood estimator for this situation.

Lemma 3.2. *For the multiple linear regression model with prior knowledge given by* (3.12) *and* (3.18), *the associated parameter region* Θ *takes the form*

$$\Theta = \{(\vartheta_0, \vartheta_1, \ldots, \vartheta_k)^\top \in \mathbb{R}^{k+1} : a \leqq \vartheta_0 \leqq b, \quad (\vartheta_1, \ldots, \vartheta_k)^\top \in \bar{\Theta}(\vartheta_0)\},$$

where $\bar{\Theta}(\vartheta_0) = \{\bar{\boldsymbol{\vartheta}} \in \mathbb{R}^k : \bar{\boldsymbol{\vartheta}}' H^{-1} \bar{\boldsymbol{\vartheta}} \leqq \min((\vartheta_0 - a)^2, (b - \vartheta_0)^2)\}$ (cp. OMAN (1983), Lemma 2.1).

4. Conjugate prior distributions

As we have already mentioned, a natural way to determine a prior distribution is to choose first a family of probability distributions depending on a number of free parameters and then to specify them such that the resulting distribution yields the best fit to the prior knowledge within the chosen family. There appear to be three possibilities for the choice of such a family of distributions:

a) We make an unrestricted choice. Then, in general, the determination of the posterior distribution and of the Bayes estimator will have to proceed (approximately) by help of numerical integration methods.

b) We use the latitude which we have in the specification of the final prior and demand for certain restrictions on the family in favour of analytical tractability.

c) We choose so-called noninformative (vague, diffuse) prior distributions. These distributions are constructed according to certain invariance and information theoretic principles and can be used as standards in such cases where we have only very little prior knowledge (see Section 9.1).

We are now going to deal with a special method of choice concerning the second of the above mentioned possibilities and construct so-called conjugate prior distributions $P_{\tilde{\vartheta}, \tilde{w}}$. These are constructed in such a way that the posterior distributions are of the same type as the priors which will simplify the Bayesian analysis substantially. The conjugate prior distributions have been introduced by RAIFFA/SCHLAIFER (1961) for a general statistical decision problem.

4.1. Definition of a conjugate prior

We will start our investigations with some simplifying assumptions concerning the classes

$$\mathcal{P}_{\tilde{y}}(v_n) = \{P_{\tilde{y}|\vartheta, w} : \vartheta \in \Theta, \quad w \in W\}, \qquad v_n \in V \tag{4.1}$$

of probability distributions of the observation vectors which are induced by the error distributions from $\mathcal{P}_{\tilde{e}} = \{P_{\tilde{e}}^{(w)}\}$ (see Section 2.1).

Assumption 4.1. (i) *For each $n \in \mathbb{N}$ such that $v_n \in V$ let the distributions from $\mathcal{P}_{\tilde{y}}(v_n)$ have densities $p_{\tilde{y}|\vartheta, w}$ with respect to Lebesgue measure.*
(ii) *For each $v_n \in V$ let exist a sufficient statistic T for $\mathcal{P}_{\tilde{y}}(v_n)$.*

Let

$$C_T(v_n) = \{t \in \mathbb{R}^{r+s} : \exists y \in C(v_n) : t = T(y)\} \tag{4.2}$$

denote the sample space of the statistic $T \mid C(v_n) \to \mathbb{R}^{r+s}$, $v_n \in V$. Further, let $l(\cdot, \cdot; y, v_n) \mid \Theta \times W \to \mathbb{R}^+$ denote the likelihood function of the observation $y \in C(v_n)$, which for all $(\vartheta, w) \in \Theta \times W$, is defined by

$$l(\vartheta, w; y, v_n) = p_{\tilde{y}|\vartheta, w}(y).$$

We assume the likelihood function to be piecewise continuous and integrable over $\Theta \times W$.

Assumption 4.2. $\forall v_n \in V \, \forall y \in C(v_n) : \int_{\Theta \times W} l(\vartheta, w; y, v_n) \, \mathrm{d}w \, \mathrm{d}\vartheta < \infty$.

Further, let $\mathcal{P}$ be a class of (prior) probability distributions $P_{\tilde{\vartheta}, \tilde{w}}$ over $\Theta \times W$ having densities $p_{\tilde{\vartheta}, w}$ with respect to Lebesgue measure λ_{r+s}. Then, for a given observation y, the posterior distribution $P_{\tilde{\vartheta}, \tilde{w}|y}$ corresponding to a prior $P_{\tilde{\vartheta}, \tilde{w}} \in \mathcal{P}$ has also a density with respect to λ_{r+s} which, according to Bayes' formula (see A.15), is defined by

$$p_{\tilde{\vartheta}, \tilde{w}|y}(\cdot, \cdot) = \frac{l(\cdot, \cdot; y, v_n) \, p_{\tilde{\vartheta}, \tilde{w}}(\cdot, \cdot)}{\int_{\Theta \times W} l(\vartheta, w; y, v_n) \, p_{\tilde{\vartheta}, \tilde{w}}(\vartheta, w) \, \mathrm{d}w \, \mathrm{d}\vartheta}. \tag{4.3}$$

We are interested in such classes of prior distributions for which the posterior distributions are contained in $\mathcal{P}$ again.

Definition 4.1. The class $\mathcal{P}$ is called *closed under sampling* in V if it holds

$$\forall v_n \in V \, \forall y \in C(v_n) \, \forall P_{\tilde{\vartheta}, \tilde{w}} \in \mathcal{P} : P_{\tilde{\vartheta}, \tilde{w}|y} \in \mathcal{P}.$$

We note that this concept goes back to Wetherill (1961).

By help of the following characterization a sufficient condition can be formulated for the closedness of $\mathcal{P}$ under sampling.

Definition 4.2. The class $\mathcal{P}$ is called *closed under multiplication* if for each pair of priors $(P_1, P_2) \in \mathcal{P} \times \mathcal{P}$ there exists a prior $P_3 \in \mathcal{P}$ such that for the corresponding densities it holds: $p_1 p_2 \propto p_3$. (Here the symbol $\propto$ stands for the proportionality of functions.)

Lemma 4.1. *If $\mathcal{P}$ is closed under multiplication and if for each $v_n \in V$ and $y \in C(v_n)$ the likelihood function satisfies the relationship*

$$l(\cdot, \cdot; y, v_n) \propto p_{\tilde{\vartheta}, \tilde{w}}(\cdot, \cdot) \tag{4.4}$$

with $p_{\tilde{\vartheta}, \tilde{w}}$ being the density of some prior distribution $P_{\tilde{\vartheta}, \tilde{w}} \in \mathcal{P}$ then the class $\mathcal{P}$ is closed under sampling.

Proof. Let $\bar{P}_{\tilde{\vartheta}, \tilde{w}} \in \mathcal{P}$ be a prior with density $\bar{p}_{\tilde{\vartheta}, \tilde{w}}$ and $\bar{P}_{\tilde{\vartheta}, \tilde{w}|y}$ the corresponding posterior distribution with density $\bar{p}_{\tilde{\vartheta}, \tilde{w}|y}$. Then we have

$$\bar{p}_{\tilde{\vartheta}, \tilde{w}|y}(\cdot, \cdot) \propto \bar{p}_{\tilde{\vartheta}, \tilde{w}}(\cdot, \cdot)\, l(\cdot, \cdot; y, v_n) \qquad \text{(from (4.3))}$$

$$\propto \bar{p}_{\tilde{\vartheta}, \tilde{w}}(\cdot, \cdot)\, p_{\tilde{\vartheta}, \tilde{w}}(\cdot, \cdot) \qquad \text{(from (4.4))}$$

so that $\bar{P}_{\tilde{\vartheta}, \tilde{w}|y} \in \mathcal{P}$ because of the assumption of closedness under multiplication. □

Because of the sufficiency Assumption 4.1 (ii) the likelihood function can be factorized (see A.9) by some function k such that

$$l(\cdot, \cdot; y, v_n) \propto k(\cdot, \cdot; \boldsymbol{T}(y), v_n)$$

for all $v_n \in V$ and $y \in C(v_n)$. Such a function k we call a *kernel* of the likelihood function.

Now we construct a special class $\mathcal{P}$ of priors with densities satisfying (4.4). To do so we require the following assumption on the kernel k of the likelihood function.

Assumption 4.3. *Let exist an integer a and functions $H|V \to \mathbb{R}^a$ and $k|\Theta \times W \times \mathbb{R}^{r+s} \times \mathbb{R}^a \to \mathbb{R}^+$ such that*

$$\forall v_n \in V\, \forall y \in C(v_n): l(\cdot, \cdot; y, v_n) \propto k(\cdot, \cdot; \boldsymbol{T}(y), \boldsymbol{H}(v_n)).$$

Thus the kernel k of the likelihood function is uniquely determined by the sufficient statistic $\boldsymbol{T}$ and the function $\boldsymbol{H}$.

Remark 4.1. The essential point in Assumption 4.3 is that the dimension a of the sample space of the function $\boldsymbol{H}$ does not depend on the size of the design. Assumption 4.3 then means that $\boldsymbol{H}$ contains a sufficient information on the design used for observation.

In the following let

$$T_H = \bigcup_{v_n \in V} (C_T(v_n) \times \{\boldsymbol{H}(v_n)\}) \tag{4.5}$$

denote the joint sample space generated by the maps $\boldsymbol{T}$ and $\boldsymbol{H}$.

Definition 4.3. The function $p_{\tilde{\vartheta}, \tilde{w}}|\Theta \times W \to \mathbb{R}^+$ is called a *conjugate density* of the kernel k if there exists a pair $(\boldsymbol{t}, \boldsymbol{h}) \in T_H$ such that

$$p_{\tilde{\vartheta}, \tilde{\boldsymbol{w}}}(\cdot, \cdot) = k(\cdot, \cdot; \boldsymbol{t}, \boldsymbol{h}) \Big/ \int_{\Theta \times W} k(\boldsymbol{\vartheta}, \boldsymbol{w}; \boldsymbol{t}, \boldsymbol{h}) \, \mathrm{d}\boldsymbol{w} \, \mathrm{d}\boldsymbol{\vartheta}.$$

Because of our Assumptions 4.2 and 4.3, for all values $(\boldsymbol{t}, \boldsymbol{h}) \in T_H$ the kernel $k(\cdot, \cdot; \boldsymbol{t}, \boldsymbol{h})$ is an integrable function on $\Theta \times W$ and the conjugate densities are probability densities with respect to Lebesgue measure λ_{r+s}. The corresponding set of distributions

$$\mathcal{P}_k = \{P_{\tilde{\vartheta}, \tilde{\boldsymbol{w}}} : \mathrm{d}P_{\tilde{\vartheta}, \tilde{\boldsymbol{w}}}/\mathrm{d}\lambda_{r+s} \propto k(\cdot, \cdot; \boldsymbol{t}, \boldsymbol{h})\}_{(\boldsymbol{t}, \boldsymbol{h}) \in T_H} \tag{4.6}$$

generated by these densities is called the *class of conjugate prior distributions.*

4.2. Properties of conjugate priors

We now give condtions under which the class of conjugate prior distributions is closed under sampling and make statements on the form of the corresponding posterior distributions.

From the definition of $\mathcal{P}_k$ it is clear that the relationship (4.4) is satisfied for every prior $P_{\tilde{\vartheta}, \tilde{\boldsymbol{w}}} \in \mathcal{P}_k$. Hence, from Lemma 4.1 follows immediately

Lemma 4.2. *The class $\mathcal{P}_k$ of conjugate prior distributions is closed under sampling if it is closed under multiplication. This is the case if it holds*

$$\begin{aligned} &\forall (\boldsymbol{t}_1, \boldsymbol{h}_1) \in T_H \qquad \forall (\boldsymbol{t}_2, \boldsymbol{h}_2) \in T_H \qquad \exists (\boldsymbol{t}_0, \boldsymbol{h}_0) \in T_H: \\ &k(\cdot, \cdot; \boldsymbol{t}_1, \boldsymbol{h}_1) \, k(\cdot, \cdot; \boldsymbol{t}_2, \boldsymbol{h}_2) \propto k(\cdot, \cdot; \boldsymbol{t}_0, \boldsymbol{h}_0). \end{aligned} \tag{4.7}$$

This means that the posterior distributions have the same structure as the prior distributions if Condition (4.7) is satisfied.

Assumption 4.4. *Let the observations be independent, i.e.,*

$$l(\cdot, \cdot; \boldsymbol{y}, v_n) = \prod_{i=1}^{n} l(\cdot, \cdot; y_i, \boldsymbol{x}_i)$$

for every $v_n = (\boldsymbol{x}_1, \ldots, \boldsymbol{x}_n) \in V$ and $\boldsymbol{y} = (y_1, \ldots, y_n) \in C(v_n)$.

It will turn out that in the case of independent observations the kernel k of the likelihood function has the Property (4.7) if the set V is closed under addition of designs as defined subsequently.

Definition 4.4. Let be $v_n = (\boldsymbol{x}_1, \ldots, \boldsymbol{x}_n) \in X_{\mathrm{E}}^n$ and $v_m = (\boldsymbol{x}'_1, \ldots, \boldsymbol{x}'_m) \in X_{\mathrm{E}}^m$. The $(n+m)$-tuple

$$v_n + v_m := (\boldsymbol{x}_1, \ldots, \boldsymbol{x}_n, \boldsymbol{x}'_1, \ldots, \boldsymbol{x}'_m)$$

is called the *sum* of the designs v_n and v_m.

Let

$$\mathbb{N}_0 = \{n \in \mathbb{N} : X_{\mathrm{E}}^n \cap V \neq \varnothing\} \tag{4.8}$$

denote the set of possible sample sizes.

Assumption 4.5. $\forall (n, m) \in \mathbb{N}_0 \times \mathbb{N}_0 \, \forall (v_n, v_m) \in V \times V$:

$$n + m \in \mathbb{N}_0 \Rightarrow v_n + v_m \in V.$$

Now we introduce binary algebraic operations in the parameter region T_H of the class of conjugate priors. These operations will help to simplify the computation of the posterior distributions.

Definition 4.5. (i) For all pairs $(\boldsymbol{t}_1, \boldsymbol{h}_1) = (\boldsymbol{T}(\boldsymbol{y}_1), \boldsymbol{H}(v_n)) \in T_H$ and $(\boldsymbol{t}_2, \boldsymbol{h}_2) = (\boldsymbol{T}(\boldsymbol{y}_2), \boldsymbol{H}(v_m)) \in T_H$ let be

$$\boldsymbol{h}_1 \ast \boldsymbol{h}_2 := \boldsymbol{H}(v_n + v_m) \quad \text{and} \quad \boldsymbol{t}_1 \oplus \boldsymbol{t}_2 := \boldsymbol{T}(\boldsymbol{y}_1, \boldsymbol{y}_2).$$

(ii) The operation $\odot$ is called *composition* in T_H iff

$$\forall (\boldsymbol{t}_1, \boldsymbol{h}_1) \in T_H \, \forall (\boldsymbol{t}_2, \boldsymbol{h}_2) \in T_H : (\boldsymbol{t}_1, \boldsymbol{h}_1) \odot (\boldsymbol{t}_2, \boldsymbol{h}_2) = (\boldsymbol{t}_1 \oplus \boldsymbol{t}_2, \boldsymbol{h}_1 \ast \boldsymbol{h}_2).$$

Because of Assumption 4.5, the set T_H is closed under the operation $\odot$, i.e., the composition $(\boldsymbol{t}_1 \oplus \boldsymbol{t}_2, \boldsymbol{h}_1 \ast \boldsymbol{h}_2)$ of any two pairs $(\boldsymbol{t}_1, \boldsymbol{h}_1)$ and $(\boldsymbol{t}_2, \boldsymbol{h}_2)$ from T_H again yields an element of T_H.

Lemma 4.3. *Under the Assumptions* 4.3–4.5 *the kernel k of the likelihood function satisfies, for all parameters* $(\boldsymbol{t}_1, \boldsymbol{h}_1)$ *and* $(\boldsymbol{t}_2, \boldsymbol{h}_2)$ *from* T_H, *the relation*

$$k(\cdot, \cdot; \boldsymbol{t}_1, \boldsymbol{h}_1) \, k(\cdot, \cdot; \boldsymbol{t}_2, \boldsymbol{h}_2) \propto k(\cdot, \cdot; \boldsymbol{t}_1 \oplus \boldsymbol{t}_2, \boldsymbol{h}_1 \ast \boldsymbol{h}_2).$$

Proof. Let be $v_n, v_m \in V$; $\boldsymbol{y}_1 \in C(v_n)$, $\boldsymbol{y}_2 \subset C(v_m)$; $\boldsymbol{h}_1 = \boldsymbol{H}(v_n)$, $\boldsymbol{h}_2 = \boldsymbol{H}(v_m)$ and $\boldsymbol{t}_i = \boldsymbol{T}(\boldsymbol{y}_i)$; $i = 1, 2$. With Assumption 4.4 it follows for the likelihood function of the compound observation $\boldsymbol{y} = (\boldsymbol{y}_1^\top, \boldsymbol{y}_2^\top)^\top$:

$$\begin{aligned} l(\cdot, \cdot; \boldsymbol{y}, v_n + v_m) &= l(\cdot, \cdot; \boldsymbol{y}_1, v_n) \, l(\cdot, \cdot; \boldsymbol{y}_2, v_m) \\ &\propto k(\cdot, \cdot; \boldsymbol{t}_1, \boldsymbol{h}_1) \, k(\cdot, \cdot; \boldsymbol{t}_2, \boldsymbol{h}_2). \end{aligned}$$

On the other hand we have with Assumption 4.3

$$\begin{aligned} l(\cdot, \cdot; \boldsymbol{y}, v_n + v_m) &\propto k(\cdot, \cdot; \boldsymbol{T}(\boldsymbol{y}), \boldsymbol{H}(v_n + v_m)) \\ &= k(\cdot, \cdot; \boldsymbol{t}_1 \oplus \boldsymbol{t}_2, \boldsymbol{h}_1 \ast \boldsymbol{h}_2). \quad \square \end{aligned}$$

From this we get with Lemma 4.2 the following result:

Theorem 4.1. *Let the Assumptions* 4.1–4.5 *be satisfied. Then the class* $\mathcal{P}_k$ *of conjugate prior distributions is closed under sampling.*

Thus, Theorem 4.1 states that

$$\forall P_{\tilde{\vartheta}, \tilde{w}} \in \mathcal{P}_k \, \forall v_n \in V \, \forall \boldsymbol{y} \in C(v_n) : P_{\tilde{\vartheta}, \tilde{w} | \boldsymbol{y}} \in \mathcal{P}_k.$$

The structure of the posterior distributions corresponding to conjugate priors is determined by the kernel k of the likelihood function. The parameters of the posterior distribution can be obtained by the composition of the parameters of the prior distribution and the likelihood function.

Theorem 4.2. *Let the Assumptions* 4.1–4.5 *be valid and let* $p_{\tilde{\vartheta}, \tilde{w}}$ *be the density of a conjugate prior with parameter* $(t, \boldsymbol{h}) \in T_H$, *i.e.,* $p_{\tilde{\vartheta}, \tilde{w}} \propto k(\cdot, \cdot; t, \boldsymbol{h})$. *Then the posterior density for given design* $v_n \in V$ *and observation* $\boldsymbol{y} \in C(v_n)$ *is determined by*

$$p_{\tilde{\vartheta}, w|y}(\cdot, \cdot) \propto k(\cdot, \cdot; \boldsymbol{t} \oplus \boldsymbol{T}(\boldsymbol{y}), \boldsymbol{h} \circledast \boldsymbol{H}(v_n)) .$$

Proof. Under the Assumptions 4.1–4.3 we have, according to Bayes' formula (cp. (4.3)),

$$\begin{aligned} p_{\tilde{\vartheta}, \tilde{w}|y}(\cdot, \cdot) &\propto p_{\tilde{\vartheta}, \tilde{w}}(\cdot, \cdot)\, l(\cdot, \cdot; \boldsymbol{y}, v_n) \\ &\propto k(\cdot, \cdot; \boldsymbol{t}, \boldsymbol{h})\, k(\cdot, \cdot; \boldsymbol{T}(\boldsymbol{y}), \boldsymbol{H}(v_n)) , \end{aligned}$$

so that the assertion follows from Lemma 4.3. □

Thus, for given parameters of the prior distribution, the posterior distribution is uniquely characterized by the values of the sufficient statistic $\boldsymbol{T}$ and the function $\boldsymbol{H}$.

Remark 4.2. The closedness of the class of conjugate priors under sampling and the composition operation in the parameter region of this class have been investigated here only for the case of independent observations. In the general case, the Condition (4.7) must be examined to prove the closedness of the class $\mathcal{P}_k$ (see Lemma 4.2). According to this condition, we can then define as composition of the parameters $(\boldsymbol{t}_1, \boldsymbol{h}_1) \in T_H$ and $(\boldsymbol{t}_2, \boldsymbol{h}_2) \in T_H$ that parameter $(\boldsymbol{t}_0, \boldsymbol{h}_0) \in T_H$ for which $k(\cdot, \cdot; \boldsymbol{t}_1, \boldsymbol{h}_1)\ k(\cdot, \cdot; \boldsymbol{t}_2, \boldsymbol{h}_2) \propto k(\cdot, \cdot; \boldsymbol{t}_0, \boldsymbol{h}_0)$.

In general, the class $\mathcal{P}_k$ of conjugate priors will be rich enough, so that the prior knowledge can be represented by some probability distribution from $\mathcal{P}_k$ with a suitably chosen parameter out of T_H. On the other hand, the class $\mathcal{P}_k$ can still be enlarged by extending the parameter region T_H. According to Assumptions 4.1 and 4.3, T_H is a subset of $\mathbb{R}^{r+s} \times \mathbb{R}^a$. Now, let T_H^0 be a set with the property

$$T_H \subset T_H^0 \subseteq \mathbb{R}^{r+s} \times \mathbb{R}^a .$$

Definition 4.6. The function $k^0 | \Theta \times W \times T_H^0 \to \mathbb{R}^+$ is called an *extended kernel* of the likelihood function if k^0 is a continuous extension of $k | \Theta \times W \times T_H \to \mathbb{R}^+$ and $k^0(\cdot, \cdot; \boldsymbol{t}, \boldsymbol{h})$ is integrable even for all parameters $(\boldsymbol{t}, \boldsymbol{h}) \in T_H^0 \backslash T_H$.

If the operation $\odot$ can be set forth continuously on $T_H^0 \times T_H^0$ and if T_H^0 is closed under this operation, then we call

$$\mathcal{P}_{k^0} = \{P_{\tilde{\vartheta}, \tilde{w}} : \mathrm{d}P_{\tilde{\vartheta}, \tilde{w}}/\mathrm{d}\lambda_{r+s} \propto k^0(\cdot, \cdot; \boldsymbol{t}, \boldsymbol{h}), (\boldsymbol{t}, \boldsymbol{h}) \in T_H^0\} \tag{4.9}$$

an *extended class of conjugate prior distributions.*

The extended class $\mathcal{P}_{k^0}$ is closed under sampling, too, and the posterior distributions can be obtained via Theorem 4.2 with the composition being set forth on $T_H^0 \times T_H^0$.

With the concept of the conjugate priors we have available a family of prior distributions which allows for sufficient latitude in the choice of a distribution expressing our prior information and which is at the same time analytically well tractable. All we have to do is the adequate transformation of prior knowledge into the specification of the free parameters $\boldsymbol{t}$ and $\boldsymbol{h}$ such that the resulting distribution fits the prior knowledge best. The posterior distribution is of the same type as the prior, which assures a good comparability, and the posterior parameters can be obtained by relatively simple algebraic operations. All this leads to a simplification of the analytical computation of the Bayes estimator which will be based on the statistic $\boldsymbol{T}$ and the function $\boldsymbol{H}$.

The application of conjugate priors is very promising for a sequential estimation procedure, where the posterior distribution, obtained from the prior and a first sample, serves as prior distribution for Bayes estimation in connection with a subsequent sample. The posterior distribution at each stage of the sequential procedure can be computed by successive composition of the posterior parameters obtained at the preceding stage and the data of the actual sample.

Essentially, the concept of conjugate priors assumes the prior information to arise from previous samples which are generated by the same stochastic process as the actual observations. This is an obvious disadvantage because it forces the prior information to satisfy certain constraints. Thus, in practical applications it is necessary to prove the assumption of a conjugate prior carefully and to guard against deviations from it (cp. Section 4.4).

4.3. Conjugate prior and posterior distributions for normal errors

We consider the linear regression model with independently and identically distributed errors of unknown variance and determine an extended class of conjugate prior distributions and, by help of the algebraic operations introduced in Definition 4.5, we find the corresponding posterior distributions.

Thus, let the class of error distributions be given by

$$\mathcal{P}_{\tilde{e}} = \left\{\mathrm{N}\left(\mathbf{0}, \frac{1}{w}\boldsymbol{I}_n\right)\right\}_{w \in \mathbb{R}^+},$$

where $W = \mathbb{R}^+$ is the index set of this class with elements w standing for the possible precisions (reciprocal variances) of the observations. The class of distributions of the observation vector induced by $\mathcal{P}_{\tilde{e}}$ is given by

$$\mathcal{P}_{\tilde{y}}(v_n) = \left\{ P_{\tilde{y}|\vartheta,w} = \mathrm{N}\left(\boldsymbol{F}(v_n)\boldsymbol{\vartheta}, \frac{1}{w}\boldsymbol{I}_n \right) : \boldsymbol{\vartheta} \in \Theta = \mathbb{R}^r, w \in W = \mathbb{R}^+ \right\}$$

and the likelihood function of the observation $\boldsymbol{y} \in C(v_n)$ has the form

$$\forall \boldsymbol{\vartheta} \in \Theta \, \forall w \in W : l(\boldsymbol{\vartheta}, w; \boldsymbol{y}, v_n) = (w/2\pi)^{n/2} \exp\left(-\frac{w}{2} \| \boldsymbol{y} - \boldsymbol{F}(v_n)\boldsymbol{\vartheta} \|^2_{\boldsymbol{I}_n} \right). \quad (4.10)$$

Let $\boldsymbol{T}_1$ define the r-dimensional statistic

$$\boldsymbol{T}_1(\tilde{\boldsymbol{y}}(v_n)) = (\boldsymbol{F}(v_n)^\top \boldsymbol{F}(v_n))^+ \boldsymbol{F}(v_n)^\top \tilde{\boldsymbol{y}}(v_n), \quad (4.11)$$

where $(\boldsymbol{F}(v_n)^\top \boldsymbol{F}(v_n))^+$ represents the MP-inverse (see A.36) of the matrix $\boldsymbol{F}(v_n)^\top \boldsymbol{F}(v_n)$. Then for arbitrary $\boldsymbol{y} \in C(v_n)$ and $\boldsymbol{\vartheta} \in \Theta$ it holds

$$\| \boldsymbol{y} - \boldsymbol{F}(v_n)\boldsymbol{\vartheta} \|^2_{\boldsymbol{I}_n} = \| \boldsymbol{y} - \boldsymbol{F}(v_n)\boldsymbol{T}_1(\boldsymbol{y}) \|^2_{\boldsymbol{I}_n} + \| \boldsymbol{\vartheta} - \boldsymbol{T}_1(\boldsymbol{y}) \|^2_{\boldsymbol{F}(v_n)^\top \boldsymbol{F}(v_n)} .$$

With the statistic

$$T_2(\tilde{\boldsymbol{y}}(v_n)) = \| \tilde{\boldsymbol{y}}(v_n) - \boldsymbol{F}(v_n)\boldsymbol{T}_1(\tilde{\boldsymbol{y}}(v_n)) \|^2_{\boldsymbol{I}_n} \quad (4.12)$$

the likelihood function can be written as

$$l(\boldsymbol{\vartheta}, w; \boldsymbol{y}, v_n) = (w/2\pi)^{n/2} \exp\left(-\frac{w}{2} [T_2(\boldsymbol{y}) + \| \boldsymbol{\vartheta} - \boldsymbol{T}_1(\boldsymbol{y}) \|^2_{\boldsymbol{F}(v_n)^\top \boldsymbol{F}(v_n)}] \right).$$

This means that

$$\boldsymbol{T}(\tilde{\boldsymbol{y}}(v_n)) := (\boldsymbol{T}_1(\tilde{\boldsymbol{y}}(v_n)), T_2(\tilde{\boldsymbol{y}}(v_n))), \qquad v_n \in V$$

is a sufficient statistic for $\mathcal{P}_{\tilde{y}}(v_n)$.

Further, let $\boldsymbol{H}|V \to \mathcal{M}_r^{\geqq} \times \mathbb{N}$ be a function defined by

$$\boldsymbol{H}(v_n) = (\boldsymbol{H}_1(v_n), H_2(v_n)) := ((\boldsymbol{F}(v_n)^\top \boldsymbol{F}(v_n))^+, n - r). \quad (4.13)$$

Then the function $k(\cdot, \cdot; \boldsymbol{T}(\boldsymbol{y}), \boldsymbol{H}(v_n))$ defined on $\Theta \times W$ by

$$k(\boldsymbol{\vartheta}, w; \boldsymbol{T}(\boldsymbol{y}), \boldsymbol{H}(v_n)) = w^{r/2} w^{H_2(v_n)/2} \exp\left(-\frac{w}{2} [T_2(\boldsymbol{y}) + \| \boldsymbol{\vartheta} - \boldsymbol{T}_1(\boldsymbol{y}) \|^2_{\boldsymbol{H}_1^+(v_n)}] \right) \quad (4.14)$$

is a kernel of the likelihood function $l(\cdot, \cdot; \boldsymbol{y}, v_n)$.

The function $\boldsymbol{H}$ can be written as a map into the Euclidean space $\mathbb{R}^{r^2+1}$ when identifying $\boldsymbol{H}(v_n) = ((\boldsymbol{F}(v_n)^\top \boldsymbol{F}(v_n))^+, n - r) =: ((f_{ij}(v_n))_{r \times r}, n - r)$ with the vector $(f_{11}(v_n), \ldots, f_{r1}(v_n), \ldots, f_{rr}(v_n), n - r)^\top$. Thus, particularly, Assumption 4.3 is satisfied.

Since $\mathcal{P}_{\tilde{y}}(v_n)$ satisfies Assumption 4.4 the composition operation is applicable (see Definition 4.5).

Lemma 4.4. *Let Assumption* 4.5 *be satisfied and denote* $\boldsymbol{t}_i = (\boldsymbol{t}_{1i}, t_{2i})$

$\in \mathbb{R}^r \times \mathbb{R}^+$, $\boldsymbol{h}_i = (\boldsymbol{h}_{1i}, h_{2i}) \in \mathscr{M}_r^{\geqq} \times \mathbb{N}$, $i = 1, 2$. *Then it holds*

$$\boldsymbol{h}_1 \ast \boldsymbol{h}_2 = ((\boldsymbol{h}_{11}^+ + \boldsymbol{h}_{12}^+)^+, h_{21} + h_{22} - r)$$

$$\boldsymbol{t}_1 \oplus \boldsymbol{t}_2 = \left(\boldsymbol{t}_0, \sum_{i=1}^{2} (t_{2i} + \|\boldsymbol{t}_{1i} - \boldsymbol{t}_0\|^2_{\boldsymbol{h}_{1i}^+}) \right),$$

where $\boldsymbol{t}_0 = (\boldsymbol{h}_{11}^+ + \boldsymbol{h}_{12}^+)^+(\boldsymbol{h}_{11}^+ \boldsymbol{t}_{11} + \boldsymbol{h}_{12}^+ \boldsymbol{t}_{12})$.

(For a proof see PILZ (1979b).)

The function $k(\cdot, \cdot; \boldsymbol{T}(\boldsymbol{y}), \boldsymbol{H}(v_n))$ from (4.14) is integrable with respect to Lebesgue measure on $\Theta \times W = \mathbb{R}^r \times \mathbb{R}^+$ if $\boldsymbol{H}_1^+(v_n) = \boldsymbol{F}(v_n)^\top \boldsymbol{F}(v_n)$ is positive definite. This is the case if and only if V is restricted to the class of regular designs, i.e., we require

$$V \subseteqq V_{\text{reg}} := \left\{ v_n \in \bigcup_{m \geqq r} X_{\mathrm{E}}^m : \operatorname{rank} \boldsymbol{F}(v_n) = r \right\}.$$

With $V \subseteqq V_{\text{reg}}$ the sample spaces $C_T(v_n)$ of $\boldsymbol{T}(\tilde{\boldsymbol{y}}(v_n))$, $v_n \in V$, are subsets of $\mathbb{R}^r \times \mathbb{R}^+$, in case of $C(v_n) = \mathbb{R}^n$ we have $C_T(v_n) = \mathbb{R}^r \times \mathbb{R}^+$. The values of $\boldsymbol{H}_1(\cdot)$ and $H_2(\cdot)$ are positive definite matrices of order r and integers, respectively. Consequently, with $V \subseteqq V_{\text{reg}}$ it holds:

$$T_H \subseteqq \bigcup_{v_n \in V} (C_T(v_n) \times \{\boldsymbol{H}(v_n)\}) \subset \mathbb{R}^r \times \mathbb{R}^+ \times \mathscr{M}_r^{>} \times \mathbb{N}.$$

The function $k | \Theta \times W \times T_H \to \mathbb{R}^+$ can be set forth continuously to an extended kernel $k^0 | \Theta \times W \times T_H^0 \to \mathbb{R}^+$ with

$$T_H^0 = \mathbb{R}^r \times \mathbb{R}^+ \times \mathscr{M}_r^{>} \times \mathbb{R}^+$$

since k^0 is positive and integrable with respect to Lebesgue measure for all parameters $(\boldsymbol{t}, \boldsymbol{h}) = ((\boldsymbol{t}_1, t_2), (\boldsymbol{h}_1, h_2)) \in T_H^0$. Further, from Lemma 4.4 it is obvious that the composition operation can be set forth on $T_H^0 \times T_H^0$. The operation $\ast$ is the usual addition under which the set $\mathscr{M}_r^{>} \times \mathbb{R}^+$ is closed. Likewise, extension of the $\oplus$-operation to values from $\mathbb{R}^r \times \mathbb{R}^+$ does not lead out of this set. Hence, T_H^0 is closed under composition.

The conjugate density $p_{\tilde{\boldsymbol{\vartheta}}, \tilde{w}}(\cdot, \cdot; \boldsymbol{t}_1, t_2, \boldsymbol{h}_1, h_2) \propto k^0(\cdot, \cdot; \boldsymbol{t}_1, t_2, \boldsymbol{h}_1, h_2)$ of the extended kernel k^0 takes the form

$$p_{\tilde{\boldsymbol{\vartheta}}, \tilde{w}}(\boldsymbol{\vartheta}, w; \boldsymbol{t}_1, t_2, \boldsymbol{h}_1, h_2) = c w^{r/2} w^{h_2/2} \exp\left(-\frac{w}{2} (t_2 + \|\boldsymbol{\vartheta} - \boldsymbol{t}_1\|^2_{\boldsymbol{h}_1^+}) \right),$$

where

$$c^{-1} = \int_{\mathbb{R}^r} \int_{\mathbb{R}^+} k^0(\boldsymbol{\vartheta}, w; \boldsymbol{t}_1, t_2, \boldsymbol{h}_1, h_2) \, \mathrm{d}w \, \mathrm{d}\boldsymbol{\vartheta}$$

is the normalizing constant. This is precisely the density of the normal-gamma-distribution $\mathrm{NG}(\boldsymbol{t}_1, t_2, \boldsymbol{h}_1, h_2)$ (see A.3).

Theorem 4.3. *Let be* $\Theta = \mathbb{R}^r$, $W = \mathbb{R}^+$, $V \subseteqq V_{\text{reg}}$ *and* $\mathscr{P}_{\tilde{e}} = \left\{ \mathrm{N}\left(\boldsymbol{0}, \frac{1}{w} \boldsymbol{I}_n\right) \right\}_{w \in W}$.

Then the family of normal-gamma-distributions

$$\mathcal{P}_k^0 = \{P_{\tilde{\vartheta}, \tilde{w}} = \mathrm{NG}(\boldsymbol{\mu}, \alpha, \boldsymbol{\Phi}, \beta) : \boldsymbol{\mu} \in \mathbb{R}^r, \alpha \in \mathbb{R}^+, \boldsymbol{\Phi} \in \mathcal{M}_r^>, \beta \in \mathbb{R}^+\}$$

is an extended class of conjugate priors for $(\tilde{\vartheta}, \tilde{w})$.

Because the class $\mathcal{P}_{\tilde{y}}(v_n)$ of independently and identically normally distributed observations satisfies the Assumptions 4.1–4.4 and T_H^0 is closed under composition, the posterior distributions corresponding to priors from $\mathcal{P}_{k^0}$ are normal-distributions, too (cp. Theorem 4.1). Its parameters can be obtained via Theorem 4.2 replacing k by k^0 and extending the composition operations given in Lemma 4.5.

Theorem 4.4. *Let be* Θ, W, $\mathcal{P}_{\tilde{e}}$ *as in Theorem 4.3 and* V *arbitrary, not necessarily contained in* V_{reg}. *Further, let* $P_{\tilde{\vartheta}, \tilde{w}} = \mathrm{NG}(\boldsymbol{\mu}, \alpha, \boldsymbol{\Phi}, \beta)$ *be a normal-gamma-prior with parameters* $\boldsymbol{\mu} \in \mathbb{R}^r$, $\alpha \in \mathbb{R}^+$, $\boldsymbol{\Phi} \in \mathcal{M}_r^>$ *and* $\beta \in \mathbb{R}^+$. *Then the posterior distribution of* $(\tilde{\vartheta}, \tilde{w})$ *for given design* $v_n \in V$ *and observation* $\boldsymbol{y} \in C(v_n)$ *is normal gamma,* $P_{\tilde{\vartheta}, \tilde{w} | y} = \mathrm{NG}(\boldsymbol{\mu}_0, \alpha_0, \boldsymbol{\Phi}_0, \beta_0)$, *where*

$$\boldsymbol{\mu}_0 = \boldsymbol{\Phi}_0(\boldsymbol{F}(v_n)^\top \boldsymbol{y} + \boldsymbol{\Phi}^{-1}\boldsymbol{\mu}), \qquad \alpha_0 = \alpha + \|\boldsymbol{\mu} - \boldsymbol{\mu}_0\|^2_{\boldsymbol{\Phi}^{-1}} + \|\boldsymbol{y} - \boldsymbol{F}(v_n)\boldsymbol{\mu}_0\|^2$$
$$\boldsymbol{\Phi}_0^{-1} = \boldsymbol{F}(v_n)^\top \boldsymbol{F}(v_n) + \boldsymbol{\Phi}^{-1}, \qquad \beta_0 = \beta + n - r.$$

Proof. The posterior parameters can be derived from Lemma 4.4 replacing $(\boldsymbol{t}_{11}, t_{21}, \boldsymbol{h}_{11}, h_{21})$ by $(\boldsymbol{\mu}, \alpha, \boldsymbol{\Phi}, \beta)$ and $(\boldsymbol{t}_{12}, t_{22}, \boldsymbol{h}_{12}, h_{22})$ by $(\boldsymbol{T}_1(\boldsymbol{y}), T_2(\boldsymbol{y}), \boldsymbol{H}_1(v_n), H_2(v_n))$ and then observing that

$$\boldsymbol{\Phi}_0^+ = \boldsymbol{\Phi}_0^{-1}, (\boldsymbol{F}^\top \boldsymbol{F})^+ \boldsymbol{F}^\top \boldsymbol{F}(\boldsymbol{F}^\top \boldsymbol{F})^+ = (\boldsymbol{F}^\top \boldsymbol{F})^+ \quad \text{and}$$
$$\boldsymbol{F}^\top \boldsymbol{F}(\boldsymbol{F}^\top \boldsymbol{F})^+ \boldsymbol{F}^\top \boldsymbol{y} = \boldsymbol{F}^\top \boldsymbol{y}. \ \square$$

It should be noted that we do not require the design v_n to be regular when deriving the posterior distribution. This is due to the fact that $\boldsymbol{\Phi}$ is positive definite and $\boldsymbol{F}(v_n)^\top \boldsymbol{F}(v_n)$ is positive semidefinite for arbitrary $v_n \in V$, so that $\boldsymbol{\Phi}_0^{-1} = \boldsymbol{F}(v_n)^\top \boldsymbol{F}(v_n) + \boldsymbol{\Phi}^{-1}$ will always be of full rank and, a posteriori, a proper probability distribution comes out.

Let us still formulate the corresponding results for the special case of known precision $w_0 > 0$ of observation. In this case the statistic $\boldsymbol{T}_1$ from (4.11) is sufficient for

$$\bar{\mathcal{P}}_{\tilde{y}}(v_n) = \{P_{\tilde{y}|\vartheta} = \mathrm{N}(\boldsymbol{F}(v_n)\boldsymbol{\vartheta}, w_0^{-1}\boldsymbol{I}_n) : \boldsymbol{\vartheta} \in \Theta = \mathbb{R}^r\}$$

and the function $k(\cdot; \boldsymbol{T}_1(\boldsymbol{y}), \boldsymbol{H}_1(v_n))$ defined on Θ by

$$\bar{k}(\boldsymbol{\vartheta}; \boldsymbol{T}_1(\boldsymbol{y}), \boldsymbol{H}_1(v_n)) = \exp\left(-\frac{1}{2} w_0 \|\boldsymbol{\vartheta} - \boldsymbol{T}_1(\boldsymbol{y})\|^2_{\boldsymbol{H}_1^+(v_n)}\right),$$

where $\boldsymbol{H}_1$ is as defined in (4.13), is a kernel of the likelihood function. Thus, in analogy with the preceding investigations, we get the following result:

Theorem 4.5. *Let be* $\Theta = \mathbb{R}^r$ *and* $\mathcal{P}_{\tilde{y}}(v_n) = \bar{\mathcal{P}}_{\tilde{y}}(v_n)$ *as given above.*

(i) *If* $V \subseteqq V_{\text{reg}}$ *then the family*

$$\mathcal{P}_{\bar{k}^0} = \{P_{\tilde{\vartheta}} = \mathrm{N}(\boldsymbol{\mu}, \boldsymbol{\Phi}) : \boldsymbol{\mu} \in \mathbb{R}^r, \boldsymbol{\Phi} \in \mathcal{M}_r^>\}$$

of normal distributions is an extended class of conjugate prior distributions for $\tilde{\vartheta}$.

(ii) *Let be* $P_{\tilde{\vartheta}} = \mathrm{N}(\boldsymbol{\mu}, w_0^{-1}\boldsymbol{\Phi}) \in \mathcal{P}_{\bar{k}}^0$ *and* $\boldsymbol{\mu}_0$, $\boldsymbol{\Phi}_0$ *as in Theorem* 4.4. *Then the posterior distribution of* $\tilde{\vartheta}$ *for given observation* $y \in C(v_n)$ *is* $P_{\tilde{\vartheta}|y} = \mathrm{N}(\boldsymbol{\mu}_0, w_0^{-1}\boldsymbol{\Phi}_0)$.

Remark 4.3. Conjugate prior distributions for normal linear models involving a restricted parameter set Θ originating, for example, from linear or quadratic inequality constraints are derived in BUNKE/BUNKE (1986), Section 7.1.

4.4. Some remarks on the specification of a normal-gamma-prior

In this section we will give some advices how to specify a conjugate normal-gamma-prior distribution in case of normally distributed observations with likelihood function given by (4.10).

First we deal with the case of objective prior knowledge based on historical frequencies making available approximations of the prior parameters

$$\boldsymbol{\mu} = \mathbf{E}\,\tilde{\vartheta}, \qquad \boldsymbol{\Lambda} = \mathbf{Cov}\,\tilde{\vartheta}, \qquad \sigma_0^2 = \mathbf{E}\,\tilde{w}^{-1}$$

and

$$\tau = \operatorname{Var} \tilde{w}^{-1} := \frac{1}{m-1} \sum_{i=1}^{m} (\hat{\sigma}_i^2 - \sigma_0^2)^2$$

by one of the methods discussed in Section 3. Now, if the common prior distribution of $(\tilde{\vartheta}, \tilde{w})$ is normal-gamma,

$$P_{\tilde{\vartheta}, \tilde{w}} = \mathrm{NG}(\boldsymbol{\mu}, \alpha, \boldsymbol{\Phi}, \beta) \tag{4.15}$$

then, according to A.4, the marginal prior distribution of $\tilde{\vartheta}$ is multivariate Student, $P_{\tilde{\vartheta}} = t(\alpha, \boldsymbol{\mu}, \beta\boldsymbol{\Phi})$, with α degrees of freedom, mean $\mathbf{E}\,\tilde{\vartheta} = \boldsymbol{\mu}$ and covariance matrix

$$\mathbf{Cov}\,\tilde{\vartheta} = \frac{\alpha\beta}{\alpha - 2}\boldsymbol{\Phi}$$

provided that $\alpha > 2$. The marginal prior of $\tilde{w}$ is a gamma distribution $P_{\tilde{w}} = \mathrm{G}(\alpha/2, \alpha\beta/2)$, so that $\tilde{w}^{-1}$ has an inverted gamma distribution with mean and variance given by

$$\mathbf{E}\,\tilde{w}^{-1} = \alpha\beta/(\alpha - 2) \qquad \text{if} \quad \alpha > 2,$$

$$\operatorname{Var} \tilde{w}^{-1} = 2\alpha^2\beta^2/(\alpha - 4)(\alpha - 2)^2 \qquad \text{if} \quad \alpha > 4.$$

Thus, if we are given the values of $\boldsymbol{\mu}$, $\boldsymbol{\Lambda}$, σ_0^2 and τ as defined above, then we can specify the parameters of the normal-gamma-prior distribution (4.15) by $\boldsymbol{\mu}$ and

$$\boldsymbol{\Phi} = \sigma_0^{-2}\boldsymbol{\Lambda}, \qquad \alpha = 2\sigma_0^4/\tau + 4,$$

$$\beta = \frac{\alpha - 2}{\alpha}\sigma_0^2 = \frac{\sigma_0^4 + \tau}{\sigma_0^4 + 2\tau}\sigma_0^2.$$

Now we deal with the case that we are confined to stating subjective probabilities in order to specify the normal-gamma-prior. General rules for assigning subjective probabilities cannot be given. An illustrative example how to tackle this problem can be found in RAIFFA (1968), where first the median is specified, then the odd quartiles, and finally the odd octiles.

It will be difficult to make a priori statements about the magnitude of the precision w or the variance $\sigma^2 = w^{-1}$ of observations. Following LEMPERS (1971), it seems easier to specify the prior information on the basis of the unconditional distribution of the measurement error $\tilde{e}$. Subjective prior information concerning $\tilde{e}$ could be derived by imagining a hypothetical experiment. Since we have assumed that, given w, the error variable has normal distribution $\mathrm{N}(0, w^{-1})$ which does not depend on the point of observation, the joint distribution of $(\tilde{e}, \tilde{w})$ is normal-gamma. The marginal distribution of $\tilde{e}$ is Student,

$$P_{\tilde{e}} = t(\alpha, 0, \beta),$$

with α degrees of freedom, zero mean and variance

$$\operatorname{Var} \tilde{e} = \alpha\beta/(\alpha - 2).$$

Now two ways are possible for the specification of the prior parameters on the basis of the marginal distributions $P_{\tilde{\vartheta}}$ and $P_{\tilde{e}}$: Specifying first $P_{\tilde{\vartheta}}$ and then $P_{\tilde{e}}$ or the converse way of specifying $P_{\tilde{\vartheta}} = t(\alpha, \boldsymbol{\mu}, \beta\boldsymbol{\Phi})$ after determining α and β for $P_{\tilde{e}} = t(\alpha, 0, \beta)$.

Consider the first approach starting with specification of $P_{\tilde{\vartheta}}$ which enables us to formulate more probability statements on $\tilde{\boldsymbol{\vartheta}}$. Let $\boldsymbol{\mu} = (\mu_1, \ldots, \mu_r)^\top$ and $\boldsymbol{\Phi} = (\varphi_{ij})_{r \times r}$. According to A.4, the marginal distribution of every component $\tilde{\vartheta}_i$ of $\boldsymbol{\vartheta}$ is Student,

$$P_{\tilde{\vartheta}_i} = t(\alpha, \mu_i, \beta\varphi_{ii}), \qquad i = 1, \ldots, r,$$

i.e., $\tilde{\vartheta}_i$ is symmetrically distributed around its median μ_i with variance equal to $\alpha\beta\varphi_{ii}/(\alpha - 2)$. The values of μ_i then have to be determined such that they are, in our opinion, most likely to occur for $\tilde{\vartheta}_i$. Hereafter we can proceed further by assigning 50 per cent confidence intervals

$$P(\mu_i - c_i < \tilde{\vartheta}_i < \mu_i + c_i) = \frac{1}{2}, \qquad i = 1, \ldots, r$$

for $\tilde{\vartheta}_i$ which are symmetric around the median with $c_i > 0$. This way we have assigned the upper and lower quartiles of the marginal distribution of $\tilde{\vartheta}_i$. Furthermore, a conditional probability statement will be needed, for example, on the median of $\tilde{\vartheta}_i$ given that $\tilde{\vartheta}_i > \mu_i + c_i$. This is done by determining positive numbers $d_i > c_i$ such that

$$P(\mu_i + c_i < \tilde{\vartheta}_i < \mu_i + d_i \mid \tilde{\vartheta}_i > \mu_i + c_i) = \frac{1}{2}.$$

Combining both probability statements, we obtain

$$P(\mu_i - d_i < \tilde{\vartheta}_i < \mu_i + d_i) = \frac{3}{4}$$

so that with the above conditional probability statement we have implicitly defined the upper and lower octiles of $P_{\tilde{\vartheta}_i}$. If the components of $\tilde{\boldsymbol{\vartheta}} = (\tilde{\vartheta}_1, \dots, \tilde{\vartheta}_r)^\top$ can be assumed to be uncorrelated a-priori, the above specifications allow for a unique determination of $\beta\varphi_{11}, \dots, \beta\varphi_{rr}$ and α. This can be done by help of tables of the cumulative Student distribution as can be found e.g. in FISHER/YATES (1957) and, for the odd octiles, in LEMPERS/LOUTER (1971). Finally, one additional probability statement on the unconditionally distributed error $\tilde{e}$ yields the value of β, so that $\boldsymbol{\Phi} = \operatorname{diag}(\varphi_{11}, \dots, \varphi_{rr})$ also comes available. If the components of $\tilde{\boldsymbol{\vartheta}}$ are not uncorrelated, then the off-diagonal elements of $\beta\boldsymbol{\Phi}$ still have to be specified. This could be done on the basis of the joint distributions of the pairs $(\tilde{\vartheta}_i, \tilde{\vartheta}_j)$, $i, j = 1, \dots, r$, which are also Student:

$$P_{\tilde{\vartheta}_i, \tilde{\vartheta}_j} = t(\alpha, \boldsymbol{\mu}_{ij}, \beta\boldsymbol{\Phi}_{ij}), \qquad \boldsymbol{\mu}_{ij} = \begin{pmatrix} \mu_i \\ \mu_j \end{pmatrix}, \qquad \boldsymbol{\Phi}_{ij} = \begin{pmatrix} \varphi_{ii} & \varphi_{ij} \\ \varphi_{ij} & \varphi_{jj} \end{pmatrix}.$$

These distributions can be specified via conditional probability statements evaluating, for example, $P(\tilde{\vartheta}_i > \mu_i + e_i \mid \tilde{\vartheta}_j > \mu_j + e_j)$, $e_i, e_j > 0$, $i, j = 1, \dots, r$, and combining them with the unconditional probabilities $P(\tilde{\vartheta}_j > \mu_j + e_j)$.

A similar method for the elicitation of the prior distribution $P_{\tilde{\boldsymbol{\vartheta}}}$ was given by KADANE et al. (1980) who proposed to elicit quantiles of the predictive distribution for a set of future (hypothetical) observations $\tilde{\boldsymbol{Y}}_m = (\tilde{Y}_1, \dots, \tilde{Y}_m)^\top$ conditional on $m \geqq r$ values $(\boldsymbol{x}_1, \dots, \boldsymbol{x}_m)^\top = \boldsymbol{X}_m$ of the vector of regressor variables. Under the above assumptions, $\boldsymbol{Y}_m$ has a multivariate t-distribution with

$$\mathbf{E}\,\tilde{\boldsymbol{Y}}_m = \boldsymbol{X}_m\boldsymbol{\vartheta} \quad \text{and} \quad \mathbf{Cov}\,\tilde{\boldsymbol{Y}}_m = \frac{\alpha\beta}{\alpha - 2}\boldsymbol{X}_m\boldsymbol{\Phi}\boldsymbol{X}_m + \sigma_0^2\boldsymbol{I}_m$$

provided that $\alpha > 2$. The authors describe the theory underlying an interactive computer program for eliciting the hyperparameters α, β, σ_0^2, $\boldsymbol{\Phi}$, report on its implementation and give an application to a problem in the design of highway pavements. For a further discussion see also OMAN (1985).

For the construction of improved methods of elicitation models would have to be developed of how errors in elicitation occur, the probabilistic structure of elicitation errors is discussed e.g. in DICKEY (1979). For further discussions of the problem of elicitation of prior distributions and its applications to Bayesian regression estimation see LINDLEY/TVERSKY/BROWN (1979), KADANE (1980), WINKLER (1980), KAHNEMAN et al. (1982), DALAL/HALL (1983), DICKEY/CHEN (1985) and DIACONIS/YLVISAKER (1985).

Though the conjugate normal-gamma-distribution is analytically convenient to deal with and is a good choice for incorporating prior knowledge, some critical comments must be made about the disadvantages of this prior distribution. First, the marginal distribution of $\tilde{\vartheta}$ is symmetric, which, in some cases, may be in contradiction with the prior ideas of the investigator. Second, the normal-gamma distribution causes a prior dependence between $\tilde{\vartheta}$ and $\tilde{w}$. This is reflected by the fact that both the marginal prior of $\tilde{\vartheta}$ and $\tilde{w}$ depend on the parameters α and β. It implies that with the specification of one of the distributions $P_{\tilde{\vartheta}}$, $P_{\tilde{w}}$ or $P_{\tilde{e}}$ we implicitly make statements about the others. However, if the number of degrees of freedom α increases, the marginal Student distributions of $\tilde{\vartheta}$ and $\tilde{e}$ converge to normal distributions, thus being of the same type as the conditional distributions.

Despite of the disadvantages just discussed, the normal-gamma-prior has but the great merit of analytical tractability. If we use other prior distributions, the posterior distribution and the Bayes estimator cannot be obtained analytically and numerical integration methods must be employed.

Having specified a normal-gamma-prior distribution we should try somehow to check whether it is consistent with the actual states of the parameters supplied by the experiment. Particularly, we should guard against strong misspecifications of the values of $\boldsymbol{\mu}$ and $\boldsymbol{\Phi}$ which will play a central role in estimation and design.

Test on compatibility of prior knowledge and actual data

THEIL (1963) has developed a test on the compatibility of actual data and prior knowledge given by a stochastic prior estimator for $\tilde{\vartheta}$. This test can be modified and used to check whether we can rely upon the specifications concerning the prior distribution of $\tilde{\vartheta}$. To do so, consider the random vector

$$\begin{aligned}\tilde{\boldsymbol{\gamma}} &:= \boldsymbol{T}_1(\tilde{\boldsymbol{y}}(v_n)) - \boldsymbol{\mu} \\ &= (\boldsymbol{F}(v_n)^\top \boldsymbol{F}(v_n))^{-1} \boldsymbol{F}(v_n)^\top \tilde{\boldsymbol{y}}(v_n) - \boldsymbol{\mu},\end{aligned}$$

where $\boldsymbol{F}(v_n)$ is assumed to be of full rank r. Under the assumption of independently and identically normally distributed errors and a normal-gamma-prior for $(\tilde{\vartheta}, w)$ we have $P_{\tilde{e}}^{(w)} = \mathrm{N}\left(\boldsymbol{0}, \frac{1}{w}\boldsymbol{I}_n\right)$ and $P_{\tilde{\vartheta}|w} = \mathrm{N}\left(\boldsymbol{\mu}, \frac{1}{w}\boldsymbol{\Phi}\right)$, so that with $\boldsymbol{y}(v_n) = \boldsymbol{F}(v_n)\boldsymbol{\vartheta} + \tilde{\boldsymbol{e}}(v_n)$ and the stochastic independence of $\tilde{\vartheta}$ and $\tilde{\boldsymbol{e}}(v_n)$ it holds

$$P_{\tilde{\boldsymbol{y}}|w} = \mathrm{N}\left(\boldsymbol{F}(v_n)\boldsymbol{\mu}, \frac{1}{w}\boldsymbol{F}(v_n)\boldsymbol{\Phi}\boldsymbol{F}(v_n)^\top + \frac{1}{w}\boldsymbol{I}_n\right).$$

This implies that $\mathbf{E}(\tilde{\boldsymbol{\gamma}}|w) = \boldsymbol{0}$ and

$$\begin{aligned}\mathbf{Cov}(\tilde{\boldsymbol{\gamma}}|w) &= (\boldsymbol{F}(v_n)^\top \boldsymbol{F}(v_n))^{-1} \boldsymbol{F}(v_n)^\top \mathbf{Cov}(\tilde{\boldsymbol{y}}|w)\, \boldsymbol{F}(v_n)\, (\boldsymbol{F}(v_n)^\top \boldsymbol{F}(v_n))^{-1} \\ &= \frac{1}{w}(\boldsymbol{\Phi} + (\boldsymbol{F}(v_n)^\top \boldsymbol{F}(v_n))^{-1}).\end{aligned}$$

Hence we have $P_{\tilde{\gamma}|w} = \mathrm{N}\left(\mathbf{0}, \frac{1}{w}\boldsymbol{\Phi} + \frac{1}{w}(\boldsymbol{F}(v_n)^\top \boldsymbol{F}(v_n))^{-1}\right)$ and the prior specifications of $\boldsymbol{\mu}$ and $\boldsymbol{\Phi}$ can be tested on the basis of the statistic

$$\chi_r^2(w) := \tilde{\boldsymbol{\gamma}}^\top (\mathbf{Cov}(\tilde{\boldsymbol{\gamma}}|w))^{-1}\tilde{\boldsymbol{\gamma}}$$

$$= \frac{1}{w}(\boldsymbol{T}_1(\tilde{\boldsymbol{y}}(v_n)) - \boldsymbol{\mu})^\top (\boldsymbol{\Phi} + (\boldsymbol{F}(v_n)^\top \boldsymbol{F}(v_n))^{-1})^{-1}(\boldsymbol{T}_1(\tilde{\boldsymbol{y}}(v_n)) - \boldsymbol{\mu})$$

which, according to A.2, is distributed as chi-square with r degress of freedom. The trouble, however, is that the precision w will not be known. One way to overcome this difficulty would be first to test a hypothesis concerning w and then to replace it by the value adopted by the hypothesis. With regard to the fact that $\mathbf{E}\,\tilde{w} = 1/\beta$, this could be accomplished by testing the null hypothesis

$$H_0: w = 1/\beta$$

on the basis of the test statistic

$$T_2(\tilde{\boldsymbol{y}}(v_n)) = \|\tilde{\boldsymbol{y}}(v_n) - \boldsymbol{F}(v_n)\, T_1(\tilde{\boldsymbol{y}}(v_n))\|^2$$

observing that

$$wT_2(\tilde{\boldsymbol{y}}(v_n)) \sim \chi^2_{n-r}.$$

Hereafter, in the test statistic $\chi_r^2(w)$ we can replace w^{-1} by β or some alternative value if H_0 is rejected. Another way would be the direct replacement of $\sigma^2 = w^{-1}$ by the unbiased estimator

$$\hat{\sigma}^2 = (n-r)^{-1}\, T_2(\tilde{\boldsymbol{y}}(v_n)).$$

The distribution of the resulting test statistic $\chi_r^2(\hat{\sigma}^{-2})$ may be found in MEHTA/SWAMY (1974), for large values of n it is approximately distributed as χ_r^2.

II. Bayesian estimation

5. Bayes estimation of the regression parameter

Now we turn to the estimation problem G_η^e as introduced in Section 2.2 and look for a Bayes estimator with respect to a given prior distribution. First we reduce G_η^e to an estimation problem for the regression parameter and deal with Bayes estimation for $\boldsymbol{\vartheta}$ in case of quadratic loss. Our main concern then is with the particular Bayes estimator obtained under normally distributed observations and conjugate normal-gamma-prior. We investigate the sampling properties of this Bayes estimator and show the extent to which an improvement in risk over the least squares estimator is possible. The robustness of optimality of the Bayes estimator under a change of error and prior distributions and of the loss function is investigated in Section 6.

In the sequel let be $v_n \in V$ a design which is chosen fixed. To simplify notation, we will omit the dependence on v_n in all expressions where it occurs (loss, risk, estimators etc.), for example, we will write shortly $\tilde{\boldsymbol{y}}$, $\tilde{\boldsymbol{e}}$, $\boldsymbol{F}$, C, D_η instead of $\tilde{\boldsymbol{y}}(v_n)$, $\tilde{\boldsymbol{e}}(v_n)$, $\boldsymbol{F}(v_n)$, $C(v_n)$, $D_\eta(v_n)$.

5.1. Bayes estimation with quadratic loss

Let $P_{\tilde{\vartheta}, \tilde{w}}$ be a given prior distribution with density $p_{\tilde{\vartheta}, \tilde{w}}$ and let the conditional distributions $P_{\tilde{y}|\vartheta, w}$ of the observation vector have densities $p_{\tilde{y}|\vartheta, w}$ (cp. Assumption 4.1 (i)).

Because of our Assumption 1.1 of a true linear regression setup for the response surface we can confine ourselves to estimation of the regression parameter. For this let

$$D = \{\hat{\boldsymbol{\vartheta}} \mid C \to \Theta : \hat{\eta}(\cdot, \tilde{\boldsymbol{y}}) = \boldsymbol{f}(\cdot)^\top \hat{\boldsymbol{\vartheta}}(\tilde{\boldsymbol{y}}) \in D_\eta\} \tag{5.1}$$

a set of estimators for $\boldsymbol{\vartheta}$ generating a subclass of the class D_η of estimators for η. Then, for fixed $\boldsymbol{x} \in X_P$ and $v_n \in V$ let

$$L(\boldsymbol{\vartheta}, \boldsymbol{w}; \hat{\boldsymbol{\vartheta}}(\boldsymbol{y})) := L_\eta(\boldsymbol{\vartheta}, \boldsymbol{w}, \boldsymbol{x}; \boldsymbol{f}(\boldsymbol{x})^\top \hat{\boldsymbol{\vartheta}}(\boldsymbol{y}), v_n) \tag{5.2}$$

denote the loss arising in estimating ϑ by $\hat{\vartheta}(y)$ and $R|\Theta \times W \times D \to \mathbb{R}^+$ the corresponding risk function, i.e.,

$$R(\vartheta, w; \hat{\vartheta}) = \mathbf{E}_{\tilde{y}|\vartheta, w} L(\vartheta, w; \hat{\vartheta}(\tilde{y}))$$

for all $(\vartheta, w, \hat{\vartheta}) \in \Theta \times W \times D$. This way we pass from G_η^e to the estimation problem

$$G = [\Theta \times W, D, R] \tag{5.3}$$

for the regression parameter ϑ and it holds

Lemma 5.1. *Provided that $\hat{\vartheta}^* \in D$ is a Bayes estimator for ϑ in G with respect to $P_{\tilde{\vartheta}, \tilde{w}}$, so $\hat{\eta}^*(\cdot, \tilde{y}) = f(\cdot)^\top \hat{\vartheta}^*(y)$ is a Bayes estimator for $\eta(\cdot)$ in G_η^e.*

Denote $\mathbf{E}_{\tilde{\vartheta}, \tilde{w}|y}$ the expectation operator with respect to the posterior distribution $P_{\tilde{\vartheta}, \tilde{w}|y}$ of $(\tilde{\vartheta}, \tilde{w})$ for given observation $y \in C$. The posterior distribution, the density of which is given by (4.3), combines the prior knowledge expressed by $P_{\tilde{\vartheta}, \tilde{w}}$ with the information y of the actual sample.

Definition 5.1. The quantity

$$\mathbf{E}_{\tilde{\vartheta}, \tilde{w}|y} L(\tilde{\vartheta}, \tilde{w}; \hat{\vartheta}(y)) = \int_{\Theta \times W} L(\vartheta, w; \hat{\vartheta}(y))\, \mathrm{d}P_{\tilde{\vartheta}, \tilde{w}|y}(\vartheta, w)$$

is called the *posterior loss* of the estimation $\hat{\vartheta}(y)$, $\hat{\vartheta} \in D$.

A Bayes estimator for ϑ can be obtained by minimization of the posterior loss (see A.16).

Lemma 5.2. *The estimator $\hat{\vartheta}^* \in D$ is Bayesian in G with respect to $P_{\tilde{\vartheta}, \tilde{w}}$ if and only if for all $y \in C$:*

$$\mathbf{E}_{\tilde{\vartheta}, \tilde{w}|y} L(\tilde{\vartheta}, \tilde{w}; \hat{\vartheta}^*(y)) = \inf_{\hat{\vartheta} \in D} \mathbf{E}_{\tilde{\vartheta}, \tilde{w}|y} L(\tilde{\vartheta}, \tilde{w}; \hat{\vartheta}(y))\,.$$

According to the special form of loss, the Bayes estimator will represent some characteristic value of the posterior distribution, for example its mean or some other moment or quantile.

Remark 5.1. If there exists a sufficient statistic T for $\mathcal{P}_{\tilde{y}}$ as we had assumed in the preceding section (see Assumption 4.1), then it holds

$$\forall t \in C_T\, \forall y (y \in C \wedge T(y) = t) : P_{\tilde{\vartheta}, \tilde{w}|y} = P_{\tilde{\vartheta}, \tilde{w}|t} \tag{5.4}$$

for arbitrary prior distributions. A statistic T having this property is called *Bayes sufficient* for $\mathcal{P}_{\tilde{y}}$; under mild assumptions the Bayes sufficiency is equivalent to usual sufficiency in the sense of A.9 (see RAIFFA/SCHLAIFER (1961)). Under the Assumption 4.1 it follows from (5.4) that the search for Bayes estimators can be restricted to estimators of the type $\hat{\vartheta}(\tilde{y}) = g(T(\tilde{y}))$, where g is some Borel-measurable function. Moreover, if Assumption 4.3 is satisfied then the Bayes estimator depends on the design v_n only through the function $\boldsymbol{H}(v_n)$ (see PILZ (1979b)).

Now we will especially assume the risk R in G to be generated by a quadratic loss function.

Assumption 5.1. *For each $\boldsymbol{w} \in W$, $\hat{\boldsymbol{\vartheta}} \in D$ and $\boldsymbol{y} \in C$ let*

$$L(\boldsymbol{\vartheta}, \boldsymbol{w}; \hat{\boldsymbol{\vartheta}}(\boldsymbol{y})) = \|\boldsymbol{\vartheta} - \hat{\boldsymbol{\vartheta}}(\boldsymbol{y})\|^2_{U(\boldsymbol{w})},$$

where $\boldsymbol{U}(\boldsymbol{w}) \in \mathcal{M}_r^{\geqq}$ is some positive semidefinite matrix.

A more general class of loss functions will be considered in Section 6.

Remark 5.2. Observing (5.2), Assumption 5.1 means that the loss for estimating the expected response $\eta(\boldsymbol{x}) = \boldsymbol{f}(\boldsymbol{x})^\top \boldsymbol{\vartheta}$ is of the form

$$L_\eta(\boldsymbol{\vartheta}, \boldsymbol{w}, \boldsymbol{x}; \hat{\eta}(\boldsymbol{x}, \boldsymbol{y}), v_n) = h(\boldsymbol{w})\,(\boldsymbol{f}(\boldsymbol{x})^\top \boldsymbol{\vartheta} - \hat{\eta}(\boldsymbol{x}, \boldsymbol{y}))^2,$$

where, for fixed $\boldsymbol{x} \in X_P$, $\boldsymbol{U}(\boldsymbol{w}) = h(\boldsymbol{w})\boldsymbol{f}(\boldsymbol{x})\boldsymbol{f}(\boldsymbol{x})^\top$ and $h(\cdot)$ is a nonnegative function on W. In case of independent observations of the same precision $w > 0$ it is common use to set $h(w) = w$, leading to so-called *normalized loss.* Thereby it is possible to incorporate the expense of experimentation by adding a cost function to L_η (see Section 14.2). In the estimation problem, however, the cost of experimentation can be dropped because, for fixed design, it plays the role of an additive constant.

Let $P_{\tilde{\vartheta}|y,w}$ denote the conditional posterior distribution of $\tilde{\boldsymbol{\vartheta}}$ for fixed $\boldsymbol{w} \in W$ and given observation $\boldsymbol{y} \in C$. Its density is given by

$$p_{\tilde{\vartheta}|y,w}(\boldsymbol{\vartheta}) = l(\boldsymbol{\vartheta}, \boldsymbol{w}; \boldsymbol{y})\, p_{\tilde{\vartheta}|w}(\boldsymbol{\vartheta}) \Big/ \int_\Theta l(\boldsymbol{\vartheta}, \boldsymbol{w}; \boldsymbol{y})\, p_{\tilde{\vartheta}|w}(\boldsymbol{\vartheta})\, \mathrm{d}\boldsymbol{\vartheta}, \tag{5.5}$$

where $p_{\tilde{\vartheta}|w}$ represents the conditional prior density of $\tilde{\boldsymbol{\vartheta}}$ given $\boldsymbol{w} \in W$. It can be shown that the expected value of the conditional posterior distribution $P_{\tilde{\vartheta}|y,w}$ is a Bayes estimate for $\boldsymbol{\vartheta}$ which depends, however, on the unknown parameter $\boldsymbol{w}$ of the error distribution.

Theorem 5.1. *Let Assumption* 5.1 *be satisfied and $\mathcal{P}_{\tilde{y}}$ and $P_{\tilde{\vartheta},\tilde{w}}$ be such that the moment $\mathsf{E}_{\tilde{\vartheta}|y,w}\|\tilde{\boldsymbol{\vartheta}} - \mathsf{E}(\tilde{\boldsymbol{\vartheta}}|\boldsymbol{y}, \boldsymbol{w})\|^2_{U(w)}$ exists for each $\boldsymbol{y} \in C$. Then*

$$\hat{\boldsymbol{\vartheta}}^*_w(\tilde{\boldsymbol{y}}) = \mathsf{E}(\tilde{\boldsymbol{\vartheta}}|\tilde{\boldsymbol{y}}, \boldsymbol{w})$$

is a Bayes estimator for $\boldsymbol{\vartheta}$ with respect to $P_{\tilde{\vartheta},\tilde{w}}$.

Proof. We show that $\hat{\boldsymbol{\vartheta}}^*_w$ has minimum posterior loss under all estimators from D. For this let $\boldsymbol{y} \in C$ and $P_{\tilde{w}|y}$ be the marginal posterior distribution of $\tilde{\boldsymbol{w}}$. According to our assumption, there exists the expected value $\mathsf{E}_{\tilde{\vartheta}|y,w}\|\tilde{\boldsymbol{\vartheta}} - \mathsf{E}(\tilde{\boldsymbol{\vartheta}}|\boldsymbol{y}, \boldsymbol{w})\|^2_{U(w)} = \mathsf{E}_{\tilde{\vartheta}|y,w}\|\tilde{\boldsymbol{\vartheta}} - \hat{\boldsymbol{\vartheta}}^*_w(\boldsymbol{y})\|^2_{U(w)}$, so that it follows for the posterior loss of any arbitrary estimator $\hat{\vartheta}$:

$$\begin{aligned}\mathsf{E}_{\tilde{\vartheta}, \tilde{w}|y} L(\tilde{\vartheta}, \tilde{w}; \hat{\vartheta}(y)) &= \int_W \mathsf{E}_{\tilde{\vartheta}|y, w} \|\tilde{\vartheta} - \hat{\vartheta}(y)\|^2_{U(w)} \, \mathrm{d}P_{\tilde{w}|y}(w) \\ &= \int_W \mathsf{E}_{\tilde{\vartheta}|y, w} \|(\hat{\vartheta}^*_w(y) - \hat{\vartheta}(y)) + (\tilde{\vartheta} - \hat{\vartheta}^*_w(y))\|^2_{U(w)} \, \mathrm{d}P_{\tilde{w}|y}(w) \\ &= \int_W \{\|\hat{\vartheta}^*_w(y) - \hat{\vartheta}(y)\|^2_{U(w)} + \mathsf{E}_{\tilde{\vartheta}|y, w} \|\tilde{\vartheta} - \hat{\vartheta}^*_w(y)\|^2_{U(w)}\} \, \mathrm{d}P_{\tilde{w}|y}(w) \\ &\geqq \int_W \mathsf{E}_{\tilde{\vartheta}|y, w} \|\tilde{\vartheta} - \hat{\vartheta}^*_w(y)\|^2_{U(w)} \, \mathrm{d}P_{\tilde{w}|y}(w)\end{aligned}$$

the last integral being precisely the posterior loss of $\hat{\vartheta}^*_w$. □

This means that, for given $y \in C$, a Bayes estimate for ϑ can be determined as

$$\hat{\vartheta}^*_w(y) = \int_\Theta \vartheta \, p_{\tilde{\vartheta}|y, w}(\vartheta) \, \mathrm{d}\vartheta \tag{5.6}$$

with the density $p_{\tilde{\vartheta}|y, w}$ given by (5.5).

From the proof of Theorem 5.1 it can be seen that, in case of positive definiteness of the matrix $U(w)$ occuring in the loss function, $\hat{\vartheta}^*_w$ is the only Bayes estimator to exist for ϑ, i.e., if $\hat{\vartheta} \in D$ is a Bayes estimator for ϑ then this necessarily implies $P_{\tilde{y}}(\hat{\vartheta}(\tilde{y}) = \hat{\vartheta}^*_w(y)) = 1$.

Corollary 5.1. *Let the assumptions of Theorem* 5.1 *be satisfied and* $U(w)$ *be positive definite for each* $w \in W$. *Then, up to* $P_{\tilde{y}}$*- equivalence,* $\hat{\vartheta}^*_w(\tilde{y}) = \mathsf{E}(\tilde{\vartheta}\,|\,\tilde{y}, w)$ *is the unique Bayes estimator for* ϑ (see PILZ (1979b)).

The trouble with the estimator $\hat{\vartheta}^*_w$ is its dependence on the unknown parameter $w \in W$. In a number of cases, however, the structure of the prior and error distributions or of the matrix $U(\cdot)$ is such that a Bayes estimator can be determined explicitly. Often, for example, the loss for estimating ϑ will not depend on the error distribution, or the mean of the conditional posterior $P_{\tilde{\vartheta}|y, w}$ will be independent of w. In these case the mean of the marginal posterior distribution $P_{\tilde{\vartheta}|y}$ of $\tilde{\vartheta}$ yields a Bayes estimate.

Corollary 5.2. *Let Assumption* 5.1 *and either one of the following conditions be satisfied:*

(i) $\mathsf{E}(\tilde{\vartheta}|y, w) = \mathsf{E}(\tilde{\vartheta}|y)$ *for all* $w \in W$ *and* $y \in C$.

(ii) $\mathsf{E}(U(\tilde{w})|y, \vartheta) = \mathsf{E}(U(\tilde{w})|y)$ *for all* $\vartheta \in \Theta$ *and* $y \in C$, *i.e., the matrix* $\int_W U(w) \, \mathrm{d}P_{\tilde{w}|y, \vartheta}(w)$ *does not depend on* ϑ.

Then $\hat{\vartheta}^*(y) = \mathsf{E}(\tilde{\vartheta}\,|\,y)$ *is a Bayes estimator for* ϑ *with respect to* $P_{\tilde{\vartheta}, \tilde{w}}$.

Proof. For the Condition (i) the assertion follows immediately from Theorem 5.1. If Condition (ii) comes true then we have $\mathsf{E}(U(\tilde{w})|y, \vartheta)$

$= \int_W U(w)\, dP_{\tilde{w}|y,\vartheta}(w) = \bar{U}(y)$, say, and it holds for all $\hat{\vartheta} \in D$:

$$\begin{aligned} \mathsf{E}_{\tilde{\vartheta},\tilde{w}|y} \|\tilde{\vartheta} - \hat{\vartheta}(y)\|^2_{U(\tilde{w})} &= \mathsf{E}_{\tilde{\vartheta}|y}\, \mathsf{E}_{\tilde{w}|y,\vartheta} \|\tilde{\vartheta} - \hat{\vartheta}(y)\|^2_{U(\tilde{w})} \\ &= \mathsf{E}_{\tilde{\vartheta}|y} \|\tilde{\vartheta} - \hat{\vartheta}(y)\|^2_{\bar{U}(y)} . \end{aligned}$$

Now, along the same lines of the proof of Theorem 5.1 we obtain

$$\mathsf{E}_{\tilde{\vartheta}|y} \|\tilde{\vartheta} - \hat{\vartheta}(y)\|^2_{\bar{U}(y)} \geqq \mathsf{E}_{\tilde{\vartheta}|y} \|\tilde{\vartheta} - \hat{\vartheta}^*(y)\|^2_{\bar{U}(y)} ,$$

i.e., $\hat{\vartheta}^*$ minimizes the posterior loss. □

Under the above conditions, the Bayes estimate for given observation y can be calculated as

$$\hat{\vartheta}^*(y) = \int_\Theta \vartheta p_{\tilde{\vartheta}|y}(\vartheta)\, d\vartheta = \int_\Theta \vartheta \left\{ \int_W p_{\tilde{\vartheta},\tilde{w}|y}(\vartheta, w)\, dw \right\} d\vartheta . \tag{5.7}$$

The Conditions (i) and (ii) of Corollary 5.2 are satisfied, naturally, if $\tilde{\vartheta}$ and $\tilde{w}$ are independent a-posteriori, i.e., if $P_{\tilde{\vartheta}|y,w} = P_{\tilde{\vartheta}|y}$ and $P_{\tilde{w}|y,\vartheta} = P_{\tilde{w}|y}$. It should be noted, however, that a prior independence of ϑ and w does not necessarily imply their posterior independence and thus validity of the Conditions (i) and (ii). But, in any case they are valid if, in addition to prior independence of $\tilde{\vartheta}$ and $\tilde{w}$, the likelihood function can be factorized as $l(\vartheta, w; y) = l_1(\vartheta; y) \times l_2(w; y)$. Also, Condition (ii) is satisfied, of course, if the loss is independent of w, i.e., if $U(w) = U$ is a matrix of constants. Furthermore, if the observations were independently distributed with constant precision $w > 0$ and we were to assume normalized quadratic loss, i.e., $U(w) = wU$ with some $U \in \mathcal{M}_r^{\geqq}$, then Condition (ii) would become the counterpart to Condition (i), requiring the mean $\mathsf{E}(\tilde{w}|y, \vartheta)$ of the conditional posterior distribution of $\tilde{w}$ to be independent of ϑ.

If the Condition (i) of Corollary 5.2 is satisfied then Bayes estimation of ϑ does not require the specification of the prior distribution of w, it suffices to specify the conditional prior $P_{\tilde{\vartheta}|y}$ of $\tilde{\vartheta}$ and combining (5.5), (5.6) and (5.7) yields

$$\hat{\vartheta}^*_w(y) = \hat{\vartheta}^*(y) = \int_\Theta \vartheta\, l(\vartheta, w; y) p_{\tilde{\vartheta}|w}(\vartheta)\, d\vartheta \Big/ \int_{\Theta \times W} l(\vartheta, w; y) p_{\tilde{\vartheta}|w}(\vartheta)\, d\vartheta$$

which then does not depend on w.

From the above formulae it can be seen that the computation of a Bayes estimate requires several (multidimensional) integrations. In case of a conjugate prior distribution the required posterior distributions can be obtained by relatively simple operations (see Theorem 4.3). Moreover, the posterior distribution is of the same type as the prior, so that if this is some standard distribution then the required expectations will be known explicitly or will be analytically computable. In case of a non-conjugate prior, however, the computation of the posterior density $p_{\tilde{\vartheta}|y}$ will cause some numerical effort. Particularly, the normalizing constant

$$\int_{\Theta \times W} l(\boldsymbol{\vartheta}, \boldsymbol{w}; \boldsymbol{y})\, p_{\tilde{\vartheta}, \tilde{w}}(\boldsymbol{\vartheta}, \boldsymbol{w})\, \mathrm{d}\boldsymbol{w}\, \mathrm{d}\boldsymbol{\vartheta}$$

rarely can be given analytically.

To avoid numerical difficulties of integration, some authors propose to determine an approximate Bayes estimate $\hat{\boldsymbol{\vartheta}}_{\mathrm{M}}$ as the mode of the posterior distribution, i.e.,

$$\hat{\boldsymbol{\vartheta}}_{\mathrm{M}}(\boldsymbol{y}) = \arg\sup_{\boldsymbol{\vartheta} \in \Theta} p_{\tilde{\vartheta}|y}(\boldsymbol{\vartheta}) \tag{5.8}$$

(cp. De Groot (1970), Lindley/Smith (1972)).

For the practical computation of $\hat{\boldsymbol{\vartheta}}_{\mathrm{M}}$ an iterative procedure can be recommended which maximizes the joint posterior density $p_{\tilde{\vartheta}, \tilde{w}|y}$ starting with an initial estimate $\boldsymbol{w}^1$ for $\boldsymbol{w}$ and then successively reestimates $\boldsymbol{\vartheta}$ and $\boldsymbol{w}$ by

$$\begin{aligned} \boldsymbol{\vartheta}^i &= \arg\sup_{\boldsymbol{\vartheta} \in \Theta} p_{\tilde{\vartheta}, \tilde{w}|y}(\boldsymbol{\vartheta}, \boldsymbol{w}^i) \\ \boldsymbol{w}^{i+1} &= \arg\sup_{\boldsymbol{w} \in W} p_{\tilde{\vartheta}, \tilde{w}|y}(\boldsymbol{\vartheta}^i, \boldsymbol{w}), \qquad i = 1, 2, \ldots \end{aligned}$$

Under mild regularity assumptions the sequence $\{\boldsymbol{\vartheta}^i\}$ converges to $\hat{\boldsymbol{\vartheta}}_{\mathrm{M}}(\boldsymbol{y})$. The approximation $\hat{\boldsymbol{\vartheta}}_{\mathrm{M}}(\boldsymbol{y})$ to the Bayes estimate $\hat{\boldsymbol{\vartheta}}^*(\boldsymbol{y})$ will be good if the posterior density $p_{\tilde{\vartheta}|y}$ is nearly symmetric around its maximum value $\hat{\boldsymbol{\vartheta}}_{\mathrm{M}}(\boldsymbol{y})$. In case of exact symmetry, $\hat{\boldsymbol{\vartheta}}_{\mathrm{M}}(\boldsymbol{y})$ and $\hat{\boldsymbol{\vartheta}}^*(\boldsymbol{y})$ will coincide, but in case of a strong kurtosis of the posterior distribution $P_{\tilde{\vartheta}|y}$ the approximation may be fairly distant from the exact Bayes estimate.

5.2. The conjugate Bayes estimator

We will now give explicitly a Bayes estimator for the parameter vector $\boldsymbol{\vartheta}$ in case of independently and identically normally distributed errors of the same precision $w > 0$, i.e.,

$$\mathcal{P}_{\tilde{e}} = \mathcal{P}_{\tilde{e}}^{(\mathrm{N})} := \left\{ \mathrm{N}\left(\mathbf{0}, \frac{1}{w} \boldsymbol{I}_n\right) \right\}_{w \in \mathbb{R}^+} \tag{5.9}$$

whereby we assume to have prior knowledge about the regression and precision parameter which can be expressed by a conjugate prior distribution

$$\begin{aligned} &P_{\tilde{\vartheta}, \tilde{w}} = \mathrm{NG}(\boldsymbol{\mu}, \alpha, \boldsymbol{\Phi}, \beta), \\ &\boldsymbol{\mu} \in \mathbb{R}^r, \quad \boldsymbol{\Phi} \in \mathcal{M}_r^{\geqq}, \quad \alpha \in \mathbb{R}^+, \quad \beta \in \mathbb{R}^+. \end{aligned} \tag{5.10}$$

Furthermore, as before we assume quadratic loss

$$L(\boldsymbol{\vartheta}, w; \hat{\boldsymbol{\vartheta}}(\boldsymbol{y})) = \|\boldsymbol{\vartheta} - \hat{\boldsymbol{\vartheta}}(\boldsymbol{y})\|^2_{U(w)}, \qquad \boldsymbol{U}(w) \in \mathcal{M}_r^{\geqq} \tag{5.11}$$

for estimating $\boldsymbol{\vartheta}$ by some $\hat{\boldsymbol{\vartheta}} \in D$. In Section 6 we will demonstrate then that the Bayes estimator obtained under the above Assumptions (5.9) to (5.11) is relatively robust under a change to more general error and prior distributions and more general loss functions. This gives good justification to solve the problem of optimum experimental design for estimation of the response surface on the basis of this estimator.

Theorem 5.2. *Let* $\Theta = \mathbb{R}^r$, $\mathscr{P}_{\tilde{e}} = \mathscr{P}_{\tilde{e}}^{(N)}$, $P_{\tilde{\vartheta},\tilde{w}} = \mathrm{NG}(\boldsymbol{\mu}, \alpha, \boldsymbol{\Phi}, \beta)$ *and the loss function be quadratic as given in* (5.11). *Then*

$$\hat{\vartheta}_B(\tilde{\boldsymbol{y}}) = (\boldsymbol{F}^\top \boldsymbol{F} + \boldsymbol{\Phi}^{-1})^{-1}(\boldsymbol{F}^\top \tilde{\boldsymbol{y}} + \boldsymbol{\Phi}^{-1}\boldsymbol{\mu}) \tag{5.12}$$

is a Bayes estimator for $\boldsymbol{\vartheta}$ *with respect to* $P_{\tilde{\vartheta},\tilde{w}}$.

Proof. With $P_{\tilde{\vartheta},\tilde{w}} = \mathrm{NG}(\boldsymbol{\mu}, \alpha, \boldsymbol{\Phi}, \beta)$ the posterior distribution $P_{\tilde{\vartheta},\tilde{w}|y}$ is normal-gamma with parameters $\boldsymbol{\mu}_0$, α_0, $\boldsymbol{\Phi}_0$, β_0 as defined in Theorem 4.4. Thus, we have $P_{\tilde{\vartheta}|y,w} = \mathrm{N}\left(\boldsymbol{\mu}_0, \frac{1}{w}\boldsymbol{\Phi}_0\right)$ (see A.3) and the expectation $\boldsymbol{\mu}_0 = \mathsf{E}(\tilde{\boldsymbol{\vartheta}}|\boldsymbol{y}, w) = \hat{\boldsymbol{\vartheta}}_B(\boldsymbol{y})$ does not depend on w. Hence, Condition (i) of Corollary 5.2 is satisfied which proves the Bayes optimality of $\hat{\boldsymbol{\vartheta}}_B$. □

Corollary 5.3. *Let* $\Theta = \mathbb{R}^r$, $\mathscr{P}_{\tilde{e}} = \mathscr{P}_{\tilde{e}}^{(N)}$ *and L be given by* (5.11) *with the additional condition that* $U(w)$ *be positive definite for every* $w \in \mathbb{R}^+$. *Then* $\hat{\boldsymbol{\vartheta}}_B$ *from* (5.12) *is an admissible estimator for* $\boldsymbol{\vartheta}$.

This follows with Theorem 5.2 and A.16 from the fact that in case of positive definiteness of $\boldsymbol{U}(w)$ the estimator $\hat{\boldsymbol{\vartheta}}_B(\tilde{\boldsymbol{y}}) = \mathsf{E}(\tilde{\boldsymbol{\vartheta}}|\tilde{\boldsymbol{y}}, w) = \mathsf{E}(\tilde{\boldsymbol{\vartheta}}|\boldsymbol{y})$ is the unique Bayes estimator for $\boldsymbol{\vartheta}$ with respect to $P_{\tilde{\vartheta},\tilde{w}} = \mathrm{NG}(\boldsymbol{\mu}, \alpha, \boldsymbol{\Phi}, \beta)$ (see Corollary 5.1).

The result of Theorem 5.2 can obviously be generalized to the case $\mathscr{P}_{\tilde{e}} = \left\{\mathrm{N}\left(\boldsymbol{0}, \frac{1}{w}\boldsymbol{\Sigma}\right)\right\}_{w \in \mathbb{R}^+}$ with some $\boldsymbol{\Sigma} \in \mathscr{M}_n^>$, i.e., correlated observations with variances and covariances known up to a common factor. Then, under the same remaining assumptions as in Theorem 5.2,

$$\hat{\boldsymbol{\vartheta}}_B^{\Sigma}(\tilde{\boldsymbol{y}}) = (\boldsymbol{F}^\top \boldsymbol{\Sigma}^{-1}\boldsymbol{F} + \boldsymbol{\Phi}^{-1})^{-1}(\boldsymbol{F}^\top \boldsymbol{\Sigma}^{-1}\tilde{\boldsymbol{y}} + \boldsymbol{\Phi}^{-1}\boldsymbol{\mu}) \tag{5.13}$$

is a Bayes estimator for $\boldsymbol{\vartheta}$ (see Bandemer et al. (1977), Section 4.3.3). For the case that $\mathscr{P}_{\tilde{e}} = \{\mathrm{N}(\boldsymbol{0}, \boldsymbol{\Sigma}) : \boldsymbol{\Sigma} \in \mathscr{M}_n^>\}$, where nothing is known about the covariance structure, the conjugate prior for $(\tilde{\boldsymbol{\vartheta}}, \tilde{\boldsymbol{\Sigma}}^{-1})$ is a so-called normal-Wishart-distribution which means that the conditional prior distribution of $\tilde{\boldsymbol{\vartheta}}$ given $\boldsymbol{\Sigma}$ is normal and $\tilde{\boldsymbol{\Sigma}}^{-1}$ has a Wishart distribution. In case of quadratic loss $\|\boldsymbol{\vartheta} - \hat{\boldsymbol{\vartheta}}(\boldsymbol{y})\|^2_{U(\Sigma)}$ with $\boldsymbol{U}(\boldsymbol{\Sigma}) \in \mathscr{M}_r^{\geqq}$, the posterior expectation

$$\mathsf{E}(\tilde{\boldsymbol{\vartheta}}|\tilde{\boldsymbol{y}}) = (\boldsymbol{F}^\top \boldsymbol{\Sigma}_0^{-1}\boldsymbol{F} + \boldsymbol{\Phi}^{-1})^{-1}(\boldsymbol{F}^\top \boldsymbol{\Sigma}_0^{-1}\tilde{\boldsymbol{y}} + \boldsymbol{\Phi}^{-1}\boldsymbol{\mu}),$$

where

$$\boldsymbol{\mu} = \mathsf{E}\,\tilde{\boldsymbol{\vartheta}}, \qquad \boldsymbol{\Phi} = \mathsf{Cov}\,\tilde{\boldsymbol{\vartheta}}, \qquad \boldsymbol{\Sigma}_0 = \mathsf{E}\,\tilde{\boldsymbol{\Sigma}},$$

is a Bayes estimator for $\boldsymbol{\vartheta}$ (see De Groot (1970), Bunke/Bunke (1986)).

We note that the Bayes estimator $\hat{\boldsymbol{\vartheta}}_B$ from (5.12) explicitly refers only to the first and second order moments $\boldsymbol{\mu}$ and $\boldsymbol{\Phi}$ of the conditional prior distribution $P_{\tilde{\vartheta}|w} = \mathrm{N}\left(\boldsymbol{\mu}, \frac{1}{w}\boldsymbol{\Phi}\right)$ whereas the parameters α and β occuring in the marginal prior distribution $P_{\tilde{w}} = \mathrm{G}\left(\frac{\alpha}{2}+1, \frac{\beta}{2}\right)$ (see A.3) are not involved in (5.12). However, these parameters enter the analysis implicitly, since they are needed in the specification of $\boldsymbol{\Phi}$ via $\mathbf{Cov}\,\tilde{\boldsymbol{\vartheta}} = (\mathbf{E}\,\tilde{w}^{-1})\,\boldsymbol{\Phi}$ and $\mathbf{E}\,\tilde{w}^{-1} = \alpha\beta/(\alpha-2)$ (see Section 4.4).

The Bayes estimator (5.12) can be represented in the form

$$\hat{\boldsymbol{\vartheta}}_{\mathrm{B}}(\tilde{\boldsymbol{y}}) = (\boldsymbol{F}^{\top}\boldsymbol{F} + \boldsymbol{\Phi}^{-1})^{-1}(\boldsymbol{F}^{\top}\boldsymbol{F}\hat{\boldsymbol{\vartheta}}_{\mathrm{LS}}(\tilde{\boldsymbol{y}}) + \boldsymbol{\Phi}^{-1}\boldsymbol{\mu}) \tag{5.14}$$

as a function of the sufficient statistic

$$\hat{\boldsymbol{\vartheta}}_{\mathrm{LS}}(\tilde{\boldsymbol{y}}) = (\boldsymbol{F}^{\top}\boldsymbol{F})^{-1}\boldsymbol{F}^{\top}\tilde{\boldsymbol{y}}. \tag{5.15}$$

This is the least squares estimator for $\boldsymbol{\vartheta}$ in case of homoscedastic errors, i.e., $\mathbf{Cov}\,\tilde{\boldsymbol{e}} = \frac{1}{w}\boldsymbol{I}_n$. (If $\boldsymbol{F}$ is not of full rank then $(\boldsymbol{F}^{\top}\boldsymbol{F})^{-1}$ is to be replaced by $(\boldsymbol{F}^{\top}\boldsymbol{F})^{+}$, see Rao (1973)). Thus the Bayes estimator $\hat{\boldsymbol{\vartheta}}_{\mathrm{B}}$ can be interpreted as a weighted mean of $\hat{\boldsymbol{\vartheta}}_{\mathrm{LS}}$ and the prior expectation $\boldsymbol{\mu} = \mathbf{E}\,\tilde{\boldsymbol{\vartheta}}$. Thereby the weighting matrices are the corresponding precision matrices of the sampling and prior distribution of $\hat{\boldsymbol{\vartheta}}_{\mathrm{LS}}$ and $\tilde{\boldsymbol{\vartheta}}$,

$$\begin{aligned} &\mathbf{Cov}\,(\hat{\boldsymbol{\vartheta}}_{\mathrm{LS}}|w)^{-1} = w\boldsymbol{F}^{\top}\boldsymbol{F} \\ \text{and}\quad &\mathbf{Cov}\,(\tilde{\boldsymbol{\vartheta}}|w)^{-1} = w\boldsymbol{\Phi}^{-1}, \end{aligned} \tag{5.16}$$

respectively, the sum of which forms the precision matrix of the posterior distribution,

$$\mathbf{Cov}\,(\tilde{\boldsymbol{\vartheta}}|\boldsymbol{y}, w)^{-1} = w(\boldsymbol{F}^{\top}\boldsymbol{F} + \boldsymbol{\Phi}^{-1}). \tag{5.17}$$

If the prior precision matrix and the information matrix of the experimental design happen to be diagonal, i.e.,

$$\boldsymbol{F}^{\top}\boldsymbol{F} = \mathrm{diag}\,(a_1, \ldots, a_r),$$
$$\boldsymbol{\Phi}^{-1} = \mathrm{diag}\,(b_1, \ldots, b_r),$$

then (5.14) becomes

$$\hat{\boldsymbol{\vartheta}}_{\mathrm{B}}(\tilde{\boldsymbol{y}}) = \mathrm{diag}\left(\frac{a_1}{a_1+b_1}, \ldots, \frac{a_r}{a_r+b_r}\right)\hat{\boldsymbol{\vartheta}}_{\mathrm{LS}}(\tilde{\boldsymbol{y}}) + \mathrm{diag}\left(\frac{b_1}{a_1+b_1}, \ldots, \frac{b_r}{a_r+b_r}\right)\boldsymbol{\mu}.$$

Hence, in this case the components $\hat{\vartheta}^i_{\mathrm{B}}$ of $\hat{\boldsymbol{\vartheta}}_{\mathrm{B}}$ are convex combinations of the corresponding components $\hat{\vartheta}^i_{\mathrm{LS}}$ and μ_i of $\hat{\boldsymbol{\vartheta}}_{\mathrm{LS}}$ and $\boldsymbol{\mu}$, respectively,

$$\hat{\vartheta}^i_{\mathrm{B}} = c_i\hat{\vartheta}^i_{\mathrm{LS}} + (1-c_i)\mu_i, \qquad c_i = a_i/(a_i+b_i), \qquad i = 1, \ldots, r.$$

This is a Sclove-type estimator when $\boldsymbol{\mu} = \boldsymbol{0}$ (see Sclove (1968)).

The estimator $\hat{\boldsymbol{\vartheta}}_{\mathrm{B}}$ can be rewritten as

$$\hat{\boldsymbol{\vartheta}}_{\mathrm{B}}(\tilde{\boldsymbol{y}}) = \hat{\boldsymbol{\vartheta}}_{\mathrm{LS}}(\tilde{\boldsymbol{y}}) + (\boldsymbol{F}^{\top}\boldsymbol{F} + \boldsymbol{\Phi}^{-1})^{-1}\,\boldsymbol{\Phi}^{-1}(\boldsymbol{\mu} - \hat{\boldsymbol{\vartheta}}_{\mathrm{LS}}(\tilde{\boldsymbol{y}})) \tag{5.18}$$

so that the term $(\boldsymbol{F}^{\top}\boldsymbol{F} + \boldsymbol{\Phi}^{-1})^{-1}\boldsymbol{\Phi}^{-1}(\boldsymbol{\mu} - \hat{\boldsymbol{\vartheta}}_{\mathrm{LS}})$ can be viewed as a correction which is to be added to the least squares estimator if we have available prior knowledge on $\boldsymbol{\vartheta}$ via $P_{\tilde{\vartheta}|w} = \mathrm{N}\left(\boldsymbol{\mu}, \frac{1}{w}\boldsymbol{\Phi}\right)$. Of course, if for a given observation $\boldsymbol{y}$ we have coincidence $\boldsymbol{\mu} = \hat{\boldsymbol{\vartheta}}_{\mathrm{LS}}(\boldsymbol{y})$ then the Bayes and least squares estimate will coincide too, $\hat{\boldsymbol{\vartheta}}_{\mathrm{B}}(\boldsymbol{y}) = \hat{\boldsymbol{\vartheta}}_{\mathrm{LS}}(\boldsymbol{y})$. Moreover, under the condition that $\boldsymbol{J} := (\boldsymbol{\Phi}\boldsymbol{F}^\top\boldsymbol{F})^{-1}$ has norm

$$\|\boldsymbol{J}\| = (\mathrm{tr}\,\boldsymbol{J}^\top\boldsymbol{J})^{1/2} < 1\,,$$

$\hat{\boldsymbol{\vartheta}}_{\mathrm{B}}$ can be expanded in a series

$$\hat{\boldsymbol{\vartheta}}_{\mathrm{B}}(\tilde{\boldsymbol{y}}) = \sum_{k=0}^{\infty} [(-1)^k \boldsymbol{J}^k]\,(\hat{\boldsymbol{\vartheta}}_{\mathrm{LS}}(\tilde{\boldsymbol{y}}) + \boldsymbol{J}\boldsymbol{\mu})$$

which makes it possible to compute the Bayes estimator approximately, starting with the LSE (see NÄTHER/PILZ (1980), Section 3). A sufficient condition for $\|\boldsymbol{J}\| < 1$ to hold is that $\|\boldsymbol{\Phi}^{-1}\| < \|\boldsymbol{F}^\top\boldsymbol{F}\|$ which, in a sense, means a smaller precision of prior knowledge provided by $\boldsymbol{\mu}$ than the precision of sampling knowledge provided by $\hat{\boldsymbol{\vartheta}}_{\mathrm{LS}}$. In the limiting case $\|\boldsymbol{\Phi}^{-1}\| \to 0$, which implies that $\boldsymbol{\Phi}^{-1} = 0$ and thus reflects a state of prior ignorance (see Section 8), the Bayes and least squares estimates coincide.

The compromise which the Bayes estimator makes between the least squares estimate and the prior expectation becomes clear if we observe that $\hat{\boldsymbol{\vartheta}}_{\mathrm{B}}(\boldsymbol{y})$ minimizes the sum

$$\begin{aligned} S(\boldsymbol{\vartheta}) &= \|\boldsymbol{y} - \boldsymbol{F}\boldsymbol{\vartheta}\|^2 + \|\boldsymbol{\vartheta} - \boldsymbol{\mu}\|^2_{\boldsymbol{\Phi}^{-1}} \\ &= \|\boldsymbol{\vartheta} - \hat{\boldsymbol{\vartheta}}_{\mathrm{LS}}(\boldsymbol{y})\|^2_{\boldsymbol{F}^\top\boldsymbol{F}} + \|\boldsymbol{\vartheta} - \boldsymbol{\mu}\|^2_{\boldsymbol{\Phi}^{-1}} + s^2(\boldsymbol{y})\,, \end{aligned} \tag{5.19}$$

where $s^2(\boldsymbol{y}) = \|\boldsymbol{y} - \boldsymbol{F}\hat{\boldsymbol{\vartheta}}_{\mathrm{LS}}(\boldsymbol{y})\|^2$ stands for the residual variance of observation,

$$\hat{\boldsymbol{\vartheta}}_{\mathrm{B}}(\boldsymbol{y}) = \arg\min_{\boldsymbol{\vartheta} \in \mathbb{R}^r} S(\boldsymbol{\vartheta})\,. \tag{5.20}$$

An obvious generalization leads to the introduction of a "smoothing factor" $\lambda > 0$ and the minimization of

$$S_\lambda(\boldsymbol{\vartheta}) = \|\boldsymbol{y} - \boldsymbol{F}\boldsymbol{\vartheta}\|^2 + \lambda\|\boldsymbol{\vartheta} - \boldsymbol{\mu}\|^2_{\boldsymbol{\Phi}^{-1}}$$

instead of $S(\boldsymbol{\vartheta})$, which offers a formal similarity to "smoothing" problems (see, e.g. WAHBA (1977), VAN DER LINDE (1985)) and to (Tichonov-) regularization methods known from numerical mathematics (see, e.g. MOROZOV (1985)). The resulting estimate $\hat{\boldsymbol{\vartheta}}_\lambda(\boldsymbol{y}) = (\boldsymbol{F}^\top\boldsymbol{F} + \lambda\boldsymbol{\Phi}^{-1})^{-1}(\boldsymbol{F}^\top\boldsymbol{y} + \lambda\boldsymbol{\Phi}^{-1}\boldsymbol{\mu})$, viewed as a function of $\lambda > 0$, leads to the so-called "curve décolletage" introduced by DICKEY (1975), see also Section 6.2.

Numerically, the Bayes estimate can be obtained as a solution of the linear equation system

$$(\boldsymbol{F}^\top\boldsymbol{F} + \boldsymbol{\Phi}^{-1})\,\hat{\boldsymbol{\vartheta}}_{\mathrm{B}}(\boldsymbol{y}) = \boldsymbol{F}^\top\boldsymbol{y} + \boldsymbol{\Phi}^{-1}\boldsymbol{\mu}$$

which has a symmetric and positive definite coefficient matrix so that, for example, Choleski-type procedures can be applied to compute $\hat{\boldsymbol{\vartheta}}_{\mathrm{B}}(\boldsymbol{y})$.

Moreover, with the denotations

$$\bar{F} = \begin{pmatrix} F \\ \hline F_0 \end{pmatrix}, \qquad \bar{\Sigma} = \left(\begin{array}{c|c} \Sigma & 0 \\ \hline 0 & \Sigma_0 \end{array}\right), \qquad \bar{y} = \begin{pmatrix} y \\ y_0 \end{pmatrix},$$

where $\Sigma = I_n$ and F_0, Σ_0, y_0 are such that

$$F_0^\top \Sigma_0^{-1} F_0 = \Phi^{-1}, \qquad F_0^\top \Sigma_0^{-1} y_0 = \Phi^{-1}\mu,$$

the Bayes estimate can be written as

$$\begin{aligned} \hat{\vartheta}_B(y) &= (F^\top \Sigma^{-1} F + F_0^\top \Sigma_0^{-1} F_0)^{-1} (F^\top \Sigma^{-1} y + F_0^\top \Sigma^{-1} y_0) \\ &= (\bar{F}^\top \bar{\Sigma}^{-1} \bar{F})^{-1} \bar{F}^\top \bar{\Sigma}^{-1} \bar{y} \end{aligned} \tag{5.22}$$

which is in the form of a generalized least squares estimate. Thus, all the numerical procedures which have been developed for this type of estimates (see, e.g. BANDEMER/NÄTHER (1980)) may be applied to compute $\hat{\vartheta}_B(y)$.

A simple choice for y_0, F_0 and Σ_0 would be the following:

$$F_0 = I_r, \qquad \Sigma_0 = \Phi, \qquad y_0 = \mu.$$

From the fact that $\hat{\vartheta}_B$ coincides with the generalized least squares estimator in the enlarged experiment $(\bar{y}, \bar{F}, \bar{\Sigma})$ it follows that the prior knowledge available through the first and second order moments of $P_{\tilde{\vartheta}|w} = N\left(\mu, \frac{1}{w}\Phi\right)$ has the same effect as an additional experiment (y_0, F_0, Σ_0) performed independently of the actual experiment (y, F, Σ). In other words, use of prior knowledge helps to reduce experimental efforts. Finally, from (5.21) and (5.22) arises the possibility of a recursive representation of the Bayes estimator, in particular, the applicability of known recursion formulae for the generalized least squares estimator.

The following alternative representations of $\hat{\vartheta}_B$, which can be derived by help of the matrix identities A.33, may be valuable for numerical computation too:

$$\begin{aligned} \hat{\vartheta}_B &= \mu + \Phi((F^\top F)^{-1} + \Phi)^{-1} (\hat{\vartheta}_{LS} - \mu) \\ &= \mu + (F^\top F + \Phi^{-1})^{-1} F^\top (\tilde{y} - F\mu) \\ &= \mu + \Phi F^\top (F \Phi F^\top + I_n)^{-1} (\tilde{y} - F\mu). \end{aligned} \tag{5.23}$$

5.3. Bayes linear estimation

Consider the following classes

$$D_l = \{\hat{\vartheta} : \hat{\vartheta}(\tilde{y}) = Z\tilde{y}, \quad Z \in \mathcal{M}_{r \times n}\} \tag{5.24}$$

and

$$D_a = \{\hat{\vartheta} : \hat{\vartheta}(\tilde{y}) = Z\tilde{y} + z, \quad Z \in \mathcal{M}_{r \times n}, \quad z \in \mathbb{R}^r\} \tag{5.25}$$

of linear and affine-linear estimators for ϑ. Clearly, $D_l \subset D_a$. If we confine ourselves to these classes then it suffices to have knowledge of the first and second moments of the error and prior distributions to construct the Bayes estimator, and no distributional assumptions are required.

Let $\mathcal{P}^0_{\tilde{e}}$ denote the class of all error distributions with uncorrelated components of the same (unknown) precision, i.e.,

$$\mathcal{P}^0_{\tilde{e}} = \left\{ P_{\tilde{e}} : \mathbf{E}\,\tilde{e} = \mathbf{0}, \quad \mathbf{Cov}\,\tilde{e} = \frac{1}{w} I_n \right\}_{w \in \mathbb{R}^+}. \tag{5.26}$$

Correspondingly, let us consider a normalized quadratic loss function, i.e.,

$$L(\boldsymbol{\vartheta}, w; \hat{\boldsymbol{\vartheta}}(\boldsymbol{y})) = w\|\boldsymbol{\vartheta} - \hat{\boldsymbol{\vartheta}}(\boldsymbol{y})\|^2_U, \quad U \in \mathcal{M}^{\geqq}_r. \tag{5.27}$$

We assume to have available prior knowledge about the regression and precision parameters which can be represented by some prior distribution out of the family

$$\mathcal{P}(\boldsymbol{\mu}, \boldsymbol{\Phi}, w_0) := \left\{ P_{\tilde{\vartheta}, \tilde{w}} : \mathbf{E}(\tilde{\boldsymbol{\vartheta}}|w) = \boldsymbol{\mu}, \quad \mathbf{Cov}(\tilde{\boldsymbol{\vartheta}}|w) = \frac{1}{w}\boldsymbol{\Phi}, \quad \mathbf{E}\,\tilde{w} = w_0 \right\} \tag{5.28}$$

with fixed first and second moments $\boldsymbol{\mu} \in \mathbb{R}^r$, $\boldsymbol{\Phi} \in \mathcal{M}^{\geqq}_r$ and $w_0 > 0$. For all estimators out of D_l and D_a, the Bayes risk is constant on the family $\mathcal{P}$ from (5.28). Any estimator minimizing the Bayes risk within the classes D_l and D_a is said to be *Bayes linear* and *Bayes affine*, respectively. Before determining such estimators, we will solve a general quadratic matrix optimization problem which has also some importance for finding robust Bayes and minimax linear estimators.

Lemma 5.3. *Let* $A \in \mathcal{M}^{>}_n$, $B \in \mathcal{M}_{n \times r}$, $C \in \mathcal{M}_{r \times r}$ *and* $U \in \mathcal{M}^{\geqq}_r$. *Then it holds*

$$\inf_{Z \in \mathcal{M}_{r \times n}} \operatorname{tr} U(ZAZ^\top - ZB - B^\top Z^\top + C) = \operatorname{tr} U(C - B^\top A^{-1} B)$$

and the minimum is achieved for any matrix of the form

$$Z^* = U^+ U(B^\top A^{-1} - G) + G$$

with $G \in \mathcal{M}_{r \times n}$ *chosen arbitrarily.*

Proof. Let $Z \in \mathcal{M}_{r \times n}$. Differentiating $h(Z) := \operatorname{tr} U(ZAZ^\top - ZB - B^\top Z^\top + C)$ with respect to Z we obtain

$$\frac{1}{2} \frac{\partial}{\partial Z} h(Z) = UZA - UB^\top$$

(see A.41). Thus, in order to have a minimum, it is necessary that $UZ = UB^\top A^{-1}$. This equation has the general solution

$$Z = U^+ UB^\top A^{-1} + (I_r - U^+ U)G = Z^*$$

with arbitrary $G \in \mathcal{M}_{r \times n}$; see Rao/Mitra (1971), p. 24. Further, we easily verify that $h(Z^*) = \operatorname{tr} U(C - B^\top A^{-1} B)$ and

$$h(Z) = h(Z^*) + \operatorname{tr} U(Z - Z^*)A(Z - Z^*)^\top.$$

Since $(Z - Z^*) A (Z - Z^*)^\top$ is positive semidefinite, it follows with A.28 that $h(Z) \geqq h(Z^*)$ for any $Z \in \mathcal{M}_{r\times n}$. □

Choosing, particularly, $G = B^\top A^{-1}$ we obtain the so-called principal solution $Z^* = B^\top A^{-1}$, which represents the unique solution in case of the positive definiteness of the matrix U.

Now, let $\hat{\vartheta} = Z\tilde{y} \in D_1$. Then $\hat{\vartheta}$ has Bayes risk

$$\begin{aligned}\varrho(P_{\tilde{\vartheta},\tilde{w}}; \hat{\vartheta}) &= \mathbf{E}_{\tilde{y},\tilde{\vartheta},\tilde{w}} w\|\tilde{\vartheta} - Z\tilde{y}\|_U^2 \\ &= \operatorname{tr} U[Z(\mathbf{E}\,\tilde{w}\tilde{y}\tilde{y}^\top)Z^\top - Z(\mathbf{E}\,\tilde{w}\tilde{y}\tilde{\vartheta}^\top) - (\mathbf{E}\,\tilde{w}\tilde{\vartheta}\tilde{y}^\top)Z^\top + \mathbf{E}\,\tilde{w}\tilde{\vartheta}\tilde{\vartheta}^\top].\end{aligned} \tag{5.29}$$

Defining

$$\begin{aligned}A_w &= \mathbf{E}(\tilde{y}\tilde{y}^\top|w) = F(\mathbf{E}\,\tilde{\vartheta}\tilde{\vartheta}^\top|w)F^\top + \frac{1}{w}I_n = FM_wF^\top + \frac{1}{w}I_n, \\ B_w &= \mathbf{E}(\tilde{y}\tilde{\vartheta}^\top|w) = F(\mathbf{E}\,\tilde{\vartheta}\tilde{\vartheta}^\top|w) = FM_w,\end{aligned} \tag{5.30}$$

where $M_w = \mathbf{E}(\tilde{\vartheta}\tilde{\vartheta}^\top|w) = \frac{1}{w}\Phi + \mu\mu^\top$ represents the second order moment matrix of the conditional prior $P_{\tilde{\vartheta}|w}$, then Lemma 5.3 leads to the following result.

Theorem 5.3. *Let $M = M_{w_0} = \mathbf{E}(\tilde{\vartheta}\tilde{\vartheta}^\top|w_0)$ and $\Sigma = w_0^{-1} I_n = \mathbf{Cov}(\tilde{e}|w_0)$. Then any estimator of the form*

$$\hat{\vartheta}^1_{\mathrm{B}}(\tilde{y}) = [U^+UMF^\top(FMF^\top + \Sigma)^{-1} + (I_r - U^+U)\,G]\,\tilde{y}, \quad G \in \mathcal{M}_{r\times n}$$

is a Bayes linear estimator for ϑ with respect to quadratic loss (5.27). The Bayes risk is given by

$$\varrho(P_{\tilde{\vartheta},\tilde{w}}; \hat{\vartheta}^1_{\mathrm{B}}) = w_0 \operatorname{tr}\{UM - UMF^\top(FMF^\top + \Sigma)^{-1}FM\}.$$

Proof. Using the denotations from (5.30), the Bayes risk (5.29) takes the form

$$\varrho(P_{\tilde{\vartheta},\tilde{w}}; \hat{\vartheta}) = w_0 \operatorname{tr} U[ZA_{w_0}Z^\top - ZB_{w_0} - B_{w_0}^\top Z^\top + M_{w_0}]$$

and the result follows immediately from Lemma 5.3 setting

$$A := A_{w_0} = FMF^\top + \Sigma, \quad B := B_{w_0} = FM, \quad C := M_{w_0} = M. \quad \square$$

In case of a regular weight matrix U we obtain the uniquely determined Bayes linear estimator

$$\hat{\vartheta}^1_{\mathrm{B}}(\tilde{y}) = MF^\top(FMF^\top + \Sigma)^{-1}\tilde{y}, \tag{5.31}$$

which then is independent of the particular choice of U.

Now, consider affine estimators $\hat{\vartheta}(\tilde{y}) = Z\tilde{y} + z \in D_a$. Their risk

$$R(\vartheta, w; \hat{\vartheta}) = w\,\mathbf{E}_{\tilde{y}|\vartheta,w}\|\vartheta - \hat{\vartheta}(\tilde{y})\|_U^2 = w\,\mathbf{E}_{\tilde{y}|\vartheta,w}\|(\vartheta - z) - Z\tilde{y}\|_U^2$$

can be written as

$$R(\vartheta, w; \hat{\vartheta}) = w \operatorname{tr} \boldsymbol{U}[\bar{\boldsymbol{Z}} \boldsymbol{A}_{\vartheta, w} \bar{\boldsymbol{Z}}^\top - \bar{\boldsymbol{Z}} \boldsymbol{B}_\vartheta - \boldsymbol{B}_\vartheta^\top \bar{\boldsymbol{Z}}^\top + \vartheta\vartheta^\top],$$

where

$$\bar{\boldsymbol{Z}} = [z \vdots \boldsymbol{Z}], \quad \boldsymbol{A}_{\vartheta, w} = \left[\begin{array}{c|c} 1 & \vartheta^\top \boldsymbol{F}^\top \\ \hline \boldsymbol{F}\vartheta & \mathsf{E}(\tilde{y}\tilde{y}^\top | \vartheta, w) \end{array}\right], \quad \boldsymbol{B}_\vartheta = \left[\begin{array}{c} \vartheta^\top \\ \hline \boldsymbol{F}\vartheta\vartheta^\top \end{array}\right]. \tag{5.32}$$

Hence, the Bayes risk is given by

$$\varrho(P_{\tilde{\vartheta}, \tilde{w}}; \hat{\vartheta}) = w_0 \operatorname{tr} \boldsymbol{U}[\bar{\boldsymbol{Z}} \bar{\boldsymbol{A}} \bar{\boldsymbol{Z}}^\top - \bar{\boldsymbol{Z}} \bar{\boldsymbol{B}} - \bar{\boldsymbol{B}}^\top \bar{\boldsymbol{Z}}^\top + \boldsymbol{M}]$$

with

$$\bar{\boldsymbol{A}} = \mathsf{E}\, \boldsymbol{A}_{\tilde{\vartheta}, \tilde{w}} = \left[\begin{array}{c|c} 1 & \boldsymbol{\mu}^\top \boldsymbol{F}^\top \\ \hline \boldsymbol{F}\boldsymbol{\mu} & \boldsymbol{F}\boldsymbol{M}\boldsymbol{F}^\top + \boldsymbol{\Sigma} \end{array}\right], \qquad \bar{\boldsymbol{B}} = \mathsf{E}\, \boldsymbol{B}_{\tilde{\vartheta}} = \left[\begin{array}{c} \boldsymbol{\mu}^\top \\ \hline \boldsymbol{F}\boldsymbol{M} \end{array}\right] \tag{5.33}$$

and $\boldsymbol{M} = \boldsymbol{M}_{w_0} = w_0^{-1}\boldsymbol{\Phi} + \boldsymbol{\mu}\boldsymbol{\mu}^\top$, $\boldsymbol{\Sigma} = w_0^{-1} \boldsymbol{I}_n$ as before. The minimization of the Bayes risk with respect to $\bar{\boldsymbol{Z}}$ again requires the solution of a matrix optimization problem of the type considered in Lemma 5.3, which leads to the following result.

Theorem 5.4. *Let the loss function be quadratic according to* (5.27) *and prior knowledge be given through the family* $\mathcal{P}$ *from* (5.28). *Then any estimator of the form*

$$\hat{\vartheta}_{\mathrm{B}}^{\mathrm{a}}(\tilde{y}) = [\boldsymbol{U}^+ \boldsymbol{U}\boldsymbol{\Phi}\boldsymbol{F}^\top (\boldsymbol{F}\boldsymbol{\Phi}\boldsymbol{F}^\top + \boldsymbol{I}_n)^{-1} + (\boldsymbol{I}_r - \boldsymbol{U}^+ \boldsymbol{U})\, \boldsymbol{G}](\tilde{y} - \boldsymbol{F}\boldsymbol{\mu}) + \boldsymbol{U}^+ \boldsymbol{U}(\boldsymbol{\mu} - \boldsymbol{g}) + \boldsymbol{g}$$

with arbitrary $\boldsymbol{G} \in \mathcal{M}_{r \times n}$, $\boldsymbol{g} \in \mathbb{R}^r$ *is a Bayes affine estimator for* ϑ. *The Bayes risk, which does not depend on the particular choice of* $\boldsymbol{G}$ *and* $\boldsymbol{g}$, *is given by*

$$\varrho(P_{\tilde{\vartheta}, \tilde{w}}; \hat{\vartheta}_{\mathrm{B}}^{\mathrm{a}}) = \operatorname{tr}\{\boldsymbol{U}\boldsymbol{\Phi} - \boldsymbol{U}\boldsymbol{\Phi}\boldsymbol{F}^\top (\boldsymbol{F}\boldsymbol{\Phi}\boldsymbol{F}^\top + \boldsymbol{I}_n)^{-1} \boldsymbol{F}\boldsymbol{\Phi}\}.$$

Proof. Using the block inversion formula A.37 we get

$$\bar{\boldsymbol{A}}^{-1} = \left[\begin{array}{c|c} (1 - \boldsymbol{\mu}^\top \boldsymbol{F}^\top \boldsymbol{A}^{-1} \boldsymbol{F}\boldsymbol{\mu})^{-1} & -w_0 \boldsymbol{\mu}^\top \boldsymbol{F}^\top (\boldsymbol{F}\boldsymbol{\Phi}\boldsymbol{F}^\top + \boldsymbol{I}_n)^{-1} \\ \hline -w_0 (\boldsymbol{F}\boldsymbol{\Phi}\boldsymbol{F}^\top + \boldsymbol{I}_n)^{-1} \boldsymbol{F}\boldsymbol{\mu} & w_0 (\boldsymbol{F}\boldsymbol{\Phi}\boldsymbol{F}^\top + \boldsymbol{I}_n)^{-1} \end{array}\right]$$

with $\boldsymbol{A} = \boldsymbol{F}\boldsymbol{M}\boldsymbol{F}^\top + \boldsymbol{\Sigma}$. Observing that, by A.34,

$$(1 - \boldsymbol{\mu}^\top \boldsymbol{F}^\top \boldsymbol{A}^{-1} \boldsymbol{F}\boldsymbol{\mu})^{-1} = 1 + w_0 \boldsymbol{\mu}^\top \boldsymbol{F}^\top (\boldsymbol{F}\boldsymbol{\Phi}\boldsymbol{F}^\top + \boldsymbol{I}_n)^{-1} \boldsymbol{F}\boldsymbol{\mu}$$

we have

$$\bar{\boldsymbol{B}}^\top \bar{\boldsymbol{A}}^{-1} = [\boldsymbol{\mu} - \boldsymbol{\Phi}\boldsymbol{F}^\top (\boldsymbol{F}\boldsymbol{\Phi}\boldsymbol{F}^\top + \boldsymbol{I}_n)^{-1} \boldsymbol{F}\boldsymbol{\mu} \;\vdots\; \boldsymbol{\Phi}\boldsymbol{F}^\top (\boldsymbol{F}\boldsymbol{\Phi}\boldsymbol{F}^\top + \boldsymbol{I}_n)^{-1}].$$

According to Lemma 5.3, the Bayes risk (5.33) is minimized by choosing $\bar{\boldsymbol{Z}}^* = \boldsymbol{U}^+ \boldsymbol{U} \bar{\boldsymbol{B}}^\top \bar{\boldsymbol{A}}^{-1} + (\boldsymbol{I}_r - \boldsymbol{U}^+ \boldsymbol{U})\, \bar{\boldsymbol{G}} =: [z^* \vdots \boldsymbol{Z}^*]$ with arbitrary $\bar{\boldsymbol{G}} = [\boldsymbol{g} \vdots \mathrm{G}]$, $\mathrm{g} \in \mathbb{R}^r$, $\boldsymbol{G} \in \mathcal{M}_{r \times n}$. The corresponding Bayes affine estimator is given by $\hat{\vartheta}_{\mathrm{B}}^{\mathrm{a}}(\tilde{y}) = \boldsymbol{Z}^* \tilde{y} + z^*$ and the Bayes risk $\varrho(P_{\tilde{\vartheta}, \tilde{w}}; \hat{\vartheta}_{\mathrm{B}}^{\mathrm{a}}) = w_0 \operatorname{tr} \boldsymbol{U}(\boldsymbol{M} - \bar{\boldsymbol{B}}^\top \bar{\boldsymbol{A}}^{-1} \bar{\boldsymbol{B}})$ takes the form indicated in Theorem 5.4. □

In case of a regular weight matrix U the Bayes affine estimator is uniquely determined by

$$\hat{\vartheta}_{\mathrm{B}}^{\mathrm{a}}(\tilde{y}) = \Phi F^{\top}(F\Phi F^{\top}+I_n)^{-1}(\tilde{y}-F\mu)+\mu, \tag{5.34}$$

which coincides with the representation (5.23) of the conjugate Bayes estimator derived in the preceding section. Hence, in case of an additional assumption of normality of $P_{\tilde{e}}$ and of a conjugate normal gamma prior distribution $P_{\tilde{\vartheta},\tilde{w}}$, we do not restrict generality if we confine ourselves to the class of affine estimators (see also Section 6.1).

Remark 5.3. (i) In the special case where $\mathsf{E}\,\tilde{\vartheta}=\mu=0$ we have $M=w_0^{-1}\Phi$ and the Bayes affine estimator coincides with the Bayes linear estimator.

(ii) From our preceding derivations it is clear that the general representation of the Bayes affine estimator holds for arbitrary, not necessarily regular covariance matrices Φ. If either one of the matrices Φ or $F^{\top}F$ is regular, then $\hat{\vartheta}_{\mathrm{B}}^{\mathrm{a}}$ can be written more conveniently as $\hat{\vartheta}_{\mathrm{B}}^{\mathrm{a}}(\tilde{y})=(F^{\top}F+\Phi^{-1})^{-1}(F^{\top}\tilde{y}+\Phi^{-1}\mu)$ or $\hat{\vartheta}_{\mathrm{B}}^{\mathrm{a}}(\tilde{y})=\mu+\Phi(\Phi+V)^{-1}(\hat{\vartheta}_{\mathrm{LS}}(\tilde{y})-\mu)$, where $V=(F^{\top}F)^{-1}$ and $\hat{\vartheta}_{\mathrm{LS}}(\tilde{y})=VF^{\top}\tilde{y}$ (cp. (5.12) and (5.23)).

Remark 5.4. An obvious generalization to the case of unknown covariance structure, i.e.,

$$\mathcal{P}_{\tilde{e}}=\{P_{\tilde{e}}:\mathsf{E}_{\tilde{e}}=0,\quad \mathsf{Cov}\,\tilde{e}=\Sigma\}_{\Sigma\in\mathcal{M}_n^{>}}, \tag{5.35}$$

yields that

$$\hat{\vartheta}_{\mathrm{B}}^{\Sigma_0}(\tilde{y})=\Phi F^{\top}(F\Phi F^{\top}+\Sigma_0)^{-1}(\tilde{y}-F\mu)+\mu \tag{5.36}$$

is a Bayes affine estimator with respect to any prior $P_{\tilde{\vartheta},\tilde{\Sigma}}$ such that $\mathsf{E}\,\tilde{\vartheta}=\mu$, $\mathsf{Cov}\,\tilde{\vartheta}=\Phi$, $\mathsf{E}\,\tilde{\Sigma}=\Sigma_0$ and with respect to quadratic loss $\|\vartheta-\hat{\vartheta}(\tilde{y})\|_U^2$, $U\in\mathcal{M}_r^{\geqq}$.

5.4. Risk improvement over the LSE

From (5.18) it is clear that, in general, the Bayes estimator $\hat{\vartheta}_{\mathrm{B}}$ will not be unbiased; its bias is given by

$$b(\vartheta)=\mathsf{E}\,\hat{\vartheta}_{\mathrm{B}}(\tilde{y})-\vartheta=(F^{\top}F+\Phi^{-1})^{-1}(\mu-\vartheta).$$

In the following let denote

$$\psi=(F^{\top}F+\Phi^{-1})^{-1} \tag{5.37}$$

which, apart from the precision factor w, is identical with the covariance matrix of the posterior distribution of $\tilde{\vartheta}$ (cp. (5.17)). The squared bias of $\hat{\vartheta}_{\mathrm{B}}$ then takes the form

$$\|b(\vartheta)\|^2=\|\vartheta-\mu\|_Q^2,\quad \text{where}\quad Q=\Phi^{-1}\psi^2\Phi^{-1}. \tag{5.38}$$

Notice that this is a continuous, increasing function of the weighted distance between the true parameter and its prior expectation.

It can be shown that the Bayes estimator has smaller covariance matrix (in the semiordering sense A.IV) than the LSE.

Lemma 5.4. (i) $\mathbf{Cov}\,\hat{\boldsymbol{\vartheta}}_B(\tilde{\boldsymbol{y}}) = \frac{1}{w}\boldsymbol{\psi}\boldsymbol{F}^\top\boldsymbol{F}\boldsymbol{\psi}$,

(ii) $\mathbf{Cov}\,\hat{\boldsymbol{\vartheta}}_B(\tilde{\boldsymbol{y}}) < \mathbf{Cov}\,\hat{\boldsymbol{\vartheta}}_{LS}(\tilde{\boldsymbol{y}})$.

Proof. Assertion (i) follows directly from the representation

$$\hat{\boldsymbol{\vartheta}}_B(\tilde{\boldsymbol{y}}) = \boldsymbol{\psi}\boldsymbol{F}^\top\tilde{\boldsymbol{y}} + \boldsymbol{\psi}\boldsymbol{\Phi}^{-1}\boldsymbol{\mu}.$$

The second assertion follows by verifying the equivalent statement that $(\mathbf{Cov}\,\hat{\boldsymbol{\vartheta}}_{LS})^{-1} < (\mathbf{Cov}\,\hat{\boldsymbol{\vartheta}}_B)^{-1}$. But this is true since

$$\begin{aligned}(\mathbf{Cov}\,\hat{\boldsymbol{\vartheta}}_{LS})^{-1} &= w\boldsymbol{F}^\top\boldsymbol{F} < w\boldsymbol{F}^\top\boldsymbol{F} + 2w\boldsymbol{\Phi}^{-1} + w\boldsymbol{\Phi}^{-1}(\boldsymbol{F}^\top\boldsymbol{F})^{-1}\boldsymbol{\Phi}^{-1}\\ &= w\boldsymbol{\psi}^{-1}(\boldsymbol{F}^\top\boldsymbol{F})^{-1}\boldsymbol{\psi}^{-1} = (\mathbf{Cov}\,\hat{\boldsymbol{\vartheta}}_B)^{-1}\end{aligned}$$

observing that $\boldsymbol{\Phi}^{-1}$ and $\boldsymbol{\Phi}^{-1}(\boldsymbol{F}^\top\boldsymbol{F})^{-1}\boldsymbol{\Phi}^{-1}$ are positive definite. □

From the above result it is clear that the sampling variance of any linear combination $\boldsymbol{c}^\top\boldsymbol{\vartheta}$ is smaller if estimated by $\boldsymbol{c}^\top\hat{\boldsymbol{\vartheta}}_B$ instead of $\boldsymbol{c}^\top\hat{\boldsymbol{\vartheta}}_{LS}$, i.e.,

$$\mathrm{Var}_{\tilde{y}|\vartheta,w}\,\boldsymbol{c}^\top\hat{\boldsymbol{\vartheta}}_B(\tilde{\boldsymbol{y}}) < \mathrm{Var}_{\tilde{y}|\vartheta,w}\,\boldsymbol{c}^\top\hat{\boldsymbol{\vartheta}}_{LS}(\tilde{\boldsymbol{y}}) \quad \text{for all} \quad \boldsymbol{c} \neq \boldsymbol{0}$$

and arbitrary $\boldsymbol{\vartheta} \in \mathbb{R}^r$, $w \in \mathbb{R}^+$. Particularly, any component of the Bayes estimator $\hat{\boldsymbol{\vartheta}}_B$ has smaller variance than the corresponding component of the LSE. From this the possibility arises that the Bayes estimator has smaller risk than the LSE for certain parameter values $\boldsymbol{\vartheta}$ and w varying in some subregion $\Theta \times W_0$ of the parameter space $\Theta \times W = \mathbb{R}^r \times \mathbb{R}^+$. Before demonstrating this fact we need explicit expressions for the usual risk and the Bayes risk of $\hat{\boldsymbol{\vartheta}}_B$.

Lemma 5.5 *Under the assumptions of Theorem* 5.2 *the risk function and the Bayes risk are given by*

$$R(\boldsymbol{\vartheta}, w; \hat{\boldsymbol{\vartheta}}_B) = \frac{1}{w}\,\mathrm{tr}\,\{\boldsymbol{U}(w)\boldsymbol{\psi}\boldsymbol{F}^\top\boldsymbol{F}\boldsymbol{\psi}\} + \|\boldsymbol{\vartheta} - \boldsymbol{\mu}\|^2_{A(w)}$$

and

$$\varrho(P_{\tilde{\vartheta},\tilde{w}}; \hat{\boldsymbol{\vartheta}}_B) = \mathbf{E}_{\tilde{\vartheta},\tilde{w}} R(\tilde{\boldsymbol{\vartheta}}, \tilde{w}; \hat{\boldsymbol{\vartheta}}_B) = \mathrm{tr}\,\boldsymbol{U}_0\boldsymbol{\psi},$$

where $\quad \boldsymbol{A}(w) = \boldsymbol{\Phi}^{-1}\boldsymbol{\psi}\boldsymbol{U}(w)\boldsymbol{\psi}\boldsymbol{\Phi}^{-1}$ *and* $\boldsymbol{U}_0 = \mathbf{E}_{\tilde{w}}\{\tilde{w}^{-1}\boldsymbol{U}(\tilde{w})\}$.

Proof. With the above assumptions we have

$$\begin{aligned}R(\boldsymbol{\vartheta}, w; \hat{\boldsymbol{\vartheta}}_B) &= \mathbf{E}_{\tilde{y}|\vartheta,w}\|\boldsymbol{\vartheta} - \hat{\boldsymbol{\vartheta}}_B(\tilde{\boldsymbol{y}})\|^2_{U(w)}\\ &= \mathrm{tr}\{\boldsymbol{U}(w)\,[\mathbf{Cov}\,\hat{\boldsymbol{\vartheta}}_B + \boldsymbol{b}(\boldsymbol{\vartheta})\,\boldsymbol{b}(\boldsymbol{\vartheta})^\top]\}\\ &= \frac{1}{w}\,\mathrm{tr}\{\boldsymbol{U}(w)\boldsymbol{\psi}\boldsymbol{F}^\top\boldsymbol{F}\boldsymbol{\psi}\} + \boldsymbol{b}(\boldsymbol{\vartheta})^\top\boldsymbol{U}(w)\,\boldsymbol{b}(\boldsymbol{\vartheta})\end{aligned}$$

and inserting the bias $\boldsymbol{b}(\boldsymbol{\vartheta}) = \boldsymbol{\psi}\boldsymbol{\Phi}^{-1}(\boldsymbol{\mu} - \boldsymbol{\vartheta})$ gives the required result for the risk. Further, taking expectation with respect to $P_{\tilde{\vartheta}, \tilde{w}}$ yields

$$\begin{aligned}\varrho(P_{\tilde{\vartheta}, \tilde{w}}; \hat{\vartheta}_B) &= \mathbf{E}_{\tilde{\vartheta}, \tilde{w}} \operatorname{tr}\{\tilde{w}^{-1} U(\tilde{w}) \psi F^\top F \psi\} + \mathbf{E}_{\tilde{\vartheta}, \tilde{w}} (\tilde{\vartheta} - \mu)^\top A(\tilde{w}) (\tilde{\vartheta} - \mu) \\ &= \operatorname{tr} U_0 \psi F^\top F \psi + \mathbf{E}_{\tilde{w}} \operatorname{tr} A(\tilde{w}) \operatorname{\mathbf{Cov}}(\tilde{\vartheta} | \tilde{w}).\end{aligned}$$

But, $\mathbf{E}_{\tilde{w}} \operatorname{tr} A(\tilde{w}) \operatorname{\mathbf{Cov}}(\tilde{\vartheta}|\tilde{w}) = \mathbf{E}_{\tilde{w}} \operatorname{tr} \tilde{w}^{-1} A(\tilde{w}) \Phi = \operatorname{tr} U_0 \psi \Phi^{-1} \psi$ and thus

$$\begin{aligned}\varrho(P_{\tilde{\vartheta}, \tilde{w}}; \hat{\vartheta}_B) &= \operatorname{tr} U_0 \psi F^\top F \psi + \operatorname{tr} U_0 \psi \Phi^{-1} \psi = \operatorname{tr} U_0 \psi [F^\top F + \Phi^{-1}] \psi \\ &= \operatorname{tr} U_0 \psi. \quad \square\end{aligned}$$

Remark 5.5. In case of normalized quadratic loss (5.27) we have $U_0 = U$ and thus

$$\begin{aligned}R(\vartheta, w; \hat{\vartheta}_B) &= \operatorname{tr} U \psi F^\top F \psi + w (\vartheta - \mu)^\top \Phi^{-1} \psi U \psi \Phi^{-1} (\vartheta - \mu), \\ \varrho(P_{\tilde{\vartheta}, \tilde{w}}; \hat{\vartheta}_B) &= \operatorname{tr} U \psi.\end{aligned}$$

More general expressions for R and ϱ, which also admit a non-regular covariance matrix Φ, were given in the preceding section using the equivalence between $\hat{\vartheta}_B$ and the Bayes affine estimator.

The following result demonstrates that there exists a nonempty subregion in the parameter space for which the Bayes estimator $\hat{\vartheta}_B$ has smaller risk than the LSE and shows the extent to which such an improvement is possible.

Theorem 5.5. *Let the assumptions of Theorem 5.2 be satisfied and assume* F *to have full rank r. Then* $\hat{\vartheta}_B$ *has a smaller risk than* $\hat{\vartheta}_{LS}$ *for all parameters* $(\vartheta, w) \in \mathbb{R}^r \times \mathbb{R}^+$ *for which it holds*

$$\|\vartheta - \mu\|^2_{A(w)} \leq \operatorname{tr} U(w) [\operatorname{\mathbf{Cov}} \hat{\vartheta}_{LS} - \operatorname{\mathbf{Cov}} \hat{\vartheta}_B]. \tag{5.39}$$

Proof. With $R(\vartheta, w; \hat{\vartheta}_{LS}) = \operatorname{tr} U(w) \operatorname{\mathbf{Cov}} \hat{\vartheta}_{LS}$ and $R(\vartheta, w; \hat{\vartheta}_B)$ as given in Lemma 5.5 we obtain

$$R(\vartheta, w; \hat{\vartheta}_{LS}) - R(\vartheta, w; \hat{\vartheta}_B) = \operatorname{tr} U(w) [\operatorname{\mathbf{Cov}} \hat{\vartheta}_{LS} - \operatorname{\mathbf{Cov}} \hat{\vartheta}_B] - \|\vartheta - \mu\|^2_{A(w)}.$$

This yields the result by virtue of the fact that $\|\vartheta - \mu\|^2_{A(w)}$ is a continuous, nonnegative and increasing function of the weighted distance between ϑ and μ, which can be made arbitrarily small in a neighbourhood of μ, and the fact that $\operatorname{tr} U(w) [\operatorname{\mathbf{Cov}} \hat{\vartheta}_{LS} - \operatorname{\mathbf{Cov}} \hat{\vartheta}_B] > 0$ due to the positive definite character of the difference matrix $\operatorname{\mathbf{Cov}} \hat{\vartheta}_{LS} - \operatorname{\mathbf{Cov}} \hat{\vartheta}_B$ (see Lemma 5.4 (ii)). $\square$

Obviously, the subregion of parameter values leading to an improvement in risk describes an ellipsoid with center point μ provided that $U(w)$ is regular. For a fixed design, the size of this ellipsoid and the magnitude of the improvement depend on the "precision of prior knowledge" expressed by Φ^{-1}.

For the special case that $U(w) \equiv I_r$, for which the risk function is identical with the usual mean square error (MSE) defined by

$$\mathrm{MSE}(\hat{\vartheta}) = \mathbf{E}_{\tilde{y}|\vartheta, w}\|\hat{\vartheta}(\tilde{y}) - \vartheta\|^2, \quad \hat{\vartheta} \in D$$

Theorem 5.5 reads: $\mathrm{MSE}(\hat{\vartheta}_{\mathrm{B}}) \leqq \mathrm{MSE}(\hat{\vartheta}_{\mathrm{LS}})$ if and only if

$$(\vartheta - \mu)^\top \Phi^{-1}\psi^2\Phi^{-1}(\vartheta - \mu) \leqq \mathrm{tr}\,(I_r - F^\top F\psi^2 F^\top F)\,\mathbf{Cov}\,\hat{\vartheta}_{\mathrm{LS}} \tag{5.40}$$

(cp. SMITH (1973)).

Further, setting $U(w) \equiv cc^\top$ with some constant vector $c \in \mathbb{R}^r$, from Theorem 5.5 may be derived a condition under which the Bayes estimator of the linear combination $c^\top\vartheta$ has smaller MSE than the corresponding least squares estimation $c^\top\hat{\vartheta}_{\mathrm{LS}}$. Particularly, choosing c as the ith unit vector gives the corresponding result on the improvement in MSE for the single components ϑ_i of ϑ, $i = 1, \ldots, r$. GILES and RAYNER (1979) derived a similar condition which guarantees that $\mathrm{MSE}(c^\top\hat{\vartheta}_{\mathrm{B}}) \leqq \mathrm{MSE}(c^\top\hat{\vartheta}_{\mathrm{LS}})$ for every $c \neq \mathbf{0}$. Their condition reads

$$(\vartheta - \mu)^\top((F^\top F)^{-1} + 2\Phi)^{-1}(\vartheta - \mu) \leqq w. \tag{5.41}$$

This is equivalent to stating that $\hat{\vartheta}_{\mathrm{B}}$ has a smaller matrix valued risk than $\hat{\vartheta}_{\mathrm{LS}}$.

In general, the region of parameters for which this inequality comes true will be considerably smaller than the region defined by (5.40) since the former takes account of the whole (overcountable) variety of possible linear combinations.

The result of Theorem 5.5 indicates that an improvement in risk over the LSE is possible if the prior expectation μ is chosen sufficiently close to the true regression parameter ϑ. By virtue of A.24, a sufficient condition for an improvement to hold is that

$$\|\vartheta - \mu\|^2 \leqq \lambda_{\max}^{-1}(A(w))\,\mathrm{tr}\,U(w)\,[\mathbf{Cov}\,\hat{\vartheta}_{\mathrm{LS}} - \mathbf{Cov}\,\hat{\vartheta}_{\mathrm{B}}]. \tag{5.42}$$

The region of all parameters ϑ satisfying (5.42) describes the largest sphere in the parameter region $\Theta = \mathbb{R}^r$ within which the Bayes estimator has smaller risk than $\hat{\vartheta}_{\mathrm{LS}}$.

Clearly, since the Conditions (5.39)–(5.42), within the sampling theoretic framework, depend on the true parameters ϑ and w, which are unknown to us, we cannot be sure that an improvement in risk over the LSE will be achieved in a particular case at hand. However, with respect to the prior distribution for $\tilde{\vartheta}$, we can always achieve an improvement in the corresponding expected risk. This is clear from the fact that this is precisely the Bayes risk which, by definition, is minimized by the Bayes estimator.

Theorem 5.6. *Let the assumptions of Theorem* 5.5 *be satisfied and* U_0 *as defined in Lemma* 5.5. *Then it holds*

$$\varrho(P_{\tilde{\vartheta}, \tilde{w}}; \hat{\vartheta}_{\mathrm{LS}}) - \varrho(P_{\tilde{\vartheta}, \tilde{w}}; \hat{\vartheta}_{\mathrm{B}}) = \mathrm{tr}\,U_0((F^\top F)^{-1} - \psi) > 0.$$

Proof. First we observe that

$$\varrho(P_{\tilde{\vartheta},\tilde{w}};\hat{\vartheta}_{LS}) = \mathbf{E}_{\tilde{\vartheta},\tilde{w}} R(\tilde{\vartheta},\tilde{w};\hat{\vartheta}_{LS}) = \mathbf{E}_{\tilde{w}} \operatorname{tr} \tilde{w}^{-1} U(\tilde{w}) (F^\top F)^{-1} = \operatorname{tr} U_0 (F^\top F)^{-1}$$

which together with $\varrho(P_{\tilde{\vartheta},\tilde{w}};\hat{\vartheta}_B) = \operatorname{tr} U_0 \psi$ (see Lemma 5.5) yields the above difference $\operatorname{tr} U_0((F^\top F)^{-1} - \psi)$. Now, since $F^\top F < \psi^{-1} = F^\top F + \Phi^{-1}$, it follows with A.29 that $(F^\top F)^{-1} - \psi$ is positive definite. Thus we have $\operatorname{tr} U_0((F^\top F)^{-1} - \psi) > 0$. □

Particularly, the expected improvement in MSE is given by

$$\mathbf{E}\{\mathrm{MSE}(\hat{\vartheta}_{LS}) - \mathrm{MSE}(\hat{\vartheta}_B)\} = \sigma_0^2 \operatorname{tr}((F^\top F)^{-1} - \psi), \tag{5.43}$$

where $\sigma_0^2 = \mathbf{E}\,\tilde{w}^{-1}$ stands for the prior expected variance of observation.

The degree of improvement in expected risk depends crucially on the prior covariance matrix Φ and the expected variance of observation (via $U_0 = \mathbf{E}\,\tilde{w}^{-1} U(\tilde{w})$), the latter will but be immaterial when measuring the relative improvement $\operatorname{tr} U_0((F^\top F)^{-1} - \psi)/\operatorname{tr} U_0(F^\top F)^{-1}$. The expected improvement

$$\operatorname{tr} U_0((F^\top F)^{-1} - \psi) = \operatorname{tr} U_0((F^\top F)^{-1} - (F^\top F + \Phi^{-1})^{-1})$$

will be small if the prior knowledge becomes vague, i.e., if $\|\Phi^{-1}\|$ approaches zero. On the other hand, if we know that the prior covariance matrix does not exceed a certain upper bound (in the semiordering sense) then we can guarantee the following least expected improvement over the LSE.

Corollary 5.4. *Assume that $\Phi \leqq \Phi_0$ for some given matrix $\Phi_0 \in \mathcal{M}_r^>$. Then it holds*

$$\varrho(P_{\tilde{\vartheta},\tilde{w}};\hat{\vartheta}_{LS}) - \varrho(P_{\tilde{\vartheta},\tilde{w}};\hat{\vartheta}_B) \geqq \operatorname{tr} U_0((F^\top F)^{-1} - \psi_0) > 0,$$

where $\psi_0 = (F^\top F + \Phi_0^{-1})^{-1}$ is an upper bound for the posterior covariance matrix of $\tilde{\vartheta}$.

Proof. With $\Phi \leqq \Phi_0$ we have $\psi = (F^\top F + \Phi^{-1})^{-1} \leqq (F^\top F + \Phi_0^{-1})^{-1} = \psi_0$ and thus $\operatorname{tr} U_0 \psi \leqq \operatorname{tr} U_0 \psi_0$ by virtue of A.28. Hence, $\operatorname{tr} U_0((F^\top F)^{-1} - \psi) \geqq \operatorname{tr} U_0((F^\top F)^{-1} - \psi_0)$ and the result follows from Theorem 5.6. □

A further improvement in risk can be achieved by a suitable choice of the experimental design (via $F^\top F$), in our case such that $\operatorname{tr} U_0 \psi = \operatorname{tr} U_0 (F^\top F + \Phi^{-1})^{-1}$ becomes minimal. This is a design criterion which, among others, will be discussed in detail in Part III.

Remark 5.6. As can be seen from the proof of Lemma 5.5, the computation of the Bayes risks of $\hat{\vartheta}_B$ and $\hat{\vartheta}_{LS}$ does not make use of the distributional assumptions on $\tilde{y}$, $\tilde{\vartheta}$ and $\tilde{w}$, but only refers to the first and second order moments, which is due to the fact that $\hat{\vartheta}_B$ and $\hat{\vartheta}_{LS}$ belong to the class of affine-linear estimators. This means, as long as quadratic loss (5.11) is concerned, all the statements given above remain true even if the assumptions of normally distributed observations and a conjugate normal-gamma-prior for $(\tilde{\boldsymbol{\vartheta}}, \tilde{w})$ are dropped. We shall be concerned with such and related robustness properties of the Bayes estimator in Sections 6.1 and 6.2.

Example 5.1. Consider the simple linear regression model

$$\mathbf{E}\tilde{y}(x) = \vartheta_1 + \vartheta_2 x, \quad x \in [-1, 1],$$

assume the errors $\tilde{e}_1, \tilde{e}_2$ to be normally (independently and identically) distributed with the same precision w, i.e.,

$$p_{\tilde{e}_i}^{(w)}(t_i) = (w/2\pi)^{1/2} \exp\left(-\frac{w}{2} t_i^2\right), \qquad i = 1, 2$$

$$p_{\tilde{e}}^{(w)}(t) = p_{\tilde{e}_1}^{(w)}(t_1) p_{\tilde{e}_2}^{(w)}(t_2), \qquad t = (t_1, t_2) \in \mathbb{R}^2.$$

Further, let us assume a conditional normal prior $P_{\tilde{\vartheta}|w} = \mathrm{N}\left(\boldsymbol{\mu}, \frac{1}{w}\boldsymbol{\Phi}\right)$ for $\tilde{\boldsymbol{\vartheta}} = (\tilde{\vartheta}_1, \tilde{\vartheta}_2)$ with

$$\boldsymbol{\mu} = \begin{pmatrix} 1/2 \\ 1/2 \end{pmatrix} \quad \text{and} \quad \boldsymbol{\Phi} = \begin{pmatrix} 1/12 & 0 \\ 0 & 3/4 \end{pmatrix}.$$

With the design $v_2 = (-1, 0)$ and the observation $\boldsymbol{y} = (2, 1)^\top$ we have

$$\boldsymbol{F} = \begin{pmatrix} 1 & -1 \\ 1 & 0 \end{pmatrix}, \quad \boldsymbol{F}^\top \boldsymbol{F} = \begin{pmatrix} 2 & -1 \\ -1 & 1 \end{pmatrix}, \quad \boldsymbol{F}^\top \boldsymbol{y} = \begin{pmatrix} 3 \\ -2 \end{pmatrix}, \quad \hat{\boldsymbol{\vartheta}}_{\mathrm{LS}}(\boldsymbol{y}) = \begin{pmatrix} 1 \\ -1 \end{pmatrix}$$

$$\boldsymbol{\psi} = \begin{pmatrix} 14 & -1 \\ -1 & \dfrac{7}{3} \end{pmatrix}^{-1} = \begin{pmatrix} 0.074 & 0.032 \\ 0.032 & 0.442 \end{pmatrix}, \qquad \boldsymbol{\Phi}^{-1}\boldsymbol{\mu} = \begin{pmatrix} 6 \\ \dfrac{2}{3} \end{pmatrix}.$$

Thus the Bayes estimate for $\boldsymbol{\vartheta} = (\vartheta_1, \vartheta_2)$ becomes

$$\hat{\boldsymbol{\vartheta}}_{\mathrm{B}}(\boldsymbol{y}) = \boldsymbol{\psi}(\boldsymbol{F}^\top \boldsymbol{y} + \boldsymbol{\Phi}^{-1}\boldsymbol{\mu}) = (\boldsymbol{F}^\top \boldsymbol{F} \hat{\boldsymbol{\vartheta}}_{\mathrm{LS}}(\boldsymbol{y}) + \boldsymbol{\Phi}^{-1}\boldsymbol{\mu})$$
$$= \begin{pmatrix} 0.074 & 0.032 \\ 0.032 & 0.442 \end{pmatrix} \begin{pmatrix} 9 \\ -1.333 \end{pmatrix} = \begin{pmatrix} 0.621 \\ -0.305 \end{pmatrix}.$$

We see that the components of $\hat{\boldsymbol{\vartheta}}_{\mathrm{B}}(\boldsymbol{y})$ range between the corresponding components of $\boldsymbol{\mu}$ and $\hat{\boldsymbol{\vartheta}}_{\mathrm{LS}}(\boldsymbol{y})$. Thereby the first component is pulled stronger to the prior expected value 0.5 than the second component which is due to the smaller variance of $\tilde{\vartheta}_1$.

Let us compare the expected mean square errors of $\hat{\boldsymbol{\vartheta}}_{\mathrm{B}}$ and $\hat{\boldsymbol{\vartheta}}_{\mathrm{LS}}$. For the least squares estimator we have MSE $(\hat{\boldsymbol{\vartheta}}_{\mathrm{LS}}) = w^{-1} \operatorname{tr}(\boldsymbol{F}^\top \boldsymbol{F})^{-1} = 3w^{-1} = 3\sigma^2$. In order to find the region of parameters for which the Bayes estimator has smaller MSE than $\hat{\vartheta}_{LS}$ we need to compute

$$A = \Phi^{-1}\psi^2\Phi^{-1} = \begin{pmatrix} 0.925 & 0.261 \\ 0.261 & 0.349 \end{pmatrix}$$

$$\operatorname{tr}(I_r - F^\top F\psi^2 F^\top F)\,\mathbf{Cov}\,\hat{\vartheta}_{LS} = \frac{1}{w}(\operatorname{tr}(F^\top F)^{-1} - \operatorname{tr} F^\top F\psi^2)$$
$$= 2.824\sigma^2.$$

Then, according to (5.40), $\hat{\vartheta}_B$ has smaller MSE than $\hat{\vartheta}_{LS}$ for all parameters ϑ and w such that $\|\vartheta - \mu\|_A^2 \leqq 2.824 w^{-1}$, or equivalently,

$$0.328(\vartheta_1 - 0.5)^2 + 0.185(\vartheta_1 - 0.5)(\vartheta_2 - 0.5) + 0.124(\vartheta_2 - 0.5)^2 \leqq w^{-1}.$$

For comparison, the subregion (5.41) which guarantees that $c^\top\hat{\vartheta}_B$ has smaller MSE than $c^\top\hat{\vartheta}_{LS}$ for every $c \neq 0$ takes the form

$$1.017(\vartheta_1 - 0.5)^2 + 2(\vartheta_1 - 0.5)(\vartheta_2 - 0.5) + 3.5(\vartheta_2 - 0.5)^2 \leqq w.$$

The figure below displays both regions for the case that $w^{-1} = \sigma_0^2 = 1$. It shows that the region of ϑ-values for which MSE $(\hat{\vartheta}_B) \leqq$ MSE $(\hat{\vartheta}_{LS})$ is considerably large whereas the second ellipsoid is much smaller for the reasons already discussed above.

Finally, comparing the expected MSE's we obtain

$$\operatorname{EMSE}(\hat{\vartheta}_{LS}) = \sigma_0^2 \operatorname{tr}(F^\top F)^{-1} = 3\sigma_0^2,$$
$$\operatorname{EMSE}(\hat{\vartheta}_B) = \sigma_0^2 \operatorname{tr}\psi = 0.516\sigma_0^2.$$

Thus, taking the Bayes estimator with respect to the above prior distribution we can expect a substantial improvement corresponding to 83 per cent of the expected MSE of $\hat{\vartheta}_{LS}$.

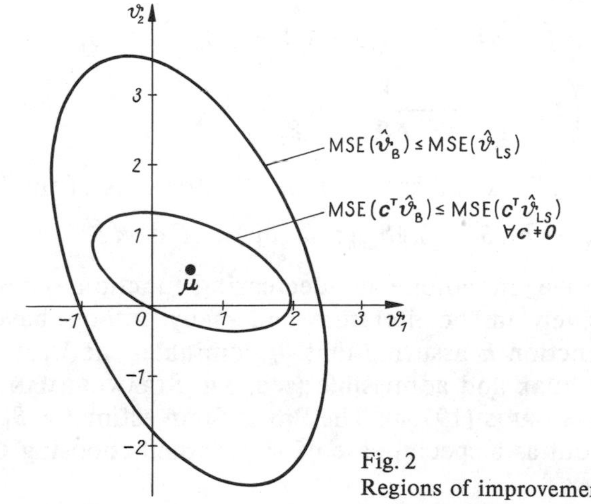

Fig. 2
Regions of improvement in MSE over the LSE

The application of the Bayes estimator crucially depends on the knowledge of the prior moments μ, Φ and, implicitly via Φ, on $\sigma_0^2 = \mathbf{E}\,\tilde{w}^{-1}$. In the Sections 6.2 and 6.3 we investigate the sensitivity of $\hat{\vartheta}_B$ when there is some uncertainty in the specification of these quantities, that is if we are only able to specify certain bounds or classes within which the prior moments can vary.

5.5. Stein-type modifications of the conjugate Bayes estimator

We have already seen that the conjugate Bayes (affine) estimator effects an improvement in risk over the LSE in some subregion of the parameter space $\Theta = \mathbb{R}^r$, the crucial problem remains, however, the specification of the prior "inputs" μ and Φ. We will now demonstrate that a simple bootstrap version of the Bayes linear estimator

$$\hat{\vartheta}^1_B(\tilde{y}) = M((F^\top \Sigma^{-1} F)^{-1} + M)^{-1} \hat{\vartheta}_{LS}(y) \tag{5.44}$$

(cp. (5.23) with $M = \mathsf{E}\,\tilde{\vartheta}\tilde{\vartheta}^\top$ instead of Φ, and Section 5.3) results in a (nonlinear) Stein-type estimator which guarantees a uniform improvement over the LSE on the whole parameter space. We consider such versions which arise from (5.44) using a data-based estimate

$$\hat{M} = a\,\mathsf{Cov}\,\hat{\vartheta}_{LS} + b\hat{\vartheta}_{LS}\vartheta^T_{LS} = a(F^\top \hat{\Sigma}^{-1} F)^{-1} + b\hat{\vartheta}_{LS}\hat{\vartheta}^T_{LS} \tag{5.45}$$

of the second order moment matrix $M = \mathsf{Cov}\,\tilde{\vartheta} + (\mathsf{E}\,\tilde{\vartheta})(\mathsf{E}\,\tilde{\vartheta})^\top$, where a, b are real numbers and $\hat{\Sigma} = \hat{\sigma}^2 I_n$ with

$$\hat{\sigma}^2 = \frac{1}{n-r}(\tilde{y} - F\hat{\vartheta}_{LS})^\top (\tilde{y} - F\hat{\vartheta}_{LS})$$

an estimate of $W = \mathsf{E}_{\tilde{w}}\,\mathsf{Cov}\,(\tilde{e}|\tilde{w}) = w_0^{-1} I_n$. Now, replacing M in (5.44) by $\hat{M}$ yields the Bayes-Stein estimator

$$\begin{aligned}\hat{\vartheta}_{BS}(\tilde{y}) &= (a\hat{V} + b\hat{\vartheta}_{LS}\hat{\vartheta}^\top_{LS})\,[(a+1)\hat{V} + b\hat{\vartheta}_{LS}\hat{\vartheta}^\top_{LS}]^{-1}\hat{\vartheta}_{LS} \\ &= \left(1 - \frac{1}{1 + a + b\hat{\vartheta}^\top_{LS}\hat{V}^{-1}\hat{\vartheta}_{LS}}\right)\hat{\vartheta}_{LS},\end{aligned} \tag{5.46}$$

where $\hat{V} = (F^\top \hat{\Sigma}^{-1} F)^{-1}$. This is a special type of estimators of the form

$$\hat{\vartheta}_h(\tilde{y}) = (1 - h(\|\hat{\vartheta}_{LS}\|^2_C)/\|\hat{\vartheta}_{LS}\|^2_C)\hat{\vartheta}_{LS}(\tilde{y}), \quad C \in \mathcal{M}_r^> \tag{5.47}$$

with h being a positive, monotone nondecreasing function. These have been studied extensively in the literature and many authors have given conditions on the function h assuring that $\hat{\vartheta}_h$ dominates the least squares estimator and is minimax and admissible (see, e.g. STRAWDERMAN (1971), ALAM (1973), EFRON/MORRIS (1973)). The Bayes-Stein estimator $\hat{\vartheta}_{BS}$ introduced above comes out as a special case of (5.47) when choosing $C = \hat{V}^{-1}$ and

$$h(t) = t/(1 + a + bt).$$

For $t > 0$ and a, $b > 0$ this function is positive, monotone nondecreasing and the shrinkage factor satisfies

$$a/(a+1) \leqq 1 - h(t)/t < 1,$$

in particular, it is nonnegative. Abbreviating

$$G(\tilde{y}) = \hat{\vartheta}_{LS}^{\top} F^{\top} F \hat{\vartheta}_{LS} / m\hat{\sigma}^2, \qquad m = n - r,$$

where $mG(\tilde{y})/r$ represents the usual F-statistics, the estimator $\hat{\vartheta}_{BS}$ takes the form

$$\hat{\vartheta}_{BS}(\tilde{y}) = \left(1 - \frac{1}{1 + a + mbG(\tilde{y})}\right) \hat{\vartheta}_{LS}(\tilde{y}). \tag{5.48}$$

By an appropriate choice of the constants a and b we can construct estimators having desired optimality properties.

Theorem 5.7. *Let* $\Theta = \mathbb{R}^r$, $\mathcal{P}_{\tilde{e}} = \mathcal{P}_{\tilde{e}}^{(N)}$ *and L according to* (5.27). *Further, assume* $r \geqq 3$, $n \geqq r + 2$ *and*

$$c := \operatorname{tr}(F^{\top}F)^{-1}/\lambda_{\max}(U(F^{\top}F)^{-1}) > 2.$$

Then any estimator $\hat{\vartheta}_{BS}$ *of the form* (5.48) *with* $a \geqq -1$ *and* $b \geqq (m+2)/2m(c-2)$ *is uniformly better than* $\hat{\vartheta}_{LS}$, *i.e.,*

$$\forall \vartheta \in \Theta \; \forall w \in \mathbb{R}^+ : R(\vartheta, w; \hat{\vartheta}_{BS}) \leqq R(\vartheta, w; \hat{\vartheta}_{LS}).$$

Proof. The estimators defined by (5.48) may be written

$$\hat{\vartheta}_{BS}(\tilde{y}) = (1 - h(G)/G)\hat{\vartheta}_{LS}, \quad \text{where}$$
$$h(G) = G/(1 + a + mbG).$$

Observing that $\mathrm{d}h(G)/\mathrm{d}G = (1+a)/(1+a+mbG)^2 \geqq 0$, $h(\cdot)$ is monotone nondecreasing on $[0, \infty]$. Furthermore, it holds

$$0 \leqq h(G) \leqq 1/mb \leqq 2(c-2)/(m+2).$$

Thus, the assertion follows immediately from Theorem 4.3.1 in Bunke/Bunke (1986) (cp. also Bock (1975), Bunke (1977)). □

We want to point at two special cases of the Bayes-Stein estimator (5.48): for $a = -1$, $b = (n - r + 3)/(r - 2)$ we obtain the James-Stein estimator, choosing $a = -1$, $b = 1/r$ we arrive at Mallows's "minimum-C_p estimator". It should be noted that in case of $-1 \leqq a < 0$ the shrinkage factor $1 - 1/(1 + a + mbG)$ in (5.48) may become negative. Then $\hat{\vartheta}_{BS}$ can be further improved by an appropriate "positive-part" modification

$$\hat{\vartheta}_{BS}^{+}(\tilde{y}) = [1 - 1/(1 + a + mbG(\tilde{y})]^{+} \hat{\vartheta}_{LS}(\tilde{y}).$$

A similar interpretation or heuristical motivation could be given for a number of contraction estimators studied in the recent literature, which contract towards linear subspaces $\mathcal{L}$ of $\mathbb{R}^r$ and which arise for example when replacing $G(\tilde{y})$ in (5.48) by the likelihood ratio statistic for testing the hypothesis $H_0: \boldsymbol{\vartheta} \in \mathscr{L}$ (or $H_0: \boldsymbol{\vartheta} - \boldsymbol{\mu} \in \mathscr{L}$ when contracting into the direction of $\boldsymbol{\mu} \neq \mathbf{0}$), see, e.g. JUDGE/BOCK (1978), BOCK (1982), OMAN (1982).

A more general frame for the treatment of such Stein-type modifications as considered above provides the empirical Bayes estimation principle, which includes the possibility of using additional information from model replications for estimation of the prior distribution. However, except of the case of model replications with a fixed design matrix $\boldsymbol{F}$ and constant error covariance matrix, we can then derive, essentially, only asymptotical optimality properties (see BUNKE/GLADITZ (1974)). Empirical Bayes versions of the Bayes (affine) estimator will be considered in Section 9.2.

5.6. Hierarchical Bayes estimation

The application of the conjugate Bayes or Bayes affine estimators requires the specification of the hyperparameter $\boldsymbol{\gamma} = (\boldsymbol{\mu}, \boldsymbol{\Phi})$ of the prior distribution of $\tilde{\boldsymbol{\vartheta}}$. If we have no additional information and are uncertain about the exact value of $\boldsymbol{\gamma}$ then this uncertainty can, be expressed by assigning another prior distribution to the possible values of $\boldsymbol{\gamma}$. This is termed a *three-stage hierarchical model*, the first stage refers to the distribution of the observations and the second and third stage, respectively, to the distributions of the regression coefficients and its hyperparameters. Clearly, the hierarchical model can be extended to include further stages with new hyperparameters at each stage and a completely specified prior at the final stage. But it will be enough to go to three stages since, in general, the gain in information decreases as one moves to higher levels of hyperparameters (see GOEL/DE GROOT (1981)).

Let us consider the following particular form of a three-stage hierarchy of normal distributions as investigated in LINDLEY/SMITH (1972) and SMITH (1973):

$$
\begin{aligned}
&P_{\tilde{y}|\vartheta} = \mathrm{N}(\boldsymbol{F}\boldsymbol{\vartheta}, \boldsymbol{\Sigma}), \\
&P_{\tilde{\vartheta}|\mu_1} = \mathrm{N}(\boldsymbol{F}_1 \boldsymbol{\mu}_1, \boldsymbol{\Phi}_1), \quad \boldsymbol{F}_1 \in \mathscr{M}_{r \times s}, \quad \boldsymbol{\Phi}_1 \in \mathscr{M}_r^{>}, \quad s \leqq r, \qquad (5.49)\\
&P_{\tilde{\mu}_1|\mu_2} = \mathrm{N}(\boldsymbol{F}_2 \boldsymbol{\mu}_2, \boldsymbol{\Phi}_2), \quad \boldsymbol{F}_2 \in \mathscr{M}_{s \times m}, \quad \boldsymbol{\Phi}_2 \in \mathscr{M}_s^{>}, \quad m \leqq s.
\end{aligned}
$$

Here $\boldsymbol{\mu}_2 \in \mathbb{R}^m$, $\boldsymbol{F}$, $\boldsymbol{F}_1$, $\boldsymbol{F}_2$ and the covariance matrices are assumed to be known. The second stage describes the form of relationship between the regression coefficients in the first stage, while the third stage describes knowledge about this form of relationship. The matrices $\boldsymbol{F}_1$ and $\boldsymbol{F}_2$ have been introduced to allow for exchangeability assumptions on the parameters. For example, if all the regression coefficients were believed to be exchangeable and independent, i.e. particularly $\mathbf{E}\tilde{\vartheta}_1 = \ldots = \mathbf{E}\tilde{\vartheta}_r$, $\operatorname{Var}\tilde{\vartheta}_1 = \ldots = \operatorname{Var}\tilde{\vartheta}_r$ and zero covariances, then we would have $s = m = 1$, $\boldsymbol{F}_1 = (1, \ldots, 1)^\top$, $\boldsymbol{\Phi}_1 = c\boldsymbol{I}_r$ with some $c > 0$, $\boldsymbol{F}_2 = 1$ and $\tilde{\mu}_1$ a random variable with expectation $\mu_2 \in \mathbb{R}^1$ and variance $\boldsymbol{\Phi}_2 > 0$.

In general, an exchangeability assumption means that the distribution of certain subsets of the parameters is invariant under permutations. The concept of prior exchangeability goes back to De Finetti (1964). In one sense, the hierarchical fashion of specification of the prior distribution is equivalent to choosing a completely specified prior distribution for $\tilde{\vartheta}$ since the second and third stage priors can be combined to an overall prior distribution for $\tilde{\vartheta}$:

$$P_{\tilde{\vartheta}} = \mathrm{N}(F_1 F_2 \mu_2,\ \Phi_1 + F_1 \Phi_2 F_1^\top). \tag{5.50}$$

However, as was noted by Goel and De Groot (1981), the hierarchical model offers the possibility of Bayesian learning about the hyperparameters. This means, instead of having to specify a particular value for μ_1 in the prior distribution of $\tilde{\vartheta}$, we can actually learn about its value when taking an observation y and observing that

$$P_{\tilde{y}|\mu_1} = \mathrm{N}(FF_1 \mu_1,\ \Sigma + F\Phi_1 F^\top) \tag{5.51}$$

and

$$P_{\tilde{y}|\mu_2} = \mathrm{N}(FF_1 F_2 \mu_2,\ \Sigma + F\Phi_1 F^\top + FF_1 \Phi_2 F_1^\top F^\top). \tag{5.52}$$

Of course, we will have to assume that μ_2 has a particular value, but the estimations for ϑ will often be less sensitive to the values of the higher-level hyperparameter μ_2 than to the values of μ_1. In other words, the introduction of an additional prior distribution for the hyperparameter μ_1 offers some protection against the consequences of a misspecification of this parameter in the two-stage Bayesian model.

We now proceed to drive Bayes estimators for ϑ and μ_1 from the hierarchical prior structure (5.49).

Theorem 5.8. *Under the three-stage hierarchy* (5.49), $\tilde{\vartheta}$ *has posterior distribution*

$$P_{\tilde{\vartheta}|y} = \mathrm{N}(\vartheta^*, \Phi^*),$$

where

$$\Phi^* = \{F^\top \Sigma^{-1} F + (F_1 \Phi_2 F_1^\top + \Phi_1)^{-1}\}^{-1},$$
$$\vartheta^* = \Phi^* F^\top \Sigma^{-1} y + \Phi^* (F_1 \Phi_2 F_1^\top + \Phi_1)^{-1} F_1 F_2 \mu_2$$

(cp. Lindley/Smith (1972)).

The proof follows from the form (5.50) of the marginal distribution of $\tilde{\vartheta}$ which, by virtue of Bayes' formula, $p_{\tilde{\vartheta}|y} \propto p_{\tilde{y}|\vartheta} \cdot p_{\tilde{\vartheta}}$, leads to the above given posterior distribution.

If we confine ourselves to the class D_a of affine estimators then we can drop the normality assumption at every stage of the model (5.49), i.e., $\hat{\vartheta}_{HB}$ is Bayes optimal in D_a for the hierarchical second order moment model

$$\begin{aligned} \mathsf{E}(\tilde{y}|\vartheta) &= F\vartheta, & \mathsf{Cov}(\tilde{y}|\vartheta) &= \Sigma, \\ \mathsf{E}(\tilde{\vartheta}\,|\,\mu_1) &= F_1 \mu_1, & \mathsf{Cov}(\tilde{\vartheta}\,|\,\mu_1) &= \Phi_1, \\ \mathsf{E}(\tilde{\mu}_1|\,\mu_2) &= F_2 \mu_2, & \mathsf{Cov}(\tilde{\mu}_1|\,\mu_2) &= \Phi_2. \end{aligned} \tag{5.53}$$

We call

$$\hat{\vartheta}_{HB} = \{F^\top \Sigma^{-1} F + [\Phi_1 + F_1 \Phi_2 F_1^\top]^{-1}\}^{-1} \{F^\top \Sigma^{-1} \tilde{y} + [\Phi_1 + F_1 \Phi_2 F_1^\top]^{-1} F_1 F_2 \mu_2\} \tag{5.54}$$

the (three-stage) *hierarchical Bayes (affine) estimator* for ϑ. The form of $\hat{\vartheta}_{HB}$ is a generalization of the form of the Bayes estimator $\hat{\vartheta}_B^\Sigma = (F^\top\Sigma^{-1}F + \Phi_1^{-1})^{-1}(F^\top\Sigma^{-1}\tilde{y} + \Phi_1^{-1}F_1\mu_1)$ obtained by omitting the third stage prior, where Φ_1 is replaced by the enlarged covariance matrix $\Phi_1 + F_1\Phi_2F_1^\top \geqq \Phi_1$ and μ_1 is replaced by its expected value $\mathbf{E}\tilde{\mu}_1 = F_2\mu_2$. Likewise, $\hat{\vartheta}_{HB}$ has the form of a weighted average of the generalized LSE $\hat{\vartheta}_{LS}$ and the prior mean $\mathbf{E}\tilde{\vartheta} = F_1F_2\mu_2$ with weights equal to the corresponding precision matrices $F^\top\Sigma^{-1}F = (\mathbf{Cov}\,\hat{\vartheta}_{LS}^\Sigma)^{-1}$ and $\Phi_1 + F_1\Phi_2F_1^\top = (\mathbf{Cov}\,\tilde{\vartheta})^{-1}$ (see Formula (5.50)).

The enlargement of the covariance matrix Φ_1 by the term $F_1\Phi_2F_1^\top$ is due to the uncertainty relative to the specification of the hyperparameter μ_1. If there is complete certainty about the value of μ_1 then this can be expressed through a one-point distribution for $\tilde{\mu}_1$ at the third stage, which would imply that $P_{\tilde{\mu}_1|\mu_2}(\tilde{\mu}_1 = F_2\mu_2) = 1$, $\Phi_2 = \mathbf{0}_{s,s}$ and $\hat{\vartheta}_{HB}$ coincides with the usual conjugate Bayes estimator $\hat{\vartheta}_B^\Sigma(\tilde{y}) = (F^\top\Sigma^{-1}F + \Phi_1^{-1})^{-1}(F^\top\Sigma^{-1}\tilde{y} + \Phi_1^{-1}F_1F_2\mu_2)$.

Remark 5.7. A Bayes estimate for the hyperparameter μ_1 would be

$$\mu_1^* = \Phi_1^*\{F_1F^\top(\Sigma + F\Phi_1F^\top)^{-1}y + \Phi_2^{-1}F_2\mu_2\},$$

where $$(\Phi_1^*)^{-1} = F_1F^\top(\Sigma + F\Phi_1F^\top)^{-1}FF_1^\top + \Phi_2^{-1}.$$

This follows by combining $P_{\tilde{y}|\mu_1}$ from (5.51) with $P_{\tilde{\mu}_1} = \mathrm{N}(F_2\mu_2, \Phi_2)$ via Bayes' formula to yield the posterior distribution $P_{\tilde{\mu}_1|y} = \mathrm{N}(\mu_1^*, \Phi_1^*)$.

When our third-stage knowledge of the second-stage hyperparameter μ_1 is vague then a Bayes estimator for ϑ can be obtained by using the improper limit of $\hat{\vartheta}_{HB}$ when Φ_2^{-1} approaches the matrix of zeroes.

Corollary 5.4. *Let* $\hat{\mu}_1 = [F_1F^\top(\Sigma + F\Phi_1F^\top)^{-1}FF_1^\top]^{-1}F_1^\top F^\top(\Sigma + F\Phi_1F^\top)^{-1}\tilde{y}$ *denote the generalized LSE for* μ_1 *obtained from* (5.51). *Then*

$$\hat{\vartheta}_{HB}^* = (F^\top\Sigma^{-1}F + \Phi_1^{-1})^{-1}(F^\top\Sigma^{-1}\tilde{y} + \Phi_1^{-1}F_1\hat{\mu}_1)$$

is a Bayes estimator for ϑ *relative to third-stage prior ignorance.*

The proof follows by rewriting Φ^* from Theorem 5.8 as

$$(\Phi^*)^{-1} = F^\top\Sigma^{-1}F + \Phi_1^{-1} - \Phi_1^{-1}F_1(F_1^\top\Phi_1^{-1}F_1 + \Phi_2^{-1})^{-1}F_1^\top\Phi_1^{-1},$$

setting $\Phi_2^{-1} = \mathbf{0}$ and then using A.33 again to rearrange $\hat{\vartheta}_{HB}$ from (5.54) as indicated above (cp. SMITH (1973)).

Note that $\hat{\vartheta}_{HB}$ does not involve the final hyperparameter μ_2. Interestingly, the form of $\hat{\vartheta}_{HB}^*$ precisely corresponds to using an empirical Bayes estimator in a two-stage hierarchy $P_{\tilde{y}|\vartheta} = \mathrm{N}(F\vartheta, \Sigma)$, $P_{\tilde{\vartheta}} = \mathrm{N}(F_1\mu_1, \Phi_1)$ with known Φ_1 but with unknown prior mean estimated by least squares, $F_1\hat{\mu}_1$ (cp. Section 9.2). Particularly, in case that $F_1 = I_r$, which means that there is no exchangeability between the regression coefficients, both $\hat{\mu}_1$ and $\hat{\vartheta}_{HB}^*$ coincide with the first-stage generalized LSE, $\hat{\mu}_1 = \hat{\vartheta}_{HB}^* = (F^\top\Sigma^{-1}F)^{-1}F^\top\Sigma^{-1}\tilde{y}$.

The crucial point with the application of $\hat{\vartheta}_{HB}$ is that it requires knowledge of the covariance matrices $\boldsymbol{\Sigma}$ and $\boldsymbol{\Phi}_1$. One way to overcome this difficulty would be to consider $\boldsymbol{\omega} = (\boldsymbol{\Sigma}, \boldsymbol{\Phi}_1)$ as a nuisance parameter, to assign a joint prior density $p_{\tilde{\vartheta}, \tilde{\omega}}$ and then to employ a modal approximation procedure, starting with an initial value of $\boldsymbol{\omega}$.

On principle, the hierarchical modelling could also be extended to include a distributional hierarchy for the covariance matrices $\boldsymbol{\Sigma}$ and $\boldsymbol{\Phi}_1$, as it is usually done for example in the two-stage Bayesian model assuming a (conjugate) Wishart distribution for $\tilde{\boldsymbol{\Sigma}}$ (see, e.g. De Groot (1970)). For Bayes estimation of the regression parameter $\boldsymbol{\vartheta}$ this means only that the covariance matrices are to be replaced by its expectations $\mathbf{E}\tilde{\boldsymbol{\Sigma}}$, $\tilde{\boldsymbol{\Phi}}_1$. But then immediately arises the specification problem for these moments. Another way would be to assign a noninformative (ignorance) prior (see Section 9.1) to the hyperparameters, for example according to some entropy based criterion as proposed by Akaike (1980), and then computing a Bayes estimator for $\boldsymbol{\vartheta}$ with respect to this ignorance prior.

6. Optimality and robustness of the Bayes estimator

In this chapter we give a collection of optimality properties of the estimator $\hat{\boldsymbol{\vartheta}}_B$ from (5.12), which was proved to be a Bayes estimator for $\boldsymbol{\vartheta}$ with respect to normally distributed errors, conjugate normal-gamma-prior and quadratic loss. Particularly we investigate the extent to which this estimator is robust under a change to larger classes of error and prior distributions and more general loss functions. This is of great importance, since in practical situations our prior knowledge will be more or less incomplete so that, for example, only some moments or fractiles of the required probability distributions can be specified and the loss function rarely can be characterized by more than some qualitative properties such as continuity, monotonicity and convexity.

Moreover, we investigate the sensitivity of $\hat{\boldsymbol{\vartheta}}_B$ against a misspecification of the prior moments and indicate robustifications of the Bayes estimator taking account of possible misspecifications and of an uncertain error covariance structure.

First we investigate the robustness of Bayes optimality of $\hat{\boldsymbol{\vartheta}}_B$, then we introduce some Bayesian robustness measures and indicate some minimax properties of the Bayes and hierarchical Bayes estimators.

6.1. Robustness of Bayes optimality

The Bayes optimality of the estimator

$$\hat{\boldsymbol{\vartheta}}_B(\tilde{\boldsymbol{y}}) = (\boldsymbol{F}^\top \boldsymbol{F} + \boldsymbol{\Phi}^{-1})^{-1} (\boldsymbol{F}^\top \tilde{\boldsymbol{y}} + \boldsymbol{\Phi}^{-1} \boldsymbol{\mu})$$

has been established under the assumptions of

(a) quadratic loss, i.e., $L(\boldsymbol{\vartheta}, w; \hat{\boldsymbol{\vartheta}}(\boldsymbol{y})) = \|\boldsymbol{\vartheta} - \hat{\boldsymbol{\vartheta}}(\boldsymbol{y})\|^2_{U(w)}$,

(b) normal errors, i.e., $\mathscr{P}_{\tilde{e}} = \mathscr{P}^{(N)}_{\tilde{e}} = \left\{ \mathrm{N}\left(\boldsymbol{0}, \frac{1}{w} \boldsymbol{I}_n\right)\right\}_{w \in \mathbb{R}^+}$,

(c) conjugate normal-gamma-prior for $(\tilde{\boldsymbol{\vartheta}}, \tilde{w})$, i.e.,

$$P_{\tilde{\vartheta}|w} = \mathrm{N}\left(\boldsymbol{\mu}, \frac{1}{w}\boldsymbol{\Phi}\right), \qquad P_{\tilde{w}} = G\left(\frac{\alpha}{2} + 1, \frac{\beta}{2}\right). \tag{6.1}$$

We have already seen that $\hat{\boldsymbol{\vartheta}}_{\mathrm{B}}$ is also Bayesian in the subclass D_{a} of affine linear estimators when dropping the distributional assumptions in (b) and (c) and only requiring $P_{\tilde{e}}$ and $P_{\tilde{\vartheta}, \tilde{w}}$ to be such that $\mathbf{E}\, \tilde{\boldsymbol{e}} = \boldsymbol{0}$, $\mathbf{Cov}\,(\tilde{\boldsymbol{e}}|w) = \frac{1}{w}\boldsymbol{I}_n$, $\mathbf{E}\, \tilde{\boldsymbol{\vartheta}} = \boldsymbol{\mu}$ and $\mathbf{Cov}\,(\tilde{\boldsymbol{\vartheta}}|w) = \frac{1}{w}\boldsymbol{\Phi}$ (see Section 5.3).

We will now give some further results demonstrating that the Bayes optimality of $\hat{\boldsymbol{\vartheta}}_{\mathrm{B}}$ is still preserved if one or more of the assumptions (a)–(c) are weakened.

We first formulate a general condition on the error and prior distributions which, in case of quadratic loss, assures Bayes optimality of $\hat{\boldsymbol{\vartheta}}_{\mathrm{B}}$. To this let $\mathscr{P}_{\tilde{e}} = \{P^{(w)}_{\tilde{e}}\}_{w \in W}$ be a general class of error distributions with some nonempty index set W.

Theorem 6.1. *Let $\Theta = \mathbb{R}^r$ and L be quadratic. The error distributions $P^{(w)}_{\tilde{e}} \in \mathscr{P}_{\tilde{e}}$ and the prior distribution $P_{\tilde{\vartheta}, \tilde{w}}$ are assumed to have densities $p^{(w)}_{\tilde{e}}$ and $p_{\tilde{\vartheta}, \tilde{w}}$, respectively, with respect to Lebesgue measure. If for every $w \in W$ there exists a symmetric function $g_w | \Theta \to \mathbb{R}^+$ such that for all $\boldsymbol{\vartheta} \in \Theta$ and all observations $\boldsymbol{y} \in C$ it holds*

$$p^{(w)}_{\tilde{e}}(\boldsymbol{y} - \boldsymbol{F}\boldsymbol{\vartheta})\, p_{\tilde{\vartheta}|w}(\boldsymbol{\vartheta}) = g_w(\boldsymbol{\vartheta} - \hat{\boldsymbol{\vartheta}}_{\mathrm{B}}(\boldsymbol{y})),$$

then $\hat{\boldsymbol{\vartheta}}_{\mathrm{B}}$ is Bayesian with respect to $P_{\tilde{\vartheta}, \tilde{w}}$.

Proof. Let $w \in W$ and $\boldsymbol{y} \in C$. The density of the conditional posterior distribution $P_{\tilde{\vartheta}|\boldsymbol{y}, w}$ is given by

$$\begin{aligned} p_{\tilde{\vartheta}|\boldsymbol{y}, w}(\boldsymbol{\vartheta}) &= p^{(w)}_{\tilde{e}}(\boldsymbol{y} - \boldsymbol{F}\boldsymbol{\vartheta})\, p_{\tilde{\vartheta}|w}(\boldsymbol{\vartheta}) \Big/ \int_\Theta p^{(w)}_{\tilde{e}}(\boldsymbol{y} - \boldsymbol{F}\boldsymbol{\vartheta})\, p_{\tilde{\vartheta}|w}(\boldsymbol{\vartheta})\, \mathrm{d}\boldsymbol{\vartheta} \\ &= g_w(\boldsymbol{\vartheta} - \hat{\boldsymbol{\vartheta}}_{\mathrm{B}}(\boldsymbol{y})) \Big/ \int_\Theta g_w(\boldsymbol{\vartheta} - \hat{\boldsymbol{\vartheta}}_{\mathrm{B}}(\boldsymbol{y}))\, \mathrm{d}\boldsymbol{\vartheta}. \end{aligned}$$

Substituting $\boldsymbol{s} = \boldsymbol{\vartheta} - \hat{\boldsymbol{\vartheta}}_{\mathrm{B}}(\boldsymbol{y})$, the corresponding expectation can be written as

$$\begin{aligned} \mathbf{E}\,(\tilde{\boldsymbol{\vartheta}}|\boldsymbol{y}, w) &= \int_\Theta (\boldsymbol{s} + \hat{\boldsymbol{\vartheta}}_{\mathrm{B}}(\boldsymbol{y}))\, g_w(\boldsymbol{s})\, \mathrm{d}\boldsymbol{s} \Big/ \int_\Theta g_w(\boldsymbol{s})\, \mathrm{d}\boldsymbol{s} \\ &= \hat{\boldsymbol{\vartheta}}_{\mathrm{B}}(\boldsymbol{y}) + \int_\Theta \boldsymbol{s}\, g_w(\boldsymbol{s})\, \mathrm{d}\boldsymbol{s} \Big/ \int_\Theta g_w(\boldsymbol{s})\, \mathrm{d}\boldsymbol{s}. \end{aligned}$$

Since g_w is symmetric, by assumption, we have $\int_\Theta s g_w(s)\,\mathrm{d}s = \mathbf{0}$. Hence, $\mathbf{E}(\tilde{\boldsymbol{\vartheta}}|\boldsymbol{y}, w) = \hat{\boldsymbol{\vartheta}}_{\mathrm{B}}(\boldsymbol{y})$ and the expectation does not depend on w. The assertion then follows from Corollary 5.2. □

Theorem 6.1 tells us that $\hat{\boldsymbol{\vartheta}}_{\mathrm{B}}$ is a Bayes estimator for $\boldsymbol{\vartheta}$ if the conditional posterior density of $\tilde{\boldsymbol{\vartheta}}$ for given w and $\boldsymbol{y}$ is a symmetric function of the difference of $\boldsymbol{\vartheta}$ and its estimate:

$$p_{\tilde{\vartheta}|y,w}(\boldsymbol{\vartheta}) \propto g_w(\boldsymbol{\vartheta} - \hat{\boldsymbol{\vartheta}}_{\mathrm{B}}(\boldsymbol{y}))\,.$$

Restricting $\mathcal{P}_{\tilde{e}}$ to the class of normal distributions, we obtain

Corollary 6.1. *Let $\Theta = \mathbb{R}^r$, L quadratic and $\mathcal{P}_{\tilde{e}} = \left\{\mathrm{N}\left(\mathbf{0}, \frac{1}{w}\boldsymbol{I}_n\right)\right\}_{w \in \mathbb{R}^+}$. Then $\hat{\boldsymbol{\vartheta}}_{\mathrm{B}}$ is Bayesian with respect to every prior $P_{\tilde{\vartheta},\tilde{w}}$ for which $P_{\tilde{\vartheta}|w} = \mathrm{N}\left(\boldsymbol{\mu}, \frac{1}{w}\boldsymbol{\Phi}\right)$.*

The proof follows from the fact that, under the assumptions of Corollary 6.1, the optimality conditions in Theorem 6.1 are met. Particularly, we have

$$p_{\tilde{e}}^{(w)}(\boldsymbol{y} - \boldsymbol{F}\boldsymbol{\vartheta})\, p_{\tilde{\vartheta}|w}(\boldsymbol{\vartheta}) \propto \exp\left\{-\frac{w}{2}\,[\|\boldsymbol{y} - \boldsymbol{F}\boldsymbol{\vartheta}\|^2 + \|\boldsymbol{\vartheta} - \boldsymbol{\mu}\|^2_{\boldsymbol{\Phi}^{-1}}]\right\}$$

$$\propto \exp\left\{-\frac{w}{2}\,\|\boldsymbol{\vartheta} - \hat{\boldsymbol{\vartheta}}_{\mathrm{B}}(\boldsymbol{y})\|^2_{\boldsymbol{F}^\top\boldsymbol{F} + \boldsymbol{\Phi}^{-1}}\right\} \tag{6.2}$$

which is obviously a symmetric function of $\boldsymbol{\vartheta} - \hat{\boldsymbol{\vartheta}}_{\mathrm{B}}(\boldsymbol{y})$.

This means that $\hat{\boldsymbol{\vartheta}}_{\mathrm{B}}$ is not only a Bayes estimator with respect to the conjugate normal-gamma-distribution. In case of normally distributed errors it suffices to have a conditional normal prior for $\boldsymbol{\vartheta}$, and no distributional assumptions on the variance parameter are necessary.

Another characterization of the classes of error and prior distributions leading to Bayes optimality of $\hat{\boldsymbol{\vartheta}}_{\mathrm{B}}$ can be obtained via characteristic functions. Let $\varphi_{\tilde{e}}^{(w)}$ and $\varphi_{F\tilde{\vartheta}}^{(w)}$ denote the characteristic functions corresponding to $\mathrm{P}_{\tilde{e}}^{(w)}$ and the conditional distribution $P_{F\tilde{\vartheta}|w}$ of $\boldsymbol{F}\tilde{\boldsymbol{\vartheta}}$ for given $w \in W$, respectively.

Theorem 6.2. *Let L be quadratic, $\boldsymbol{S} = \boldsymbol{F}(\boldsymbol{F}^\top\boldsymbol{F} + \boldsymbol{\Phi}^{-1})^{-1}\boldsymbol{F}^\top$, $\boldsymbol{q} = \boldsymbol{F}(\boldsymbol{F}^\top\boldsymbol{F} + \boldsymbol{\Phi}^{-1})^{-1}\boldsymbol{\Phi}^{-1}\boldsymbol{\mu}$ and assume $\boldsymbol{F}$ to have full rank. Then $\hat{\boldsymbol{\vartheta}}_{\mathrm{B}}$ is Bayesian with respect to $P_{\tilde{\vartheta},\tilde{w}}$ if for all $w \in W$ it holds:*

$$(\boldsymbol{I}_n - \boldsymbol{S})\nabla \ln \varphi_{F\tilde{\vartheta}}^{(w)}(\cdot) = \boldsymbol{S}\nabla \ln \varphi_{\tilde{e}}^{(w)}(\cdot) + \mathrm{i}\boldsymbol{q}\,. \tag{6.3}$$

The proof can be done analogously to that of Theorem 3.2 and Lemma 3.1 in Rao (1976 a) (see Pilz (1979 b)).

Corollary 6.2. *Let the assumptions of Theorem 6.2 be satisfied with quadratic* *loss such that $U(w)$ is positive definite for every $w \in W$. Then $\hat{\vartheta}_B$ is Bayesian if and only if Equation (6.3) holds.*

Proof. According to Lemma 3.1 from RAO (1976 a) and Theorem 6.2 the equality $\mathbf{E}(\tilde{\vartheta} | \tilde{y}, w) = \hat{\vartheta}_B(\tilde{y})$ holds $P_{\tilde{y}}$ – a.e. if and only if (6.3) is satisfied. The result then follows from Corollary 5.1. □

Corollary 6.2 enables us to sharpen the result of Corollary 6.1 to say that in the special case of normal errors the normality of $P_{\tilde{\vartheta}|w}$ is not only sufficient but also necessary for Bayes optimality of $\hat{\vartheta}_B$.

Corollary 6.3. *Let $\Theta = \mathbb{R}^r$, L quadratic, $\mathcal{P}_{\tilde{e}} = \left\{ \mathrm{N}\left(\mathbf{0}, \frac{1}{w} \boldsymbol{I}_n\right) \right\}_{w \in \mathbb{R}^+}$ and assume $\boldsymbol{F}$ and $\boldsymbol{U}(\cdot)$ to have full rank. Then $\hat{\vartheta}_B$ is Bayesian with respect to $P_{\tilde{\vartheta}, \tilde{w}}$ if and only if $P_{\tilde{\vartheta}|w} = \mathrm{N}\left(\boldsymbol{\mu}, \frac{1}{w} \boldsymbol{\Phi}\right)$.*

Proof. The error distribution $P_{\tilde{e}} = \mathrm{N}\left(\mathbf{0}, \frac{1}{w} \boldsymbol{I}_n\right)$ has characteristic function $\varphi_{\tilde{e}}^{(w)}(\boldsymbol{t}) = \exp\left(-\frac{w}{2} \boldsymbol{t}^\top \boldsymbol{t}\right)$, $\boldsymbol{t} \in \mathbb{R}^n$ (see A.6). Inserting $\varphi_{\tilde{e}}^{(w)}$ into (6.3) yields:

$$(\boldsymbol{I}_n - \boldsymbol{S}) \nabla \ln \varphi_{\boldsymbol{F}\tilde{\vartheta}}^{(w)}(\boldsymbol{t}) = -w\boldsymbol{S}\boldsymbol{t} + \mathrm{i}\boldsymbol{q} \quad \text{for all} \quad \boldsymbol{t} \in \mathbb{R}^n.$$

Observing that $(\boldsymbol{I}_n - \boldsymbol{S})^{-1} = \boldsymbol{I}_n + \boldsymbol{F\Phi F}^\top$ this equation reads

$$\ln \varphi_{\boldsymbol{F}\tilde{\vartheta}}^{(w)}(\boldsymbol{t}) = -w\boldsymbol{F\Phi F}^\top \boldsymbol{t} + \mathrm{i}\boldsymbol{F\mu}$$

which implies that $\varphi_{\boldsymbol{F}\tilde{\vartheta}}^{(w)}(\boldsymbol{t}) = \exp\left(-\frac{w}{2} \boldsymbol{t}^\top \boldsymbol{F\Phi F}^\top \boldsymbol{t} + \mathrm{i}\boldsymbol{t}^\top \boldsymbol{F\mu}\right)$. Thus, by virtue of A.6 we have that $P_{\boldsymbol{F}\tilde{\vartheta}|w} = \mathrm{N}\left(\boldsymbol{F\mu}, \frac{1}{w} \boldsymbol{F\Phi F}^\top\right)$. From this the result follows by observing that $\boldsymbol{F}$ has full rank. □

The statement of Corollary 6.3 gives a characterization of the class of normally, independently and identically distributed errors: If $\hat{\vartheta}_B(\tilde{y}) = (\boldsymbol{F}^\top \boldsymbol{F} + \boldsymbol{\Phi}^{-1})^{-1} (\boldsymbol{F}^\top \tilde{y} + \boldsymbol{\Phi}^{-1} \boldsymbol{\mu})$ is a Bayes estimator with respect to every prior $P_{\tilde{\vartheta}, \tilde{w}}$ such that $P_{\tilde{\vartheta}|w} = \mathrm{N}\left(\boldsymbol{\mu}, \frac{1}{w} \boldsymbol{\Phi}\right)$ then it necessarily holds $P_{\tilde{e}}^{(w)} = \mathrm{N}\left(\mathbf{0}, \frac{1}{w} \boldsymbol{I}_n\right)$.

Within the given context, this characterization is in close agreement with the general results of DIACONIS/YLVISAKER (1979). For a general statistical model they proved that, under very mild assumptions, the linearity of the posterior expectation implies that the parameter vector has a conjugate prior distribution if the probability distribution of the observations belongs to the regular exponential family.

The results of GOEL/DE GROOT (1980) indicate that the above characterization can be sharpened such that, under different assumptions on the design matrix $\boldsymbol{F}$ and on the posterior distribution $P_{\tilde{\vartheta}|y}$, the Bayes optimality of the estimator $\hat{\boldsymbol{\vartheta}}_{\mathrm{B}}$ implies that both $P_{\tilde{\vartheta}|w}$ and $P_{\tilde{e}}^{(w)}$ (or the distributions of linear combinations of $\tilde{\boldsymbol{\vartheta}}$ and $\tilde{\boldsymbol{e}}$) are normal.

The following Theorems 6.3 and 6.4, with obvious reinterpretations, are essentially identical with Theorem 1 and Corollaries 1, 2, 3 in GOEL/DE GROOT (1980) observing that in case of quadratic loss the posterior expectation $\mathbf{E}(\tilde{\boldsymbol{\vartheta}}|\tilde{\boldsymbol{y}})$ is the unique Bayes estimator for $\boldsymbol{\vartheta}$ (see Corollaries 5.1 and 5.2).

Theorem 6.3. *Assume that $\tilde{\boldsymbol{\vartheta}}$ and $\tilde{\boldsymbol{e}}$ have finite second moments and the components of $\tilde{\boldsymbol{e}}$ be independent. Further, assume that $\boldsymbol{F}$ has full rank $r < n$ and no column of $\boldsymbol{F}$ is proportional to any column of $\boldsymbol{I}_n$. Then $\hat{\boldsymbol{\vartheta}}_{\mathrm{B}}$ from* (5.12) *can be Bayesian only if $P_{\tilde{\vartheta}|w}$ and $P_{\tilde{e}}^{(w)}$ are normal.*

We remark that the assumption of independent components can be weakened, it suffices to have a linear structure for $\tilde{\boldsymbol{e}}$, i.e., $\tilde{\boldsymbol{e}}$ can be represented as $\tilde{\boldsymbol{e}} = \boldsymbol{a} + \boldsymbol{B}\tilde{\boldsymbol{e}}_0$, where $\boldsymbol{a}$ is a vector of constants, $\boldsymbol{B}$ is a matrix of constants and $\tilde{\boldsymbol{e}}_0$ a random vector with independent, nondegenerate components. Furthermore, there holds an analogous result to that of Theorem 6.3 for the case that $n \leqq r$ if, additionally, the components of $\tilde{\boldsymbol{\vartheta}}$ are assumed to be independent (see GOEL/DE GROOT (1980), Theorem 2).

The above assumptions on $\tilde{\boldsymbol{e}}$ and $\boldsymbol{F}$ can be replaced by the assumption of weak homoscedasticity of the posterior distribution as indicated in the following theorem.

Theorem 6.4. *Assume that* $\mathbf{Cov}(\tilde{\boldsymbol{\vartheta}}|\boldsymbol{y}) =$ *constant for all $\boldsymbol{y}$, i.e., the posterior covariance does not depend on the observation. Then it holds*

(i) *If $n > r = \mathrm{rank}(\boldsymbol{F})$ the Bayes optimality of $\hat{\boldsymbol{\vartheta}}_{\mathrm{B}}$ implies that $\tilde{\boldsymbol{\vartheta}}|w$ and $\boldsymbol{F}^{\top}\tilde{\boldsymbol{e}}$ are normally distributed.*

(ii) *If $r \geqq n = \mathrm{rank}(\boldsymbol{F})$ then $\hat{\boldsymbol{\vartheta}}_{\mathrm{B}}$ can be Bayesian only if $\tilde{\boldsymbol{\vartheta}}|w$ and $\tilde{\boldsymbol{e}}$ are normally distributed.*

Up to now we assumed quadratic loss in our investigations on the robustness of $\hat{\boldsymbol{\vartheta}}_{\mathrm{B}}$. We can relax this assumption and allow for more general loss functions. We first formulate an analogue to Theorem 6.1 guaranteeing Bayes optimality of an estimator $\hat{\boldsymbol{\vartheta}}^*$ under monotone loss functions and a general class $\mathcal{P}_{\tilde{e}} = \{P_{\tilde{e}}^{(w)}\}_{w \in W}$ of error probability distributions.

Theorem 6.5. *Let $\Theta = \mathbb{R}^r$ and assume the distributions $P_{\tilde{e}}^{(w)} \in \mathcal{P}_e$ and the conditional prior $P_{\tilde{\vartheta}|w}$ to have densities $p_{\tilde{e}}^{(w)}$ and $p_{\tilde{\vartheta}|w}$, respectively, with respect to Lebesgue measure. Further, for all $w \in W$ let exist a strictly monotone decreasing function $g_w|\mathbb{R}^+ \to \mathbb{R}^+$ and a positive definite matrix $\boldsymbol{Q}(w)$ such that*

$$p_{\tilde{e}}^{(w)}(\boldsymbol{y} - \boldsymbol{F}\boldsymbol{\vartheta})\, p_{\tilde{\vartheta}|w}(\boldsymbol{\vartheta}) = g_w(\|\boldsymbol{\vartheta} - \hat{\boldsymbol{\vartheta}}^*(\boldsymbol{y})\|^2_{Q(w)})$$

for all observations $\boldsymbol{y}$. If the loss function L is of the form

$$L(\boldsymbol{\vartheta}, w; \hat{\boldsymbol{\vartheta}}(\boldsymbol{y})) = L_m(\|\boldsymbol{\vartheta} - \hat{\boldsymbol{\vartheta}}(\boldsymbol{y})\|^2_{U(w)}),$$

where $U(w) \in \mathcal{M}_r^{\geqq}$ and $L_m|\mathbb{R}^+ \to \mathbb{R}^+$ is continuous and montone increasing, then $\hat{\vartheta}^*$ is a Bayes estimator with respect to $P_{\tilde{\vartheta}, \tilde{w}}$.

Proof. First we observe that

$$p_{\tilde{\vartheta}|y,w}(\vartheta) = g_w(\|\vartheta - \hat{\vartheta}^*(y)\|^2_{Q(w)}) \Big/ \int_\Theta g_w(\|t - \hat{\vartheta}^*(y)\|^2_{Q(w)})\, dt$$

and therefore $\hat{\vartheta}^*(y)$ is the unique mode of $P_{\tilde{\vartheta}|y,w}$ since g_w is strictly decreasing. Thus, it follows from A.7 that for arbitrary estimators $\hat{\vartheta} \in D$ it holds

$$\begin{aligned}
\mathbf{E}_{\tilde{\vartheta}|y,w} L(\tilde{\vartheta}, w; \hat{\vartheta}(y)) &= \int_\Theta L_m(\|\vartheta - \hat{\vartheta}(y)\|^2_{U(w)})\, p_{\tilde{\vartheta}|y,w}(\vartheta)\, d\vartheta \\
&\geqq \int_\Theta L_m(\|\vartheta - \hat{\vartheta}^*(y)\|^2_{U(w)})\, p_{\tilde{\vartheta}|y,w}(\vartheta)\, d\vartheta \\
&= \mathbf{E}_{\tilde{\vartheta}|y,w} L(\tilde{\vartheta}, w; \hat{\vartheta}^*(y)).
\end{aligned}$$

From this it follows that

$$\begin{aligned}
\mathbf{E}_{\tilde{\vartheta},\tilde{w}|y} L(\tilde{\vartheta}, \tilde{w}; \hat{\vartheta}(y)) &= \int_W \mathbf{E}_{\tilde{\vartheta}|y,w} L(\tilde{\vartheta}, w; \hat{\vartheta}(y))\, dP_{\tilde{w}|y}(w) \\
&\geqq \int_W \mathbf{E}_{\tilde{\vartheta}|y,w} L(\tilde{\vartheta}, w; \hat{\vartheta}^*(y))\, dP_{\tilde{w}|y}(w) \\
&= \mathbf{E}_{\tilde{\vartheta},\tilde{w}|y} L(\tilde{\vartheta}, \tilde{w}; \hat{\vartheta}^*(y)),
\end{aligned}$$

i.e., $\hat{\vartheta}^*$ minimizes the posterior loss and thus is Bayesian. □

Remark 6.1. The assumption of strict monotonocity of the functions g_w can be dropped if, additionally, the function L_m is required to be convex (see DEUTSCH (1965)).

In constrast with the quadratic loss situation, where the conditional posterior density $p_{\tilde{\vartheta}|y,w}$ was only required to be a symmetric function of the difference $\vartheta - \hat{\vartheta}^*(y)$, in case of the above more general type of loss functions we have, additionally, to require this density to be a strictly decreasing function of the distance between ϑ and $\hat{\vartheta}^*(y)$.

Corollary 6.4. *Let $\Theta = \mathbb{R}^r$ and $\mathcal{P}_{\tilde{e}} = \left\{ \mathrm{N}\left(\mathbf{0}, \frac{1}{w} \boldsymbol{I}_n\right) \right\}_{w \in \mathbb{R}^+}$. If the loss function meets the same conditions as in Theorem 6.5 then $\hat{\vartheta}^*(\tilde{y}) = \hat{\vartheta}_B(\tilde{y}) = (\boldsymbol{F}^\top \boldsymbol{F} + \boldsymbol{\Phi}^{-1})^{-1}(\boldsymbol{F}^\top \tilde{y} + \boldsymbol{\Phi}^{-1}\boldsymbol{\mu})$ is a Bayes estimator with respect to all priors $P_{\tilde{\vartheta},\tilde{w}}$ for which $P_{\tilde{\vartheta}|w} = \mathrm{N}\left(\boldsymbol{\mu}, \frac{1}{w}\boldsymbol{\Phi}\right)$.*

This follows from the fact that the product $p_{\tilde{e}}^{(w)}(y - F\vartheta)\, p_{\tilde{\vartheta}|w}(\vartheta)$ given by (6.2) meets the conditions of Theorem 6.5.

By help of Corollary 6.4 we can generalize the result of Corollary 5.3 to say that $\hat{\vartheta}_B$ is not only admissible in case of quadratic loss but also in case of monotone loss functions (see Bandemer et al. (1977), Section 4.5.1).

To summarize our considerations we can say that, as long as we are concerned with quadratic loss and linear estimators then Bayes optimality of $\hat{\vartheta}_B$ holds without any distributional assumption except of homoscedasticity and uncorrelatedness of the errors and knowledge of the first and second order prior moments, but Bayes optimality of $\hat{\vartheta}_B$ under arbitrary estimators can be guaranteed only in case of normality of $P_{\tilde{e}}^{(w)}$ and $P_{\tilde{\vartheta}|w}$. This can be viewed an analogue to the optimality behaviour of the LSE, which (in case of quadratic loss) is the best linear unbiased estimator under homoscedastic and uncorrelated errors, but is best unbiased under arbitrary estimators only in case of additional normality of the errors.

6.2. Bayesian robustness measures

The application of conjugate Bayes or Bayes affine estimators requires knowledge of the first and second order moments of the prior distribution of the regression parameter and of the error distribution. On the one hand, an exact specification of these quantities is impossible, because of lack of knowledge, time or money. But, on the other hand, it is not at all necessary to do this in order to derive "good" estimators. Thus, we assume to have only approximate knowledge of the indicated moments and proceed to appropriate families of prior and error distributions. Our efforts are then directed to finding robust estimators showing satisfactory Bayes risk with respect to every prior from the chosen family. In particular, the (hierarchical) Bayes estimator is shown to have some minimax properties with regard to appropriate families.

Up to now, we can only record the very beginnings of robustness investigations from the Bayesian viewpoint, they are not comparable to the advanced state of investigations on the robustness of "classical" statistical methods as presented, for example, in Huber's (1981) monograph. The present state of Bayesian approaches to robustness theory is documented in Berger (1984), where also an extensive literature survey is given. Important recent contributions are due to Berger/Berliner (1986), Berger (1987) and Das Gupta/Studden (1988 a, b, c).

We first consider local Bayesian robustness measures based on the a-posteriori evaluation of the sensitivity of an estimate against the choice of a prior distribution for a fixed observation. The corresponding global, a-preposteriori evaluation of an estimator leads to the restricted minimax principle, which plays a central role in our robustness investigations. It will be seen in the next section that this principle is equivalent to hierarchical Bayes estimation with a least favourable prior distribution for the regression hyperparameter $\mu = \mathbf{E}\,\tilde{\vartheta}$ at the third stage.

Let us consider general families $\mathcal{P}_{\tilde{e}}$ of error distributions of the type

$$\mathcal{P}_{\tilde{e}} = \{P_{\tilde{e}}^{(w)}\}_{w \in W}, \qquad w = (w_1, w_2) \in W := W_1 \times W_2 . \tag{6.4}$$

Here W_1 and W_2 are arbitrary index sets which need not be parametric. Further, let $\Theta \subseteqq \mathbb{R}^r$ and suppose we have (incomplete) prior knowledge about the regression parameter and the error distribution which can be represented by some family $\mathcal{P}_1 = \mathcal{P}_{\tilde{\vartheta}, \tilde{w}_1}$ of prior distributions. We choose this general setup to model situations, where we are only able or willing to specify prior knowledge concerning some basic characteristics of the error distribution which may be summarized by some subindex w_1, say, of $w = (w_1, w_2) \in W$, where w_2 plays the role of a nuisance parameter.

This includes the extreme cases $W_1 = \emptyset$, where we have only prior knowledge about the regression parameter, and $W_2 = \emptyset$, where we have a fully specified family of prior distributions on $\Theta \times W$.

Now, for a given abservation $\boldsymbol{y} \in \mathbb{R}^n$, an estimate $\hat{\boldsymbol{\vartheta}}(\boldsymbol{y})$ is evaluated on the basis of the (partial) posterior expected loss

$$\bar{L}(P_1, w_2; \hat{\boldsymbol{\vartheta}}(\boldsymbol{y})) = \int_{\Theta \times W_1} L(\boldsymbol{\vartheta}, w; \hat{\boldsymbol{\vartheta}}(\boldsymbol{y}))\, P_{\tilde{\vartheta}, \tilde{w}_1 | \boldsymbol{y}}(\mathrm{d}(\boldsymbol{\vartheta}, w_1)), \tag{6.5}$$

where $P_1 = P_{\tilde{\vartheta}, \tilde{w}_1} \in \mathcal{P}_1$ and L is some given (measurable) loss function. We remark that $\bar{L}$ from (6.5) coincides with the usual posterior expected loss $\mathbf{E}_{\tilde{\vartheta}, \tilde{w}|\boldsymbol{y}} L(\tilde{\boldsymbol{\vartheta}}, \tilde{w}; \hat{\boldsymbol{\vartheta}}(\boldsymbol{y}))$ if $W_1 = W$, $W_2 = \emptyset$.

Definition 6.1. Let $c > 0$ and D a given class of estimators for $\boldsymbol{\vartheta}$. The estimator $\hat{\boldsymbol{\vartheta}}_0 \in D$ is said to be *c-locally robust in D with respect to $\mathcal{P}_{\tilde{e}}$ and $\mathcal{P}_1$* for the observation $\boldsymbol{y} \in \mathbb{R}^n$ if it holds

$$\bar{L}_s(\hat{\boldsymbol{\vartheta}}_0) := \sup_{w_2 \in W_2} \sup_{P_1 \in \mathcal{P}_1} |\bar{L}(P_1, w_2; \hat{\boldsymbol{\vartheta}}_0(\boldsymbol{y})) - \inf_{\hat{\vartheta} \in D} \bar{L}(P_1, w_2; \hat{\boldsymbol{\vartheta}}(\boldsymbol{y}))| \leqq c.$$

This local notion of robustness based on the concrete observation proceeds from an a-posteriori viewpoint and thus refers to the robustness of inference.

Now, let us deal with the important special case of homoscedasticity, where w_1 signifies the precision of observation, i.e.,

$$\mathcal{P}_{\tilde{e}} \subseteqq \left\{ P_{\tilde{e}}^{(w)} : \mathbf{E}(\tilde{\boldsymbol{e}} | w) = \mathbf{0}, \quad \mathbf{Cov}(\tilde{\boldsymbol{e}} | w) = \frac{1}{w_1} \boldsymbol{I}_n \right\}_{w_1 \in W_1}, \qquad W_1 = \mathbb{R}^+. \tag{6.6}$$

Correspondingly, assume, as before, normalized quadratic loss

$$L(\boldsymbol{\vartheta}, w; \hat{\boldsymbol{\vartheta}}(\boldsymbol{y})) = w_1 \|\boldsymbol{\vartheta} - \hat{\boldsymbol{\vartheta}}(\boldsymbol{y})\|_U^2, \qquad \boldsymbol{U} \in \mathcal{M}_r^{\geqq} \tag{6.7}$$

independent of $w_2 \in W_2$.

In case of normal error distributions and conjugate priors local robustness investigations for the conjugate Bayes or Bayes affine estimators, respectively, are equivalent to evaluating the local stability of these estimators. The degree of local robustness then depends exclusively on the goodness of the specification of the first and second order moments $\mathbf{E}\,\tilde{\boldsymbol{\vartheta}} = \boldsymbol{\mu}$, $\mathbf{Cov}(\tilde{\boldsymbol{\vartheta}} | w_1) = w_1^{-1} \boldsymbol{\Phi}$ and $\mathbf{E}\,\tilde{w}_1 = w_0$.

Lemma 6.1. *Let* $\Theta = \mathbb{R}^r$, $W_1 = \mathbb{R}^+$, $W_2 = \emptyset$, $\mathcal{K} \subseteqq \mathbb{R}^r$, $\mathcal{Q} \subseteqq \mathcal{M}_r^{\geqq}$ *and* $\mathcal{W}_0 \subseteqq \mathbb{R}^+$.

Further, assume $\mathcal{P}_{\tilde{e}} = \left\{ \mathrm{N}\left(\mathbf{0}, \frac{1}{w_1} \boldsymbol{I}_n\right)\right\}_{w_1 \in W_1}$,

$$\mathcal{P}_1 = \left\{ P_{\tilde{\vartheta}, \tilde{w}_1} : P_{\tilde{\vartheta}|w_1} = \mathrm{N}\left(\boldsymbol{\mu}, \frac{1}{w_1} \boldsymbol{\Phi}\right) \wedge \boldsymbol{\mu} \in \mathcal{K},\ \boldsymbol{\Phi} \in \mathcal{Q},\ \mathsf{E}\tilde{w}_1 \in \mathcal{W}_0 \right\} \qquad (6.8)$$

and normalized quadratic loss according to (6.7). *Then it holds for arbitrary estimators* $\hat{\boldsymbol{\vartheta}}$:

$$\bar{L}_s(\hat{\boldsymbol{\vartheta}}) = \sup \{ w_0 \| \hat{\boldsymbol{\vartheta}}(\boldsymbol{y}) - \hat{\boldsymbol{\vartheta}}_B^a(\boldsymbol{y})\|_U^2 : w_0 \in \mathcal{W}_0, \quad \hat{\boldsymbol{\vartheta}}_B^a \in D_B^a \},$$

i.e., the degree of local robustness is determined by the maximum distance from the class of Bayes affine estimators

$$D_B^a = \{ \hat{\boldsymbol{\vartheta}}_B^a(\tilde{\boldsymbol{y}}) = \boldsymbol{\Phi F}^\top (\boldsymbol{F\Phi F}^\top + \boldsymbol{I}_n)^{-1} (\tilde{\boldsymbol{y}} - \boldsymbol{F\mu}) + \boldsymbol{\mu} : \boldsymbol{\mu} \in \mathcal{K}, \quad \boldsymbol{\Phi} \in \mathcal{Q} \}. \qquad (6.9)$$

Proof. Let $P_1 \in \mathcal{P}_1$ and D the class of all estimators for $\boldsymbol{\vartheta}$. Since the Bayes affine estimator $\hat{\boldsymbol{\vartheta}}_B^a(\tilde{\boldsymbol{y}}) = \hat{\boldsymbol{\vartheta}}_B(\tilde{\boldsymbol{y}})$ minimizes the posterior expected loss (cp. A.16 and Section 5.3) we have

$$\inf_{\hat{\boldsymbol{\vartheta}} \in D} \bar{L}(P_1, w_2; \hat{\boldsymbol{\vartheta}}(\boldsymbol{y})) = \bar{L}(P_1, w_2; \hat{\boldsymbol{\vartheta}}_B^a(\boldsymbol{y})),$$

independent of $w_2 \in W_2 = \emptyset$. On the other hand, from the assumption of normalized quadratic loss follows that $\hat{\boldsymbol{\vartheta}}_B^a(\boldsymbol{y}) = \mathsf{E}(\tilde{\boldsymbol{\vartheta}}|\boldsymbol{y}, w) = \mathsf{E}(\tilde{\boldsymbol{\vartheta}}|\boldsymbol{y})$ and thus

$$\begin{aligned}\bar{L}(P_1, w_2; \hat{\boldsymbol{\vartheta}}(\boldsymbol{y})) &= \mathsf{E}_{\tilde{\vartheta}, \tilde{w}_1|\boldsymbol{y}} \tilde{w}_1 \|\tilde{\boldsymbol{\vartheta}} - \hat{\boldsymbol{\vartheta}}(\boldsymbol{y})\|_U^2 \\ &= w_0 \{ \operatorname{tr} \boldsymbol{UM} - 2\hat{\boldsymbol{\vartheta}}_B^a(\boldsymbol{y})^\top \boldsymbol{U}\hat{\boldsymbol{\vartheta}}(\boldsymbol{y}) + \hat{\boldsymbol{\vartheta}}(\boldsymbol{y})^\top \boldsymbol{U}\hat{\boldsymbol{\vartheta}}(\boldsymbol{y})\},\end{aligned}$$

which implies the assertion. □

Clearly, in case of a correct specification of the moments $\boldsymbol{\mu}$ and $\boldsymbol{\Phi}$ we have $\bar{L}_s(\hat{\boldsymbol{\vartheta}}_B) = 0$. The result of Lemma 6.1 visualizes the direct link between the study of the effects of a misspecification of the required moments and the study of the local robustness of Bayes (affine) estimators.

Now we make some statements concerning the sensitivity or numerical stability, respectively, of Bayes affine estimators against a misspecification of the prior inputs $\boldsymbol{\mu}$ and $\boldsymbol{\Phi}$.

a) $\boldsymbol{\Phi}$ known, $\boldsymbol{\mu}$ unknown

If the prior expectation is subject to uncertainty such that $\boldsymbol{\mu}$ can take values in the neighbourhood of some given vector $\boldsymbol{\mu}_0 \in \mathbb{R}^r$, i.e.,

$$\mathsf{E}\tilde{\boldsymbol{\vartheta}} \in \mathcal{K} = \{ \boldsymbol{\mu} \in \mathbb{R}^r : \|\boldsymbol{\mu} - \boldsymbol{\mu}_0\|^2 \leqq \delta \}$$

for some $\delta > 0$, then the Bayes estimates will be located in the neighbourhood of the Bayes affine estimate $\hat{\boldsymbol{\vartheta}}_B^0(\boldsymbol{y}) = (\boldsymbol{F}^\top \boldsymbol{F} + \boldsymbol{\Phi}^{-1})^{-1} (\boldsymbol{F}^\top \boldsymbol{y} + \boldsymbol{\Phi}^{-1}\boldsymbol{\mu}_0)$ formed with $\boldsymbol{\mu}_0$ instead of $\boldsymbol{\mu}$:

$$\begin{aligned}\|\hat{\boldsymbol{\vartheta}}_B^a(\boldsymbol{y}) - \hat{\boldsymbol{\vartheta}}_B^0(\boldsymbol{y})\|_U^2 &= \|\boldsymbol{\psi\Phi}^{-1}(\boldsymbol{\mu} - \boldsymbol{\mu}_0)\|_U^2, \qquad \boldsymbol{\psi} = (\boldsymbol{F}^\top \boldsymbol{F} + \boldsymbol{\Phi}^{-1})^{-1} \\ &\leqq \delta \lambda_{\max}(\boldsymbol{\Phi}^{-1}\boldsymbol{\psi U \psi \Phi}^{-1}), \qquad \forall \hat{\boldsymbol{\vartheta}}_B^a \in D_B^a\end{aligned}$$

provided that $\boldsymbol{\Phi} \in \mathcal{M}_r^>$. In case of a singular $\boldsymbol{\Phi}$ we may write

$$\|\hat{\boldsymbol{\vartheta}}_B^a(\boldsymbol{y}) - \hat{\boldsymbol{\vartheta}}_B^0(\boldsymbol{y})\|_U^2 \leqq \delta \lambda_{\max}(\bar{\boldsymbol{\psi}}^\top \boldsymbol{U} \bar{\boldsymbol{\psi}}), \qquad \bar{\boldsymbol{\psi}} = (\boldsymbol{I}_r + \boldsymbol{\Phi} \boldsymbol{F}^\top \boldsymbol{F})^{-1}$$

based on the more general representation (5.34) of the Bayes affine estimator. In case of $\boldsymbol{U} = \boldsymbol{I}_r$ and $\bar{\boldsymbol{\psi}}^2 < \boldsymbol{I}_r$, the latter occurs for example if $\boldsymbol{F}^\top \boldsymbol{F}$ and $\boldsymbol{\Phi}$ commute ($\boldsymbol{F}^\top \boldsymbol{F} \boldsymbol{\Phi} = \boldsymbol{\Phi} \boldsymbol{F}^\top \boldsymbol{F}$), we would have $\|\hat{\boldsymbol{\vartheta}}_B^a(\boldsymbol{y}) - \hat{\boldsymbol{\vartheta}}_B^0(\boldsymbol{y})\| < \delta$.

b) $\boldsymbol{\mu}$ known, $\boldsymbol{\Phi}$ unknown
If we only know that the prior covariance matrix is positive definite, i.e., $\boldsymbol{\Phi} \in \mathcal{Q} = \mathcal{M}_r^>$, then the Bayes estimates will be constrained to lie in an ellipsoid with midpoint 1/2 $(\hat{\boldsymbol{\vartheta}}_{LS}(\boldsymbol{y}) + \boldsymbol{\mu})$. With additional assumptions on the structure of $\boldsymbol{\Phi}$ the Bayes estimates can be further constrained. The Theorems 6.6 and 6.7 given below follow immediately from corresponding theorems stated in Chamberlain/Leamer (1976) observing that $\hat{\boldsymbol{\vartheta}}_B$ can be written as a matrix weighted average (5.14).

Theorem 6.6. *For every positive definite matrix $\boldsymbol{\Phi}$ and every possible observation $\boldsymbol{y}$ the Bayes estimate $\hat{\boldsymbol{\vartheta}}_B(\boldsymbol{y})$ is constrained to lie in the ellipsoid*

$$\|\hat{\boldsymbol{\vartheta}}_B(\boldsymbol{y}) - \boldsymbol{c}\|_{F^\top F}^2 < \frac{1}{4} \|\hat{\boldsymbol{\vartheta}}_{LS}(\boldsymbol{y}) - \boldsymbol{\mu}\|_{F^\top F}^2$$

with midpoint $\boldsymbol{c} = \dfrac{1}{2} (\hat{\boldsymbol{\vartheta}}_{LS}(\boldsymbol{y}) + \boldsymbol{\mu})$.

The goodness of the bound given in Theorem 6.6 crucially depends on the distance between $\hat{\boldsymbol{\vartheta}}_{LS}(\boldsymbol{y})$ and $\boldsymbol{\mu}$. Moreover, it also depends on the chosen design. If the maximal eigenvalue of $\boldsymbol{F}^\top \boldsymbol{F}$ is not too large as guaranteed for example by choosing the design to be nearly E-optimal, then we can expect a small bound for $\hat{\boldsymbol{\vartheta}}_B(\boldsymbol{y})$ in case of uncertain $\boldsymbol{\Phi}$.

Theorem 6.7. *Assume $\boldsymbol{\Phi} \in \mathcal{M}_r^>$ to be an arbitrary diagonal matrix.*
(i) *If the coordinate system is chosen such that $\boldsymbol{\mu}$ coincides with the origin and $\hat{\boldsymbol{\vartheta}}_{LS}(\boldsymbol{y})$ has positive components then it holds $\boldsymbol{\mu} < \hat{\boldsymbol{\vartheta}}_B(\boldsymbol{y}) < \hat{\boldsymbol{\vartheta}}_{LS}(\boldsymbol{y})$.*
(ii) *If the coordinate system is chosen such that $\hat{\boldsymbol{\vartheta}}_{LS}(\boldsymbol{y}) - \boldsymbol{\mu}$ has nonnegative components and, additionally, $\boldsymbol{F}^\top \boldsymbol{F}$ is diagonal then $\hat{\boldsymbol{\vartheta}}_B(\boldsymbol{y})$ is constrained to lie in or on the rectangular solid with diagonal* $[\boldsymbol{\mu}, \hat{\boldsymbol{\vartheta}}_{LS}(\boldsymbol{y})]$.

We remark that we have already seen in Section 5.2 that in case of diagonality of both $\boldsymbol{F}^\top \boldsymbol{F}$ and $\boldsymbol{\Phi}$ the components of $\hat{\boldsymbol{\vartheta}}_B$ lie between the corresponding components of $\hat{\boldsymbol{\vartheta}}_{LS}$ and $\boldsymbol{\mu}$.

Leamer (1982) has derived further bounds for the Bayes estimate provided that the prior covariance matrix can be bounded from above or from below such that $\boldsymbol{\Phi}_* \leqq \boldsymbol{\Phi} \leqq \boldsymbol{\Phi}^*$ with predetermined $\boldsymbol{\Phi}_*$, $\boldsymbol{\Phi}^* \in \mathcal{M}_r^>$.

The case of uncertain prior expectation $w_0 = \mathbf{E}\,\tilde{w}_1$, which is equivalent to assuming that $\boldsymbol{\Phi}$ is known up to a constant multiple, can be easily studied by displaying the behaviour of

$$\hat{\vartheta}_B(y) = (F^\top F + \lambda\Phi^{-1})^{-1}(F^\top y + \lambda\Phi^{-1}\mu)$$

against the ratio $\lambda = w_1^{-1}/\mathbf{E}\,\tilde{w}_1^{-1}$ of sampling and prior expected variance. With varying λ this r-dimensional curve is precisely DICKEY's (1975) "curve décolletage". In two dimensions for example this curve describes an hyperbola. Moreover, if the error components are no longer assumed to be homoscedastically and independently distributed, i.e., $\mathbf{Cov}\,\tilde{e} = \Sigma$, $\mathbf{E}\,\tilde{\Sigma} = \Sigma_0$ with some $\Sigma_0 \in \mathcal{M}_n^>$, then instead of $\hat{\vartheta}_B$ we have to take

$$\hat{\vartheta}_B^{\Sigma_0}(\tilde{y}) = (F^\top\Sigma_0^{-1}F + \Phi^{-1})^{-1}(F^\top\Sigma_0^{-1}\tilde{y} + \Phi^{-1}\mu).$$

The effect of a misspecification of Σ_0 then can be studied as above replacing $F^\top F$, Φ^{-1}, $\hat{\vartheta}_{LS}(y)$ and μ by Φ^{-1}, $F^\top\Sigma_0^{-1}F$, μ and $\hat{\vartheta}_{LS}^{\Sigma_0}(y) = (F^\top\Sigma_0^{-1}F)^{-1}F^\top\Sigma_0^{-1}y$, respectively, provided that μ and Φ are accurately specified. However, if both Σ_0 and Φ are subject to uncertainty and may be arbitrary positive definite matrices then $\hat{\vartheta}_B$ may assume values essentially everywhere in $\mathbb{R}^r$ (cp. CHAMBERLAIN/LEAMER (1976), Theorem 1).

A robustified Bayes (affine) estimator in case that both μ and Φ, and even more the error covariance matrix $\mathbf{Cov}\,\tilde{e}$, are subject to uncertainty and only known to vary within certain bounds will be constructed in the next section.

In any case, if we have doubts concerning the correct specification of the prior moments, we should investigate the possible effects on the Bayes risk of the resulting estimators and compare it with the risk of the LSE. The following formulae, which can be obtained after some straightforward algebraic manipulations indicate the increase in Bayes risk due to misspecification of the moments, assuming normalized quadratic loss (6.7).

(i) If $\mu = \mathbf{E}\,\tilde{\vartheta}$ is incorrectly specified by $\mu_0 \in \mathbb{R}^r$ then the Bayes risk of $\hat{\vartheta}_B$ is increased by

$$\varrho_0 = \|\mu - \mu_0\|_A^2, \qquad A = \Phi^{-1}\psi U\psi\Phi^{-1}.$$

(ii) If the prior covariance matrix Φ is incorrectly specified by $\Phi_0 \in \mathcal{M}_r^>$ then the increase in Bayes risk arising by use of

$$\hat{\vartheta}_B^0(\tilde{y}) = (F^\top F + \Phi_0^{-1})(F^\top\tilde{y} + \Phi_0^{-1}\mu) =: \psi_0(F^\top\tilde{y} + \Phi_0^{-1}\mu)$$

instead of $\hat{\vartheta}_B(\tilde{y}) = \psi(F^\top\tilde{y} + \Phi^{-1}\mu)$ takes the form

$$\varrho_1 = \operatorname{tr} U\psi_1 F^\top[F(\Phi_1^{-1} + \Phi_0^{-1})^{-1}F^\top + I_n]F\psi_1,$$

where $\Phi_1^{-1} = \Phi^{-1} - \Phi_0^{-1}$ measures the misspecification of prior precision and $\psi_1 = \psi - \psi_0 = -\psi_0(\psi_0 + \Phi_1)^{-1}\psi_0$ indicates the misspecification of the posterior covariance matrix.

We will now turn from the local notion of robustness to a global evaluation of estimators taking account of the whole variety of possible outcomes $y \in \mathbb{R}^n$ of observation. This entails an a-preposteriori viewpoint, i.e., an evaluation of local robustness according to the (unconditional) distribution

$P_{\tilde{y}}$ of the observation vector. Proceeding from the posterior expected loss (6.5) we then arrive, very naturally, at the partial Bayes risk

$$\varrho_1(P_1, w_2; \hat{\vartheta}) = \int_{\mathbb{R}^n} \bar{L}(P_1, w_2; \hat{\vartheta}(y))P_{\tilde{y}}(dy). \tag{6.10}$$

Definition 6.2. Let $c > 0$ and D a given class of estimators. The estimator $\hat{\vartheta}_0 \in D$ is said to be *c-globally robust* in D with respect to $\mathcal{P}_{\tilde{e}}$ and $\mathcal{P}_1$ if it holds

$$\varrho_s(\hat{\vartheta}_0) := \sup_{w_2 \in W_2} \sup_{P_1 \in \mathcal{P}_1} |\varrho_1(P_1, w_2; \hat{\vartheta}_0) - \inf_{\hat{\vartheta} \in D} \varrho_1(P_1, w_2; \hat{\vartheta})| \leqq c.$$

Contrary to the local notion, the global criterion refers to the robustness of the estimation procedure, BERGER (1984) called this criterion "procedure robustness". In view of procedure robustness, the so-called *$\mathcal{P}_1$-minimax-regret criterion*

$$\varrho_s(\hat{\vartheta}_0) \to \inf_{\hat{\vartheta}_0 \in D} \tag{6.11}$$

appears to be a natural criterion for the choice of an estimator. However, this criterion is very difficult to handle mathematically and, moreover, has the disadvantage that the resulting minimax-regret estimators are not necessarily Bayesian with respect to any single distribution from $\mathcal{P}_1$ (cp., e.g., WATSON (1974)).

As an alternative we consider the following criterion which makes a natural compromise between minimax and Bayes optimality.

Definition 6.3. The estimator $\hat{\vartheta}_0 \in D$ is called a *$\mathcal{P}_1$-minimax estimator* in D if it holds

$$\sup_{w_2 \in W_2} \sup_{P_1 \in \mathcal{P}_1} \varrho_1(P_1, w_2; \hat{\vartheta}_0) = \inf_{\hat{\vartheta} \in D} \sup_{w_2 \in W_2} \sup_{P_1 \in \mathcal{P}_1} \varrho_1(P_1, w_2; \hat{\vartheta}).$$

This criterion which, in the literature, is also known as *restricted minimaxity* or *Gamma-minimaxity*, respectively, was used, among others, by HODGES/LEHMANN (1952), BUNKE (1964) and SOLOMON (1972) in case that $W_1 = W$, $W_2 = \emptyset$, where the partial Bayes risk coincides with the usual Bayes risk. In the context of parameter estimation in regression models it was used by WIND (1973), WATSON (1974) and PILZ (1981 a). This criterion makes a natural compromise between minimax and Bayes optimality, according to the different levels of prior knowledge about the model parameters ϑ, w_1 and w_2. If $\mathcal{P}_1 = \{P_1\}$ and W_2 reduce to single-element subsets then any $\mathcal{P}_1$-minimax estimator is Bayesian with respect to P_1 and vice versa. On the other hand, $\mathcal{P}_1$-minimax optimality coincides with usual minimax optimality if $\mathcal{P}_{\tilde{\vartheta}, \tilde{w}_1}$ is taken to be the set $(\Theta \times W_1)^*$ of all prior distributions on the parameter space $\Theta \times W_1$.

Lemma 6.2. *Assume* $\Theta \subset \mathbb{R}^r$ *and* $W_1 \subset \mathbb{R}^s$, $s \geqq 1$, *to be compact,* $\mathcal{P}_{\tilde{\vartheta}, \tilde{w}_1} = (\Theta \times W_1)^*$, $\mathcal{P}_{\tilde{e}}$ *arbitrary and D the class of all estimators* $\hat{\vartheta}$ *for which the* *risk function* $R(\cdot, \cdot, w_2; \hat{\vartheta})$ *is continuous over* $\Theta \times W_1$, *for any* $w_2 \in W_2$. *Then* $\hat{\vartheta}_0 \in D$ *is a* $\mathcal{P}_1$*-minimax estimator in* D *if and only if it is minimax, i.e., iff it holds*

$$\sup_{(\vartheta, w) \in \Theta \times W} R(\vartheta, w; \hat{\vartheta}_0) = \inf_{\hat{\vartheta} \in D} \sup_{(\vartheta, w) \in \Theta \times W} R(\vartheta, w; \hat{\vartheta})$$

Proof. The assertion follows immediately from the fact that, under the above assumptions,

$$\sup_{w_2 \subset W_2} \sup_{P_1 \subset \mathcal{P}_1} \varrho_1(P_1, w_2; \hat{\vartheta}) = \sup_{(\vartheta, w) \subset \Theta \times W} R(\vartheta, w; \hat{\vartheta}) \tag{6.12}$$

for any $\hat{\vartheta} \in D$ (cp. A.19 (ii) and Lemma 6.5 below). □

We remark that the assertion of Lemma 6.2 also holds true in case that $\Theta = \mathbb{R}^r$, since then (6.12) holds trivially. From this viewpoint, a minimax estimator can be interpreted as a globally most robust Bayes estimator. This is also supported by the fact that in case of equality

$$\inf_{\hat{\vartheta} \in D} \sup_{w_2 \in W_2} \sup_{P_1 \in \mathcal{P}_1} \varrho_1(P_1, w_2; \hat{\vartheta}) = \sup_{w_2 \in W_2} \sup_{P_1 \in \mathcal{P}_1} \inf_{\hat{\vartheta} \in D} \varrho_1(P_1, w_2; \hat{\vartheta}) \tag{6.13}$$

every minimax estimator is Bayesian with respect to a least favourable prior distribution from $\mathcal{P}_1 = \mathcal{P}_{\tilde{\vartheta}, \tilde{w}}$. These facts explain the interest in the criterion of $\mathcal{P}_1$-minimax optimality from the robust Bayesian viewpoint.

The ($\mathcal{P}_1$-) minimax criterion bears a conservative attitude, it seeks robustness against the class of all priors on the parameter space. The desired protection against the worst case offers, however, the danger of a relatively large Bayes risk. If we have available sufficiently precise prior knowledge which allows the specification of a sufficiently reliable prior distribution $P^0_{\tilde{\vartheta}, \tilde{w}}$ over $\Theta \times W$ then the following concept of procedure robustness can be recommended.

Definition 6.4. Let $c > 0$. The estimator $\hat{\vartheta}_0 \in D$ is called a *c-constrained Bayes estimator* in D with respect to $P_{\tilde{\vartheta}, \tilde{w}}$ if it holds

(i) $R(\vartheta, w; \hat{\vartheta}_0) \leqq c \qquad \forall (\vartheta, w) \in \Theta \times W$,

(ii) $\varrho(P_{\tilde{\vartheta}, \tilde{w}}; \hat{\vartheta}_0) = \inf_{\hat{\vartheta} \in D} \varrho(P_{\tilde{\vartheta}, \tilde{w}}; \hat{\vartheta})$.

This concept goes back to Hodges/Lehmann (1952) and makes a compromise between the supremum and the expectation (with respect to $P^0_{\tilde{\vartheta}, \tilde{w}}$), respectively, of the risk function. The balance between these quantities is made through the choice of the constant $c > 0$. This is equivalent to determining a constant $a \in [0, 1]$ and minimizing

$$a\varrho(P^0_{\tilde{\vartheta}, \tilde{w}}; \hat{\vartheta}) + (1 - a) \sup_{(\vartheta, w) \in \Theta \times W} R(\vartheta, w; \hat{\vartheta}) \to \inf_{\hat{\vartheta} \in D}. \tag{6.13}$$

Here the constant a expresses the "degree of confidence" in the prior distribution $P^0_{\tilde{\vartheta}, \tilde{w}}$. Modifications of the concept of c-constrained Bayes estimation and methods for the construction of such estimators, especially with regard to the Stein effect

$$R(\vartheta, w; \hat{\vartheta}_0) \leqq R(\vartheta, w; \hat{\vartheta}_{LS}) = \operatorname{tr} U(F^\top F)^{-1} =: c,$$

may be found in STRAWDERMAN (1971), EFRON/MORRIS (1972), (1973), BERGER (1980), (1982 a, b), STEIN (1981), BOCK (1982) and DEY/BERGER (1983).

In many cases the problem of finding c-constrained Bayes estimators with respect to some prior $P^0_{\tilde{\vartheta}, \tilde{w}}$ is equivalent to seeking $\mathcal{P}_1$-minimax estimators with respect to some family of contaminated prior distributions

$$\mathcal{P}_1 = \{P_{\tilde{\vartheta}, \tilde{w}} \,|\, P_{\tilde{\vartheta}, \tilde{w}} = (1-\varepsilon) P^0_{\tilde{\vartheta}, \tilde{w}} + \varepsilon \bar{P}, \quad \bar{P} \in (\Theta \times W)^*\},$$

where $\varepsilon = \varepsilon(c) \in (0, 1)$, see, e.g. MASRELIEZ/MARTIN (1977), BERGER (1982 c), MARAZZI (1980), (1985).

6.3. Restricted minimax optimality

We now consider, in some more detail, the criterion of $\mathcal{P}_1$-minimax optimality and make assertions on the optimality of the conjugate Bayes estimator relative to this criterion which demonstrate the extent to which this estimator is robust against a violation of the distributional assumptions (n. i. i. d. errors, conjugate prior) and, additionally, against a misspecification of the prior inputs $\mathbf{E}\,\tilde{\vartheta}$, $\mathbf{Cov}\,\tilde{\vartheta}$ and $\mathbf{Cov}\,\tilde{e}$.

It will be seen that we can drop the distributional assumptions and easily modify the Bayes estimator $\hat{\vartheta}_B$ to allow even for an arbitrary covariance structure of the error and prior distributions. For the case that, besides $\mathbf{Cov}\,\tilde{\vartheta}$ and $\mathbf{Cov}\,\tilde{e}$, also $\mathbf{E}\,\tilde{\vartheta}$ is subject to uncertainty we construct a $\mathcal{P}_1$-minimax estimator which will be seen to coincide with a hierarchical Bayes estimator with respect to a least favourable prior distribution for the regression hyperparameter $\boldsymbol{\mu} = \mathbf{E}\,\tilde{\vartheta}$ at the third stage.

Before giving some specific results, we will formulate a very general condition which is sufficient for $\mathcal{P}_1$-minimax optimality of an estimator. This condition requires the construction of an estimator which is Bayesian with respect to such error and prior distributions from $\mathcal{P}_{\tilde{e}}$ and $\mathcal{P}_1 = \mathcal{P}_{\tilde{\vartheta}, \tilde{w}_1}$, respectively, which are least favourable in the sense that they lead to maximum Bayes risk within these classes.

Lemma 6.3. *Let $w_2^0 \in W_2$, $W^0 = W_1 \times \{w_2^0\}$ and $\mathcal{P}^0_{\tilde{e}} = \{P^{(w)}_{\tilde{e}}\}_{w \in W^0}$* a subclass *of $\mathcal{P}_{\tilde{e}}$. Suppose there exist an estimator $\hat{\vartheta}_0 \in D$ and a prior distribution $P_1^0 = P^0_{\tilde{\vartheta}, \tilde{w}_1}$ such that $\hat{\vartheta}_0$ is Bayesian with respect to P_1 and $\mathcal{P}^0_{\tilde{e}}$ and for which it holds*

$$\sup_{w_2 \in W_2} \sup_{P_1 \in \mathcal{P}_1} \varrho_1(P_1, w_2; \hat{\vartheta}_0) \leqq \varrho_1(P_1^0, w_2^0; \hat{\vartheta}_0). \tag{6.14}$$

Then $\hat{\vartheta}_0$ is a $\mathcal{P}_1$-minimax estimator in D.

(For a proof see PILZ (1981 a).)

The condition (6.14) means that the error distributions from $\mathscr{P}_{\tilde{e}}^0$ and the prior P_1^0 are least favourable within $\mathscr{P}_{\tilde{e}}$ and $\mathscr{P}_1$, respectively, in the sense that

$$\varrho_1(P_1^0, w_2; \hat{\vartheta}_0) \leqq \varrho_1(P_1^0, w_2^0; \hat{\vartheta}_0) \quad \text{for all} \quad w_2 \in W_2$$

and

$$\varrho_1(P_1, w_2^0; \hat{\vartheta}_0) \leqq \varrho_1(P_1^0, w_2^0; \hat{\vartheta}_0) \quad \text{for all} \quad P_1 \in \mathscr{P}_1 .$$

In the special case where $W_1 = W$, $W_2 = \varnothing$ and $\mathscr{P}_1 = (\Theta \times W)^*$ the assertion of Lemma 6.3 reduces to the well-known fact that any Bayes estimator with respect to a least favourable prior distribution is minimax (cp. A.14 and Lemma 6.2).

With the help of Lemma 6.3 we can prove that in case of homoscedastic error distributions the conjugate Bayes estimator is $\mathscr{P}_1$-minimax with $\mathscr{P}_1$ including all prior distributions with fixed first and second order conditional moments of $\tilde{\vartheta}$ (conditional on the precision of observation).

Theorem 6.8. *Let* $\Theta = \mathbb{R}^r$, $\mathscr{P}_{\tilde{e}}$ *of the type* (6.6) *and assume normalized quadratic loss* (6.7). *Then* $\hat{\vartheta}_B(\tilde{y}) = (F^\top F + \Phi^{-1})^{-1}(F^\top \tilde{y} + \Phi^{-1}\mu)$ *is a* $\mathscr{P}_1$*-minimax estimator when choosing*

$$\mathscr{P}_1 = \left\{ P_{\tilde{\vartheta}, \tilde{w}_1} : \mathsf{E}(\tilde{\vartheta}|w_1) = \mu, \quad \mathsf{Cov}(\tilde{\vartheta}|w_1) = \frac{1}{w_1}\Phi, \quad \mathsf{E}\,\tilde{w}_1 < \infty \right\}. \tag{6.15}$$

A proof of this result, formulated with a slightly more general loss function, may be found in Pilz (1981 a). The proof follows by verifying that, with $\mathscr{P}_{\tilde{e}}^0 = \left\{ \mathrm{N}\left(0, \frac{1}{w_1} I_n\right) \right\}_{w_1 \in \mathbb{R}^+}$ and any prior $P_{\tilde{\vartheta}, \tilde{w}_1}^0 \in \mathscr{P}_1$ for which $P_{\tilde{\vartheta}|w_1}^0 = \mathrm{N}\left(\mu, \frac{1}{w_1}\Phi\right)$, the assumptions of Lemma 6.3 are satisfied.

Remark 6.2. Note that in the above theorem we do not require any assumption on the type of error and prior distributions. Theorem 6.8 is a generalization of a result due to Wind (1973), who proved the $\mathscr{P}_1$-minimaxity of $\hat{\vartheta}_B$ for the special case in which $\mathscr{P}_{\tilde{e}}$ is restricted to the class of normally i. i. d. errors.

An important nontrivial example of a family $\mathscr{P}_{\tilde{e}}$ of error distributions for which Theorem 6.8 applies is the family of so-called *exponential power distributions.* These are distributions having independently and identically distributed components $\tilde{e}_i$, $i = 1, \ldots, n$, with densities $p_{\tilde{e}_i}$ with respect to Lebesgue measure given by

$$p_{\tilde{e}_i}^{(w)}(t) = c_1 \sigma^{-1} \exp\left(-c_2 \left|\frac{t}{\sigma}\right|^\beta\right), \; t \in \mathbb{R}^1, \; w = (w_1, w_2) = (\sigma^{-2}, \beta) \in \mathbb{R}^+ \times [1, \infty], \tag{6.16}$$

where $c_1 = c_2^{1/\beta}/2\Gamma(1 + 1/\beta)$, $c_2 = (\Gamma(3/\beta)/\Gamma(1/\beta))^{\beta/2}$.

This kind of error distributions has been investigated in Box/Tiao

(1962), (1973). The parameter $w_2 = \beta$ can be interpreted as a measure of the deviation from normality, which occurs for $\beta = 2$, where the parameter $w_1 = \sigma^{-2}$ characterizes the precision (inverse of the variance) of observation. For $1 \leqq \beta < 2$ the exponential power distributions are less and for $\beta > 2$ they are more concentrated around the mean $\mathbf{E}\,\tilde{e} = 0$ than the normal distribution $P_{\tilde{e}} = \mathrm{N}\left(\mathbf{0}, \frac{1}{w_1} \boldsymbol{I}_n\right)$.

From Theorem 6.8 we conclude that $\hat{\vartheta}_{\mathrm{B}}$ is $\mathcal{P}_1$-minimax, with $\mathcal{P}_1$ given by (6.15), if the error vector follows an exponential power distribution. Viewed in this light, the Bayes estimator shows a satisfying robustness against non-normality of measurement error.

A more general result on the restricted minimaxity of the conjugate Bayes estimator even holds if we drop the assumption of homoscedastic errors and allow for arbitrary covariance structure with the only exception that the error variances be bounded from above. This situation corresponds to the case, where $W_1 = \varnothing$, $W_2 = W$ is some nonempty index set and $\mathcal{P}_{\tilde{e}}$ is an arbitrary class of error distributions such that

$$\mathcal{P}_{\tilde{e}} = \{P_{\tilde{e}}^{(w)}\}_{w \in W}, \quad \forall w \in W: \mathbf{E}(\tilde{e}/w) = 0\,, \quad \mathbf{Cov}(\tilde{e}/w) \leqq \boldsymbol{\Sigma}_0 \tag{6.17}$$

for some given $\boldsymbol{\Sigma}_0 \in \mathcal{M}_n^{>}$. Correspondingly, we consider a general class $\mathcal{P}_1 = \mathcal{P}_{\tilde{\vartheta}}$ of prior distributions for $\tilde{\vartheta}$,

$$\mathcal{P}_1 = \{P_{\tilde{\vartheta}}: \mathbf{E}\,\tilde{\vartheta} \in \mathcal{K}, \quad \mathbf{Cov}\,\tilde{\vartheta} \leqq \boldsymbol{\Phi}_0\}\,, \tag{6.18}$$

where $\mathcal{K}$ is some non-empty subset of $\mathbb{R}^r$ and $\boldsymbol{\Phi}_0 \in \mathcal{M}_r^{>}$ is some given upper bound for $\mathbf{Cov}\,\tilde{\vartheta}$.

Remark 6.3. Essentially, the assumptions $\mathbf{Cov}\,\tilde{e} \leqq \boldsymbol{\Sigma}_0$ and $\mathbf{Cov}\,\tilde{\vartheta} \leqq \boldsymbol{\Phi}_0$ require knowledge of upper bounds for the variances of the components of $\tilde{e}$ and $\tilde{\vartheta}$, respectively. If it holds, for example, $\mathrm{Var}\,\tilde{e}_i \leqq \sigma_i^2 < \infty$ with known $\sigma_i^2 > 0$, $i = 1, \ldots, n$, then we have $\mathbf{Cov}\,\tilde{e} \leqq n\sigma_0^2 \boldsymbol{I}_n$ with $\sigma_0^2 = \max(\sigma_1^2, \ldots, \sigma_n^2)$. If additional knowledge is available then sharper bounds can be given, for example in case of uncorrelated observations we would have $\mathbf{Cov}\,\tilde{e} \leqq \sigma_0^2 \boldsymbol{I}_n$ instead. More generally speaking, if the covariance structure admits a Toeplitz-type covariance matrix then the results of GRENANDER/SZEGÖ (1959) on the eigenvalues of such matrices may be used to derive upper bounds $\boldsymbol{\Phi}_0$ and $\boldsymbol{\Sigma}_0$. For first order autoregressive and moving-average processes such bounds may be found in OMAN (1982).

We first consider the case that the first order moment $\mathbf{E}\,\tilde{\vartheta}$ is exactly known, i.e., $\mathcal{K}$ from (6.18) is a single-element subset $\mathcal{K} = \{\boldsymbol{\mu}\}$ with some $\boldsymbol{\mu} \in \mathbb{R}^r$. Then the modified Bayes estimator

$$\hat{\vartheta}_{\mathrm{B}}^{0}(\tilde{y}) = (\boldsymbol{F}^\top \boldsymbol{\Sigma}_0^{-1} \boldsymbol{F} + \boldsymbol{\Phi}_0^{-1})^{-1} (\boldsymbol{F}^\top \boldsymbol{\Sigma}_0^{-1} \tilde{y} + \boldsymbol{\Phi}_0^{-1} \boldsymbol{\mu})\,, \tag{6.19}$$

which arises from the generalized Bayes estimator (5.13) replacing the unknown covariance matrices of $\tilde{e}$ and $\tilde{\vartheta}$ by their upper bounds $\boldsymbol{\Sigma}_0$ and $\boldsymbol{\Phi}_0$, respectively, comes out as a $\mathcal{P}_1$-minimax estimator. The proof makes essential use of a stochastic dominance relation between distribution functions (see A.8) and of the fact that the normal distributions $P_{\tilde{e}} = \mathrm{N}(\mathbf{0}, \boldsymbol{\Sigma}_0)$ and $P_{\tilde{\vartheta}} = \mathrm{N}(\boldsymbol{\mu}, \boldsymbol{\Phi}_0)$ are least favourable distributions within the above classes $\mathcal{P}_{\tilde{e}}$ and $\mathcal{P}_1$, respectively.

Theorem 6.9. *Let $\Theta = \mathbb{R}^r$, $\mathcal{P}_{\tilde{e}}$ according to* (6.17) *and assume quadratic loss $L(\boldsymbol{\vartheta}, w; \hat{\boldsymbol{\vartheta}}(\boldsymbol{y})) = \|\boldsymbol{\vartheta} - \hat{\boldsymbol{\vartheta}}(\boldsymbol{y})\|_U^2$, $\boldsymbol{U} \in \mathcal{M}_r^{\geqq}$, independent of $w \in W$. Further, let $\mathcal{P}_1$ according to* (6.18) *with $\mathcal{K} = \{\boldsymbol{\mu}\}$. Then $\hat{\boldsymbol{\vartheta}}_{\mathrm{B}}^0$ as given above is a $\mathcal{P}_1$-minimax estimator within the class D of all estimators, i.e., it holds*

$$\sup_{w \in W} \sup_{P_{\tilde{\vartheta}} \in \mathcal{P}_1} \mathbf{E}_{\vartheta} R(\tilde{\boldsymbol{\vartheta}}, w; \hat{\boldsymbol{\vartheta}}_{\mathrm{B}}^0) = \inf_{\hat{\vartheta} \in D} \sup_{w \in W} \sup_{P_{\tilde{\vartheta}} \in \mathcal{P}_1} \mathbf{E}_{\tilde{\vartheta}} R(\tilde{\boldsymbol{\vartheta}}, w; \hat{\boldsymbol{\vartheta}}).$$

Proof. The proof follows by verifying the validity of the sufficient conditions for $\mathcal{P}_1$-minimaxity stated in Lemma 6.3. To this let $P_1^0 = P_{\tilde{\vartheta}}^0 = \mathrm{N}(\boldsymbol{\mu}, \boldsymbol{\Phi}_0)$, $W_2 = W$ and $w_2^0 = w^0$ index the normal distribution $P_{\tilde{e}}^{(w)} = \mathrm{N}(\mathbf{0}, \boldsymbol{\Sigma}_0)$. Then $\hat{\boldsymbol{\vartheta}}_{\mathrm{B}}$ is Bayesian in D with respect to P_1^0 and $\mathcal{P}_{\tilde{e}}^0 = \{\mathrm{N}(\mathbf{0}, \boldsymbol{\Sigma}_0)\}$. Thus, it only remains to verify that $\hat{\boldsymbol{\vartheta}}_0 = \hat{\boldsymbol{\vartheta}}_{\mathrm{B}}^0$ satisfies (6.14).

The distribution $P_{\tilde{e}}^{(w^0)} = \mathrm{N}(\mathbf{0}, \boldsymbol{\Sigma}_0)$ is a maximum element of $\mathcal{P}_{\tilde{e}}$ with respect to the stochastic dominance relation $>$ introduced in A.8. Thus, letting $P_{\hat{\vartheta}_{\mathrm{B}}^0}(\cdot \mid \boldsymbol{\vartheta}, w)$ denote the sampling distribution of $\hat{\boldsymbol{\vartheta}}_{\mathrm{B}}^0$ induced by $P_{\tilde{e}}^{(w)} \in \mathcal{P}_{\tilde{e}}$, then it follows from the linearity of $\hat{\boldsymbol{\vartheta}}_{\mathrm{B}}^0$ (relative to $\tilde{\boldsymbol{y}}$) that

$$R(\boldsymbol{\vartheta}, w; \hat{\boldsymbol{\vartheta}}_{\mathrm{B}}^0) = \int_{\mathbb{R}^r} \|\boldsymbol{\vartheta} - \boldsymbol{t}\|_U^2 P_{\hat{\vartheta}_{\mathrm{B}}^0}(\mathrm{d}\boldsymbol{t}) \mid \boldsymbol{\vartheta}, w)$$

$$\leqq \int_{\mathbb{R}^r} \|\boldsymbol{\vartheta} - \boldsymbol{t}\|_U^2 P_{\hat{\vartheta}_{\mathrm{B}}^0}(\mathrm{d}\boldsymbol{t} \mid \boldsymbol{\vartheta}, w^0) = R(\boldsymbol{\vartheta}, w^0; \hat{\boldsymbol{\vartheta}}_{\mathrm{B}}^0)$$

for all pairs $(\boldsymbol{\vartheta}, w) \subset \Theta \times W$. Hence, for the partial Bayes risk $\varrho_1(P_1, w; \hat{\boldsymbol{\vartheta}}_{\mathrm{B}}^0) = \mathbf{E}_{\tilde{\vartheta}} R(\tilde{\boldsymbol{\vartheta}}, w; \hat{\boldsymbol{\vartheta}}_{\mathrm{B}}^0)$ of $\hat{\boldsymbol{\vartheta}}_{\mathrm{B}}^0$ with respect to $P_1 = P_{\tilde{\vartheta}} \in \mathcal{P}_1$ it holds

$$\sup_{w \in W} \sup_{P_1 \in \mathcal{P}_1} \varrho_1(P_1, w; \hat{\boldsymbol{\vartheta}}_{\mathrm{B}}^0) \leqq \sup_{P_1 \in \mathcal{P}_1} \varrho_1(P_1, w^0; \hat{\boldsymbol{\vartheta}}_{\mathrm{B}}^0). \tag{6.20}$$

Further, $\hat{\boldsymbol{\vartheta}}_{\mathrm{B}}^0(\tilde{\boldsymbol{y}}) = (\boldsymbol{F}^\top \boldsymbol{\Sigma}_0^{-1} \boldsymbol{F} + \boldsymbol{\Phi}_0^{-1})^{-1} (\boldsymbol{F}^\top \boldsymbol{\Sigma}_0^{-1} \tilde{\boldsymbol{y}} + \boldsymbol{\Phi}_0^{-1} \boldsymbol{\mu}) =: \boldsymbol{Z}\tilde{\boldsymbol{y}} + \boldsymbol{z}$ has (partial) Bayes risk

$$\begin{aligned}\varrho_1(P_1, w^0; \hat{\boldsymbol{\vartheta}}_{\mathrm{B}}^0) &= \mathbf{E}_{\tilde{\vartheta}} R(\tilde{\boldsymbol{\vartheta}}, w^0; \hat{\boldsymbol{\vartheta}}_{\mathrm{B}}^0) \\ &= \operatorname{tr} \boldsymbol{U}[\boldsymbol{Z}\boldsymbol{\Sigma}_0\boldsymbol{Z}^\top + (\boldsymbol{I}_r - \boldsymbol{Z}\boldsymbol{F})(\mathbf{Cov}\,\tilde{\boldsymbol{\vartheta}})(\boldsymbol{I}_r - \boldsymbol{Z}\boldsymbol{F})^\top].\end{aligned}$$

Observing that $\mathbf{Cov}\,\tilde{\boldsymbol{\vartheta}} \leqq \boldsymbol{\Phi}_0$ for any $P_1 = P_{\tilde{\vartheta}} \in \mathcal{P}_1$ we thus have

$$\sup_{P_1 \in \mathcal{P}_1} \varrho_1(P_1, w^0; \hat{\boldsymbol{\vartheta}}_{\mathrm{B}}^0) \leqq \varrho_1(P_1^0, w^0, \hat{\boldsymbol{\vartheta}}_{\mathrm{B}}^0) \tag{6.21}$$

which, together with (6.20), leads to the required inequality

$$\sup_{w \in W} \sup_{P_1 \in \mathcal{P}_1} \varrho_1(P_1, w; \hat{\boldsymbol{\vartheta}}_{\mathrm{B}}^0) \leqq \varrho_1(P_1^0, w^0; \hat{\boldsymbol{\vartheta}}_{\mathrm{B}}^0). \quad \square$$

The assumption $\Theta = \mathbb{R}^r$ made in Theorem 6.9 is necessary to prove the $\mathcal{P}_1$-minimaxity of $\hat{\boldsymbol{\vartheta}}_{\mathrm{B}}^0$ within the class D of all estimators, since the Bayes optimality of the affine linear estimator $\hat{\vartheta}_B^0$ within D is guaranteed only with respect to a normal prior distribution for $\tilde{\vartheta}$ (cp. our discussion in Section 5.3 and Section 6.1).

In case of a restricted parameter set the $\mathcal{P}_1$-minimax optimality of the Bayes affine estimator $\hat{\vartheta}_B^0$ can be shown only within the subclass D_a of affine estimators.

Corollary 6.5. *Assume $\Theta \subseteqq \mathbb{R}^r$ and let L, $\mathcal{P}_{\tilde{e}}$ and $\mathcal{P}_1$ as defined in Theorem* 6.9. *Then $\hat{\vartheta}_B^0$ from* (6.19) *is $\mathcal{P}_1$-minimax in D_a.*

Proof. The assertion follows again from Lemma 6.3: $\hat{\vartheta}_B^0$ is Bayesian in D_a with respect to any $P_{\tilde{\vartheta}} \in \mathcal{P}_1$ and $P_{\tilde{e}} \in \mathcal{P}_e$ for which $\mathbf{Cov}\, \tilde{e} = \boldsymbol{\Sigma}_0$ and $\mathbf{Cov}\, \tilde{\vartheta} = \boldsymbol{\Phi}_0$, respectively. The validity of (6.14) then can be shown as in the proof of Theorem 6.9 since the verification of (6.20) and (6.21) only requires that $\mathbf{Cov}\, \tilde{e} \leq \boldsymbol{\Sigma}_0$ and $\mathbf{Cov}\, \tilde{\vartheta} \leq \boldsymbol{\Phi}_0$, respectively, but does not take account of the type of the distributions $P_{\tilde{e}}$ and $P_{\tilde{\vartheta}}$. □

Let us now consider the situation when there is additional uncertainty with regard to the specification of the first order moment $\mathbf{E}\,\tilde{\vartheta}$, i.e., we consider a general family $\mathcal{P}_1$ of priors of the form (6.18) with card $\mathcal{K} > 1$, and determine a $\mathcal{P}_1$-minimax estimator within the subclass D_a of affine-linear estimators. However, the $\mathcal{P}_1$-minimaxity then does not carry over to the class of all estimators as it was the case in the situation with $\mathcal{K} = \{\boldsymbol{\mu}\}$ considered in Theorem 6.9. The construction of estimators which are $\mathcal{P}_1$-minimax under all estimators appears to be a very complicated unsolved problem.

Assumption 6.1. *Let $\mathcal{K} \in \mathbb{R}^r$ be a compact set which is symmetric around some center point $\boldsymbol{\mu}_0 \in \mathbb{R}^r$.*

Special cases of practical importance refer to an ellipsoid-shaped set

$$\mathcal{K} = \mathcal{E}(\boldsymbol{H}, \boldsymbol{\mu}_0) := \{\boldsymbol{\mu} \in \mathbb{R}^r : (\boldsymbol{\mu} - \boldsymbol{\mu}_0)^\top \boldsymbol{H}(\boldsymbol{\mu} - \boldsymbol{\mu}_0) \leqq 1\}, \quad \boldsymbol{H} \in \mathcal{M}_r^{>} \tag{6.22}$$

or to a rectangular solid $\mathcal{K}$ with midpoint $\boldsymbol{\mu}_0$, respectively. In the given case, the above estimator $\hat{\vartheta}_B^0$ is no longer $\mathcal{P}_1$-minimax optimal. It will be seen that the $\mathcal{P}_1$-minimax estimator has the form of a hierarchical Bayes-affine estimator and the uncertainty with regard to $\mathbf{E}\,\tilde{\vartheta}$ has the same effect as an "enlargement" of the covariance matrix of $\tilde{\vartheta}$.

Lemma 6.4. *Let $P_{\tilde{e}}$ and L as assumed in Theorem* 6.9 *and $\mathcal{P}_1$ as given by* (6.18) *with $\mathcal{K}$ according to Assumption* 6.1. *Then it holds*

$$\inf_{\hat{\vartheta} \in D_a} \sup_{w \in W} \sup_{P_{\tilde{\vartheta}} \in \mathcal{P}_1} \mathbf{E}_{\tilde{\vartheta}} R(\tilde{\vartheta}, w; \hat{\vartheta}) = \inf_{Z \in \mathcal{M}_{r \times n}} \sup_{\boldsymbol{\mu} \in \mathcal{K}} h(\boldsymbol{\mu}, \boldsymbol{Z}),$$

where

$$h(\boldsymbol{\mu}, \boldsymbol{Z}) = \operatorname{tr} \boldsymbol{U}(\boldsymbol{Z}\boldsymbol{\Sigma}_0\boldsymbol{Z}^\top + \boldsymbol{Z}_0\boldsymbol{\Phi}_0\boldsymbol{Z}_0^\top) + \|\boldsymbol{Z}_0(\boldsymbol{\mu} - \boldsymbol{\mu}_0)\|_U^2 \tag{6.23}$$

and $\boldsymbol{Z}_0 = \boldsymbol{I}_r - \boldsymbol{Z}\boldsymbol{F}$.

Proof. Let $\hat{\vartheta} = Z\tilde{y} + z \in D_a$. Then, since $\mathbf{Cov}\,\tilde{\vartheta} \leqq \Phi_0$ and $\mathbf{Cov}\,\tilde{e} \leqq \Sigma_0$, it follows from the representation (5.33) of the Bayes risk of $\hat{\vartheta} \in D_a$ that

$$\sup_{w \in W} \sup_{P_{\tilde{\vartheta}} \in \mathcal{P}_1} \mathbf{E}_{\tilde{\vartheta}} R(\tilde{\vartheta}, w; \hat{\vartheta}) = \operatorname{tr} U(Z\Sigma_0 Z^\top + Z_0 \Phi_0 Z_0^\top) + \sup_{\mu \in \mathcal{K}} \|Z_0 \mu - z\|_U^2 \tag{6.24}$$

(cp. PILZ (1984), Lemma 2). Writing $z = Z_0 \mu_0 - g$ with arbitrary $g \in \mathbb{R}^r$, we conclude from the symmetry of $\bar{\mathcal{K}} = \{\bar{\mu} \in \mathbb{R}^r : \bar{\mu} = \mu - \mu_0, \mu \in \mathcal{K}\}$ around the origin that

$$\begin{aligned}\sup_{\mu \in \mathcal{K}} \|Z_0 \mu - z\|_U^2 &= \sup_{\bar{\mu} \in \bar{\mathcal{K}}} \{\|Z_0 \bar{\mu}\|_U^2 + 2g^\top U Z_0 \bar{\mu}\} + \|g\|_U^2 \\ &\geqq \sup_{\bar{\mu} \in \bar{\mathcal{K}}} \|Z_0 \bar{\mu}\|_U^2 = \sup_{\mu \in \mathcal{K}} \|Z_0(\mu - \mu_0)\|_U^2 .\end{aligned}$$

This combined with (6.24) yields the assertion. □

Remark 6.4. Together with (6.24) Lemma 6.4 states that the $\mathcal{P}_1$-minimax estimate in D_a is of the form

$$\hat{\vartheta}_{\mathcal{P}_1} = Z^* \tilde{y} + (I_r - Z^* F)\mu_0 ,$$

where Z^* minimizes the right-hand side expression of (6.24) with respect to $Z \in \mathcal{M}_{r \times n}$.

The following result proves to be very useful for the treatment of minimax problems and builds the basis for the application of ideas and results from the (convex) theory of optimum experimental design layed out in the Chapters 10 and 11.

Lemma 6.5. *Assume $h(\cdot)$ to be a continuous function defined on a compact set $\mathcal{K} \subset \mathbb{R}^r$. Then it holds*

$$\sup_{\mu \in \mathcal{K}} h(\mu) = \sup_{\xi \in \mathcal{K}^*} \int_{\mathcal{K}} h(\mu)\, \xi(\mathrm{d}\mu) ,$$

where $\mathcal{K}^*$ denotes the set of all probability measures ξ defined on the σ-algebra of the Borel subsets of $\mathcal{K}$.

This assertion follows from a more general result due to PECK/DULMAGE (1957).

Now, introducing the matrices

$$C_\xi = \int_{\mathcal{K}} (\mu - \mu_0)(\mu - \mu_0)^\top \xi(\mathrm{d}\mu) , \qquad \xi \in \mathcal{K}^* \tag{6.25}$$

for the function $h(\cdot, Z)$ defined by (6.23) we obtain

$$\begin{aligned}\int_{\mathcal{K}} h(\mu, Z)\, \xi(\mathrm{d}\mu) &= \operatorname{tr} U[Z\Sigma_0 Z^\top + Z_0(\Phi_0 + C_\xi) Z_0^\top] \\ &=: \bar{h}(C_\xi, Z) .\end{aligned} \tag{6.26}$$

The matrix C_ξ represents the centered second order moment matrix of the probability measure $\xi \in \mathcal{K}^*$. We remark that C_ξ can be interpreted as the information matrix of the "experimental design" ξ over the compact "experimental region" $\bar{\mathcal{K}} = \{\bar{\mu} \in \mathbb{R}^r : \bar{\mu} = \mu - \mu_0,\ \mu \in \mathcal{K}\}$, cp. Section 10.3.

With the above preliminary considerations we are now in a position to determine a $\mathcal{P}_1$-minimax affine estimator.

Theorem 6.10. *Let $\Theta \subseteqq \mathbb{R}^r$ arbitrary and L as assumed in Theorem 6.9. Further, let $\mathcal{P}_{\tilde{e}}$ as defined in (6.17) and $\mathcal{P}_1$ according to (6.18) with $\mathcal{K}$ satisfying Assumption 6.1. Then the $\mathcal{P}_1$-minimax estimator in D_a is uniquely determined by*

$$\hat{\vartheta}_{\mathcal{P}_1} = (F^\top \Sigma_0^{-1} F + (\Phi_0 + C_0)^{-1})^{-1} (F^\top \Sigma_0^{-1} \tilde{y} + (\Phi_0 + C_0)^{-1} \mu_0),$$

where

$$C_0 = C_{\xi_0} \quad \text{and} \quad \xi_0 = \arg \sup_{\xi \in \mathcal{K}^*} \operatorname{tr} U(F^\top \Sigma_0^{-1} F + (\Phi_0 + C_\xi)^{-1})^{-1}.$$

Proof. First note that the function $\bar{h}(C_\xi, Z)$ defined by (6.26) is convex with respect to $Z \in \mathcal{M}_{r \times n}$ and also continuous and linear with respect to C_ξ chosen from $\mathcal{C} = \{C_\xi \in \mathcal{M}_r^{\geqq} : \xi \in \mathcal{K}^*\}$ which, by virtue of A.19 (i), is a compact set. Thus, from a minimax theorem due to Sion (see A.20) we conclude that

$$\inf_{\hat{\vartheta} \in D_a} \sup_{w \in W} \sup_{P_{\tilde{\vartheta}} \in \mathcal{P}_1} \mathsf{E}_{\tilde{\vartheta}} R(\tilde{\vartheta}, w; \hat{\vartheta}) = \sup_{C_\xi \in \mathcal{C}} \inf_{Z \in \mathcal{M}_{r \times n}} \bar{h}(C_\xi, Z) \tag{6.27}$$

observing the results of Lemma 6.4 and 6.5. To minimize $\bar{h}(C_\xi, \cdot)$ for fixed $\xi \in \mathcal{K}^*$ write

$$Z = [F^\top \Sigma_0^{-1} F + (\Phi_0 + C_\xi)^{-1}]^{-1} F^\top \Sigma_0^{-1} + G =: Z_\xi + G$$

with arbitrary $G \in \mathcal{M}_{r \times n}$. With $Z_0 = I_r - ZF$ we then obtain

$$Z \Sigma_0 Z^\top + Z_0 (\Phi_0 + C_\xi) Z_0^\top = [F^\top \Sigma_0^{-1} F + (\Phi_0 + C_\xi)^{-1}]^{-1} + G_0,$$

where $G_0 = GF(\Phi_0 + C_\xi) F^\top G^\top + G \Sigma_0 G^\top$ is positive semidefinite. Therefore, we have $\operatorname{tr} UG_0 \geqq 0$ and thus

$$\inf_{Z \in \mathcal{M}_{r \times n}} \bar{h}(C_\xi, Z) = \operatorname{tr} U[F^\top \Sigma_0^{-1} F + (\Phi_0 + C_\xi)^{-1}]^{-1} \tag{6.28}$$

which proves the $\mathcal{P}_1$-minimaxity of $\hat{\vartheta}_{\mathcal{P}_1} = Z_\xi \tilde{y} + (I_r - Z_{\xi_0} F) \mu_0$, observing Remark 6.4.

The existence of a measure ξ_0 and thus of a matrix C_0 which maximizes the indicated trace functional is guaranteed by the fact that this functional is continuous over the compact set $\mathcal{C}$. □

Obviously, $\hat{\vartheta}_{\mathcal{P}_1}$ as given in Theorem 6.10 coincides with the estimator $\hat{\vartheta}_B^0$ from (6.19) if $\mathcal{K}$ collapses to a single-element subset $\mathcal{K} = \{\mu\}$. In this case we have $C_\xi \equiv C_0 = 0_{r,r}$ and $\hat{\vartheta}_{\mathcal{P}_1}$ is a Bayes affine estimator with respect to any distribution $P_{\tilde{\vartheta}, \tilde{e}}$ such that $\mathsf{E}\,\tilde{e} = 0$, $\mathsf{Cov}\,\tilde{e} = \Sigma_0$, $\mathsf{E}\,\tilde{\vartheta} = \mu$ and $\mathsf{Cov}\,\tilde{\vartheta} = \Phi_0$.

In the above case of card $\mathcal{K} > 1$ with a region $\mathcal{K}$ according to Assumption 6.1 the $\mathcal{P}_1$-minimax estimator formally coincides with a Bayes affine estimator with respect to distributions $P_{\tilde{\vartheta}, \tilde{e}}$ such that $\mathsf{E}\,\tilde{e} = 0$, $\mathsf{Cov}\,\tilde{e} = \Sigma_0$

and

$$\mathbf{E}\,\tilde{\boldsymbol{\vartheta}} = \boldsymbol{\mu}_0, \qquad \mathbf{Cov}\,\tilde{\boldsymbol{\vartheta}} = \boldsymbol{\Phi}_0 + \boldsymbol{C}_0. \tag{6.29}$$

This means that for the subclass D_a the principle of $\mathcal{P}_1$-minimax estimation is equivalent to Bayes estimation using an "enlarged" prior covariance matrix, where the enlargement by $\boldsymbol{C}_0$ is due to the uncertainty with regard to the knowledge of the first order moment of $\tilde{\boldsymbol{\vartheta}}$. But this enlargement is also effected by the hierarchical Bayes estimation principle when we express the uncertainty with regard to $\boldsymbol{\mu} = \mathbf{E}\,\tilde{\boldsymbol{\vartheta}}$ by an appropriate probability distribution for the hyperparameter $\boldsymbol{\mu}$ (cp. Section 5.6).

Corollary 6.6. *Under the assumption of Theorem 6.10 the $\mathcal{P}_1$-minimax estimator $\hat{\boldsymbol{\vartheta}}_{\mathcal{P}_1}$ in D_a is identical with the hierarchical Bayes affine estimator in the three-stage hierarchical model*

$$\begin{aligned}
\mathbf{E}(\tilde{\boldsymbol{y}}|\boldsymbol{\vartheta}) &= \boldsymbol{F}\boldsymbol{\vartheta}, & \mathbf{Cov}(\tilde{\boldsymbol{y}}|\boldsymbol{\vartheta}) &= \boldsymbol{\Sigma}_0\\
\mathbf{E}(\tilde{\boldsymbol{\vartheta}}|\boldsymbol{\mu}) &= \boldsymbol{\mu}, & \mathbf{Cov}(\tilde{\boldsymbol{\vartheta}}|\boldsymbol{\mu}) &= \boldsymbol{\mu}_0\\
\mathbf{E}(\tilde{\boldsymbol{\mu}}|\boldsymbol{\mu}_0) &= \boldsymbol{\mu}_0, & \mathbf{Cov}(\tilde{\boldsymbol{\mu}}|\boldsymbol{\mu}_0) &= \boldsymbol{C}_0
\end{aligned}$$

with least favourable covariance matrices at every stage (cp. Theorem 5.8 and (5.33), (5.54)).

If our prior knowledge becomes increasingly vague, which corresponds to an enlargement (increasing diameter) of $\mathcal{K}$, then $(\boldsymbol{\Phi}_0 + \boldsymbol{C}_0)^{-1}$ tends to the matrix of zeroes and in the limiting case $\mathcal{K} = \mathbb{R}^r$ the $\mathcal{P}_1$-minimax estimator coincides with the generalized LSE $(\boldsymbol{F}^\top\boldsymbol{\Sigma}_0^{-1}\boldsymbol{F})^{-1}\boldsymbol{F}^\top\boldsymbol{\Sigma}_0^{-1}\tilde{\boldsymbol{y}}$.

Theorem 6.10 reduces the problem of determining a $\mathcal{P}_1$-minimax affine estimator to the problem of finding a least favourable second order moment matrix $\boldsymbol{C}_0$ which maximizes the trace functional (hierarchical Bayes risk)

$$\operatorname{tr} \boldsymbol{U}[\boldsymbol{F}^\top\boldsymbol{\Sigma}_0^{-1}\boldsymbol{F} + (\boldsymbol{\Phi}_0 + \boldsymbol{C}_\xi)^{-1}]^{-1}$$

over the compact and convex set

$$\mathscr{C} = \left\{\boldsymbol{C}_\xi = \int_{\mathcal{K}} (\boldsymbol{\mu} - \boldsymbol{\mu}_0)(\boldsymbol{\mu} - \boldsymbol{\mu}_0)^\top \xi(\mathrm{d}\boldsymbol{\mu}) : \xi \in \mathcal{K}^*\right\}. \tag{6.30}$$

The following lemma demonstrates that such a matrix can be obtained as solution to a Bayesian experimental design problem (referring to the so-called criterion of L_B-optimality), which we will deal with extensively in Part III and for which we will give explicit solutions as well as methods for the construction of approximate solutions.

Lemma 6.6. *If the matrix $\boldsymbol{F}$ is of full rank then it holds*

$$\boldsymbol{C}_0 = \arg \inf_{\boldsymbol{C}_\xi \in \mathscr{C}} \operatorname{tr} \boldsymbol{V}\boldsymbol{U}\boldsymbol{V}(\boldsymbol{C}_\xi + \boldsymbol{\Phi}_0 + \boldsymbol{V}) \quad \textit{with} \quad \boldsymbol{V} = (\boldsymbol{F}^\top\boldsymbol{\Sigma}_0^{-1}\boldsymbol{F})^{-1}.$$

This follows immediately from Theorem 6.10 using the matrix identity

$$[F^\top \Sigma_0^{-1} F + (\Phi_0 + C_\xi)^{-1}]^{-1} = V - V(V + \Phi_0 + C_\xi)^{-1} V$$

(cp. A.33). The least favourable moment matrix C_0 is thus identical with the information matrix of an L_B-optimal design measure in $\mathcal{K}^*$ with respect to the weight matrix VUV (cp. Definitions 10.2 and 10.4).

For the special case where $\mathcal{K}$ is an ellipsoid we can give a partial solution as follows.

Corollary 6.7. *Let* $\mathcal{K} = \mathcal{E}(H, \mu_0)$ *according to* (6.22). *Then*

$$C_0 = aH^{-1/2}(H^{1/2}VUVH^{1/2})^{1/2}H^{-1/2} - (V + \Phi_0)$$

with (6.31)

$$a = (1 + \operatorname{tr} H(V + \Phi_0))/\operatorname{tr}(H^{1/2}VUVH^{1/2})^{1/2}$$

yields a solution to the minimization problem formulated in Lemma 6.6 *and the estimator* $\hat{\vartheta}_{\mathcal{P}_1}$ *formed with* C_0 *is* $\mathcal{P}_1$*-minimax in* D_a *provided the matrix on the right-hand side of* (6.31) *is positive semidefinite.*

This follows from Theorem 12.3 setting $f(\mu) = \mu - \mu_0$ and $U_L = VUV$.

For the ellipsoid case $\mathcal{K} = \mathcal{E}(H, \mu_0)$ it is easily verified that $C_0 \leqq H^{-1}$ (in the usual semiordering sense for positive semidefinite matrices, see the appendix on matrix algebra), which leads to the following upper bound for the maximum risk associated with $\hat{\vartheta}_{\mathcal{P}_1}$:

$$\sup_{w \in W} \sup_{P_{\tilde{\vartheta}} \in \mathcal{P}_1} \mathsf{E}_{\tilde{\vartheta}} R(\tilde{\vartheta}, w; \hat{\vartheta}_{\mathcal{P}_1}) \leqq \operatorname{tr} U[F^\top \Sigma_0^{-1} F + (\Phi_0 + H^{-1})^{-1}]^{-1}. \quad (6.32)$$

Replacing C_0 in the representation of $\hat{\vartheta}_{\mathcal{P}_1}$ given in Theorem 6.10 by the upper bound H^{-1}, the resulting estimator will be $\mathcal{P}_1$-minimax in D_a with regard to some uncertainty ellipsoid $\mathcal{K}_0 \supset \mathcal{E}(H, \mu_0)$ and could therefore be used as a simple and tractable approximate solution for the optimum $\hat{\vartheta}_{\mathcal{P}_1}$.

An explicit from for C_0 and thus for $\hat{\vartheta}_{\mathcal{P}_1}$ in the special case where $\mathcal{K}$ describes a rectangular solid in $\mathbb{R}^r$ and where U, $F^\top \Sigma_0^{-1} F$ and Φ_0 are diagonal matrices may be found in Pilz (1984).

The $\mathcal{P}_1$-minimax risk of $\hat{\vartheta}_{\mathcal{P}_1}$ is bounded from above and from below by the corresponding risks of the generalized LSE $\hat{\vartheta} = (F^\top \Sigma_0^{-1} F)^{-1} F^\top \Sigma_0^{-1} \tilde{y}$ and the $\mathcal{P}_1$-minimax estimator $\hat{\vartheta}_B^0$ from (6.19) for the extreme case $\mathcal{K} = \{\mu\}$.

Corollary 6.8. *Under the assumptions of Theorem* 6.10 *it holds*

$$\operatorname{tr} U(V^{-1} + \Phi_0^{-1})^{-1} \leqq \sup_{w \in W} \sup_{P_{\tilde{\vartheta}} \in \mathcal{P}_1} \mathsf{E}_{\tilde{\vartheta}} R(\tilde{\vartheta}, w; \hat{\vartheta}_{\mathcal{P}_1}) \leqq \operatorname{tr} UV$$

with

$$V = (F^\top \Sigma_0^{-1} F)^{-1}.$$

This follows with

$$\sup_{w \in W} \sup_{P_{\tilde{\vartheta}} \in \mathcal{P}_1} \mathsf{E}_{\tilde{\vartheta}} R(\tilde{\vartheta}, w; \hat{\vartheta}_{\mathcal{P}_1}) = \operatorname{tr} U[F^\top \Sigma_0^{-1} F + (\Phi_0 + C_0)^{-1}]^{-1}$$

immediately from the fact that $V^{-1} \leqq V^{-1} + (\Phi_0 + C_0)^{-1} \leqq V^{-1} + \Phi_0^{-1}$.

Sharper lower bounds for the $\mathcal{P}_1$-minimax risk can be obtained replacing Φ_0 by $\Phi_0 + C_\xi$ with a particularly chosen $\xi \in \mathcal{K}^*$. Sharper upper bounds can be derived by exploiting the special structure of $\mathcal{K}$, as we have demonstrated for example in (6.32) for the ellipsoid case $\mathcal{K} = \mathcal{E}(H, \mu_0)$.

7. Bayes linear and admissible linear estimation

7.1. General remarks

Besides the fact that the usual LSE does not take account of prior knowledge, two further essential drawbacks led to the rapid development of alternative estimation methods over the past twenty years: the non-admissibility of the LSE in case that the dimension of the parameter space is greater than two and the numerical instability in case of an ill-conditioned matrix $F^\top F$. According to the nature of the resulting estimators, we can distinguish, essentially, the following 3 categories of proposed alternatives to the LSE:

(i) Linear biased estimators

These include, among others, ridge estimators with fixed ridge coefficients, shrunken estimators with fixed shrinkage factors, the Kuks-Ol'man estimator, minimum MSE-estimators with respect to fixed parameters (see, e.g. Farebrother (1975)) a. s. o., which all belong to the class of Bayes (affine-) linear estimators.

(ii) Pretest and Marquardt-type estimators

Pretest estimators decide on the use of a restricted estimate on the basis of the outcome of a statistical test about the validity of a prior restriction concerning ϑ, overviews over these estimators are given in Judge/Bock (1978) and Bunke/Bunke (1986). By Marquardt-type estimators we subsume all estimation procedures for selecting "significant" regression coefficients, including for example principal component estimators and the more general fractional rank estimators (see Marquardt (1970)).

(iii) Stein-type estimators

In this category we can summarize all estimators of the form

$$\hat{\vartheta} = [I_r - A/\|\hat{\vartheta}_{LS} - \vartheta_0\|_B^2](\hat{\vartheta}_{LS} - \vartheta_0) + \vartheta_0 + o(\|\hat{\vartheta}_{LS} - \vartheta_0\|^{-1}), \quad (7.1)$$

where $A, B \in \mathcal{M}_{r \times r}$, ϑ_0 is a given centre point and $o(\|\hat{\vartheta}_{LS} - \vartheta_0\|^{-1}) = (o_1, \ldots, o_r)^\top$ means a vector such that $o_i/\|\hat{\vartheta}_{LS} - \vartheta_0\| \to 0$ for $\|\hat{\vartheta}_{LS} - \vartheta_0\| \to \infty$. This category comprises a multitude of particular minimax and generalized Bayes estimators which have been investigated since the early seventies (see, e.g. Brown (1971), Strawderman (1971), Alam

(1973), Bock (1975), Berger (1976 a, b), (1980 b), (1982 a), Ben Mansour (1984), but also Bayes estimators with respect to t-like prior distributions and empirical Bayes estimators (see Efron/Morris (1973), (1976), Zellner /Vandaele (1975), Zellner (1976), Fen Li (1982), ridge estimators with stochastic ridge coefficients (see, e.g. Ullah/Vinod (1984)), minimax ridge and minimax adaptive estimators (see Strawderman (1978), Casella (1980)) and finally minimax contraction estimators (see Oman (1982), Bock (1982)).

In the following Sections 7 and 8 we will consider in some more detail estimators of the first category, in particular we will characterize the class of all admissible (unconstrained) linear estimators. The estimators of the second category seem to have become somewhat "old-fashioned". The reasons for this are that, on the one hand, Marquardt-type estimators can be dominated by Bayes linear estimators (see La Motte (1978)) and, on the other hand, pretest estimators can be dominated by appropriate contraction estimators of the form (7.1.), see, e.g. Sclove et al. (1972). Further, estimators of the third category are very difficult to compare and to evaluate, due to their complicated risk structure. As a matter of fact, these estimators effect a significant improvement in risk over the LSE only in certain relatively small regions of the parameter space. Moreover, when choosing a (nonlinear) estimator of the third category, then it is very difficult to check whether it is admissible. The only safe way to guarantee admissibility lies in the construction of generalized Bayes estimators, i.e., Bayes estimators with respect to generalized prior distributions or limit points of sequences of proper Bayes estimators, respectively. Brown (1971) showed that the generalized Bayes property is necessary for an estimator to be admissible.

By contrast, (biased) linear estimators have the advantage that, besides their computational simplicity and easiness of risk calculations, there exist necessary and sufficient criteria for proving the admissibility of such estimators. Rao (1976 b) showed that in case of an unconstrained parameter set $\Theta = \mathbb{R}^r$ the class of admissible linear estimators consists of all estimators which are proper Bayes or limit points of sequences of proper Bayes rules, a rigorous mathematical formulation of this result will be given in the sequel. Another advantage associated with the restriction to the class of (affine) linear estimators results from the fact that we only need (approximate) knowledge of the first two moments of the error and prior distributions. In case of admitting arbitrary (measurable) estimators we would require knowledge of the functional form of these distributions to derive (generalized) Bayes estimators, quite apart from the fact that these estimators, in general, cannot be given in a closed form analytical representation.

The class of all estimators which come out as limit points of sequences of proper Bayes linear estimators (so-called almost Bayes linear estimators) is very large, it includes for example the LSE, the principal component and Marquardt (fractional rank) estimators, and similar estimators which, in their very nature, are neither Bayesian nor do they incorporate prior knowledge in any form. These estimators rather express a state of prior ignorance or of so-called noninformative prior knowledge, respectively.

The situation is quite different if we are faced with a restricted compact parameter set $\Theta \subset \mathbb{R}^r$. Then the above mentioned estimators are no longer admissible, not even in the class of (affine) linear estimators. It will be seen in Chapter 16 that in case of a compact Θ the class of admissible (affine) linear estimators coincides with the class of proper Bayes (affine) linear estimators and every limit point of a sequence of such estimators again yields a proper Bayes (affine) linear estimator.

7.2. Admissible linear estimation with unconstrained Θ

We first present the basic results of RAO (1976 b) on the admissibility of linear estimators in case that $\Theta = \mathbb{R}^r$. For the sake of simplicity, and in view of a unified representation of these results and the corresponding results in case of a compact set Θ, as will be considered in Chapter 16, we assume the error covariance matrix

$$\mathbf{Cov}\, \tilde{e} = \Sigma \in \mathcal{M}_n^>$$

to be known. Further, as before, we assume quadratic loss $\|\vartheta - \hat{\vartheta}(y)\|_U^2$, $U \in \mathcal{M}_r^{\geqq}$, for estimating ϑ, independent of the underlying error distribution from

$$\mathcal{P}_{\tilde{e}} = \{P_{\tilde{e}}^{(w)}\}_{w \in W} \subseteqq \{P_{\tilde{e}} : \mathbf{E}\, \tilde{e} = \mathbf{0}\,, \mathbf{Cov}\, \tilde{e} = \Sigma\}\,. \tag{7.2}$$

Since the risk of an affine linear estimator $\hat{\vartheta}(\tilde{y}) = Z\tilde{y} + z$, $Z \in \mathcal{M}_{r \times n}$, $z \in \mathbb{R}^r$, does not depend on $w \in W$, we will write shortly $R(\vartheta; \hat{\vartheta})$ instead of $R(\vartheta, w; \hat{\vartheta})$.

Definition 7.1. Let D a given class of estimators for ϑ, $U \in \mathcal{M}_r^{\geqq}$ and $\Theta \subseteqq \mathbb{R}^r$.

(i) The estimator $\hat{\vartheta}_1 \in D$ is said to be a (U, Θ)*-improvement* over $\hat{\vartheta}_2 \in D$ if it holds $R(\vartheta; \hat{\vartheta}_1) \leqq R(\vartheta; \hat{\vartheta}_2)$ for all $\vartheta \in \Theta$ and $R(\vartheta_0; \hat{\vartheta}_1) < R(\vartheta_0; \hat{\vartheta}_2)$ for at least one parameter $\vartheta_0 \in \Theta$.

(ii) The estimator $\hat{\vartheta} \in D$ is said to be (U, Θ)*-admissible* in D if there exists no (U, Θ)-improvement over $\hat{\vartheta}$ within D.

Lemma 7.1. *Assume $\hat{\vartheta} \in D$ to be (U_0, Θ)-admissible in D for some given positive definite matrix U_0. Then $\hat{\vartheta}$ is (U, Θ)-admissible in D for every weight matrix $U \in \mathcal{M}_r^{\geqq}$.*

(For a proof see RAO (1976b).)

In view of a desirable robustness against a misspecification of the weight matrix occuring in our quadratic loss function we are interested in checking a possible improvement and admissibility, respectively, of a given estimator irrespective of the underlying weight matrix U.

Definition 7.2. (i) The estimator $\hat{\vartheta}_1 \in D$ is called a *universal Θ-improvement* over $\hat{\vartheta}_2 \in D$ if $\hat{\vartheta}_1$ is a $(\boldsymbol{U}, \Theta)$-improvement over $\hat{\vartheta}_2$ for any $\boldsymbol{U} \in \mathcal{M}_r^{\geqq}$, $\boldsymbol{U} \neq \boldsymbol{0}_{r,r}$.

(ii) The estimator $\hat{\vartheta} \in D$ is said to be *universally Θ-admissible* in D if it is $(\boldsymbol{U}, \Theta)$-admissible in D for any $\boldsymbol{U} \in \mathcal{M}_r^{\geqq}$.

Theorem 7.1. (i) *The estimator $\hat{\vartheta}_1 \in D$ is a universal Θ-improvement over $\hat{\vartheta}_2 \in D$ if and only if $\hat{\vartheta}_1$ is a $(\boldsymbol{c}\boldsymbol{c}^\top, \Theta)$-improvement over $\hat{\vartheta}_2$ for any $\boldsymbol{c} \in \mathbb{R}^r$, $\boldsymbol{c} \neq \boldsymbol{0}$.*

(ii) *The estimator $\hat{\vartheta} \in D$ is universally Θ-admissible in D if and only if it is $(\boldsymbol{I}_r, \Theta)$-admissible in D.*

Remark upon the proof. Assertion (i) was proved by THEOBALD (1974), the second assertion follows immediately from Lemma 7.1.

In the following let be satisfied

Assumption 7.1. rank $(\boldsymbol{F}^\top \boldsymbol{\Sigma}^{-1} \boldsymbol{F}) = r$.

Further, denote $\boldsymbol{V} = (\boldsymbol{F}^\top \boldsymbol{\Sigma}^{-1} \boldsymbol{F})^{-1}$ and

$$\boldsymbol{A}_{[0,1]} = \{\boldsymbol{A} \in \mathcal{M}_r^{\geqq} : \boldsymbol{0}_{r,r} \leqq \boldsymbol{A} \leqq \boldsymbol{I}_r\} \tag{7.3}$$

the set of all positive semidefinite matrices whose eigenvalues lie in the closed interval [0, 1].

We first investigate admissibility within the class $D_1 = \{\hat{\vartheta}(\tilde{\boldsymbol{y}}) = \boldsymbol{Z}\tilde{\boldsymbol{y}}, \boldsymbol{Z} \in \mathrm{M}_{r \times n}\}$ and then proceed to the more general class D_a of affine-linear estimators. RAO gave the following very simple algebraic characterization of admissible linear estimators, which forms an analogue to COHEN's (1966) characterization of admissibility within the class of all (measurable) estimators under the additional assumption of normally distributed errors.

Theorem 7.2. (i) *The estimator $\hat{\vartheta} = \boldsymbol{Z}\tilde{\boldsymbol{y}} \in D_1$ is universally $\mathbb{R}^r$-admissible in D_1 if and only if there exists an $\boldsymbol{A} \in \mathcal{A}_{[0,1]}$ such that*

$$\boldsymbol{Z} = \boldsymbol{V}^{1/2} \boldsymbol{A} \boldsymbol{V}^{1/2} \boldsymbol{F}^\top \boldsymbol{\Sigma}^{-1}, \tag{7.4}$$

where $\boldsymbol{V}^{1/2}$ means the symmetric square root of $\boldsymbol{V}$.

(ii) *Under the additional assumption that $\tilde{\boldsymbol{y}} \sim \mathrm{N}(\boldsymbol{F}\vartheta, \boldsymbol{\Sigma})$ the linear estimator $\hat{\vartheta} = \boldsymbol{Z}\tilde{\boldsymbol{y}}$ is universally $\mathbb{R}^r$-admissible in the class of all estimators iff it holds* (7.4) *with some $\boldsymbol{A} \in \mathcal{A}_{[0,1]}$ such that* rank $(\boldsymbol{I}_r - \boldsymbol{A}) \geqq r - 2$.

(See COHEN (1966), RAO (1976b), HOFFMANN (1980).)

From Theorem 7.2 (i) we see that any admissible linear estimator can be written in the form

$$\hat{\vartheta} = \boldsymbol{A}\hat{\vartheta}_{\Sigma} = \boldsymbol{A}\boldsymbol{V}\boldsymbol{F}^\top \boldsymbol{\Sigma}^{-1} \tilde{\boldsymbol{y}}, \quad \boldsymbol{A} \in \mathcal{A}_{[0,1]}, \tag{7.5}$$

where $\hat{\vartheta}_{\Sigma}$ stands for the generalized LSE.

Now, under the Assumption 7.1, any Bayes linear estimator has a representation

$$\hat{\vartheta}_P = M_P(M_P + V)^{-1}\hat{\vartheta}_\Sigma, \qquad M_P \in \mathcal{M} \tag{7.6}$$

(cp. Section 5.3), where

$$\mathcal{M} = \left\{ M_P \in \mathcal{M}_r^{\geqq} : M_P = \int_\Theta \vartheta\vartheta^\top P(\mathrm{d}\vartheta),\ P \in \Theta^* \right\} \tag{7.7}$$

denotes the set of all second order moment matrices generated by the class Θ^* of all prior distributions over Θ. Introducing the matrices

$$A_P = I_r - V^{1/2}(V + M_P)^{-1}V^{1/2},\ P \in \Theta^* \tag{7.8}$$

for which we have $0_{r,r} \leqq A_P < I_r$ and thus $A_P \in \mathcal{A}_{[0,1]}$, the Bayes linear estimator given by (7.6) can be rewritten as

$$\hat{\vartheta}_P = V^{1/2}A_P V^{-1/2}\hat{\vartheta}_\Sigma = V^{1/2}A_P V^{1/2}F^\top\Sigma^{-1}\tilde{y}. \tag{7.9}$$

Hence, according to Theorem 7.2 (i), in case of $\Theta = \mathbb{R}^r$ any $\hat{\vartheta}_P$ with $P \in \Theta^*$ is universally Θ-admissible in D_1. Obviously, for $\Theta = \mathbb{R}^r$ the set $\mathcal{M}$ defined by (7.7) coincides with the set of all positive semidefinite matrices $\mathcal{M}_r^{\geqq}$. Accordingly, for $\Theta = \mathbb{R}^r$ the set $\{A_P : P \in \Theta^*\}$ consists of all positive semidefinite matrices whose eigenvalues lie in the semiopen intervall [0, 1), thereby $A_P = 0_{r,r}$ corresponds to $M_P = 0_{r,r}$.

Corollary 7.1. *Let $\Theta = \mathbb{R}^r$. Then the class $D_1^B = \{\hat{\vartheta}_P : P \in \Theta^*\}$ of Bayes linear estimators can be represented as*

$$D_1^B = \{\hat{\vartheta} = V^{1/2}AV^{-1/2}\hat{\vartheta}_\Sigma : A \in \mathcal{A}_{[0,1)}\}, \tag{7.10}$$

where $\mathcal{A}_{[0,1)} = \{A \in \mathcal{M}_r^{\geqq} : 0 \leqq \lambda_{\max}(A) < 1\}$, and D_1^B is a proper subset of the class

$$D_1^{ad} = \{\hat{\vartheta} = V^{1/2}AV^{-1/2}\hat{\vartheta}_\Sigma : A \in \mathcal{A}_{[0,1]}\} \tag{7.11}$$

of universally $\mathbb{R}^r$-admissible linear estimators.

Now, if we choose a sequence of prior distributions $\{P_i\}_{i\in\mathbb{N}} \subset \Theta^*$ for which

$$\lim_{i\to\infty} \lambda_{\max}(M_{P_i}) = \infty,$$

then the corresponding sequence $\{A_{P_i}\}_{i\in\mathbb{N}}$ satisfies

$$\lim_{i\to\infty} \lambda_{\max}(A_{P_i}) = 1. \tag{7.12}$$

In this sense, any matrix $A \in \mathcal{A}_{[0,1]}$ represents a limit point of the set $\{A_P : P \in \Theta^*\} = \mathcal{A}_{[0,1)}$, however, there exists no finite M_P such that $\hat{\vartheta} = M_P(M_P + V)^{-1}\hat{\vartheta}_\Sigma = V^{1/2}AV^{-1/2}\hat{\vartheta}_\Sigma$ would hold for some $A \in \mathcal{M}_r^{\geqq}$ with $\lambda_{\max}(A) = 1$. Such estimators can only be obtained as limit points of a corresponding sequence of Bayes linear estimators $\{\hat{\vartheta}_{P_i}\}_{i\in\mathbb{N}}$ (in the sense of convergence defined by (7.12)). These estimators correspond to situations in which certain linear combinations of $\hat{\vartheta}$ have infinite variance, i.e., they are generalized Bayes estimators. Accordingly, RAO characterized the class of universally $\mathbb{R}^r$-admissible estimators D_l as the union of Bayes linear and generalized Bayes linear estimators. We wish to make this characterization more precise. At first, it is easily seen that there exists a one-to-one correspondence between the admissible linear estimator (7.5) and the underlying matrix A i.e., the mapping

$$\varphi | D_l^{\mathrm{ad}} \to \mathscr{A}_{[0,1]} : \varphi(\hat{\vartheta}) = A \tag{7.13}$$

yields a bijection from the class (7.11) of universally $\mathbb{R}^r$-admissible linear estimators onto the set $\mathscr{A}_{[0,1]}$. This suggests the introduction of a metric m in D_l^{ad} according to

$$m(\hat{\vartheta}_1, \hat{\vartheta}_2) = \|\varphi(\hat{\vartheta}_1) - \varphi(\hat{\vartheta}_2)\| = \|A_1 - A_2\|,$$

where

$$\hat{\vartheta}_i = V^{1/2} A_i V^{-1/2} \hat{\vartheta}_\Sigma, \quad A_i \in \mathscr{A}_{[0,1]}, \quad i = 1, 2. \tag{7.14}$$

Here $\|\cdot\|$ denotes the usual Euclidean norm, i.e., $\|M\| = (\operatorname{tr} M^2)^{1/2}$ for $M \in \mathscr{M}_r^{\geqq}$.

Definition 7.3. Let D a given class of estimators for ϑ. The estimator $\hat{\vartheta} \in D$ is said to be *almost Bayesian* in D if there exists a sequence $\{P_i\}_{i \in \mathbb{N}} \subset \Theta^*$ of prior distributions for $\hat{\vartheta}$ such that for the corresponding sequence $\{\hat{\vartheta}_i\}_{i \in \mathbb{N}}$ of Bayes estimators it holds

$$\hat{\vartheta}_i \xrightarrow[P_{\bar{y}|\vartheta} - \text{a.e.}]{} \hat{\vartheta} \quad \text{for all} \quad \vartheta \in \Theta.$$

Remark 7.1. Under relatively mild assumptions this convergence implies the convergence of the corresponding Bayes risks, which refers to the notion of an almost Bayes estimator (see, e.g. BANDEMER et al. (1977), Lemma 4.4.1, and PILZ (1983), pp. 81). In case that $D = D_a(D_l)$ both notions of convergence coincide.

Lemma 7.2. *Let* $\Theta = \mathbb{R}^r$. *The universally* $\mathbb{R}^r$*-admissible linear estimator* $\hat{\vartheta} = V^{1/2} A V^{-1/2} \hat{\vartheta}_\Sigma$, $A \in \mathscr{A}_{[0,1]}$, *is almost Bayesian in* D_l *iff there exists a sequence* $\{P_i\}_{i \in \mathbb{N}} \subset \Theta^*$ *such that*

$$\lim_{i \to \infty} m(\hat{\vartheta}, \hat{\vartheta}_{P_i}) = \lim_{i \to \infty} \|A_{P_i} - A\| = 0.$$

Proof. The Bayes linear estimators with respect to P_i, $i = 1, 2, \ldots$, are given by $\hat{\vartheta}_{P_i} = V^{1/2} A_{P_i} V^{-1/2} \hat{\vartheta}_\Sigma$ (cp. (7.9)). The sequence of these estimators converges $P_{\bar{y}|\vartheta}$ – a.e. to $\hat{\vartheta}$ iff $A_{P_i} \to A$ for $i \to \infty$ in the sense of the metric m from (7.14). □

With account to the representation (7.8) we get as an immediate consequence of Lemma 7.2 the following assertion.

Corollary 7.2. (i) *The universally* $\mathbb{R}^r$*-admissible estimator* $\hat{\vartheta} = V^{1/2} A V^{-1/2} \hat{\vartheta}_\Sigma$ *is* *almost Bayesian in D_1 iff there exists a sequence of priors $\{P_i\}_{i\in\mathbb{N}}$ for $\hat{\vartheta}$ such that*

$$\lim_{i\to\infty}(M_{P_i}+V)^{-1}=V^{-1/2}(I_r-A)V^{-1/2}.$$

(ii) *The almost Bayes estimators in D_1 are boundary points of the convex and compact set D_1^{ad} with respect to the metric m.*

Proof. We only prove the second assertion, since the former is obvious from (7.8).

Let $A\in\mathscr{A}_{[0,1]}$ with $\lambda_{\max}(A)=1$. Further, let $\{k_i\}_{i\in\mathbb{N}}$ a sequence of positive real numbers with $k_i\to\infty$ for $i\to\infty$ and $\{P_i\}_{i\in\mathbb{N}}$ a sequence of priors for $\hat{\vartheta}$ with

$$M_{P_i}=V^{1/2}\left\{\left(I_r-\frac{k_i}{k_{i+1}}A\right)^{-1}-I_r\right\}V^{1/2},$$

$i=1,2,\ldots$. Then we have $A_P=k_i(k_i+1)^{-1}A$ and from $0_{r,r}\leqq A\leqq I_r$ it follows that

$$\{A_{P_i}\}_{i\in\mathbb{N}}\subset\mathscr{A}_{[0,1)}.$$

For the corresponding sequence of Bayes linear estimators we obtain

$$\hat{\vartheta}_{P_i}=V^{1/2}A_{P_i}V^{-1/2}\hat{\vartheta}_\Sigma\xrightarrow[P_{\bar{y}|\vartheta}-\text{a.e.}]{}\hat{\vartheta}=V^{1/2}AV^{-1/2}\hat{\vartheta}_\Sigma,\qquad\forall\vartheta\in\mathbb{R}^r$$

observing that $\|A_{P_i}-A\|\to 0$ for $i\to\infty$. Hence, $\hat{\vartheta}$ is almost Bayesian in D_1 and it holds $\hat{\vartheta}\in D_1^{ad}$. However, $\hat{\vartheta}\notin D_1^{B}$ since $A\notin\mathscr{A}_{[0,1)}$. Finally, the convexity and compactness of D_1^{ad} is clear from the fact that $\mathscr{A}_{[0,1]}$ is convex and compact (with respect to the metric induced by $\|A\|=(\operatorname{tr}A^2)^{1/2}$). □

Summarizing, we get the following result.

Theorem 7.3. (i) *The class of universally $\mathbb{R}^r$-admissible estimators in D_1 is the closure of the set of (proper) Bayes linear estimators with respect to the metric m: $D_1^{ad}=\operatorname{clo}(D_1^{B})$.*

(ii) *The estimator $\hat{\vartheta}\in D_1$ is universally $\mathbb{R}^r$-admissible in D_1 if and only if it is Bayesian or almost Bayesian in D_1.*

Now, let us briefly consider some well-known estimators which do not make use of actual prior knowledge, but prove to be admissible linear estimators in case that $\Theta=\mathbb{R}^r$. According to Theorem 7.2 (i), in order that an estimator be admissible in D_1 it is necessary and sufficient to have a representation $\hat{\vartheta}=V^{1/2}AV^{1/2}F^\top\Sigma^{-1}\bar{y}$ with some $A\in\mathscr{A}_{[0,1]}$. Simultaneously, on the basis of Theorem 7.2 (ii), we investigate admissibility among all estimators in case of normally distributed observations.

(a) Generalized least squares estimator

The generalized LSE $\hat{\vartheta}_\Sigma=VF^\top\Sigma^{-1}\bar{y}$ has the required form with $A=I_r\in\mathscr{A}_{[0,1]}$. Hence, $\hat{\vartheta}_\Sigma\in D_1^{ad}$. However, since rank $(I_r-A)=0$, the $\mathbb{R}^r$-admissibility of $\hat{\vartheta}_\Sigma$ among all estimators is only guaranteed as long as $r\leqq 2$.

The existence of nonlinear improvements over $\hat{\vartheta}_\Sigma$ in case that the dimensionality of Θ is greater than 2 was already proved by JAMES/STEIN (1961).

(b) Principal component and Marquardt estimators
Let $d_1 \geqq d_2 \geqq \ldots \geqq d_r > 0$ and $s_1, \ldots, s_k$ the eigenvalues and eigenvectors, respectively, of $V^{-1} = F^\top \Sigma^{-1} F$. Further, let m an integer with $0 \leqq m < r$ and $0 \leqq c < 1$. The so-called Marquardt (fractional rank) estimator

$$\hat{\vartheta}_{m,c} = \left(\sum_{i=1}^{m} d_i^{-1} s_i s_i^\top + c d_{m+1}^{-1} s_{m+1} s_{m+1}^\top \right) F^\top \Sigma^{-1} \tilde{y} \tag{7.15}$$

represents a smoothing of the principal component estimator $\hat{\vartheta}_{m,0}$, which is known to be the generalized LSE under the restriction $\vartheta \in \Theta = \mathscr{L}(\{s_1, \ldots, s_m\})$, the linear space generated by $s_1, \ldots, s_m$. The Marquardt estimator can be written in the form $\hat{\vartheta}_{m,c} = V^{1/2} A_{m,c} V^{-1/2} \hat{\vartheta}_\Sigma$ with

$$A_{m,c} = V^{-1/2} \left(\sum_{i=1}^{m} d_i^{-1} s_i s_i^\top + c d_{m+1}^{-1} s_{m+1} s_{m+1}^\top \right) V^{-1/2}.$$

Obviously, $A_{m,c} \in \mathscr{A}_{[0,1]}$, and thus $\hat{\vartheta}_{m,c} \in D_1^{\text{ad}}$. Moreover, since $\lambda_{\max}(A_{0,c}) \leqq \operatorname{tr} A_{0,c} = c$, we have $A_{0,c} \in \mathscr{A}_{[0,1)}$ so that for $m = 0$ the Marquardt estimator turns out to be a (proper) Bayes linear estimator. For $m > 1$, however, we have rank $(I_r - A_{m,c}) \leqq r - m < r$ and thus $\lambda_{\max}(A_{m,c}) = 1$; i.e., in this case $\hat{\vartheta}_{m,c}$ does not belong to the class of proper Bayes linear estimators. Finally, admissibility of $\hat{\vartheta}_{m,c}$ is guaranteed at most for $0 \leqq m \leqq 2$.

(c) Shrunken estimators
The so-called shrunken estimators

$$\hat{\vartheta}_g = g \hat{\vartheta}_\Sigma, \qquad g \in (0, 1) \tag{7.16}$$

originally introduced by MAYER/WILLKE (1973) and DEMPSTER (1973) dominate the generalized LSE $\hat{\vartheta}_\Sigma$ over relatively large subregions of $\Theta = \mathbb{R}^r$ (see LOWERRE (1974)). They can be written in the form (7.4) with $A = g \cdot I_r \in \mathscr{A}_{[0,1)}$, hence they belong to the class of proper Bayes linear estimators and are universally $\mathbb{R}^r$-admissible in D_1. Under the additional assumption of normally distributed observations, shrunken estimators are even admissible under all estimators, since rank $(I_r - A) = r$ for all $g \in (0, 1)$.

(d) Ridge estimators
Let S denote the matrix of (orthonormal) eigenvectors of $V^{-1} = F^\top \Sigma^{-1} F$ and $H = \operatorname{diag}(h_1, \ldots, h_r)$ a diagonal matrix with elements $h_i \geqq 0$, $i = 1, \ldots, r$. The so-called *ridge estimators*

$$\hat{\vartheta}_R = (F^\top \Sigma^{-1} F + SHS^\top)^{-1\,\top} \Sigma^{-1} \tilde{y} \tag{7.17}$$

are Bayes linear estimators with respect to any prior distribution for $\tilde{\vartheta}$ with $M_P = SH^{-1}S^\top$, and thus also universally $\mathbb{R}^r$-admissible in D_1, provided that $h_1, \ldots, h_r$ are all positive numbers. In particular, this holds for the ordinary ridge estimators $\hat{\vartheta}_R$ having $H = hI_r$, $h > 0$, which were originally introduced by HOERL/KENNARD (1970) in order to stabilize the (generalized) LSE $\hat{\vartheta}_\Sigma = (F^\top \Sigma^{-1} F)^{-1} F^\top \Sigma^{-1} \tilde{y}$ in case of an ill-conditioned matrix $F^\top \Sigma^{-1} F$. The main justification for ridge regression was but their theorem that a positive $h > 0$ (depending, however, on ϑ) exists, for which $\mathrm{MSE}(\hat{\vartheta}_R) < \mathrm{MSE}(\hat{\vartheta}_\Sigma)$, see also GOLDSTEIN/SMITH (1974), LOWERRE (1974) and THEOBALD (1974). More generally, we see from Theorem 5.5 that with $\hat{\vartheta}_R$ we can achieve a risk reduction with respect to arbitrary quadratic risk.

Under the Assumption 7.1 the ridge estimator can be written as

$$\hat{\vartheta}_R = V^{1/2} A_R V^{-1/2} \hat{\vartheta}_\Sigma \quad \text{with} \quad A_R^{-1} = I_r + V^{1/2} S H S^\top V^{1/2}$$

so that $A_R \in \mathscr{A}_{[0,1]}$ and thus $\hat{\vartheta}_R \in D_1^{\mathrm{ad}}$ even if not all diagonal elements of H are positive. The admissibility under all estimators in case of $\tilde{y} \sim \mathrm{N}(F\vartheta, \Sigma)$ is but only guaranteed if at least $r - 2$ out of the values $h_1, \ldots, h_r$ are positive, since

$$I_r - A_R = V^{1/2} S (I_r + H S^\top V S)^{-1} H S^\top V^{1/2}$$

(cp. A.22) and therefore rank $(I_r - A_R) \leqq \operatorname{rank} H$.

A further Bayesian interpretation of ridge and shrunken estimators as well as an interpretation in terms of non-Bayesian prior knowledge will be given in the next section.

Finally, we indicate corresponding characterizations of admissibility within the class of affine estimators $D_a = \{\hat{\vartheta} = Z\tilde{y} + z : Z \in \mathscr{M}_{r \times n},\ z \in \mathbb{R}^r\}$, analogously to the above considered situation with D_1. According to RAO (1976 b), the class of universally $\mathbb{R}^r$-admissible estimators in D_a is given by

$$D_a^{\mathrm{ad}} = \{\hat{\vartheta} = V^{1/2} A V^{-1/2} (\hat{\vartheta}_\Sigma - a) + a : A \in \mathscr{A}_{[0,1]},\ a \in \mathbb{R}^r\}. \quad (7.18)$$

In case of normally distributed observations an estimator from D_a^{ad} is admissible under all estimators if, additionally, rank $(I_r - A) \geqq r - 2$.

We identify every estimator from D_a^{ad} with the augmented matrix $\bar{A} := [A \,\vdots\, a] \in \mathscr{M}_{r \times (r+1)}$ and introduce a metric $\bar{m}$ in this class according to

$$\begin{aligned} \bar{m}(\hat{\vartheta}_1, \hat{\vartheta}_2) &:= (\|A_1 - A_2\|^2 + (a_1 - a_2)^\top (a_1 - a_2))^{1/2}, \\ \hat{\vartheta}_i &= V^{1/2} A_i V^{-1/2} (\hat{\vartheta}_\Sigma - a_i) + a_i \in D_a^{\mathrm{ad}}, \quad i = 1, 2. \end{aligned} \quad (7.19)$$

We remark that the convex set D_a is both open and closed, but not compact with respect to $\bar{m}$.

Now, for $\Theta = \mathbb{R}^r$ let $P \in \Theta^*$ a prior distribution with $\mathbf{E}\tilde{\vartheta} = \mu_P$ and $\mathbf{Cov}\,\tilde{\vartheta} = C_P$. Then the Bayes affine estimator, which is uniquely determined in case that U is positive definite, takes the form

$$\hat{\vartheta}_P^{\mathrm{a}} = C_P (C_P + V)^{-1} (\hat{\vartheta}_\Sigma - \mu_P) + \mu_P \quad (7.20)$$

(cp. Section 5.3). Defining the matrices

$$\bar{A}_P = I_r - V^{1/2} (C_P + V)^{-1} V^{1/2}, \qquad P \in \Theta^* \quad (7.21)$$

the Bayes affine estimator from (7.20) can be written equivalently as

$$\hat{\boldsymbol{\vartheta}}_P^{\mathrm{a}} = \boldsymbol{V}^{1/2}\bar{\boldsymbol{A}}_P\boldsymbol{V}^{-1/2}(\hat{\boldsymbol{\vartheta}}_\Sigma - \boldsymbol{\mu}_P) + \boldsymbol{\mu}_P. \tag{7.22}$$

For $\Theta = \mathbb{R}^r$ we have $\{\boldsymbol{C}_P : P \in \Theta^*\} = \mathcal{M}_r^{\geqq}$ and thus $\{\bar{\boldsymbol{A}}_P : P \in \Theta^*\} = \mathcal{A}_{[0,1)}$, which leads to the following algebraic characterization of the class $D_{\mathrm{a}}^{\mathrm{B}} = \{\hat{\boldsymbol{\vartheta}}_P^{\mathrm{a}} : P \in \Theta^*\}$ of Bayes affine estimators:

$$D_{\mathrm{a}}^{\mathrm{B}} = \{\hat{\boldsymbol{\vartheta}} = \boldsymbol{V}^{1/2}\boldsymbol{A}\boldsymbol{V}^{-1/2}(\hat{\boldsymbol{\vartheta}}_\Sigma - \boldsymbol{a}) + \boldsymbol{a} : \boldsymbol{A} \in \mathcal{A}_{[0,1)},\ \boldsymbol{a} \in \mathbb{R}^r\}, \tag{7.23}$$

$$D_{\mathrm{a}}^{\mathrm{B}} \subset D_{\mathrm{a}}^{\mathrm{ad}}.$$

Every estimator $\hat{\boldsymbol{\vartheta}} = \boldsymbol{V}^{1/2}\boldsymbol{A}\boldsymbol{V}^{-1/2}(\hat{\boldsymbol{\vartheta}}_\Sigma - \boldsymbol{a}) + \boldsymbol{a} \in D_{\mathrm{a}}^{\mathrm{ad}}$ which is not contained in $D_{\mathrm{a}}^{\mathrm{B}}$ represents an almost Bayes estimator in D_{a} and can be obtained as a limit point of the sequence of Bayes affine estimators with respect to $\{P_i\}_{i \in \mathbb{N}}$ such that

$$\boldsymbol{\mu}_{P_i} = \boldsymbol{a}, \quad \bar{\boldsymbol{A}}_{P_i} = k_i(k_i + 1)^{-1}\boldsymbol{A}, \quad k_i \in (0, \infty), \quad \lim_{i \to \infty} k_i = \infty,$$

i.e.,

$$\lim_{i \to \infty} \bar{m}(\hat{\boldsymbol{\vartheta}}_{P_i}^{\mathrm{a}}, \hat{\boldsymbol{\vartheta}}) = \lim_{i \to \infty} \|\bar{\boldsymbol{A}}_{P_i} - \boldsymbol{A}\| = 0.$$

Summarizing, we can state the following result.

Theorem 7.4. (i) *The set of universally $\mathbb{R}^r$-admissible estimators in D_{a} is convex and turns out to be the closure of $D_{\mathrm{a}}^{\mathrm{B}}$ with respect to the metric $\bar{m}$: $D_{\mathrm{a}}^{\mathrm{ad}} = \mathrm{clo}\,(D_{\mathrm{a}}^{\mathrm{B}})$.*

(ii) *The estimator $\hat{\boldsymbol{\vartheta}} \in D_{\mathrm{a}}$ is universally $\mathbb{R}^r$-admissible in D_{a} if it is Bayesian or almost Bayesian in D_{a}. The almost Bayes estimators in D_{a} are boundary points of $D_{\mathrm{a}}^{\mathrm{B}}$ (and thus of $D_{\mathrm{a}}^{\mathrm{ad}}$) with respect to the metric $\bar{m}$.*

7.3. Bayesian interpretation of estimators using non-Bayesian prior knowledge

In this section we shall briefly reconsider, within a Bayesian framework, some well-known estimators using prior knowledge which does not come available in form of some prior distribution for the model parameters. As a rule, the corresponding estimation procedures represent some modification of the method of least squares, where the prior knowledge enters the optimization in form of some restrictions on the parameter space.

In particular, we consider estimators having the same structure as the Bayes affine estimator $\hat{\boldsymbol{\vartheta}}_{\mathrm{B}}$ from (5.12): estimators using linear or compact parameter constraints and prior estimates, respectively. This way we obtain, on the one side, a Bayesian interpretation of these estimators and, on the other side, further optimality properties for the Bayes estimator $\hat{\boldsymbol{\vartheta}}_{\mathrm{B}}$.

For a detailed treatment of the estimation problem involving different kind of non-Bayesian prior knowledge the reader is referred to Bibby/Toutenburg (1977) and Toutenburg (1982).

7.3.1. Linear restrictions

As in the preceding section, we assume a linear regression model

$$\begin{aligned} &\tilde{y} = F\vartheta + \tilde{e}, \\ &\mathsf{E}\,\tilde{e} = 0, \qquad \mathsf{Cov}\,\tilde{e} = \Sigma \in \mathcal{M}_n^{>} \end{aligned} \tag{7.24}$$

with Assumption 7.1 holding ($F^\top \Sigma^{-1} F$ has full rank).

First, let prior knowledge be given, about the regression parameter which can be represented in form of a linear constraint

$$K\vartheta = q, \quad \text{where} \quad K \in \mathcal{M}_{s\times r}, \quad q \in \mathbb{R}^s, \quad \text{rank} \quad K = s < r. \tag{7.25}$$

In this case the parameter space is restricted to an $(r-s)$-dimensional subspace

$$\Theta = \{t \in \mathbb{R}^r : Kt = q\} \subset \mathbb{R}^r. \tag{7.26}$$

Definition 7.4. The estimator $\hat{\vartheta}_{K,q}$ is called a *linearly constrained* least squares estimator for ϑ if it holds for all observations y:

$$\|y - F\hat{\vartheta}_{K,q}(y)\|^2_{\Sigma^{-1}} = \inf_{\vartheta \in \Theta} \|y - F\vartheta\|^2_{\Sigma^{-1}}.$$

Lemma 7.3. *Let* $\hat{\vartheta}_\Sigma(\tilde{y}) = (F^\top \Sigma^{-1} F)^{-1} F^\top \Sigma^{-1} \tilde{y}$ *be the unconstrained generalized LSE for* ϑ *and define*

$$K^- = (F^\top \Sigma^{-1} F)^{-1} K^\top [K (F^\top \Sigma^{-1} F)^{-1} K^\top]^{-1}.$$

Then the linearly constrained LSE for ϑ *is given by*

$$\hat{\vartheta}_{K,q}(\tilde{y}) = \hat{\vartheta}_\Sigma(\tilde{y}) + K^-(q - K\hat{\vartheta}_\Sigma(\tilde{y}))$$

(see, e.g. GOLDBERGER (1964)).

Thus, the term $K^-(q - K\hat{\vartheta}_\Sigma(\tilde{y}))$ can be interpreted as a correction term to be added to the unconstrained generalized LSE if we have prior knowledge that ϑ is constrained to lie in the subspace $\Theta \subset \mathbb{R}^r$. Clearly, the constrained estimate $\hat{\vartheta}_{K,q}(y)$ coincides with the unconstrained estimate $\hat{\vartheta}_\Sigma(y)$ if the latter happens to satisfy the constraint (7.25), i.e., if $\hat{\vartheta}_\Sigma(y) \in \Theta$.

The estimator $\hat{\vartheta}_{K,q}$ has covariance matrix

$$\begin{aligned} \mathsf{Cov}\,\hat{\vartheta}_{K,q} &= (I_r - K^-K)(\mathsf{Cov}\,\hat{\vartheta}_\Sigma)(I_r - K^-K)^\top \\ &= \mathsf{Cov}\,\hat{\vartheta}_\Sigma - (\mathsf{Cov}\,\hat{\vartheta}_\Sigma)K^\top[K(\mathsf{Cov}\,\hat{\vartheta}_\Sigma)K^\top]^{-1}K(\mathsf{Cov}\,\hat{\vartheta}_\Sigma) \end{aligned} \tag{7.27}$$

so that $\mathsf{Cov}\,\hat{\vartheta}_{K,q} \leqq \mathsf{Cov}\,\hat{\vartheta}_\Sigma$, since the matrix subtracted from $\mathsf{Cov}\,\hat{\vartheta}_\Sigma$ is positive semidefinite. Therefore, the prior knowledge $\vartheta \in \Theta$ leads to an improvement in efficiency of estimation over the generalized LSE which corresponds to the noninformative case $\vartheta \in \mathbb{R}^r$.

The estimator $\hat{\vartheta}_{K,q}$ can be given a Bayesian interpretation in terms of an almost Bayes estimator.

Corollary 7.3. *Assume quadratic loss* $\|\vartheta - \hat{\vartheta}(y)\|^2_U$, $U \in \mathcal{M}_r^{\geqq}$.

(i) *The linearly constrained LSE* $\hat{\boldsymbol{\vartheta}}_{K,q}$ *is almost Bayesian and universally* $\mathbb{R}^r$-*admissible in* D_a *for any* $\boldsymbol{K} \in \mathcal{M}_{s \times r}$ *and* $\boldsymbol{q} \in \mathbb{R}^s$.

(ii) *If, additionally,* $\tilde{\boldsymbol{y}} \sim \mathrm{N}(\boldsymbol{F\vartheta}, \boldsymbol{\Sigma})$ then $\hat{\boldsymbol{\vartheta}}_{K,q}$ *is almost Bayesian in the class of all estimators, the universal* $\mathbb{R}^r$-*admissibility in this class is but only guaranteed for* $s \in \{r-2, r-1\}$.

Proof. (i) Rewriting $\hat{\boldsymbol{\vartheta}}_{K,q}$ as

$$\hat{\boldsymbol{\vartheta}}_{K,q} = \boldsymbol{V}^{1/2}\boldsymbol{A}_K\boldsymbol{V}^{-1/2}(\hat{\boldsymbol{\vartheta}}_\Sigma - \boldsymbol{K}^-\boldsymbol{q}) + \boldsymbol{K}^-\boldsymbol{q},$$

where

$$\boldsymbol{A}_K = \boldsymbol{I}_r - \boldsymbol{V}^{1/2}\boldsymbol{K}^\top(\boldsymbol{KVK}^\top)^{-1}\boldsymbol{KV}^{1/2} \in \mathcal{A}_{[0,1]},$$

it follows immediately from (7.18) that $\hat{\boldsymbol{\vartheta}}_{K,q} \in D_a^{ad}$. Further, it can be shown that $\hat{\boldsymbol{\vartheta}}_{K,q}$ is a limit point of the sequence of Bayes affine estimators with respect to any sequence $\{P_i\}_{i \in \mathrm{N}}$ of priors for $\tilde{\boldsymbol{\vartheta}}$ having expectations $\boldsymbol{\mu}_i = \boldsymbol{K}^-\boldsymbol{q}$ and covariance matrices

$$\boldsymbol{C}_i = k_i[\boldsymbol{I}_r - (1 - k_i^{-2})\boldsymbol{K}^-\boldsymbol{K}]\boldsymbol{V}, \quad i = 1, 2, \ldots,$$

where $\{k_i\}_{i \in \mathrm{N}}$ is a sequence of positive real numbers tending to infinity (cp. Bandemer et al. (1977), Theorem 4.4.3).

(ii) The almost Bayes optimality of $\hat{\boldsymbol{\vartheta}}_{K,q}$ under all estimators is immediately clear from the preceding result when assuming a sequence of normal priors $\{P_i = \mathrm{N}(\boldsymbol{\mu}_i, \boldsymbol{C}_i)\}_{i \in \mathrm{N}}$.

Finally, observing that rank $(\boldsymbol{I}_r - \boldsymbol{A}_K) \leq s$, the stated admissibility condition follows from Theorem 7.2 (ii). □

Clearly, $\hat{\boldsymbol{\vartheta}}_{K,q}$ is not a proper Bayes (affine) estimator, since $\lambda_{\max}(\boldsymbol{A}_K) = 1$. With the sequence of priors specified in the above proof we have

$$\mathbf{E}\,\boldsymbol{K}\tilde{\boldsymbol{\vartheta}} = \boldsymbol{KK}^-\boldsymbol{q} = \boldsymbol{q}$$

and

$$\mathbf{Cov}\,(\boldsymbol{K}\tilde{\boldsymbol{\vartheta}}) = \boldsymbol{KC}_i\boldsymbol{K}^\top = k_i^{-1}\boldsymbol{KVK}^\top,$$

the covariance matrix vanishes when i approaches infinity. Thus, the exakt constraint $\boldsymbol{K\vartheta} = \boldsymbol{q}$ can be viewed as the limiting case of a prior distribution on $\tilde{\boldsymbol{\vartheta}}$ that is concentrated along the manifold (7.26).

Furthermore, a posteriori we have

$$\begin{aligned}\lim_{i\to\infty} \mathbf{Cov}\,(\tilde{\boldsymbol{\vartheta}}|\boldsymbol{y}) &= \lim_{i\to\infty} (\boldsymbol{V} + \boldsymbol{C}_i^{-1})^{-1} \\ &= \lim_{i\to\infty}\left\{\frac{k_i}{k_i+1}\boldsymbol{I}_r - \frac{k_i-1}{k_i+1}\boldsymbol{K}^-\boldsymbol{K}\right\}\boldsymbol{V} \\ &= (\boldsymbol{I}_r - \boldsymbol{K}^-\boldsymbol{K})\boldsymbol{V} \\ &= \mathbf{Cov}\,\hat{\boldsymbol{\vartheta}}_{K,q}\end{aligned}$$

(see (7.27)). The posterior limiting moments of $\boldsymbol{K}\tilde{\boldsymbol{\vartheta}}$ are

$$\lim_{i\to\infty} \mathbf{E}\,(\boldsymbol{K}\tilde{\boldsymbol{\vartheta}}|\boldsymbol{y}) = \boldsymbol{K}\hat{\boldsymbol{\vartheta}}_{K,q}(\boldsymbol{y}) = \boldsymbol{q}$$

and

$$\lim_{i\to\infty} \mathbf{Cov}\,(K\tilde{\vartheta}\,|\,y) = K[I_r - K^- K]VK^\top = \mathbf{0}\,.$$

This means, in the limiting case the prior and posterior information become exact.

An interesting result on the connection between Bayes affine estimators and linearly constrained least squares estimators was given by LEAMER/CHAMBERLAIN (1976) in a pretesting context. They state that the Bayes estimator $\hat{\vartheta}_B^\Sigma(\tilde{y}) = (F^\top\Sigma^{-1}F + \Phi^{-1})^{-1}(F^\top\Sigma^{-1}\tilde{y} + \Phi^{-1}\mu)$ can be written as a weighted average of 2^r linearly constrained LSE's with constraints of the form $\vartheta_i = \mu_i$ for $i \in J$ where J indexes the 2^r distinct subsets of $\{1, \ldots, r\}$. This can be interpreted such that by selecting a prior distribution for Bayes estimation of ϑ we implicitly select a set of linear constraints and also procedures for imposing them.

7.3.2. Use of prior estimates

The linear constraint (7.25) just considered refers to deterministic prior knowledge. Now we consider a special kind of stochastic prior knowledge: We assume to have a priori an estimate for some of the regression coefficients. Such knowledge may arise, for example, from previous investigations.

To be specific, let $\tilde{\vartheta}_0$ a prior estimate (defined on the same probability space as the observation vector $\tilde{y}$) for the subvector $\vartheta_0 = (\vartheta_1, \ldots, \vartheta_s)^\top$ of $\vartheta = (\vartheta_1, \ldots, \vartheta_r)^\top$, $s < r$. This knowledge may be written in the form

$$\tilde{\vartheta}_0 = \vartheta_0 + \tilde{\delta}, \qquad \mathbf{E}\,\tilde{\delta} = \mathbf{0} \tag{7.28}$$

with some random vector $\tilde{\delta}$, where we have assumed that $\tilde{\vartheta}_0$ is an unbiased estimator. Further, let us assume that its covariance matrix exists:

$$\mathbf{Cov}\,\tilde{\delta} = \mathbf{E}\,\tilde{\delta}\tilde{\delta}^\top =: B_{\tilde{\delta}} \in \mathcal{M}_s^> \,. \tag{7.29}$$

This, together with (7.24), yields the combined model

$$\bar{y} = \bar{F}\vartheta + \bar{e}, \qquad \mathbf{E}\,\bar{e} = \mathbf{0}$$

$$\bar{y} = \begin{pmatrix} \tilde{y} \\ \tilde{\vartheta}_0 \end{pmatrix}, \quad \bar{F} = \begin{pmatrix} F \\ \hline S \end{pmatrix}, \quad S = (I_s \,\vdots\, \mathbf{0}_{s,\,r-s}), \quad \bar{e} = \begin{pmatrix} \tilde{e} \\ \tilde{\delta} \end{pmatrix}. \tag{7.30}$$

The combined model is called a *linear regression model with prior estimate* $\tilde{\vartheta}_0$ for ϑ_0.

Lemma 7.4. *Assume the prior estimate and the actual observations to be uncorrelated, i.e.,* $\mathbf{E}\,\tilde{e}\tilde{\vartheta}_0^\top = \mathbf{0}_{n,\,s}$. *Then*

$$\hat{\vartheta}(\tilde{y}, \tilde{\vartheta}_0) = (F^\top\Sigma^{-1}F + S^\top B_{\tilde{\delta}}^{-1}S)^{-1}(F^\top\Sigma^{-1}\tilde{y} + S^\top B_{\tilde{\delta}}^{-1}\tilde{\vartheta}_0)$$

is the generalized LSE for ϑ *in the model* (7.30) *with prior estimate* $\tilde{\vartheta}_0$ *for* ϑ_0 (see GOLDBERGER/THEIL (1961), THEIL (1963)).

In the terminology of these authors the model (7.30) is called a *mixed linear model* and $\hat{\vartheta}(\tilde{y}, \tilde{\vartheta}_0)$ from above is called a *mixed linear estimator.* For this estimator there hold analogous optimality properties as for the generalized LSE without prior estimates (see, e.g. THEIL (1963), BANDEMER et al. (1977)). Thereby it is clear that, in case of quadratic loss, the risk $\operatorname{tr} U(F^\top \Sigma^{-1} F + S^\top B_{\tilde{\delta}}^{-1} S)^{-1}$ of $\hat{\vartheta}(\tilde{y}, \tilde{\vartheta}_0)$ is smaller than the risk $\operatorname{tr} U(F^\top \Sigma^{-1} F)^{-1}$ of $\hat{\vartheta}_\Sigma(\tilde{y})$ since $S^\top B_{\tilde{\delta}}^{-1} S$ is positive semidefinite. But, without further knowledge about the covariance structure of observation, the numerical computation and the properties of $\hat{\vartheta}(\tilde{y}, \tilde{\vartheta}_0)$ depend on the exact knowledge of Σ. For the case that $\Sigma = \sigma^2 I_n$ with unknown variance $\sigma^2 > 0$, TOUTENBURG (1975) proposed to replace σ^2 by some positive value, say $\hat{\sigma}^2 > 0$, and gave conditions under which the resulting estimator

$$\hat{\vartheta}_{\hat{\sigma}}(\tilde{y}, \tilde{\vartheta}_0) = (F^\top F + \hat{\sigma}^2 S^\top B_{\tilde{\delta}}^{-1} S)^{-1} (F^\top y + \hat{\sigma}^2 S^\top B_{\tilde{\delta}}^{-1} \tilde{\vartheta}_0)$$

has smaller risk than the LSE. Furthermore he derived estimators which, under partial prior knowledge $0 < \sigma_1^2 < \sigma^2 < \sigma_2^2 < \infty$, minimize the loss in efficiency relative to the unknown optimal estimator $\hat{\vartheta}(\tilde{y}, \tilde{\vartheta}_0)$. Finally, we remark that THEIL's (1963) proposal of replacing σ^2 by the usual residual sum of squares estimator guarantees that, asymptotically, $\hat{\vartheta}_{\hat{\sigma}}(\tilde{y}, \tilde{\vartheta}_0)$ has the same risk as the optimal $\hat{\vartheta}(\tilde{y}, \tilde{\vartheta}_0)$, but may be inefficient in small sample situations.

Remark 7.2. The linearly constrained LSE $\hat{\vartheta}_{K,q}$ considered in the preceding section turns out as a limiting case of the above estimator $\hat{\vartheta}(\tilde{y}, \tilde{\vartheta}_0)$ when the constraints $K\vartheta = q$ are reinterpreted as "dummy observations" resulting from a ficticious set of data points taken according to a design matrix K. This would correspond to a combined model (7.30) with $\tilde{\vartheta}_0$ and S replaced by q and K, respectively. The generalized LSE in this combined model then takes the form

$$\begin{aligned}\hat{\vartheta}(\tilde{y}, q) &= (F^\top \Sigma^{-1} F + K^\top B_{\tilde{\delta}}^{-1} K)^{-1} (F^\top \Sigma^{-1} \tilde{y} + K^\top B_{\tilde{\delta}}^{-1} q) \\ &= \hat{\vartheta}_\Sigma(\tilde{y}) + (F^\top \Sigma^{-1} F)^{-1} K^\top [B_{\tilde{\delta}}^{-1} + K (F^\top \Sigma^{-1} F)^{-1} K^\top]^{-1} (q - K \hat{\vartheta}_\Sigma(\tilde{y})).\end{aligned}$$

Assuming that $B_{\tilde{\delta}}^{-1} = \operatorname{diag}(b_1, \ldots, b_s)$ and $b := \max(b_1, \ldots, b_s)$ approaches zero, this estimator coincides with $\hat{\vartheta}_{K,q}$ from Lemma 7.3. The limiting case $b = 0$ but corresponds to deterministic prior information $P(K\tilde{\vartheta} = q) = 1$ (cp. ALLEN (1974)).

As it was the case with the linearly constrained LSE, the mixed linear estimator $\hat{\vartheta}(\tilde{y}, \tilde{\vartheta}_0)$ can be characterized in terms of almost Bayes optimality.

Corollary 7.4. *Let $\Theta = \mathbb{R}^r$, L quadratic and $\hat{\vartheta}_0$ some realization of $\tilde{\vartheta}_0$. Then it holds*

(i) *The estimator* $\hat{\vartheta}(\tilde{y}, \hat{\vartheta}_0) = (F^\top \Sigma^{-1} F + S^\top B_{\tilde{\delta}}^{-1} S)^{-1} (F^\top \Sigma^{-1} \tilde{y} + S^\top B_{\tilde{\delta}}^{-1} \hat{\vartheta}_0)$ *is almost Bayesian and universally $\mathbb{R}^r$-admissible in D_a.*

(ii) *In the case that $\tilde{y} \sim \mathrm{N}(F\vartheta, \Sigma)$, the estimator $\hat{\vartheta}(\tilde{y}, \hat{\vartheta}_0)$ is almost Bayesian* *in the class D of all estimators and universally* $\mathbb{R}^r$*-admissible in D provided that* $s \geqq r - 2$.

This can be proved along the same lines as in the proof of Corollary 7.4 rewriting $\hat{\boldsymbol{\vartheta}}(\tilde{\boldsymbol{y}}, \hat{\boldsymbol{\vartheta}}_0)$ as

$$\hat{\boldsymbol{\vartheta}}(\tilde{\boldsymbol{y}}, \hat{\boldsymbol{\vartheta}}_0) = \boldsymbol{V}^{1/2}\boldsymbol{A}_{\tilde{\delta}}\boldsymbol{V}^{-1/2}(\hat{\boldsymbol{\vartheta}}_\Sigma - \hat{\boldsymbol{\vartheta}}_0) + \hat{\boldsymbol{\vartheta}}_0,$$

where

$$\boldsymbol{A}_{\tilde{\delta}}^{-1} = \boldsymbol{I}_r + \boldsymbol{V}^{1/2}\boldsymbol{S}^\top \boldsymbol{B}_{\tilde{\delta}}^{-1}\boldsymbol{S}\boldsymbol{V}^{1/2} \in \mathscr{A}_{[0,1]}$$

and observing Remark 7.2. Analogous assertions hold in case of homoscedastic observations of unknown variance, i.e., where we have $\mathscr{P}_{\tilde{e}} = \left\{P_{\tilde{e}} : \mathbf{E}\,\tilde{\boldsymbol{e}} = \boldsymbol{0}, \mathbf{Cov}\,\tilde{\boldsymbol{e}} = \frac{1}{w}\boldsymbol{I}_n\right\}_{w \in \mathbb{R}^+}$, replacing $\hat{\boldsymbol{\vartheta}}(\tilde{\boldsymbol{y}}, \hat{\boldsymbol{\vartheta}}_0)$ by

$$\hat{\boldsymbol{\vartheta}}_{\hat{\sigma}}(\tilde{\boldsymbol{y}}, \hat{\boldsymbol{\vartheta}}_0) = (\boldsymbol{F}^\top\boldsymbol{F} + \hat{\sigma}^2\boldsymbol{S}^\top\boldsymbol{B}_{\tilde{\delta}}^{-1}\boldsymbol{S})(\boldsymbol{F}^\top\tilde{\boldsymbol{y}} + \hat{\sigma}^2\boldsymbol{S}^\top\boldsymbol{B}_{\tilde{\delta}}^{-1}\hat{\boldsymbol{\vartheta}}_0)$$

(see BANDEMER et al. (1977), Theorem 4.4.9).

If we have a prior estimate $\tilde{\boldsymbol{\vartheta}}_0$ for the whole parameter vector $\boldsymbol{\vartheta}_0 = \boldsymbol{\vartheta}$, which implies that $\boldsymbol{S} = \boldsymbol{I}_r$ and $s = r$, then the above estimator is not only an almost Bayes but even a proper Bayes estimator with respect to $P_{\tilde{\vartheta}} = \mathrm{N}(\hat{\boldsymbol{\vartheta}}_0, \boldsymbol{B}_{\tilde{\delta}})$ or $P_{\tilde{\vartheta}, \tilde{w}} = \mathrm{NG}(\hat{\boldsymbol{\vartheta}}_0, \alpha, \hat{\sigma}^{-2}\boldsymbol{B}_{\tilde{\delta}}, \hat{\sigma}^2)$ with arbitrary $\alpha \in \mathbb{R}^+$ in case that $P_{\tilde{e}} \in \mathscr{P}_{\tilde{e}} = \left\{\mathrm{N}\left(\boldsymbol{0}, \frac{1}{w}\boldsymbol{I}_n\right)\right\}_{w \in \mathbb{R}^+}$, respectively.

Though we have a formal similarity between the mixed linear estimator and the linear Bayes estimator it is important to emphasize the differences between the above model and the Bayesian model. First, whereas in (7.30) $\boldsymbol{\vartheta}$ is assumed to be fixed, in the Bayesian approach this parameter (or some linear combination of it) follows a nondegenerate probability distribution. Second, the Assumption (7.28) causes the a priori information to be unbiased; this assumption is not required in the Bayesian framework.

7.3.3. Minimax linear, ridge and shrunken estimators

Let us briefly consider the general situation where we have (nonstochastic) prior knowledge such that the parameter vector is known to belong to some compact subset of $\mathbb{R}^r$. Over the last years, the principle of minimax linear estimation has received increasing interest as an attractive estimation principle to deal with the situation of a truncated parameter space, since it leads to optimal and admissible (linear) improvements over the (generalized) LSE, which are also superior to other alternatives to the LSE. Thereby, $\hat{\boldsymbol{\vartheta}}_{\mathrm{M}}$ is said to be a *minimax affine* or *minimax linear estimator*, respectively, if it holds

$$\sup_{\vartheta \in \Theta} \mathbf{E}_{\tilde{y}|\vartheta}\|\boldsymbol{\vartheta} - \hat{\boldsymbol{\vartheta}}_{\mathrm{M}}(\tilde{\boldsymbol{y}})\|_U^2 = \inf_{\hat{\vartheta} \in D} \sup_{\vartheta \in \Theta} \mathbf{E}_{\tilde{y}|\vartheta}\|\boldsymbol{\vartheta} - \hat{\boldsymbol{\vartheta}}(\tilde{\boldsymbol{y}})\|_U^2, \; D \in \{D_l, D_a\}. \tag{7.31}$$

We will deal extensively with minimax linear regression estimation under compact parameter restrictions in Part IV. We will characterize minimax (affine) linear estimators as Bayes (affine) linear estimators with respect to some least favourable prior distribution over the restricted parameter set, in certain cases we are led to well-known ridge and shrunken estimators.

For illustration purposes, assume that we are given prior knowledge such that $\boldsymbol{\vartheta}$ is constrained to lie in an ellipsoid

$$\Theta = \mathscr{E}(\boldsymbol{H}, \boldsymbol{\mu}) := \{\boldsymbol{\vartheta} \in \mathbb{R}^r : \|\boldsymbol{\vartheta} - \boldsymbol{\mu}\|_{\boldsymbol{H}}^2 \leqq 1\} \tag{7.32}$$

with some given shape matrix $\boldsymbol{H} \in \mathscr{M}_r^>$ and center point $\boldsymbol{\mu} \in \mathbb{R}^r$.

If we confine ourselves to estimation of a linear combination of $\boldsymbol{\vartheta}$, which implies that rank $\boldsymbol{U} = 1$ or, equivalently, that $\boldsymbol{U} = \boldsymbol{c}\boldsymbol{c}^\top$ for some $\boldsymbol{c} \in \mathbb{R}^r$, then we can explicitly derive the following minimax affine estimator.

Lemma 7.5. *Let* $\boldsymbol{U} = \boldsymbol{c}\boldsymbol{c}^\top$ *for some* $\boldsymbol{c} \in \mathbb{R}^r$ *and* $\mathscr{P}_{\tilde{e}} = \{P_{\tilde{e}} : \mathbf{E}\,\tilde{\boldsymbol{e}} = \mathbf{0},\ \mathbf{Cov}\,\tilde{\boldsymbol{e}} = \boldsymbol{\Sigma}\}$. *Then a minimax estimator in* D_{a} *is given by*

$$\hat{\boldsymbol{\vartheta}}_{\mathrm{M}}(\tilde{\boldsymbol{y}}) = (\boldsymbol{F}^\top\boldsymbol{\Sigma}^{-1}\boldsymbol{F} + \boldsymbol{H})^{-1}(\boldsymbol{F}^\top\boldsymbol{\Sigma}^{-1}\tilde{\boldsymbol{y}} + \boldsymbol{H}\boldsymbol{\mu}) \tag{7.33}$$

(see KUKS/OL'MAN (1972)).

Remark 7.3. In case of homoscedastic observations of unknown variance, where we have $\mathscr{P}_{\tilde{e}} = \left\{P_{\tilde{e}} : \mathbf{E}\,\tilde{\boldsymbol{e}} = \mathbf{0},\ \mathbf{Cov}\,\tilde{\boldsymbol{e}} = \frac{1}{w}\boldsymbol{I}_n\right\}_{w \in W}$ with $W = (w_0, \infty)$ and $w_0^{-1} > 0$ being an upper bound for the variance of observation, we have to modify $\hat{\boldsymbol{\vartheta}}_{\mathrm{M}}$ from (7.33) to

$$\hat{\boldsymbol{\vartheta}}_{\mathrm{M}}(\tilde{\boldsymbol{y}}) = (\boldsymbol{F}^\top\boldsymbol{F} + w_0^{-1}\boldsymbol{H})^{-1}(\boldsymbol{F}^\top\tilde{\boldsymbol{y}} + w_0^{-1}\boldsymbol{H}\boldsymbol{\mu})$$

(see NÄTHER/PILZ (1980)).

It is easily verified that the Kuks-Ol'man estimator is universally $\mathbb{R}^r$-admissible in D_{a}. However, it will be seen in Chapter 16 that this estimator is not minimax and universally Θ-admissible in D_{a} if rank $\boldsymbol{U} > 1$ and $r \geqq 2$, which is due to the fact that this estimator then does not belong to the class of Bayes affine estimators with regard to the restricted parameter set $\Theta = \mathscr{E}(\boldsymbol{H}, \boldsymbol{\mu})$.

For $\boldsymbol{\mu} = \mathbf{0}$ and special choices of $\boldsymbol{H}$ the Kuks-Ol'man estimator reduces to the well-known ridge and shrunken estimators which we already introduced in Section 7.2. Choosing $\boldsymbol{H} = h\boldsymbol{I}_r$ with some $h > 0$, we obtain the Hoerl-Kennard-(ordinary) ridge regression estimator

$$\hat{\boldsymbol{\vartheta}}_{\mathrm{R}}(\tilde{\boldsymbol{y}}) = (\boldsymbol{F}^\top\boldsymbol{\Sigma}^{-1}\boldsymbol{F} + h\boldsymbol{I}_r)^{-1}\boldsymbol{F}^\top\boldsymbol{\Sigma}^{-1}\tilde{\boldsymbol{y}}. \tag{7.34}$$

Corollary 7.5. (i) *Under the assumption of Lemma* 7.5 $\hat{\boldsymbol{\vartheta}}_{\mathrm{R}}$ *is a minimax linear estimator subject to a length constraint* $\boldsymbol{\vartheta}^\top\boldsymbol{\vartheta} \leqq 1/h$ *on* $\boldsymbol{\vartheta}$.

(ii) *If* $\Theta = \mathbb{R}^r$ *then* $\hat{\boldsymbol{\vartheta}}_{\mathrm{R}}$ *from* (7.34) *is admissible and Bayesian in* D_{l} *with respect to any prior* $P_{\tilde{\vartheta}}$ *such that* $\mathbf{E}\,\tilde{\boldsymbol{\vartheta}} = \mathbf{0}$ *and* $\mathbf{Cov}\,\tilde{\boldsymbol{\vartheta}} = \frac{1}{h}\boldsymbol{I}_r$. *If, additionally, the observations are normally distributed then* $\hat{\boldsymbol{\vartheta}}_{\mathrm{R}}$ *is admissible and Bayesian with respect to* $P_{\tilde{\vartheta}} = \mathrm{N}\left(\mathbf{0}, \frac{1}{h}\boldsymbol{I}_r\right)$ *under all estimators.*

The second assertion follows immediately from Theorem 7.2, Corollary 7.1 (cp. our considerations in Section 7.2 (d)) and from Corollary 6.4, where we remark that Bayes optimality of $\hat{\boldsymbol{\vartheta}}_{\mathrm{R}}$ not only holds with respect to quadratic loss but more general with respect to loss functions of the type $L(\boldsymbol{\vartheta}, \hat{\boldsymbol{\vartheta}}(\boldsymbol{y})) = L_m(\|\boldsymbol{\vartheta} - \hat{\boldsymbol{\vartheta}}(\boldsymbol{y})\|_U^2)$, $L_m(\cdot)$ being nonnegative, continuous and monotonically increasing.

Thus, from the Bayesian viewpoint, ordinary ridge regression implicitly assumes that the components of the regression parameter are "exchangeable" a priori, in the sense that they are all identically distributed according to $P_{\tilde{\vartheta}_i} = \mathrm{N}\left(0, \dfrac{1}{h}\right)$, $i = 1, \ldots, r$.

Following MARQUARDT (1970), the ridge estimator could also be given a Bayesian interpretation by augmenting the regression model by a ficticious set of data points taken according to an orthogonal experiment (i.e., information matrix proportional to $\boldsymbol{I}_r$) the response for each of the supplementary observations being equal to zero and considering them as the "prior information".

SWINDEL (1976) proposed a transformed ridge estimator

$$\hat{\boldsymbol{\vartheta}}_{\mathrm{R}}^{\mu}(\tilde{\boldsymbol{y}}) = (\boldsymbol{F}^{\top}\boldsymbol{\Sigma}^{-1}\boldsymbol{F} + h\boldsymbol{I}_r)^{-1}(\boldsymbol{F}^{\top}\boldsymbol{\Sigma}^{-1}\tilde{\boldsymbol{y}} + h\boldsymbol{\mu})$$

to allow for shrinkage towards an arbitrary vector $\boldsymbol{\mu}$ reflecting the a priori notions about the magnitudes of the regression coefficients. It can be shown that, for any $h \in \mathbb{R}^+$ and any linear combination $\boldsymbol{c}^{\top}\boldsymbol{\vartheta}$ with $\boldsymbol{c} \in \mathbb{R}^r$, the estimator $\boldsymbol{c}^{\top}\hat{\boldsymbol{\vartheta}}_{\mathrm{R}}^{\mu}(\tilde{\boldsymbol{y}})$ has smaller mean square error than $\boldsymbol{c}^{\top}\hat{\boldsymbol{\vartheta}}_{\Sigma}(\tilde{\boldsymbol{y}})$ whenever $\|\boldsymbol{\vartheta} - \boldsymbol{\mu}\|^2 \leqq 1/\lambda_{\max}(\boldsymbol{F}^{\top}\boldsymbol{\Sigma}^{-1}\boldsymbol{F})$. If this condition is not satisfied then there still exists an upper bound $k_0 > 0$ such that the improvement in MSE holds for every $k \in (0, k_0)$. Obviously, $\hat{\boldsymbol{\vartheta}}_{\mathrm{R}}^{\mu}(\tilde{\boldsymbol{y}})$ is a Bayes estimator for $\boldsymbol{\vartheta}$ with respect to $P_{\tilde{\vartheta}} = \mathrm{N}\left(\boldsymbol{\mu}, \dfrac{1}{h}\boldsymbol{I}_r\right)$, i.e., normally and independently distributed regression coefficients which must but not necessarily be exchangeable. RAO (1976 b) and HAITOWSKY/WAX (1980) proposed further generalizations of ridge regression,

$$\hat{\boldsymbol{\vartheta}}_{\mathrm{GR}}^{\mu}(\tilde{\boldsymbol{y}}) = (\boldsymbol{F}^{\top}\boldsymbol{\Sigma}^{-1}\boldsymbol{F} + \boldsymbol{G})^{-1}(\boldsymbol{F}^{\top}\boldsymbol{\Sigma}^{-1}\tilde{\boldsymbol{y}} + \boldsymbol{G}\boldsymbol{\mu}),$$

where $\boldsymbol{\mu}$ is as before and $\boldsymbol{G}$ is a positive definite matrix with known values. This generalized ridge estimate can be derived by minimizing the sum of weighted squared errors $\|\boldsymbol{y} - \boldsymbol{F}\boldsymbol{\vartheta}\|_{\Sigma^{-1}}^2$ subject to the constraint $\|\boldsymbol{\vartheta} - \boldsymbol{\mu}\|_G^2 = \text{constant}$ or, alternatively, by minimizing $\|\boldsymbol{\vartheta} - \boldsymbol{\mu}\|_G^2$ subject to $\|\boldsymbol{y} - \boldsymbol{F}\boldsymbol{\vartheta}\|_{\Sigma^{-1}}^2 = \text{constant}$. Thus, the matrix $\boldsymbol{G}$ can be considered a weight matrix by which we can attach corresponding weights according to our specific interest in the different coefficients. In a Bayesian context, $\hat{\boldsymbol{\vartheta}}_{\mathrm{GR}}^{\mu}$ is precisely the Bayes estimator with respect to monotone loss and normal prior $P_{\tilde{\vartheta}} = \mathrm{N}(\boldsymbol{\mu}, \boldsymbol{G}^{-1})$. Thus, in a Bayesian interpretation the elements of the matrix $\boldsymbol{G}$ reflect the precision of our prior knowledge assuming that the magnitudes of the coefficients are concentrated around $\boldsymbol{\mu}$.

If we let $\boldsymbol{\mu} = \boldsymbol{0}$ and $\boldsymbol{G} = g\boldsymbol{F}^{\top}\boldsymbol{\Sigma}^{-1}\boldsymbol{F}$, $g \in \mathbb{R}^+$, proportional to the information matrix of the experimental design, then from $\hat{\vartheta}_{\mathrm{GR}}$ we obtain the shrunken estimator

$$\hat{\vartheta}_c(\tilde{y}) = c\hat{\vartheta}_\Sigma(\tilde{y}), \qquad c = (1+g)^{-1} \in (0, 1) \tag{7.35}$$

introduced by MAYER/WILLKE (1973) and DEMPSTER (1973), cp. Section 7.2 (c).

Alternatively, under the assumptions of Lemma 7.5, $\hat{\vartheta}_c$ comes out as a minimax linear estimator subject to the quadratic constraint $\vartheta^\top F^\top \Sigma^{-1} F \vartheta \leqq c/(1-c)$.

A Bayesian interpretation with respect to an implicit prior distribution gives

Corollary 7.6. *Under the assumptions of Corollary* 7.5 (ii) *the shrunken estimator* $\hat{\vartheta}_c$ *with* $c \in (0, 1)$ *is Bayesian with respect to*

$$P_{\tilde{\vartheta}} = \mathrm{N}\left(\mathbf{0}, \frac{c}{1-c}(F^\top \Sigma^{-1} F)^{-1}\right).$$

Hence, the shrunken estimator can be interpreted as a Bayes estimator using prior knowledge which arises from previous or additional observations according to a design whose information matrix is proportional to that of the actual design.

An important form of a shrunken estimator is the *James-Stein estimator* (see JAMES/STEIN (1961))

$$\hat{\vartheta}_{\mathrm{JS}}(\tilde{y}) = \left(1 - \frac{g\hat{\sigma}^2}{\|\hat{\vartheta}_{\mathrm{LS}}(\tilde{y})\|^2_{F^\top F}}\right)\hat{\vartheta}_{\mathrm{LS}}(\tilde{y}), \tag{7.36}$$

where $\hat{\sigma}^2$ is an estimate of $\hat{\sigma}^2 = w^{-1}$ in case that $\mathscr{P}_{\tilde{e}} = \left\{\mathrm{N}\left(\mathbf{0}, \frac{1}{w} I_n\right)\right\}_{w \in \mathbf{R}^+}$, based, for example, on the residual sum of squares, and $g = (r-2)/(n-r+3)$. Here the shrinkage factor c is stochastic. The properties of this estimator were studied, among others, by BARANCHIK (1970), BERGER/BOCK (1976), EFRON/MORRIS (1976), YANCEY/JUDGE (1976) and BUNKE/BUNKE (1986). Note that whenever $r > 2$, $\hat{\vartheta}_{\mathrm{JS}}$ dominates the ordinary LSE for quadratic loss with $U(w) = wF^\top F$.

An informative survey of ridge regression, shrinkage estimation and related techniques for improvements over least squares estimation may be found in VINOD (1978) and SMITH/CAMPBELL (1980).

8. Bayes estimation in case of prior ignorance

Let us consider the problem of constructing suitable estimators for the regression coefficients in such cases in which our prior knowledge is rather vague and uncertain compared to the knowledge that can possibly be gained by sampling. The search for a powerful mathematical tool to represent such a state of prior ignorance has become a subject of growing attention in statistical literature over the last twenty five years. Within the Bayesian framework, the lack of prior knowledge is usually expressed by particular, infinite prior distributions. As a rule, these distributions have a simple analytical form and lead to a substantial simplification of the complicated task of specification of prior parameters and the computation of the posterior distribution, which is of particular importance in models involving a large number of (nuisance) parameters.

Moreover, the concept of prior ignorance distributions is also useful in certain problems of structural and likelihood inference, particularly in the elimination of nuisance parameters (see DAWID/STONE/ZIDEK (1973) and the literature cited there). This concept yields results which are in close accordance with those of conventional statistical methods and thus provides a link between the Bayesian and the frequency statistical concepts.

In the sequel we will develop a formal approach to estimation in case of prior ignorance which allows for the application of the Bayesian apparatus without any reference to an explicit prior distribution.

Assumption 8.1. (i) *Let the error distribution from $\mathcal{P}_{\tilde{e}} = \{P_{\tilde{e}}^{(w)}\}_{w \in W}$ be absolutely continuous with respect to Lebesgue measure.*

(ii) *For all $\boldsymbol{\vartheta} \in \Theta$ and $\boldsymbol{w} \in W$ it holds $P_{\tilde{y}|\vartheta, w}$* – a.e.

$$\int_{\Theta} \int_{W} l(\boldsymbol{\vartheta}, \boldsymbol{w}; \boldsymbol{y}) \, \mathrm{d}\boldsymbol{w} \, \mathrm{d}\boldsymbol{\vartheta} \in \mathbb{R}^+ \quad \text{and} \quad \int_{\Theta} \boldsymbol{\vartheta} \int_{W} l(\boldsymbol{\vartheta}, \boldsymbol{w}; \boldsymbol{y}) \, \mathrm{d}\boldsymbol{w} \, \mathrm{d}\boldsymbol{\vartheta} \in \mathbb{R}^r.$$

Under this assumption,

$$p^*_{\tilde{\vartheta}, \tilde{w}|y}(\cdot, \cdot) := l(\cdot, \cdot; \boldsymbol{y}) \Big/ \int_{\Theta} \int_{W} l(\boldsymbol{\vartheta}, \boldsymbol{w}; \boldsymbol{y}) \, \mathrm{d}\boldsymbol{w} \, \mathrm{d}\boldsymbol{\vartheta} \tag{8.1}$$

defines the density of a posterior distribution $P^*_{\tilde{\vartheta}, \tilde{w}|y}$ which is absolutely continuous with respect to λ_{r+s} ($s = \dim W$). The quantity

$$\mathbf{E}^*_{\tilde{\vartheta}, \tilde{w}|y} L(\tilde{\boldsymbol{\vartheta}}, \tilde{\boldsymbol{w}}; \hat{\boldsymbol{\vartheta}}(\boldsymbol{y})) = \int_{\Theta} \int_{W} L(\boldsymbol{\vartheta}, \boldsymbol{w}; \hat{\boldsymbol{\vartheta}}(\boldsymbol{y})) \, p^*_{\tilde{\vartheta}, \tilde{w}|y}(\boldsymbol{\vartheta}, \boldsymbol{w}) \, \mathrm{d}\boldsymbol{w} \, \mathrm{d}\boldsymbol{\vartheta} \tag{8.2}$$

is called the *posterior loss relative to prior ignorance.* This posterior loss does not involve any prior knowledge about the parameters $\tilde{\boldsymbol{\vartheta}}$ and $\tilde{\boldsymbol{w}}$ but only refers to the actual observation $\boldsymbol{y}$.

Definition 8.1. The estimator $\hat{\boldsymbol{\vartheta}}^* \in D$ is called a *Bayes estimator relative to* *prior ignorance* if it holds

$$\forall y \in C: \mathbf{E}^*_{\tilde{\vartheta}, \tilde{w}|y} L(\tilde{\vartheta}, \tilde{w}; \hat{\vartheta}^*(y)) = \inf_{\hat{\vartheta} \in D} \mathbf{E}^*_{\tilde{\vartheta}, \tilde{w}|y} L(\tilde{\vartheta}, \tilde{w}; \hat{\vartheta}(y)).$$

Let us consider monotone loss functions as indicated in Theorem 6.6 but which do not depend on $w \in W$. Then it can be shown that the so-called Pitman estimator $\hat{\vartheta}_P$ defined by

$$\hat{\vartheta}_P(y) = \frac{\int\limits_{\Theta} \vartheta \int\limits_{W} l(\vartheta, w; y)\, dw\, d\vartheta}{\int\limits_{\Theta} \int\limits_{W} l(\vartheta, w; y)\, dw\, d\vartheta} \tag{8.3}$$

is a Bayes estimator relative to prior ignorance if the likelihood function satisfies a certain symmetry condition.

Theorem 8.1. *Assume monotone loss* $L(\vartheta, w; \hat{\vartheta}(y)) = L_m(\|\vartheta - \hat{\vartheta}(y)\|^2_U)$, $U \in \mathcal{M}^{\geqq}_r$, *and let* $\mathcal{P}_{\tilde{e}}$ *be a class of error distributions for which Assumption* 8.1 *is satisfied and the marginal likelihood functions can be written as*

$$\int\limits_{W} l(\vartheta, w; y)\, dw = \bar{l}(\|\vartheta - t(y)\|^2_T),$$

where $t(y) \in \mathbb{R}^r$, $T \in \mathcal{M}^{\geqq}_r$ *and* $\bar{l}$ *is a nonnegative, strictly decreasing function on* $[0, \infty)$. *Then the Pitman estimator* $\hat{\vartheta}_P$ *from* (8.3) *is a Bayes estimator relative to prior ignorance and it holds* $\hat{\vartheta}_P(y) = t(y)$ *for each* $y \in C$.

Proof. Because of our assumption on l, the marginal posterior distribution $P^*_{\tilde{\vartheta}|y}$ given by the density

$$p^*_{\tilde{\vartheta}|y}(\vartheta) = \int\limits_{W} p^*_{\tilde{\vartheta}, \tilde{w}|y}(\vartheta, w)\, dw$$

$$= \bar{l}(\|\vartheta - t(y)\|^2_T) \Big/ \int\limits_{\Theta} \int\limits_{W} l(\vartheta, w; y)\, dw\, d\vartheta$$

has the (unique) mode $t(y)$. Further, $t(y)$ coincides with the expectation of $P^*_{\tilde{\vartheta}|y}$ since $p^*_{\tilde{\vartheta}|y}$ is symmetric around $t(y)$. Therefore,

$$t(y) = \int\limits_{\Theta} \vartheta p^*_{\tilde{\vartheta}|y}(\vartheta)\, d\vartheta = \int\limits_{\Theta} \vartheta \int\limits_{W} p^*_{\tilde{\vartheta}, \tilde{w}|y}(\vartheta, w)\, dw\, d\vartheta = \hat{\vartheta}_P(y)$$

and according to A.7 it holds

$$\int\limits_{\Theta} L_m(\|\vartheta - \hat{\vartheta}_P(y)\|^2_U)\, p^*_{\tilde{\vartheta}|y}(\vartheta)\, d\vartheta \leqq \int\limits_{\Theta} L_m(\|\vartheta - \hat{\vartheta}(y)\|^2_U)\, p^*_{\tilde{\vartheta}|y}(\vartheta)\, d\vartheta$$

for every estimator $\hat{\vartheta} \in D$. From this it follows that $\hat{\vartheta}_P$ minimizes the posterior loss relative to prior ignorance (8.2). □

Corollary 8.1. *Assume monotone loss as in Theorem* 8.1 *and*

$\mathcal{P}_{\tilde{e}} = \left\{ \mathrm{N}\left(\mathbf{0}, \frac{1}{w} \boldsymbol{I}_n\right)\right\}_{w \in \mathbb{R}^+}$. *Then the least squares estimator* $\hat{\boldsymbol{\vartheta}}_{\mathrm{LS}}(\tilde{\boldsymbol{y}}) = (\boldsymbol{F}^\top \boldsymbol{F})^{-1} \boldsymbol{F}^\top \tilde{\boldsymbol{y}}$ *is Bayesian relative to prior ignorance.*

Proof. With the normality assumption, the likelihood function is given by (4.10) and we obtain

$$\int_0^\infty l(\boldsymbol{\vartheta}, w; \boldsymbol{y})\, \mathrm{d}w \propto \int_0^\infty w^{n/2} \exp\left(-\frac{w}{2}\left[\|\boldsymbol{y} - \boldsymbol{F}\hat{\boldsymbol{\vartheta}}_{\mathrm{LS}}(\boldsymbol{y})\|^2 + \|\boldsymbol{\vartheta} - \hat{\boldsymbol{\vartheta}}_{\mathrm{LS}}(\boldsymbol{y})\|^2_{\boldsymbol{F}^\top \boldsymbol{F}}\right]\right) \mathrm{d}w$$

$$\propto \left(\|\boldsymbol{y} - \boldsymbol{F}\hat{\boldsymbol{\vartheta}}_{\mathrm{LS}}(\boldsymbol{y})\|^2 + \|\boldsymbol{\vartheta} - \hat{\boldsymbol{\vartheta}}_{\mathrm{LS}}(\boldsymbol{y})\|^2_{\boldsymbol{F}^\top \boldsymbol{F}}\right)^{-(n/2+1)}.$$

Hence, the result follows from Theorem 8.1 since the assumption required there on $\bar{l}$ is satisfied with $\boldsymbol{t}(\boldsymbol{y}) = \hat{\boldsymbol{\vartheta}}_{\mathrm{LS}}(\boldsymbol{y})$ and $\boldsymbol{T} = \boldsymbol{F}^\top \boldsymbol{F}$. □

Up to now we have only implicitly defined the concept of a prior ignorance distribution via the posterior density (8.1) which is completely determined by the likelihood function of observation. In the literature there have been made several attempts for the construction of ignorance priors which are based on invariance and information-theoretic considerations (see, e.g. JEFFREYS (1961), ZELLNER (1971), BOX/TIAO (1973)). But yet there appears to be no unified concept.

In case of uncorrelated and homoscedastic errors

$$\mathcal{P}_{\tilde{e}} = \{P_{\tilde{e}}^{(\sigma)} : \mathsf{E}\,\tilde{e} = \mathbf{0},\ \mathsf{Cov}\,\tilde{e} = \sigma^2 \boldsymbol{I}_n\}_{\sigma \in \mathbb{R}^+}$$

it is generally accepted to use

$$p^*_{\tilde{\vartheta}, \tilde{\sigma}}(\boldsymbol{\vartheta}, \sigma) = 1/\sigma \quad \text{for} \quad \boldsymbol{\vartheta} \in \Theta, \quad \sigma \in \mathbb{R}^+ \tag{8.4}$$

as density of an ignorance prior distribution. This corresponds to an infinite prior measure which spreads uniform weight over the whole space $\mathbb{R}^r \times \mathbb{R}^+$ of possible parameter values $(\boldsymbol{\vartheta}, \sigma)$ and which considers $\tilde{\boldsymbol{\vartheta}}$ and $\tilde{\sigma}$ to be independent. We note that the ignorance prior $P^*_{\tilde{\vartheta}, \tilde{\sigma}}$ with density (8.4) is invariant under power transformations $\varphi | \sigma \to \sigma^m$ with integer valued $m \neq 0$, namely

$$\mathrm{d}P^*_{\tilde{\vartheta}, \tilde{\sigma}^m}(\boldsymbol{\vartheta}, \sigma^m) = \sigma^{-m}\, \mathrm{d}(\sigma^m)\, \mathrm{d}\boldsymbol{\vartheta} = m\sigma^{-1} \propto \mathrm{d}P^*_{\tilde{\vartheta}, \tilde{\sigma}}(\boldsymbol{\vartheta}, \sigma).$$

Thus the above ignorance prior distribution does not depend on the chosen parameterization; it does not play any role whether attention is confined to the precision $w = \sigma^{-2}$, the variance $w = \sigma^2$, the standard deviation $w = \sigma$ or some other integer power of σ. By inspection of Bayes' formula it can easily be seen that the corresponding posterior distributions are also invariant under such transformations. More general considerations on invariance properties of prior and posterior distributions may be found, for example, in JEFFREYS (1961), HARTIGAN (1964) and VILLEGAS (1971).

The global use of the density (8.4) may but be in contradiction with a state of prior ignorance when we have further detailed knowledge of the type of error distributions; there may be situations (subfamilies of $\mathcal{P}_{\tilde{e}}$) in which (8.4) contains a relatively large amount of prior information (for a discussion of an example see, e.g., RAIFFA/SCHLAIFER (1961), Chapter 3). In such situations it may be better to closely associate the ignorance prior distribution with the special type of error distributions. This can be done, for example, by considering the ignorance priors as limiting distributions within the family of conjugate priors, an idea which was expressed by NOVICK (1969), ROTHENBERG (1973) and, implicitly, by RAIFFA/SCHLAIFER (1961). But so far this idea has not found wide application. We note that for the subfamily $\mathcal{P}_{\tilde{e}}^{(N)} = \{N(\mathbf{0}, \sigma^2 \boldsymbol{I}_n)\}_{\sigma \in \mathbb{R}^+}$ of normal distributions the density (8.4) comes out as a limiting distribution of a conjugate normal-gamma-prior

$$P^*_{\tilde{\boldsymbol{\vartheta}}, \tilde{w}} = \mathrm{NG}(\boldsymbol{\mu}, \alpha, \boldsymbol{\Phi}, \beta), \quad \mathrm{d}P^*_{\tilde{\boldsymbol{\vartheta}}, \tilde{w}}(\boldsymbol{\vartheta}, w) \propto w^{-1} \mathrm{d}w \, \mathrm{d}\boldsymbol{\vartheta}, \quad \tilde{w} = \tilde{\sigma}^{-2} \tag{8.5}$$

if we formally let

$$\boldsymbol{\Phi}^{-1} = \mathbf{0}, \qquad \alpha = 1 - r, \qquad \beta = 0, \qquad \boldsymbol{\mu} \in \mathbb{R}^r \text{ arbitrary.} \tag{8.6}$$

In particular, since with (8.5) $\boldsymbol{\Phi}$ is proportional to the covariance matrix of $\tilde{\boldsymbol{\vartheta}}$, the state of prior ignorance about the regression coefficients corresponds to assuming zero prior precisions for them.

Although, in general, an ignorance prior distribution $P^*_{\tilde{\boldsymbol{\vartheta}}, \tilde{w}}$ does not represent a probability distribution, the posterior distribution associated with it will always be a proper probability distribution. If $P^*_{\tilde{\boldsymbol{\vartheta}}, \tilde{w}}$ has density $p^*_{\tilde{\boldsymbol{\vartheta}}, \tilde{w}}$ then $P^*_{\tilde{\boldsymbol{\vartheta}}, \tilde{w}|\boldsymbol{y}}$ has probability density

$$p^*_{\tilde{\boldsymbol{\vartheta}}, \tilde{w}|\boldsymbol{y}}(\boldsymbol{\vartheta}, w) = \frac{l(\boldsymbol{\vartheta}, w;\, \boldsymbol{y})\, p^*_{\tilde{\boldsymbol{\vartheta}}, \tilde{w}}(\boldsymbol{\vartheta}, w)}{\int_{\Theta} \int_{\mathbb{R}^+} l(\boldsymbol{\vartheta}, w;\, \boldsymbol{y})\, p^*_{\tilde{\boldsymbol{\vartheta}}, \tilde{w}}(\boldsymbol{\vartheta}, w) \, \mathrm{d}w \, \mathrm{d}\boldsymbol{\vartheta}} \tag{8.7}$$

provided that the denominator is finite. A Bayes estimator with respect to an ignorance prior distribution then can be obtained by minimizing the posterior expected loss associated with the density (8.7).

Corollary 8.2. *Let L be monotone as in Theorem* 8.1 *and* $\mathcal{P}_{\tilde{e}} = \mathcal{P}_{\tilde{e}}^{(N)} = \{N(\mathbf{0}, \sigma^2 \boldsymbol{I}_n)\}_{\sigma \in \mathbb{R}^+}$. *Then* $\hat{\boldsymbol{\vartheta}}_{\mathrm{LS}}(\tilde{\boldsymbol{y}}) = (\boldsymbol{F}^\top \boldsymbol{F})^{-1} \boldsymbol{F}^\top \tilde{\boldsymbol{y}}$ *is a Bayes estimator with respect to the ignorance prior* $P^*_{\tilde{\boldsymbol{\vartheta}}, \tilde{\sigma}}$ *with density* (8.4).

Proof. With $p^*_{\tilde{\boldsymbol{\vartheta}}, \tilde{\sigma}}(\boldsymbol{\vartheta}, \sigma) = \sigma^{-1}$ the marginal posterior density

$$\begin{aligned} p^*_{\tilde{\boldsymbol{\vartheta}}|\tilde{\boldsymbol{y}}}(\boldsymbol{\vartheta}) &\propto \int_0^\infty \sigma^{-1} l(\boldsymbol{\vartheta}, \sigma;\, \boldsymbol{y}) \, \mathrm{d}\sigma \\ &\propto \int_0^\infty \sigma^{-(n+1)} \exp\left\{ -\frac{1}{2\sigma^2} \left[\|\boldsymbol{\vartheta} - \hat{\boldsymbol{\vartheta}}_{\mathrm{LS}}(\boldsymbol{y})\|^2_{\boldsymbol{F}^\top \boldsymbol{F}} + \|\boldsymbol{y} - \boldsymbol{F}\hat{\boldsymbol{\vartheta}}_{\mathrm{LS}}(\boldsymbol{y})\|^2 \right] \right\} \mathrm{d}\sigma \\ &\propto \{ s^2 + \|\boldsymbol{\vartheta} - \hat{\boldsymbol{\vartheta}}_{\mathrm{LS}}(\boldsymbol{y})\|^2_{\boldsymbol{F}^\top \boldsymbol{F}} \}^{-n/2}, \qquad s^2 = \|\boldsymbol{y} - \boldsymbol{F}\hat{\boldsymbol{\vartheta}}_{\mathrm{LS}}(\boldsymbol{y})\|^2 \end{aligned}$$

is the density of a t-distribution and has expectation $\hat{\boldsymbol{\vartheta}}_{\mathrm{LS}}(\boldsymbol{y})$. Thus, by virtue of A.7, $\hat{\vartheta}_{LS}(y)$ minimizes the posterior expected loss

$$\int_{\Theta} L_m(\|\vartheta - \hat{\vartheta}(y)\|_U^2)\, p^*_{\tilde{\vartheta}|y}(\vartheta)\, d\vartheta$$

under arbitrary estimates $\hat{\vartheta}(y) \in \mathbb{R}^r$. □

In conjunction with the conjugate prior family concept it is tempting to apply an adaptive estimation procedure for ϑ where, starting with an ignorance prior distribution, a first sample is used to construct a corresponding posterior distribution for $\tilde{\vartheta}$ which serves as a prior distribution for the analysis of a second sample etc. A Bayes estimate for ϑ then can be derived in the usual way from the posterior distribution obtained at some final stage $k > 2$. For the case of $k = 2$ normally distributed samples

$$\tilde{y}_i \sim \mathrm{N}(F_i\vartheta, \sigma_i^2 I_{n_i}), \quad i = 1, 2$$

this has been considered by TIAO/ZELLNER (1964). If it can be assumed that $\sigma_1^2 = \sigma_2^2$, which may be the case when the experiments are made sequentially under well-controlled conditions guaranteeing the constancy of the error variances, then the combined posterior density at the second stage is

$$p_{\vartheta, \tilde{\sigma}|y}(\vartheta, \sigma) \propto \sigma^{-(n+1)} \exp\left(-\frac{1}{2\sigma^2}[s^2 + \|\vartheta - \hat{\vartheta}(y)\|^2_{F^\top F}]\right),$$

where

$$n = n_1 + n_2, \quad s^2 = s_1^2 + s_2^2, \quad F^\top F = F_1^\top F_1 + F_2^\top F_2$$

and

$$\hat{\vartheta}(y) = (F_1^\top F_1 + F_2^\top F_2)^{-1}(F_1^\top y_1 + F_2^\top y_2), \tag{8.8}$$

$s_i^2 = \|y_i - F_i\hat{\vartheta}_i(y_i)\|^2$ with $\hat{\vartheta}_i(y_i) = (F_i^\top F_i)^{-1}F_i^\top y_i$ are the residual variances at the first and second stage; $i = 1, 2$. The marginal posterior density is then determined by

$$p_{\tilde{\vartheta}|y}(\vartheta) \propto [s^2 + \|\vartheta - \hat{\vartheta}(y)\|^2_{F^\top F}]^{-n/2}$$

which is a t-density with expectation equal to $\hat{\vartheta}(y)$ (cp. proof of Corollary 8.2). Referring to monotone loss, the Bayes estimator after the second stage would be $\hat{\vartheta}$ from (8.8); this is precisely the LSE for the combined observation $y = (y_1^\top, y_2^\top)^\top$ and design matrix $F^\top = (F_1^\top \,|\, F_2^\top)$. The situation becomes but more complicated when σ_1 and σ_2 are unknown and do not depend on each other. Then, assuming a prior density

$$p(\vartheta, \sigma_1, \sigma_2) \propto 1/\sigma_1\sigma_2, \quad (\vartheta, \sigma_1, \sigma_2) \in \mathbb{R}^r \times \mathbb{R}^+ \times \mathbb{R}^+,$$

i.e., an ignorance prior with $\tilde{\vartheta}$, $\tilde{\sigma}_1$ and $\tilde{\sigma}_2$ being independent and uniformly distributed over the range of possible realizations, the posterior density of $\tilde{\vartheta}$ becomes

$$p_{\tilde{\vartheta}|y}(\vartheta) \propto \prod_{i=1}^{2} [s_i^2 + \|\vartheta - \hat{\vartheta}_i(y_i)\|^2_{F_i^\top F_i}]^{-n_i/2}$$

which is in the form of a "multivariate double-t" density. This is a nonstandard density the expectation and quantiles of which cannot be expressed in terms of simple functions. TIAO and ZELLNER (1964) have given an approximation of the corresponding distribution function which then allows to derive Bayes estimates for $\boldsymbol{\vartheta}$.

Obviously, the above considerations can be generalized to situations where more than two experiments are performed sequentially (or simultaneously) with observations $\tilde{\boldsymbol{y}}_i \sim \mathrm{N}(\boldsymbol{F}_i\boldsymbol{\vartheta}, \sigma_i^2 \boldsymbol{I}_{n_i})$, $i = 1, \ldots, k \geqq 2$, and independent parameter $(\tilde{\boldsymbol{\vartheta}}, \tilde{\sigma}_1, \ldots, \tilde{\sigma}_k)$. The posterior density of $\tilde{\boldsymbol{\vartheta}}$ given $\boldsymbol{y} = (\boldsymbol{y}_1^\top, \ldots, \boldsymbol{y}_k^\top)^\top$ again is a product of t-densities (k factors) and can be approximated in the same way. For the construction of explicit estimators within an empirical Bayes approach we refer to Section 9.1.

If it can be assumed that the variances at each stage of observation are equal then the resulting Bayes estimator with respect to prior ignorance is

$$\hat{\boldsymbol{\vartheta}}(\tilde{\boldsymbol{y}}) = \left[\sum_{i=1}^{k} \boldsymbol{F}_i^\top \boldsymbol{F}_i\right]^{-1} \left[\sum_{i=1}^{k} \boldsymbol{F}_i^\top \tilde{\boldsymbol{y}}_i\right].$$

If we can start, however, with an informative prior $P_{\tilde{\boldsymbol{\vartheta}}|\sigma} = \mathrm{N}(\boldsymbol{\mu}, \sigma^2\boldsymbol{\Phi})$ at the initial stage then the resulting Bayes estimator at the stage k modifies to

$$\hat{\boldsymbol{\vartheta}}_{\mathrm{B}}(\boldsymbol{y}) = \left(\sum_{i=1}^{k} \boldsymbol{F}_i^\top \boldsymbol{F}_i + \boldsymbol{\Phi}^{-1}\right)^{-1} \left(\sum_{i=1}^{k} \boldsymbol{F}_i^\top \tilde{\boldsymbol{y}}_i + \boldsymbol{\Phi}^{-1}\right).$$

9. Further problems

9.1. Empirical Bayes estimation

Let us consider again the case that the first and second order prior moments needed to compute the linear Bayes estimator for $\boldsymbol{\vartheta}$ are only partially known, but now suppose that we are given previous or additional observations from investigations in related regression problems. In conjunction with the actual observations, this additional information can be used to construct estimates for the prior moments and thus to determine implicitly a prior distribution from the data. This is called an empirical Bayes procedure, the Bayes estimator derived from such an implicit prior distribution is called an empirical Bayes estimator.

The empirical Bayes concept has been introduced by ROBBINS (1964), for a detailed representation of the general theory and corresponding methods we refer to MARITZ (1970).

Let us assume that, in addition to the actual model $\tilde{\boldsymbol{y}} = \boldsymbol{F}\boldsymbol{\vartheta} + \tilde{\boldsymbol{e}}$, where

$$\mathbf{E}\,\tilde{\boldsymbol{e}} = \mathbf{0}, \quad \mathbf{Cov}\,\tilde{\boldsymbol{e}} = \boldsymbol{\Sigma} \quad \text{and} \quad \mathbf{E}\,\tilde{\boldsymbol{\vartheta}} = \boldsymbol{\mu}, \quad \mathbf{Cov}\,\tilde{\boldsymbol{\vartheta}} = \boldsymbol{\Phi},$$

we have available independent sets of observations from N different models

$$\begin{aligned} &\tilde{y}_i = F_i\vartheta_i + \tilde{e}_i\,; \qquad i = 1, \ldots, N, \\ &\mathsf{E}\,\tilde{e}_i = \mathbf{0}, \qquad \mathsf{Cov}\,\tilde{e}_i = \Sigma_i, \\ &\mathsf{E}\,\tilde{\vartheta}_i = \mu, \qquad \mathsf{Cov}\,\tilde{\vartheta}_i = \Phi. \end{aligned} \tag{9.1}$$

Here the error covariance matrices $\Sigma, \Sigma_1, \ldots, \Sigma_N$ are assumed to be known. Furthermore, let be satisfied

Assumption 9.1. (i) $\Sigma_i \in \mathcal{M}^{>}_{n_i}$, $F_i \in \mathcal{M}_{n_i \times r}$, rank $F_i = r$.
(ii) *The random vectors $\tilde{\vartheta}, \tilde{e}, \tilde{\vartheta}_1, \ldots, \tilde{\vartheta}_N, \tilde{e}_1, \ldots, \tilde{e}_N$ are totally independent.*

The information about the prior parameters μ and Φ contained in these observations via

$$\mathsf{E}\,\tilde{y}_i = F_i\,\mu \quad \text{and} \quad \mathsf{Cov}\,\tilde{y}_i = F_i\Phi F_i^\top + \Sigma_i \tag{9.2}$$

can be used to obtain estimates for the parameters. Let $\hat{\gamma} = \hat{\gamma}(\tilde{y}_1, \ldots, \tilde{y}_N)$ be an estimator for $\gamma = (\mu, \Phi)$. Replacing μ and Φ by their estimates, the Bayes estimator $\hat{\vartheta}_{\mathrm{B}} = \hat{\vartheta}_{\mathrm{B}}(\tilde{y})$ changes into an empirical Bayes estimator $\hat{\vartheta}_{\mathrm{EB}} = \hat{\vartheta}_{\mathrm{EB}}(\hat{\gamma}, \tilde{y})$ which then depends on the $N+1$ observations $\tilde{y}$, $\tilde{y}_1, \ldots, \tilde{y}_N$.

Let w and $w_1, \ldots, w_N$ index the error distributions of $\tilde{e} \sim P^{(w)}_{\tilde{e}}$ and $\tilde{e}_i \sim P^{(w_i)}_{\tilde{e}_i}$. The goodness of the empirical Bayes estimator must be evaluated by the total (Bayes) risk

$$\varrho_N(\hat{\vartheta}_{\mathrm{EB}}) = \mathsf{E}_{\tilde{\tau}}\,\mathsf{E}\,\{L(\tilde{\vartheta}, \tilde{w}; \hat{\vartheta}_{\mathrm{EB}}(\hat{\gamma}, \tilde{y}))\}, \tag{9.3}$$

where the inner expectation is taken with respect to the conditional distribution of $(\tilde{y}, \tilde{y}_1, \ldots, \tilde{y}_N)$ for given $\tau = (\vartheta, w, \vartheta_1, \ldots, \vartheta_N, w_1, \ldots, w_N)$ and $\mathsf{E}_{\tilde{\tau}}$ means expectation with respect to the compound prior

$$P_{\tilde{\tau}} = \left(\mathop{\times}_{i=1}^{N} P_{\tilde{\vartheta}_i, \tilde{w}_i}\right) \times P_{\tilde{\vartheta}, \tilde{w}}.$$

In general, the total risk cannot be computed explicitly. Under mild assumptions, the empirical Bayes estimator can be shown to be at least asymptotically optimal, i.e., the total risk ϱ_N of $\hat{\vartheta}_{\mathrm{EB}}$ converges to the Bayes risk ϱ of $\hat{\vartheta}_{\mathrm{B}}$ when N approaches infinity (see, e.g., EFRON/MORRIS (1972)). In case of finite N it is sometimes possible to evaluate the goodness of $\hat{\vartheta}_{\mathrm{EB}}$ by help of Monte-Carlo studies (see, e.g., CLEMMER/KRUTCHKOFF (1968), MARTZ/KRUTCHKOFF (1969)).

However, if the Bayes estimator can be expected to yield a large improvement in risk over the LSE then there is good hope that the corresponding empirical Bayes estimator will also do better than the LSE and will be close to the optimum Bayes risk provided that the prior estimates are not too bad.

Now we give estimates for μ and Φ which may be used for the construction of an empirical Bayes estimator for ϑ when replacing

$$\hat{\vartheta}_{B}^{\Sigma}(\tilde{y}) = (F^\top \Sigma^{-1} F + \Phi^{-1})^{-1}(F^\top \Sigma^{-1}\tilde{y} + \Phi^{-1}\mu)$$

by the corresponding version with estimated μ and Φ. To do so we recollect the actual and additional observations within the compound model

$$\tilde{y}^* = F^*\mu + \tilde{e}^*,$$

$$\tilde{y}^* = \begin{pmatrix} \tilde{y}_0 \\ \vdots \\ \tilde{y}_N \end{pmatrix}, \quad F^* = \begin{pmatrix} F_0 \\ \vdots \\ F_N \end{pmatrix}, \quad \tilde{e}^* = \begin{pmatrix} F_0 & & 0 \\ & \ddots & \\ 0 & & F_N \end{pmatrix} \begin{pmatrix} \tilde{\vartheta}_0 - \mu \\ \vdots \quad \vdots \\ \tilde{\vartheta}_N - \mu \end{pmatrix} + \begin{pmatrix} \tilde{e}_0 \\ \vdots \\ \tilde{e}_N \end{pmatrix}, \quad (9.4)$$

where, for convenience of notation, the subscript zero indicates observation in the actual model. First we deal with the case that only μ is unknown and then with the general case in which all the prior parameters are unknown.

(i) *Φ known, μ unknown*
Then we estimate μ by the generalized LSE

$$\mu^* = [F^{*\top}(\mathrm{Cov}\,\tilde{e}^*)^{-1}F^*]^{-1}F^{*\top}(\mathrm{Cov}\,\tilde{e}^*)^{-1}\tilde{y}^*$$

in the compound model (9.4) so that μ^* will be the best linear unbiased estimator for μ. After some rearrangements this yields

$$\mu^* = \sum_{k=0}^{N}\left[\sum_{i=0}^{N}(V_i + \Phi)^{-1}\right]^{-1}(V_k + \Phi)^{-1}\hat{\vartheta}_{LS}^{\Sigma_k},$$

where

$$\hat{\vartheta}_{LS}^{\Sigma_i} = V_i F_i^\top \Sigma_i^{-1}\tilde{y}_i \quad \text{and} \quad V_i = (F_i^\top \Sigma_i^{-1} F_i)^{-1} \qquad (9.5)$$

are the generalized LSE's and their covariance matrices, respectively, in the ith submodel. Taking quadratic loss $\|\vartheta - \hat{\vartheta}(y)\|_U^2$, the resulting empirical Bayes estimator for ϑ,

$$\hat{\vartheta}_{\mu^*} = (F^\top \Sigma^{-1}F + \Phi^{-1})^{-1}(F^\top \Sigma^{-1}\tilde{y} + \Phi^{-1}\mu^*) \qquad (9.6)$$

has total risk

$$\varrho_N(\hat{\vartheta}_{\mu^*}) = \varrho_N(\hat{\vartheta}_{LS}^{\Sigma}) - \mathrm{tr}\, UV_0A_0\left[A_0^{-1} - \left(\sum_{i=0}^{N}A_i\right)^{-1}\right]A_0V_0,$$

where $A_i = (V_i + \Phi)^{-1}$, $i = 0, \ldots, N$. Since $A_0^{-1} > \left(\sum_{i=0}^{N}A_i\right)^{-1}$, it follows that

$$\varrho_N(\hat{\vartheta}_{\mu^*}) < \varrho_N(\hat{\vartheta}_{LS}^{\Sigma}),$$

i.e., even with only $N = 1$ additional observation we achieve an improvement when using $\hat{\vartheta}_{\mu^*}$ instead of the generalized LSE $\hat{\vartheta}_{LS}^{\Sigma}$.

(ii) *$\boldsymbol{\Phi}$ and $\boldsymbol{\mu}$ unknown*

Then, following SWAMY (1971), we estimate $\boldsymbol{\mu}$ and $\boldsymbol{\Phi}$ by

$$\hat{\boldsymbol{\mu}} = \frac{1}{N+1} \sum_{i=0}^{N} \hat{\boldsymbol{\vartheta}}_{\mathrm{LS}}^{\Sigma_i} \quad \text{and} \tag{9.7}$$

$$\hat{\boldsymbol{\Phi}} = \frac{1}{N} \sum_{i=0}^{N} (\hat{\boldsymbol{\vartheta}}_{\mathrm{LS}}^{\Sigma_i} - \hat{\boldsymbol{\mu}})(\hat{\boldsymbol{\vartheta}}_{\mathrm{LS}}^{\Sigma_i} - \hat{\boldsymbol{\mu}})^{\top} - \frac{1}{N+1} \sum_{i=0}^{N} \boldsymbol{V}_i \tag{9.8}$$

respectively, with $\hat{\boldsymbol{\vartheta}}_{\mathrm{LS}}^{\Sigma_i}$ and $\boldsymbol{V}_i$ as above.

Obviously, $\hat{\boldsymbol{\mu}}$ and $\hat{\boldsymbol{\Phi}}$ are unbiased estimators. The estimator $\hat{\boldsymbol{\mu}}$ attaches equal weights to the estimates $\hat{\boldsymbol{\vartheta}}_{\mathrm{LS}}^{\Sigma_i}$ obtained in the submodels whereas, in general, these estimates will contain different amounts of information. As an alternative to $\hat{\boldsymbol{\mu}}$, a two-stage estimator $\hat{\boldsymbol{\mu}}^* = \boldsymbol{\mu}^*(\hat{\boldsymbol{\Phi}})$ could be used, i.e., the generalized LSE for $\boldsymbol{\mu}$ with $\boldsymbol{\Phi}$ replaced by its estimate $\hat{\boldsymbol{\Phi}}$.

An obvious disadvantage of $\hat{\boldsymbol{\Phi}}$ is that it may happen to be indefinite. It can be improved by the projection

$$\hat{\boldsymbol{\Phi}}^* = \arg \inf_{\boldsymbol{\Phi} \in \mathcal{M}_r^{\geqq}} \|\boldsymbol{\Phi} - \hat{\boldsymbol{\Phi}}\|^2$$

on the set of positive semidefinite matrices. Let $\hat{\lambda}_i$ and $\hat{\boldsymbol{u}}_i$ denote the eigenvalues and eigenvectors of $\hat{\boldsymbol{\Phi}}$, respectively, then it holds $\hat{\boldsymbol{\Phi}}^* = \Sigma \hat{\lambda}_i \hat{\boldsymbol{u}}_i \hat{\boldsymbol{u}}_i^{\top}$ where the sum goes over all $i \in \{1, \ldots, r\}$ for which $\hat{\lambda}_i > 0$ (see RAO (1973), p. 63).

In the above cases it is assumed that the error covariance matrices $\boldsymbol{\Sigma}_i$ be known. Now let us consider the case that they are only known up to scale factors,

$$\boldsymbol{\Sigma}_i = \sigma_i^2 \boldsymbol{S}_i \quad \text{and} \quad \mathbf{E}\, \tilde{\sigma}_i^2 = \sigma_0^2, \qquad i = 0, 1, \ldots, N. \tag{9.9}$$

Then, estimating σ_i^2 by help of the residual variances

$$\hat{\sigma}_i^2 = \frac{1}{n_i - r} \tilde{\boldsymbol{y}}_i^{\top} \boldsymbol{S}_i^{-1} (\tilde{\boldsymbol{y}}_i - \boldsymbol{F}_i \hat{\boldsymbol{\vartheta}}_{\mathrm{LS}}^{S_i}), \tag{9.10}$$

the estimator $\hat{\boldsymbol{\Phi}}$ is to be replaced by

$$\hat{\boldsymbol{\Phi}}_0 = \frac{1}{N} \sum_{i=0}^{N} (\hat{\boldsymbol{\vartheta}}_{\mathrm{LS}}^{S_i} - \hat{\boldsymbol{\mu}})(\hat{\boldsymbol{\vartheta}}_{\mathrm{LS}}^{S_i} - \hat{\boldsymbol{\mu}})^{\top} - \frac{\hat{\sigma}_0^2}{N+1} \sum_{i=0}^{N} (\boldsymbol{F}_i^{\top} \boldsymbol{S}_i^{-1} \boldsymbol{F}_i)^{-1}, \tag{9.11}$$

where $$\hat{\sigma}_0^2 = \frac{1}{N+1} \sum_{i=0}^{N} \hat{\sigma}_i^2$$

is an unbiased estimator for σ_0^2. Obviously, $\hat{\boldsymbol{\Phi}}_0$ is an unbiased estimator for $\boldsymbol{\Phi}$ and again it can be improved by the corresponding projection on $\mathcal{M}_r^{\geqq}$.

Particularly, given that the observations in each model are independent and homoscedastic, i.e., $\boldsymbol{\Sigma} = \sigma^2 \boldsymbol{I}_n$ and $\boldsymbol{\Sigma}_i = \sigma_i^2 \boldsymbol{I}_{n_i}$, the empirical Bayes estimator for ϑ takes the form

$$\hat{\vartheta}_{EB} = (F^\top F + \hat{\sigma}_0^2 \hat{\Phi}_0^{-1})^{-1} (F^\top \bar{y} + \hat{\sigma}_0^2 \hat{\Phi}_0^{-1} \hat{\mu}). \tag{9.12}$$

Under additional normality assumptions on $\tilde{y}_i$ and $\tilde{\vartheta}_i$ $(i = 0, \ldots, N)$, SMITH (1973), Section 5, has given alternative estimations for μ and Φ by choosing conjugate gamma priors for $\tilde{\sigma}_i^{-2}$ and a conjugate Wishart distribution for $\tilde{\Phi}^{-1}$. The essential difference to the above equations is that the classical unbiased estimations $\hat{\vartheta}_{LS}^{V_i}$ and $\hat{\sigma}_i^2$ are replaced by the corresponding posterior mode approximations (see also LINDLEY/SMITH (1972, Section 5.2)).

Under mild assumptions, the above estimations for μ and Φ guarantee the asymptotical optimality of the resulting empirical Bayes estimator $\hat{\vartheta}_{EB}$, for example in case (i) it suffices that $\lambda_{\min}(F_i^\top \Sigma_i^{-1} F_i) \geqq c > 0$ for $i = 1, \ldots, N$ and, in case (ii), we additionally need that $\lambda_{\max}(\Phi)$ and $\lambda_{\min}(\Phi)$ are bounded from above and below, respectively (see BUNKE/GLADITZ (1974), GLADITZ (1981)).

A substantial simplification of the numerical computation of an empirical Bayes estimator can be achieved if the observations at each model stage are taken according to a fixed design

$$F_i = F \in \mathcal{M}_{n \times r}, \quad \text{rank} \quad F = r, \qquad i = 0, 1, \ldots, N \tag{9.13}$$

which may be desirable for example for the reason of a better comparability of the experimental results. If we additionally assume normality,

$$\tilde{e}_i \sim \mathrm{N}(\mathbf{0}, \Sigma) \quad \text{and} \quad \tilde{\vartheta}_i \sim \mathrm{N}(\mu, \Phi), \qquad i = 0, 1, \ldots, N \tag{9.14}$$

then even the associated total risks become available in an explicit form. In case (i) the empirical Bayes estimator is again

$$\hat{\vartheta}_{\hat{\mu}} = (F^\top \Sigma^{-1} F + \Phi^{-1})^{-1} (F^\top \Sigma^{-1} \tilde{y} + \Phi^{-1} \hat{\mu}),$$

where now $\hat{\mu}$ simplifies to

$$\hat{\mu} = (F^\top \Sigma^{-1} F)^{-1} F^\top \Sigma^{-1} \tilde{y}_m \quad \text{with} \quad \tilde{y}_m = \frac{1}{N+1} \sum_{i=0}^{N} \tilde{y}_i.$$

In case (ii),

$$\hat{A} = (N - r - 1) \left\{ \sum_{i=0}^{N} (\hat{\vartheta}_i - \hat{\mu})(\hat{\vartheta}_i - \hat{\mu})^\top \right\}^{-1} \tag{9.15}$$

with $\quad \hat{\vartheta}_i = V F^\top \Sigma^{-1} \tilde{y}_i, \qquad V = (F^\top \Sigma^{-1} F)^{-1}$

can be shown to be an unbiased estimator for $A = (V + \Phi)^{-1}$ provided that $N > r + 1$. Rewriting the Bayes estimator $\hat{\vartheta}_B^\Sigma$ for ϑ as

$$\hat{\vartheta}_B^\Sigma = \hat{\vartheta}_{LS}^\Sigma - VA(\hat{\vartheta}_{LS}^\Sigma - \mu), \quad \text{where} \quad \hat{\vartheta}_{LS}^\Sigma = VF^\top \Sigma^{-1} \tilde{y}$$

is the generalized LSE in the actual model and substituting $\hat{A}$ for A we obtain the empirical Bayes estimator

$$\hat{\vartheta}_{\hat{A}} = \hat{\vartheta}_{LS}^\Sigma - V\hat{A}(\hat{\vartheta}_{LS}^\Sigma - \hat{\mu}). \tag{9.16}$$

The total risk of $\hat{\vartheta}_{\hat{A}}$ is smaller than that of $\hat{\vartheta}_{LS}^\Sigma$, the improvement over $\hat{\vartheta}_{LS}^\Sigma$

being proportional to $\operatorname{tr} \boldsymbol{UVAV}$ (relative to quadratic loss $\|\boldsymbol{\vartheta} - \hat{\boldsymbol{\vartheta}}(\boldsymbol{y})\|_U^2$). Moreover, even the usual risk is smaller when using $\hat{\boldsymbol{\vartheta}}_{\hat{A}}$ instead of $\hat{\boldsymbol{\vartheta}}_{\mathrm{LS}}^{\Sigma}$, i.e., the empirical Bayes estimator $\hat{\boldsymbol{\vartheta}}_{\hat{A}}$ is uniformly better (cp. RAO (1975)).

Again, in case of independent and homoscedastic errors of unknown variance, i.e., $\mathcal{P}_{\tilde{e}} = \{\mathrm{N}(\mathbf{0}, \sigma^2 \boldsymbol{I}_n)\}_{\sigma \in \mathbb{R}^+}$, we have to replace $\boldsymbol{\Sigma} = \sigma^2 \boldsymbol{I}_n$ by its estimate $\hat{\boldsymbol{\Sigma}} = \hat{\sigma}_0^2 \boldsymbol{I}_n$. Then, according to (9.13) and (9.11),

$$\hat{\sigma}_0^2 = \frac{(N+1)^{-1}}{n-r} \sum_{i=0}^{N} \tilde{\boldsymbol{y}}_i^\top [\boldsymbol{I}_n - \boldsymbol{F}(\boldsymbol{F}^\top \boldsymbol{F})^{-1} \boldsymbol{F}^\top] \tilde{\boldsymbol{y}}_i \tag{9.17}$$

is an unbiased estimator for $\sigma_0^2 = \mathbf{E}\,\tilde{\sigma}^2$ and the empirical Bayes estimator for $\boldsymbol{\vartheta}$ modifies to

$$\hat{\boldsymbol{\vartheta}}_{\hat{A}} = \hat{\boldsymbol{\vartheta}}_{\mathrm{LS}} - \hat{\sigma}_0^2 (\boldsymbol{F}^\top \boldsymbol{F})^{-1} \hat{\boldsymbol{A}} (\hat{\boldsymbol{\vartheta}}_{\mathrm{LS}} - \hat{\boldsymbol{\mu}}), \tag{9.18}$$

where $\hat{\boldsymbol{\vartheta}}_{\mathrm{LS}} = (\boldsymbol{F}^\top \boldsymbol{F})^{-1} \boldsymbol{F}^\top \tilde{\boldsymbol{y}}$ is the LSE based on the actual observations and $\hat{\boldsymbol{\mu}} = (\boldsymbol{F}^\top \boldsymbol{F})^{-1} \boldsymbol{F}^\top \tilde{\boldsymbol{y}}_m$ is the LSE formed with the arithmetic mean of all observation vectors.

9.2. Bayes estimation in inadequate models

In applications of regression analysis there is often uncertainty with regard to the specification of the components of the regression function $\boldsymbol{f} = (f_1, \ldots, f_r)^\top$. In the sequel we briefly consider the problem of approximation of the unknown response surface known to belong to some parametric family $\mathcal{F}$ on the basis of a regression experiment in a given family $\mathcal{G}$ of regression functions, where $\mathcal{F}$ and $\mathcal{G}$ represent finite-dimensional linear function spaces. Usually, $\mathcal{G}$ will have smaller dimension than $\mathcal{F}$ and the "ideal family" $\mathcal{G}$ is chosen such that it has sufficiently simple structure in terms of the design and analysis of the regression experiment. For example, the family $\mathcal{G}$ may arise from a truncated Taylor series expansion containing only low-degree polynomial terms and for which the remainder-term does not exceed a predetermined bound.

The first to investigate the consequences of model inadequacy for response surface estimation were BOX/DRAPER (1959) who proposed to use the inadequate LSE and to minimize the bias term by an appropriate choice of the experimental design. Extensions of this work appeared in KARSON et al. (1969), KUPPER/MEYDRECH (1973), COTE et al. (1973) and KIEFER (1973), (1980). PETERSEN (1973), LÄUTER (1975 b) and BUNKE/BUNKE (1986) discussed a minimax approach leading to a very intrinsic matrix optimization problem which is but not feasible for a solution. We will deal with the minimax approach in Section 17, where we give solutions using prior knowledge about $\mathcal{F}$.

Bayesian approaches were given by BLIGHT/OTT (1975), GOLDSTEIN (1976), (1980) and O'HAGAN (1978), these approaches require, however, the specification of (Gaussian) prior processes. Here we confine ourselves to the class of affine-linear estimators and look for a Bayes approximation in

$\mathcal{G}$ with respect to some prior defined on the parameter space associated with $\mathcal{F}$.

As before, we assume the response surface to be a member of the family

$$\mathcal{F} = \{\eta : \eta(x) = \boldsymbol{\vartheta}^\top \boldsymbol{f}(x), \quad \boldsymbol{x} \in X, \quad \boldsymbol{\vartheta} \in \Theta\} \tag{9.19}$$

of functions defined on $X \subseteqq \mathbb{R}^k$, where $\boldsymbol{f}$ is a given $\mathbb{R}^r$-valued function and $\Theta \subseteqq \mathbb{R}^r$ denotes the parameter space. For a given design of size n the vector of observations then follows the linear model

$$\mathbf{E}\tilde{\boldsymbol{y}} = \boldsymbol{F}\boldsymbol{\vartheta}, \qquad \mathbf{Cov}\,\tilde{\boldsymbol{y}} = \boldsymbol{\Sigma}. \tag{9.20}$$

We wish to approximate $\eta \in \mathcal{F}$ by some function out of the family

$$\mathcal{G} = \{\gamma : \gamma(\boldsymbol{x}) = \boldsymbol{\beta}^\top \boldsymbol{g}(\boldsymbol{x}), \quad \boldsymbol{x} \in X, \quad \boldsymbol{\beta} \in \mathbb{R}^s\}, \tag{9.21}$$

where $\boldsymbol{g}|X \to \mathbb{R}^s$ is a given function and $\boldsymbol{\beta} = (\beta_1, \ldots, \beta_s)^\top$ denotes an appropriate parameter vector. We consider the following class of linear estimators for $\eta(\cdot)$:

$$D = \{\hat{\eta}(\cdot, \tilde{\boldsymbol{y}}) = \boldsymbol{g}(\cdot)^\top \boldsymbol{Z}\tilde{\boldsymbol{y}} : \boldsymbol{Z} \in \mathcal{M}_{s \times n}\}. \tag{9.22}$$

Thus, $\hat{\eta}(\cdot, \tilde{\boldsymbol{y}}) = \boldsymbol{g}(\cdot)^\top \boldsymbol{Z}\tilde{\boldsymbol{y}}$ may be viewed as the linear estimator of an approximation $\gamma \in \mathcal{G}$ for $\eta \in \mathcal{F}$. Therefore, we call the estimators from D *linear approgression estimators.*

In the linear function space $\mathcal{H} = \mathcal{L}(f_1, \ldots, f_r, g_1, \ldots, g_s\})$ generated by the components of $\boldsymbol{f}$ and $\boldsymbol{g}$ we define a seminorm according to

$$\|\boldsymbol{h}\|^2 = \int_X \boldsymbol{h}(\boldsymbol{x})^2 \omega(\mathrm{d}\boldsymbol{x}), \qquad \boldsymbol{h} \in \mathcal{H}, \tag{9.23}$$

where ω is a given Borel measure on X such that $f_i \in L^2(\omega)$, $g_j \in L^2(\omega)$ for $i = 1, \ldots, r$ and $j = 1, \ldots, s$. Moreover, assume that the components of $\boldsymbol{f}$ and $\boldsymbol{g}$ separately are linearly independent in $L^2(\omega)$.

The goodness of an approgression estimator $\hat{\eta}$ for η will be evaluated on the basis of the L^2-(semi)metric corresponding to (9.23). Then the risk function of $\hat{n}$ is given by

$$\begin{aligned} R(\eta; \hat{\eta}) &= \mathbf{E}_{\tilde{y}|\eta} \|\eta(\boldsymbol{x}) - \hat{\eta}(\boldsymbol{x}, \tilde{\boldsymbol{y}})\|^2 \\ &= \mathbf{E}_{\tilde{y}|\eta} \int_X (\eta(\boldsymbol{x}) - \hat{\eta}(\boldsymbol{x}, \tilde{\boldsymbol{y}}))^2 \omega(\mathrm{d}\boldsymbol{x}). \end{aligned} \tag{9.24}$$

Introducing the matrices

$$\boldsymbol{S}_{ff} = \int_X \boldsymbol{f}(\boldsymbol{x})\boldsymbol{f}(\boldsymbol{x})^\top \omega(\mathrm{d}\boldsymbol{x}), \qquad \boldsymbol{S}_{gg} = \int_X \boldsymbol{g}(\boldsymbol{x})\boldsymbol{g}(\boldsymbol{x})^\top \omega(\mathrm{d}\boldsymbol{x})$$

and (9.25)

$$\boldsymbol{S}_{fg} = \int_X \boldsymbol{f}(\boldsymbol{x})\boldsymbol{g}(\boldsymbol{x})^\top \omega(\mathrm{d}\boldsymbol{x})$$

the risk comes out as

$$R(\eta; \hat{\eta}) = \operatorname{tr} \boldsymbol{S}_{gg} \boldsymbol{Z}\boldsymbol{\Sigma}\boldsymbol{Z}^\top + \boldsymbol{\vartheta}^\top \boldsymbol{G}(\boldsymbol{Z})\boldsymbol{\vartheta}, \tag{9.26}$$

where

$$G(Z) = F^\top Z^\top S_{gg} ZF - F^\top Z^\top S_{fg}^\top - S_{fg} ZF + S_{ff}.$$

The first term in (9.26) refers to the variance and the second one to the squared bias of the estimator $\hat{\eta}$. The matrix $G(Z)$ may be rewritten as

$$G(Z) = S + (ZF - S_{gg}^{-1} S_{fg}^\top)^\top S_{gg} (ZF - S_{gg}^{-1} S_{fg}^\top), \tag{9.27}$$

where

$$S = S_{ff} - S_{fg} S_{gg}^{-1} S_{fg}^\top \tag{9.28}$$

signifies the minimum bias matrix, i.e., the minimum squared bias attainable by some linear estimator out of D is given by $\vartheta^\top S \vartheta$. Note that the matrix S is positive semidefinite, since it can be written in the form

$$S = \int_X s(x) s(x)^\top \omega(\mathrm{d}x), \qquad s(x) = f(x) - S_{fg} S_{gg}^{-1} g(x).$$

Now, assume that we are given prior knowledge about ϑ representable by some prior distribution $P_{\tilde{\vartheta}}$ over Θ. Denoting

$$M = \int_\Theta \vartheta \vartheta^\top P_{\tilde{\vartheta}}(\mathrm{d}\vartheta) \tag{9.29}$$

the matrix of second order moments associated with the prior $P_{\tilde{\vartheta}}$, the Bayes risk corresponding to (9.26) takes the form

$$\varrho(P_{\tilde{\vartheta}}; \hat{\eta}) = \int_\Theta R(\eta; \hat{\eta}) P_{\tilde{\vartheta}}(\mathrm{d}\vartheta) = \operatorname{tr}[S_{gg} Z \Sigma Z^\top + G(Z) M]. \tag{9.30}$$

Obviously, the Bayes risk depends on the prior only through the matrix M. We are interested in a Bayes linear approgression estimator minimizing $\varrho(P_{\tilde{\vartheta}}, \hat{\eta})$ with respect to $\hat{\eta} \in D$.

Theorem 9.1. *With the risk function defined by* (9.24), *the Bayes linear approgression estimator with respect to* $P_{\tilde{\vartheta}}$ *is given by*

$$\hat{\eta}_{\mathrm{B}}(x, \tilde{y}) = g(x)^\top S_{gg}^{-1} S_{fg}^\top M F^\top (F M F^\top + \Sigma)^{-1} \tilde{y}$$

having Bayes risk

$$\varrho(P_{\tilde{\vartheta}}, \hat{\eta}_{\mathrm{B}}) = \operatorname{tr}\{S_{ff} M - S_{gg}^{-1} S_{fg}^\top M F^\top (F M F^\top + \Sigma)^{-1} F M S_{fg}\}.$$

Proof. The result follows immediately after rearranging terms in the Bayes risk (9.30) according to $\varrho(P_{\tilde{\vartheta}}; \hat{\eta}) = \operatorname{tr} S_{gg}[ZAZ^\top - ZB - B^\top Z^\top + S_{gg}^{-1} S_{ff} M]$, with $U = S_{gg}$, $A = \Sigma + FMF^\top$, $B = FMS_{fg} S_{gg}^{-1}$ and then applying Lemma 5.3. □

The above Bayes linear approgression estimator may be written in the form

$$\hat{\eta}_{\mathrm{B}}(x, \tilde{y}) = g(x)^\top S_{gg}^{-1} S_{fg}^\top \hat{\vartheta}^1_{\mathrm{B}}(\tilde{y}),$$

where (9.31)

$$\hat{\vartheta}^1_{\mathrm{B}}(\tilde{y}) = M F^\top (F M F^\top + \Sigma)^{-1} \tilde{y}$$

with $\boldsymbol{M} = \boldsymbol{M}_{P_{\tilde{\vartheta}}}$ represents the usual Bayes linear estimator for $\boldsymbol{\vartheta}$ in the adequate model (9.20), cp. Section 5.3, especially (5.31). Thus, $\hat{\eta}_B$ simply effects the projection of the Bayes linear estimator associated with this model onto the function space generated by the components of $\boldsymbol{g}(\boldsymbol{x})$. Again, assuming $\boldsymbol{F}^\top \boldsymbol{\Sigma}^{-1} \boldsymbol{F}$ to have full rank, the Bayes approgression estimator can be written in the form

$$\hat{\eta}_B(\boldsymbol{x}, \tilde{\boldsymbol{y}}) = \boldsymbol{g}(\boldsymbol{x})^\top \boldsymbol{S}_{gg}^{-1} \boldsymbol{S}_{fg}^\top \boldsymbol{M}(\boldsymbol{M} + \boldsymbol{V})^{-1} \hat{\boldsymbol{\vartheta}}_\Sigma(\tilde{\boldsymbol{y}}), \tag{9.32}$$

where $\boldsymbol{V}^{-1} = \boldsymbol{F}^\top \boldsymbol{\Sigma}^{-1} \boldsymbol{F}$ and $\hat{\boldsymbol{\vartheta}}_\Sigma = \boldsymbol{V} \boldsymbol{F}^\top \boldsymbol{\Sigma}^{-1} \tilde{\boldsymbol{y}}$ is the generalized LSE for $\boldsymbol{\vartheta}$.

Finally, if we have only vague prior knowledge about $\boldsymbol{\vartheta}$ then this can be modelled assuming a prior $P_{\tilde{\vartheta}}$ with $\boldsymbol{M} = c\boldsymbol{I}_r$ and then letting $c \to \infty$ as suggested by KIEFER (1973).

9.3. Further literature

1. LEONARD (1975) derives Bayes estimators for $\boldsymbol{\vartheta}$ and the variance parameters in a linear regression model with heteroscedastic errors, $\mathbf{Cov}\,\tilde{\boldsymbol{e}} = \operatorname{diag}(\sigma_1, \ldots, \sigma_n)$, where the variances are assumed to be exchangeable a priori and the transformed parameters $\alpha_i = \log \sigma_i$ are assumed to follow a hierarchical normal prior structure. For further approaches to estimation in case of heteroscedasticity see CARROLL/RUPPERT (1983). Modified Bayes linear estimators with estimated variances have been proposed by GOLDSTEIN (1983), for a coordinate-free approach to Bayes regression estimation in case of unknown covariance structure see GNOT (1983). Bayesian approaches to error covariance matrix estimation were given, for example, by DE GROOT (1970), LINDLEY/SMITH (1972) and DICKEY et al. (1985).

2. In BUNKE/BUNKE (1986) Bayes estimators are derived under equality, inequality and quadratic constraints on the regression parameter, respectively. Also, O'HAGAN (1973) and DAVIS (1978) derive Bayes estimators under inequality constraints using a truncated normal-gamma prior distribution for the regression and precision parameter.

3. For prior knowledge given in form of prior estimates, several authors have proposed two-stage estimation procedures (see KATTI (1962), KATTI/WAIKAR (1971), ARNOLD/AL-BAYYATI (1970)). MAYER/SINGH/WILLKE (1974) and AL-BAYYATI/ARNOLD (1972) study properties of two-stage estimators for multiple and simple linear regression, respectively.

4. Bayesian confidence estimators for the regression coefficients and other confidence estimators using different kind of prior knowledge may be found for example in BOX/TIAO (1973) and BUNKE/BUNKE (1986). In particular, there are given confidence regions containing the regression parameter with a prespecified posterior probability and having minimum possible size, the so-called highest posterior density (HPD)-regions.

5. The estimation problem with incompletely specified prior distributions has been considered in JACKSON et al. (1970), SOLOMON (1972), WIND (1973), WATSON (1974), GOLDSTEIN (1974) and PILZ (1981 a). The problem of the robustification of estimators under a change of error distributions, from the Bayesian viewpoint, is discussed in BIRNBAUM/LASKA (1967) and, for the special case of deviation from normality, in BOX/TIAO (1962), (1973). The problem of Bayes estimation with respect to general (including, for example, convex or monotone) loss functions has been considered, among others, by DE GROOT/RAO (1963), (1966), DEUTSCH (1965) and BRITNEY/WINKLER (1974).

6. In addition to the literature cited in Section 8, investigations on the posterior distribution with respect to ignorance prior distributions may be found in LINDLEY (1961), STEIN (1965) and BERGER (1980 a). For the multivariate linear regression model, Bayes estimation with respect to ignorance priors and proper prior distributions is considered in ZELLNER/CHETTY (1965), ZELLNER (1971), BOX/TIAO (1973) and BUNKE/BUNKE (1986).

7. Bayesian approachs to the problem of detection and treatment of outliers in the general linear model have been considered in BOX/TIAO (1968), GUTTMAN/DUTTER (1974), O'HAGAN (1979), SMITH (1983) and PETTIT/SMITH (1985). Another problem of great importance in practical applications concerns the handling of missing observations. For Bayes estimation in regression models with missing observations we refer to PRESS/SCOTT (1974) and DAGENAIS (1974).

8. For a discussion of the Bayesian approach to choosing among alternative linear models on the basis of their posterior probabilities we refer to LEMPERS (1971), SMITH/SPIEGELHALTER (1980), ATKINSON (1978), KLEIN/BROWN (1984), SAN MARTINI/SPEZZAFERRI (1984) and the literature cited there; for the special case of polynomial settings see, e.g., HALPERN (1973).

9. Bayes estimation methods for dynamic linear models have been developed in HARRISON/STEVENS (1976) using KALMAN filter techniques. These methods were further improved by SMITH (1979) and WEST/HARRISON/MIGON (1985).

10. Recently, increasing interest has been shown in the problem of making inferences from switching (or changing) regression models. Essentially, there are two problems associated with such models: detecting the changes and making inferences about the change points and all the other parameters of the model. Bayesian inferences for such models have been made e.g. in SMITH (1977), CHIN CHOY/BROEMELING (1980), BROEMELING/CHIN CHOY (1981) and BROEMELING (1985).

11. Over the last years, nonparametric and semiparametric Bayes approaches have received some attention, dating back to FERGUSON'S (1973) paper on Bayesian inference using Dirichlet prior distributions. For detailed presentations and reviews we refer to HARTIGAN (1983) and SIMAR (1984), for applications to regression estimation see POLI (1985) and BUNKE (1987).

III. Bayesian experimental design

10. The design problem for the linear Bayes estimator

We will now deal with the problem of experimental design for the conjugate Bayes estimator $\hat{\vartheta}_B$ introduced in Section 5.2. In contrast with classical experimental design for the LSE where interest centers around the information matrix $\boldsymbol{M}(v_n) = \frac{1}{n} \boldsymbol{F}(v_n)^\top \boldsymbol{F}(v_n)$, the matrix of interest here is $\boldsymbol{M}_B(v_n) = \boldsymbol{M}(v_n) + (n\boldsymbol{\Phi})^{-1} = \frac{1}{n}(\boldsymbol{F}(v_n)^\top \boldsymbol{F}(v_n) + \boldsymbol{\Phi}^{-1})$. That is, Bayes designs generally depend on the prior covariance matrix and, even in case of a continuous relaxation, on the number n of observations.

The experimental design considerations and construction methods developed here do not only apply to the Bayes estimator $\hat{\vartheta}_B$ but to any estimator with expected squared error loss or covariance structure proportional to $(\boldsymbol{F}(v_n)^\top \boldsymbol{F}(v_n) + \boldsymbol{A})^{-1}$ with some positive semidefinite matrix $\boldsymbol{A}$. This includes, for example, the minimax linear and mixed linear estimators (see Sections 7.3.2 and 7.3.3), the hierarchical Bayes estimator (see Section 5.6), but also applies to experimental design to supplement previous observations as considered by Covey-Crump/Silvey (1970) and to experimental design for response surface estimation in case of fitting a lower order model when a higher order model is appropriate (see Kiefer (1973)).

The design problem for the Bayes estimator will first be relaxed in the usual way by changing to continuous designs and then we introduce criteria which are analogous to classical design criteria for the LSE. Hereafter we will investigate the properties of the functionals of these criteria and the consequences for the construction of optimal designs which then is treated in detail in Sections 12 and 13.

10.1. The exact design problem

The starting point for our considerations is the compound estimation and design problem

$$G_\eta = [\Theta \times W \times X_P, S, R_\eta]$$

for the response surface η (see Definition 2.1). Thereby we assume the observations to be uncorrelated with the same (unknown) precision $w \in W = \mathbb{R}^+$ and the loss function to be quadratic.

Assumption 10.1. (i) *$\mathcal{P}_{\tilde{e}}$ contains only error probability distributions $P_{\tilde{e}}^{(w)}$ such that* $\mathbf{E}\,\tilde{e}(v_n) = \mathbf{0}$ and $\mathbf{Cov}(\tilde{e}(v_n)|w) = w^{-1}\boldsymbol{I}_n$.

(ii) *The loss for estimating $\eta(\boldsymbol{x}) = \boldsymbol{f}(\boldsymbol{x})^\top \boldsymbol{\vartheta}$ by $\hat{\eta}(\boldsymbol{x}, \boldsymbol{y})$ is given by*

$$L_\eta(\boldsymbol{\vartheta}, w; \boldsymbol{x}; \hat{\eta}(\boldsymbol{x}, \boldsymbol{y}), v_n) = h(w)\,(\boldsymbol{f}(\boldsymbol{x})^\top \boldsymbol{\vartheta} - \hat{\eta}(\boldsymbol{x}, \boldsymbol{y}))^2,$$

where h is some nonnegative function defined on $W = \mathbb{R}^+$.

Remark 10.1. This loss function L_η corresponds with the loss structure which we were concerned with in the estimation problem for the regression parameter (viz. Assumption 5.1 and Remark 5.2). The function h can be interpreted as a normalizing factor relating the loss due to misestimation to the precision of observation; setting $h(w) = w = \sigma^{-2}$ we get the usual normalized quadratic loss.

Further, let us suppose to have prior knowledge about the regression and precision parameters which we can specify by some prior probability distribution from the class

$$\mathcal{P} = \left\{ P_{\tilde{\vartheta}, \tilde{w}} : \mathbf{E}(\tilde{\boldsymbol{\vartheta}}|w) = \boldsymbol{\mu},\ \mathbf{Cov}(\tilde{\boldsymbol{\vartheta}}|w) = \frac{1}{w}\boldsymbol{\Phi},\ \mathbf{E}\,\tilde{w}^{-1} = \sigma_0^2 \right\} \tag{10.1}$$

with fixed first and second order moments.

The results obtained in Section 5 and 6 then show for which error and prior distributions $\hat{\boldsymbol{\vartheta}}_B$ from (5.12) is a Bayes estimator for the regression parameter. Particularly, under Assumption 10.1, $\hat{\boldsymbol{\vartheta}}_B$ is Bayesian with respect to every prior $P_{\tilde{\vartheta}, \tilde{w}} \in \mathcal{P}$ if we confine ourselves to linear estimators (see Theorem 6.5). Then, according to Lemma 5.1,

$$\begin{aligned} \hat{\eta}_B(\boldsymbol{x}, \boldsymbol{y}) &= \boldsymbol{f}(\boldsymbol{x})^\top \hat{\boldsymbol{\vartheta}}_B(\tilde{\boldsymbol{y}}) \\ &= \boldsymbol{f}(\boldsymbol{x})^\top [\boldsymbol{F}(v_n)^\top \boldsymbol{F}(v_n) + \boldsymbol{\Phi}^{-1}]^{-1} (\boldsymbol{F}(v_n)^\top \tilde{\boldsymbol{y}}(v_n) + \boldsymbol{\Phi}^{-1}\boldsymbol{\mu}) \end{aligned} \tag{10.2}$$

is a Bayes estimator for the expected response $\eta(\boldsymbol{x})$ in the estimation problem G_η^e (see (2.11)). In the following we investigate the design problem

$$\begin{aligned} &G_\eta^d = [X_P, V, R_\eta^d] \\ &R_\eta^d(\boldsymbol{x}; v_n) = \mathbf{E}_{\tilde{\vartheta}, \tilde{w}} R_\eta(\tilde{\boldsymbol{\vartheta}}, \tilde{w}, \boldsymbol{x}; \hat{\eta}_B, v_n), \qquad \boldsymbol{x} \in X_P,\ v_n \in V \end{aligned} \tag{10.3}$$

(cp. (2.13)) for this estimator. Our task is to determine a Bayes design v_n^* in G_η^d with respect to some prior distribution $P_{\tilde{x}}$ over the region X_P of points $\boldsymbol{x}$

for which $\eta(\boldsymbol{x})$ is to be estimated, i.e., v_n^* has to be a solution to

$$\mathsf{E}_{\tilde{x}} R_\eta^{\mathrm{d}}(\tilde{x}; v_n) = \int_{X_{\mathrm{P}}} R_\eta^{\mathrm{d}}(\boldsymbol{x}; v_n) P_{\tilde{x}}(\mathrm{d}\boldsymbol{x}) \overset{!}{=} \inf_{v_n \in V}. \tag{10.4}$$

The prior distribution $P_{\tilde{x}}$ can be generated, for example, by a weight function defined on X_{P} which gives higher weight to $\boldsymbol{x}_1 \in X_{\mathrm{P}}$ than to $\boldsymbol{x}_2 \in X_{\mathrm{P}}$ if the value $\eta(\boldsymbol{x}_1)$ is of greater interest or is to be predicted with a higher accuracy than the value $\eta(\boldsymbol{x}_2)$. If for every $\boldsymbol{x} \in X_{\mathrm{P}}$ the expected response $\eta(\boldsymbol{x})$ shall be predicted with the same accuracy then $P_{\tilde{x}}$ can be chosen as a uniform distribution. We now compute the risk function R_η^{d} for the Bayes estimator $\hat{\eta}_{\mathrm{B}}$.

Lemma 10.1. *Let* $h(\cdot)$ *and* $P_{\tilde{\vartheta}, \tilde{w}} \in \mathcal{P}$ *be such that*

$$h_0 := \mathsf{E}\{h(\tilde{w})/\tilde{w}\} < \infty.$$

Then it holds

$$R_\eta^{\mathrm{d}}(\boldsymbol{x}; v_n) = h_0 \boldsymbol{f}(\boldsymbol{x})^\top [\boldsymbol{F}(v_n)^\top \boldsymbol{F}(v_n) + \boldsymbol{\Phi}^{-1}]^{-1} \boldsymbol{f}(\boldsymbol{x}).$$

Proof. The Bayes risk of $\hat{\boldsymbol{\vartheta}}_{\mathrm{B}}$ in case of quadratic loss $L(\boldsymbol{\vartheta}, w; \hat{\boldsymbol{\vartheta}}(\boldsymbol{y})) = \|\boldsymbol{\vartheta} - \hat{\boldsymbol{\vartheta}}(\boldsymbol{y})\|^2_{U(w)}$ and error and prior distributions as assumed above was proved to be equal to $\operatorname{tr} \boldsymbol{U}_0[\boldsymbol{F}(v_n)^\top \boldsymbol{F}(v_n) + \boldsymbol{\Phi}^{-1}]^{-1}$ with $\boldsymbol{U}_0 = \mathsf{E}\left\{\frac{1}{w} \boldsymbol{U}(\tilde{w})\right\}$ (see Lemma 5.4). The result then follows from the fact that

$$L_\eta(\boldsymbol{\vartheta}, w; \boldsymbol{x}; \hat{\eta}_{\mathrm{B}}(\boldsymbol{x}, \boldsymbol{y}), v_n) = \|\boldsymbol{\vartheta} - \hat{\boldsymbol{\vartheta}}_{\mathrm{B}}(\boldsymbol{y})\|^2_{U(w)} \quad \text{with}$$
$$\boldsymbol{U}(w) = h(w) \boldsymbol{f}(\boldsymbol{x}) \boldsymbol{f}(\boldsymbol{x})^\top. \quad \square$$

Remark 10.2. In case of normalized quadratic loss the assumption formulated in Lemma 10.1 is satisfied naturally with $h_0 = 1$ for all priors $P_{\tilde{\vartheta}, \tilde{w}} \in \mathcal{P}$.

Thus, a Bayes design for $\hat{\eta}_{\mathrm{B}}$ has to minimize the integral

$$\int_{X_{\mathrm{P}}} \boldsymbol{f}(\boldsymbol{x})^\top [\boldsymbol{F}(v_n)^\top \boldsymbol{F}(v_n) + \boldsymbol{\Phi}^{-1}]^{-1} \boldsymbol{f}(\boldsymbol{x})\, P_{\tilde{x}}(\mathrm{d}\boldsymbol{x}) \tag{10.5}$$

over all possible designs $v_n \in V$. Then, under the above assumptions, $(\hat{\eta}_{\mathrm{B}}, v_n^*)$ is a Bayes strategy in the compound estimation and design problem G_η.

10.2. The Bayesian information matrix

The Bayes risk R_η^{d} given in Lemma 10.1 has the same structure as the corresponding minimax risk

$$\sup_{(\vartheta, w) \in \Theta \times W} R_\eta(\boldsymbol{\vartheta}, w, \boldsymbol{x}; \hat{\eta}_{\mathrm{LS}}, v_n) = \hat{h}_0 \boldsymbol{f}(\boldsymbol{x})^\top [\boldsymbol{F}(v_n)^\top \boldsymbol{F}(v_n)]^{-1} \boldsymbol{f}(\boldsymbol{x});$$

$$\hat{h}_0 = \sup_{w \in W} \{h(w)/w\};$$

of the least squares estimator

$$\hat{\eta}_{\mathrm{LS}}(\boldsymbol{x}, \tilde{\boldsymbol{y}}) = \boldsymbol{f}(\boldsymbol{x})^\top [\boldsymbol{F}(v_n)\boldsymbol{F}(v_n)]^{-1}\boldsymbol{F}(v_n)^\top \tilde{\boldsymbol{y}}(v_n) \tag{10.6}$$

for $\eta(\boldsymbol{x})$ if the design matrix $\boldsymbol{F}(v_n)$ is of full rank. This risk substantially depends on the information matrix

$$\boldsymbol{M}(v_n) = \frac{1}{n}\boldsymbol{F}(v_n)^\top \boldsymbol{F}(v_n)$$

and in the theory of optimum experimental design for $\hat{\eta}_{\mathrm{LS}}$ suitable functionals of this matrix are minimized (see Section 1), for example,

$$\sup_{x \in X_{\mathrm{P}}} \boldsymbol{f}(\boldsymbol{x})^\top [\boldsymbol{F}(v_n)^\top \boldsymbol{F}(v_n)]^{-1} \boldsymbol{f}(\boldsymbol{x})$$

which is the minimax analogue to (10.5). Now, in the design problem for the Bayes estimator, $\boldsymbol{F}(v_n)^\top \boldsymbol{F}(v_n)$ is replaced by $\boldsymbol{F}(v_n)^\top \boldsymbol{F}(v_n) + \boldsymbol{\Phi}^{-1}$ where the additional part $\boldsymbol{\Phi}^{-1}$ comes from the additional knowledge which we have prior to the actual experiment.

Definition 10.1. The matrix

$$\boldsymbol{M}_{\mathrm{B}}(v_n) := \frac{1}{n}(\boldsymbol{F}(v_n)^\top \boldsymbol{F}(v_n) + \boldsymbol{\Phi}^{-1}) = \boldsymbol{M}(v_n) + \frac{1}{n}\boldsymbol{\Phi}^{-1}$$

is called the *Bayesian information matrix* of the exact design $v_n \in V$.

Because of $\mathbf{Cov}\,\tilde{\boldsymbol{\vartheta}} = \sigma_0^2 \boldsymbol{\Phi}$, the additional part $\frac{1}{n}\boldsymbol{\Phi}^{-1}$ in the Bayesian information matrix measures the precision of prior information on $\tilde{\boldsymbol{\vartheta}}$ expressed in sampling units:

$$\frac{1}{n}\boldsymbol{\Phi}^{-1} = (\sigma_0^2/n)(\mathbf{Cov}\,\tilde{\boldsymbol{\vartheta}})^{-1}.$$

Remark 10.3. If we measure the information on the regression parameter $\hat{\boldsymbol{\vartheta}}$ through the Bayesian analogue to the usual Fisher information, i.e., through the matrix $\boldsymbol{J} = (j_{ik})$ having typical element

$$j_{ik} = -\mathbf{E}_{\tilde{\vartheta}|y,w}\{\partial^2 \log p_{\tilde{\vartheta}|y,w}(\boldsymbol{\vartheta})/\partial\vartheta_i\partial\vartheta_k\},$$

then $\boldsymbol{J}^{-1}$ is a lower bound to the posterior covariance matrix, i.e. $\mathbf{Cov}(\tilde{\boldsymbol{\vartheta}}|y, w) \geqq \boldsymbol{J}^{-1}$. This lower bound is attained if and only if $P_{\tilde{\vartheta}|y,w}$ is normal (cp. ROTHENBERG (1973)). In this case we have $\boldsymbol{J} = nw\boldsymbol{M}_{\mathrm{B}}(v_n)$.

An interesting property of the Bayesian information matrix is that its inverse coincides, apart from a constant factor, with the expected posterior covariance matrix (the so-called preposterior covariance matrix) of $\tilde{\boldsymbol{\vartheta}}$.

Lemma 10.2. *If $P_{\tilde{\vartheta},\tilde{w}}$ is a prior distribution from the class $\mathcal{P}$ given by* (10.1) *and* *if $\hat{\eta}_{\mathrm{B}}$ is a Bayes estimator with respect to $P_{\tilde{\vartheta}, \tilde{w}}$ then*

$$\mathsf{E}_{\tilde{y}, \tilde{w}} \mathbf{Cov}(\tilde{\vartheta} | \tilde{y}(v_n), \tilde{w}) = (\sigma_0^2/n) M_{\mathrm{B}}(v_n)^{-1}.$$

(For a proof see PILZ (1979c).)

By help of Lemma 10.2 we can give the following statistical interpretation for the Bayesian design criterion

$$\mathsf{E}_{\tilde{x}} R_\eta^{\mathrm{d}}(\tilde{x}; v_n) = (h_0/n) \int_{X_{\mathrm{P}}} f(x)^\top M_{\mathrm{B}}(v_n)^{-1} f(x) P_{\tilde{x}}(\mathrm{d}x) \stackrel{!}{=} \inf_{v_n \in V}.$$

A Bayes design minimizes the weighted average (weighted with $P_{\tilde{x}}$) of the expected posterior variance of the response over X_{P}, since

$$\mathsf{E}_{\tilde{x}} R_\eta^{\mathrm{d}}(\tilde{x}; v_n) = \sigma_0^{-2} h_0 \int_{X_{\mathrm{P}}} \mathsf{E}_{\tilde{y}, \tilde{w}} \operatorname{Var}(f(x)^\top \tilde{\vartheta} | \tilde{y}(v_n), \tilde{w}) P_{\tilde{x}}(\mathrm{d}x). \quad (10.7)$$

Let V_n be the set of designs of size n, i.e., $V_n = V \cap X_{\mathrm{E}}^n$, and

$$\mathbb{N}_0 = \{n \in \mathbb{N} : V_n \neq \emptyset\}.$$

In practical experiments the size of the design must be bounded. Therefore we can assume that $\mathbb{N}_0$ is a finite set.

Assumption 10.2. card $\mathbb{N}_0 < \infty$.

An optimal design then can be obtained in such a way that first, for each $n \in \mathbb{N}_0$, we look for a Bayes design within V_n and then choose among all these Bayes designs that one having the smallest risk, i.e., we decompose the optimization problem:

$$\inf_{v_n \in V} \mathsf{E}_{\tilde{x}} R_\eta^{\mathrm{d}}(\tilde{x}; v_n) = h_0 \min_{n \in \mathbb{N}_0} \left\{ \frac{1}{n} \inf_{v_n \in V_n} \int_{X_{\mathrm{P}}} f(x)^\top M_{\mathrm{B}}(v_n)^{-1} f(x) P_{\tilde{x}}(\mathrm{d}x) \right\}. \quad (10.8)$$

In the following we deal with the Bayesian design problem

$$B(v_n) := \int_{X_{\mathrm{P}}} f(x)^\top M_{\mathrm{B}}(v_n)^{-1} f(x) P_{\tilde{x}}(\mathrm{d}x) \stackrel{!}{=} \inf_{v_n \in V_n} \quad (10.9)$$

for a fixed size $n \in \mathbb{N}_0$. Having obtained a solution v_n^* to (10.9) for all $n \in \mathbb{N}_0$, then the design $v_{n_0}^*$ is a global optimum within V if the size n_0 is such that

$$\frac{1}{n_0} B(v_{n_0}^*) = \min_{n \in \mathbb{N}_0} \frac{1}{n} B(v_n^*).$$

10.3. The extended design problem

The exact design problem (10.9) can be embedded into a generalized problem which is mathematically more convenient to deal with, especially it yields nice convexity properties. Such an extension is accomplished by changing from exact designs to continuous designs in the Kiefer sense.

Let Ξ denote the set of all probability distributions ξ on the experimental region X_E, any element $\xi \in \Xi$ we call a *continuous* (or *approximate*) *design*. An exact design $v_n \in V_n$ then can be identified with a discrete probability measure $\xi = \xi_{v_n}$ from Ξ: If $v_n = (x_1, \ldots, x_n) \in V_n$ and $S(v_n) = \{x_{j_1}, \ldots, x_{j_m}\}$ represents the set of all $m \leqq n$ distinct points occuring in v_n with multipicities $n_{j_1}, \ldots, n_{j_m}$, where $n_{j_1} + \ldots + n_{j_m} = n$, so define ξ_{v_n} to be that measure which gives probability $p_i = n_{j_i}/n$ to the points $x_{j_i} \in S(v_n)$, $i = 1, \ldots, m$, and zero probability to all the remaining points $x \in X_E \backslash S(v_n)$.

After this embedding of V_n into Ξ it is obvious to extend the notion of the Bayesian information matrix and to set forth the design functional B from V_n to Ξ.

Definition 10.2. The matrix

$$M_B(\xi) := M(\xi) + \frac{1}{n} \Phi^{-1} = \int_{X_E} f(x) f(x)^\top \xi(dx) + \frac{1}{n} \Phi^{-1}$$

is called the *Bayesian information matrix* of the continuous design $\xi \in \Xi$ relative to the size $n \in \mathbb{N}_0$.

The extension of the functional B in (10.9) from V_n to Ξ leads us to the extended design problem

$$B(\xi) = \int_{X_P} f(x)^\top M_B(\xi)^{-1} f(x) P_{\bar{x}}(dx) \overset{!}{=} \inf_{\xi \in \Xi} \tag{10.10}$$

for the Bayes estimator $\hat{\eta}_B$.

Definition 10.3. The design $\xi^* \in \Xi$ is called a *Bayes design* in Ξ with respect to $P_{\bar{x}}$ if it minimizes the functional B over Ξ.

In general, the solutions $\xi^* \in \Xi$ to (10.10) cannot be identified with some exact design from V_n. In this case we still have to look for an approximation $\xi^*_{v_n}$ of ξ^* within V_n which can serve as an (approximately) exact Bayes design.

Now we define some special functionals of the Bayesian information matrix which are analogous to known design criteria for the least squares estimator.

Let $U_L \in \mathscr{M}_r^{\geqq}$, $c \in \mathbb{R}^r$ and p be some nonnegative function defined on X_P.

Definition 10.4. The design $\xi^* \in \Xi$ is called

(i) L_B-*optimal* in Ξ with respect to U_L iff

$$L_B(\xi) := \operatorname{tr} U_L M_B(\xi)^{-1} \geqq L_B(\xi^*) \quad \text{for all} \quad \xi \in \Xi,$$

(ii) A_B-*optimal* in Ξ iff

$$A_B(\xi) := \operatorname{tr} M_B(\xi)^{-1} \geqq A_B(\xi^*) \quad \text{for all} \quad \xi \in \Xi,$$

(iii) C_B-*optimal* in Ξ with respect to $\boldsymbol{c}$ iff

$$C_B(\xi) := \boldsymbol{c}^\top \boldsymbol{M}_B(\xi)^{-1}\boldsymbol{c} \geqq C_B(\xi^*) \quad \text{for all} \quad \xi \in \Xi,$$

(iv) I_B-*optimal* in Ξ with respect to p iff

$$I_B(\xi) := \int_{X_P} \boldsymbol{f}(\boldsymbol{x})^\top \boldsymbol{M}_B(\xi)^{-1}\boldsymbol{f}(\boldsymbol{x})\, p(\boldsymbol{x})\, \mathrm{d}\boldsymbol{x} \geqq I_B(\xi^*) \quad \text{for all} \quad \xi \in \Xi.$$

The functionals introduced in Definition 10.4 can be interpreted as the Bayesian analogues to classical L-, A-, C- and I-optimality.

The Bayesian design criterion appears to be a linear optimality criterion with a special structure of the matrix U_L.

Theorem 10.1. *The design $\xi^* \in \Xi$ is Bayesian in Ξ with respect to $P_{\tilde{x}}$ if and only if it is L_B-optimal in Ξ with respect to*

$$\boldsymbol{U}_L = \int_{X_P} \boldsymbol{f}(\boldsymbol{x})\boldsymbol{f}(\boldsymbol{x})^\top P_{\tilde{x}}(\mathrm{d}\boldsymbol{x}).$$

Proof. The result follows immediately from the linearity of the trace operator which assures that for any $\xi \in \Xi$ we have

$$\begin{aligned} B(\xi) &= \int_{X_P} \boldsymbol{f}(\boldsymbol{x})^\top \boldsymbol{M}_B(\xi)^{-1}\boldsymbol{f}(\boldsymbol{x}) P_{\tilde{x}}(\mathrm{d}\boldsymbol{x}) \\ &= \int_{X_P} \mathrm{tr}\,\{\boldsymbol{M}_B(\xi)^{-1}\boldsymbol{f}(\boldsymbol{x})\boldsymbol{f}(\boldsymbol{x})^\top\} P_{\tilde{x}}(\mathrm{d}\boldsymbol{x}) \\ &= \mathrm{tr}\left\{\boldsymbol{M}_B(\xi)^{-1} \int_{X_P} \boldsymbol{f}(\boldsymbol{x})\boldsymbol{f}(\boldsymbol{x})^\top P_{\tilde{x}}(\mathrm{d}\boldsymbol{x})\right\} \\ &= \mathrm{tr}\,\boldsymbol{M}_B(\xi)^{-1}\boldsymbol{U}_L = \mathrm{tr}\,\boldsymbol{U}_L\boldsymbol{M}_B(\xi)^{-1}. \quad \square \end{aligned}$$

The criteria of A_B-, C_B- and I_B-optimality are special cases of L_B-optimality which occur when choosing $\boldsymbol{U}_L = \boldsymbol{I}_r$, $\boldsymbol{U}_L = \boldsymbol{c}\boldsymbol{c}^\top$ and $\boldsymbol{U}_L = \int_{X_P} \boldsymbol{f}(\boldsymbol{x})\boldsymbol{f}(\boldsymbol{x})^\top p(\boldsymbol{x})\,\mathrm{d}\boldsymbol{x}$, respectively. By an appropriate choice of the prior distribution $P_{\tilde{x}}$, these criteria can also be characterized as specializations of the Bayes criterion.

Theorem 10.2. (i) *Let the prior $P_{\tilde{x}}$ be such that $\int_{X_P} \boldsymbol{f}(\boldsymbol{x})\boldsymbol{f}(\boldsymbol{x})^\top P_{\tilde{x}}(\mathrm{d}\boldsymbol{x}) = \boldsymbol{I}_r$. Then any A_B-optimal design in Ξ is Bayesian in Ξ with respect to $P_{\tilde{x}}$.*

(ii) *Let $X_P = \{\boldsymbol{x}_0\}$ with some $\boldsymbol{x}_0 \in \mathbb{R}^k$. Then any C_B-optimal design in Ξ with respect to $\boldsymbol{c} = \boldsymbol{f}(\boldsymbol{x}_0)$ is Bayesian in Ξ with respect to the one-point distribution $P_{\tilde{x}} = \delta_{x_0}$.*

(iii) *Let $P_{\tilde{x}}$ be absolutely continuous with respect to Lebesgue measure and p, apart from a normalizing factor, the probability density of $P_{\tilde{x}}$. Then any I_B-optimal design with respect to p is Bayesian with respect to $P_{\tilde{x}}$.*

Proof. The proof follows along the same lines of the proof of Theorem 10.1 observing that, under the assumptions made in (i), (ii) and (iii), the matrix $\boldsymbol{U}_{\mathrm{L}}$ takes the following forms:

(i) $\boldsymbol{U}_{\mathrm{L}} = \boldsymbol{I}_r$ and thus $B(\xi) = \operatorname{tr} \boldsymbol{M}_{\mathrm{B}}(\xi)^{-1} = A_{\mathrm{B}}(\xi)$,

(ii) $\boldsymbol{U}_{\mathrm{L}} = \boldsymbol{f}(\boldsymbol{x}_0)\boldsymbol{f}(\boldsymbol{x}_0)^{\top} = \boldsymbol{c}\boldsymbol{c}^{\top}$ and thus $B(\xi) = \operatorname{tr} \boldsymbol{c}\boldsymbol{c}^{\top} \boldsymbol{M}_{\mathrm{B}}(\xi)^{-1} = C_{\mathrm{B}}(\xi)$,

(iii) $\boldsymbol{U}_{\mathrm{L}} = a^{-1} \int\limits_{X_{\mathrm{P}}} \boldsymbol{f}(\boldsymbol{x})\boldsymbol{f}(\boldsymbol{x})^{\top} p(\boldsymbol{x})\, \mathrm{d}\boldsymbol{x}$, where $a = \int\limits_{X_{\mathrm{P}}} p(\boldsymbol{x})\, \mathrm{d}\boldsymbol{x} > 0$ is the normalizing factor, and thus

$$B(\xi) = a^{-1} \int\limits_{X_{\mathrm{P}}} \boldsymbol{f}(\boldsymbol{x})^{\top} \boldsymbol{M}_{\mathrm{B}}(\xi)^{-1} \boldsymbol{f}(\boldsymbol{x})\, p(\boldsymbol{x})\, \mathrm{d}\boldsymbol{x} = a^{-1} I_{\mathrm{B}}(\xi)\,. \quad \square$$

Remark 10.4. The assumption that $\int\limits_{X_{\mathrm{P}}} \boldsymbol{f}(\boldsymbol{x})\boldsymbol{f}(\boldsymbol{x})^{\top} P_{\tilde{x}}(\mathrm{d}\boldsymbol{x}) = \boldsymbol{I}_r$ in Theorem 10.2 (i) means that the components of the regression function $\boldsymbol{f}(\boldsymbol{x}) = (f_1(\boldsymbol{x}), \ldots, f_r(\boldsymbol{x}))^{\top}$ form an orthonormal system with respect to the measure $P_{\tilde{x}}$, i.e. $\int\limits_{X_{\mathrm{P}}} f_i(\boldsymbol{x}) f_j(\boldsymbol{x}) P_{\tilde{x}}(\mathrm{d}\boldsymbol{x}) = \delta_{ij}$, $i, j = 1, \ldots, r$. The assumption $X_{\mathrm{P}} = \{\boldsymbol{x}_0\}$ made in Theorem 10.2 (ii) refers to the special case of extrapolation of the expected response $\eta(\boldsymbol{x}_0)$ at some single point $\boldsymbol{x}_0 \in \mathbb{R}^k$.

Let us define two further design criteria which are analogous to classical D- and E-optimality. To this let $\det \boldsymbol{M}_{\mathrm{B}}(\xi)^{-1}$ and $\lambda_{\max}(\boldsymbol{M}_{\mathrm{B}}(\xi)^{-1})$ denote the determinant and the largest eigenvalue of the inverse of the Bayesian information matrix of a design $\xi \in \Xi$.

Definition 10.5. The design $\xi^* \in \Xi$ is called

(i) D_{B}-*optimal* in Ξ iff

$$D_{\mathrm{B}}(\xi) := \det \boldsymbol{M}_{\mathrm{B}}(\xi)^{-1} \geqq D_{\mathrm{B}}(\xi^*) \quad \text{for all} \quad \xi \in \Xi,$$

(ii) E_{B}-*optimal* in Ξ iff

$$E_{\mathrm{B}}(\xi) := \lambda_{\max}(\boldsymbol{M}_{\mathrm{B}}(\xi)^{-1}) \geqq E_{\mathrm{B}}(\xi^*) \quad \text{for all} \quad \xi \in \Xi.$$

Though the functionals D_{B} and E_{B} do not represent linear optimality criteria, i.e., they cannot be written in the form $\operatorname{tr} \boldsymbol{U}_{\mathrm{L}} \boldsymbol{M}_{\mathrm{B}}(\xi)^{-1}$ with some matrix $\boldsymbol{U}_{\mathrm{L}}$ being independent of the design, there are close relations between D_{B}-, E_{B}- and A_{B}-optimality (see Sections 11.3, 12.1 and 13.2).

Remark 10.5. According to Kiefer (1974) it can be shown that the above criteria are special cases of the more general criterion function

$$\Delta_s(\xi) := \left\{ \frac{1}{r} \operatorname{tr} \boldsymbol{M}_{\mathrm{B}}(\xi)^{-s} \right\}^{1/s}, \qquad 0 < s < \infty\,.$$

As $s \to +0$ the criterion becomes equivalent to D_{B}-optimality, for $s = 1$ we have A_{B}-optimality and as $s \to \infty$ we have E_{B}-optimality.

We now give a statistical interpretation for the optimality criteria introduced by the definitions 10.4 and 10.5. To do so we restrict the functionals defined on Ξ to the subset of designs $\xi = \xi_{v_n}$ which can be identified with an exact design $v_n \in V_n$ and make use of the result of Lemma 10.2 stating that $\boldsymbol{M}_{\mathrm{B}}(v_n)^{-1}$, apart from a constant factor, coincides with the expected posterior covariance matrix of $\tilde{\boldsymbol{\vartheta}}$.

a) Interpretation of L_{B}-optimality
Let $\boldsymbol{H} \in \mathcal{M}_{j\times r}$, $j \leqq r$, be such that $\boldsymbol{U}_{\mathrm{L}} = \boldsymbol{H}^{\top}\boldsymbol{H}$. Denoting by $\tilde{\boldsymbol{\beta}} = (\tilde{\beta}_1, \ldots, \tilde{\beta}_j)^{\top}$ the random vector $\tilde{\boldsymbol{\beta}} = \boldsymbol{H}\tilde{\boldsymbol{\vartheta}}$, we obtain

$$\begin{aligned} L_{\mathrm{B}}(\xi) &= \operatorname{tr} \boldsymbol{U}_{\mathrm{L}}\boldsymbol{M}_{\mathrm{B}}(v_n)^{-1} = \operatorname{tr} \boldsymbol{H}\boldsymbol{M}_{\mathrm{B}}(v_n)^{-1}\boldsymbol{H}^{\top} \\ &= (n/\sigma_0^2) \operatorname{tr}\{\boldsymbol{H}\, \mathbf{E}_{\tilde{y}, \tilde{w}}\, \mathbf{Cov}(\tilde{\boldsymbol{\vartheta}}|\tilde{\boldsymbol{y}}(v_n), \tilde{w})\boldsymbol{H}^{\top}\} \\ &= (n/\sigma_0^2) \operatorname{tr} \mathbf{E}_{\tilde{y}, \tilde{w}}\, \mathbf{Cov}(\tilde{\boldsymbol{\beta}}|\tilde{\boldsymbol{y}}(v_n), \tilde{w}) \\ &= (n/\sigma_0^2) \sum_{i=1}^{j} \mathbf{E}_{\tilde{y}, \tilde{w}} \operatorname{Var}(\tilde{\beta}_i|\tilde{\boldsymbol{y}}(v_n), \tilde{w}). \end{aligned}$$

Hence, L_{B}-optimal designs minimize the arithmetic mean of the expected posterior variances of the components of the linear combination $\tilde{\boldsymbol{\beta}} = \boldsymbol{H}\tilde{\boldsymbol{\vartheta}}$ of the regression parameter.

b) Interpretation of A_{B}-, C_{B}- and I_{B}-optimality
The criteria of A_{B}- and C_{B}-optimality can be written as $\operatorname{tr} \boldsymbol{U}_{\mathrm{L}}\boldsymbol{M}_{\mathrm{B}}(\cdot)^{-1}$ with $\boldsymbol{U}_{\mathrm{L}} = \boldsymbol{I}_r$ and $\boldsymbol{U}_{\mathrm{L}} = \boldsymbol{c}\boldsymbol{c}^{\top}$, i.e., $\boldsymbol{H} = \boldsymbol{I}_r$ and $\boldsymbol{H} = \boldsymbol{c}^{\top}$, respectively. This yields

$$A_{\mathrm{B}}(\xi) = (n/\sigma_0^2) \sum_{i=1}^{r} \mathbf{E}_{\tilde{y}, \tilde{w}} \operatorname{Var}(\tilde{\vartheta}_i|\tilde{\boldsymbol{y}}(v_n), \tilde{w}),$$

$$C_{\mathrm{B}}(\xi) = (n/\sigma_0^2)\, \mathbf{E}_{\tilde{y}, w} \operatorname{Var}(\boldsymbol{c}^{\top}\tilde{\boldsymbol{\vartheta}}|\tilde{\boldsymbol{y}}(v_n), \tilde{w}).$$

That is, A_{B}-optimal designs minimize the average expected posterior variance of the regression coefficients and C_{B}-optimal designs minimize the expected posterior variance of the linear combination $\tilde{\beta} = \boldsymbol{c}^{\top}\tilde{\boldsymbol{\vartheta}}$. Furthermore, we have

$$\begin{aligned} I_{\mathrm{B}}(\xi) &= (n/\sigma_0^2) \int_{X_{\mathrm{P}}} \boldsymbol{f}(\boldsymbol{x})^{\top}\, \mathbf{E}_{\tilde{y}, \tilde{w}}\, \mathbf{Cov}(\tilde{\boldsymbol{\vartheta}}|\tilde{\boldsymbol{y}}(v_n), \tilde{w})\boldsymbol{f}(\boldsymbol{x})\, p(\boldsymbol{x})\, \mathrm{d}\boldsymbol{x} \\ &= (n/\sigma_0^2) \int_{X_{\mathrm{P}}} \mathbf{E}_{\tilde{y}, \tilde{w}} \operatorname{Var}(\boldsymbol{f}(\boldsymbol{x})^{\top}\tilde{\boldsymbol{\vartheta}}|\tilde{\boldsymbol{y}}(v_n), \tilde{w})\, p(\boldsymbol{x})\, \mathrm{d}\boldsymbol{x}, \end{aligned}$$

so that I_{B}-optimal designs minimize a weighted average of the expected posterior variance of the response surface over X_{P}.

c) Interpretation of D_{B}- and E_{B}-optimality
D_{B}-optimal designs minimize the determinant of the preposterior covariance matrix (generalized variance) of the regression parameter:

$$D_{\mathrm{B}}(\xi) = \det \boldsymbol{M}_{\mathrm{B}}(v_n)^{-1} = (n/\sigma_0^2)^r \det \mathbf{E}_{\tilde{y}, \tilde{w}}\, \mathbf{Cov}(\tilde{\boldsymbol{\vartheta}}|\tilde{\boldsymbol{y}}(v_n), \tilde{w}).$$

Moreover, we can also give an information theoretic interpretation of the D_{B}-optimality criterion (see Section 14.3).

E_B-optimal designs minimize an upper bound to the expected posterior variance of arbitrary linear combinations of the regression coefficients:

$$E_B(\xi) = \lambda_{\max}(M_B(v_n)^{-1}) = \sup_{c \in \mathbb{R}^r} \left\{ \frac{1}{c^\top c} c^\top M_B(v_n)^{-1} c \right\} \quad \text{(see A.24)}$$

$$= (n/\sigma_0^2) \sup_{c \in \mathbb{R}^r} \left\{ \frac{1}{c^\top c} \mathsf{E}_{\tilde{y}, \tilde{w}} \operatorname{Var}(c^\top \tilde{\vartheta} | \tilde{y}(v_n), \tilde{w}) \right\}.$$

With appropriate choices of the matrix Φ the design criteria considered here also apply to the mixed linear estimator, the minimax linear estimator under quadratic constraints and the hierarchical Bayes estimator (see Sections 7.3.2, 7.3.3 and 5.6). In particular, the minimization of the squared error risks of these estimators amounts to the search for L_B-optimal designs.

Another field of application of the above criteria offers the experimental design problem of supplementing previous observations where the matrix of interest is $(F(v_n)^\top F(v_n) + F_0^\top F_0)$ with F_0 being the fixed design matrix of a previously chosen design. In this context several papers can be found discussing the maximization of the determinant of $(F_1(v_n)^\top F(v_n) + F_0^\top F_0)$, which coincides with D_B-optimality when setting $\Phi^{-1} = F_0^\top F_0$. This is discussed in COVEY-CRUMP/SILVEY (1960), DYKSTRA (1971), MAYER/HENDRICKSON (1973) and EVANS (1979).

Moreover, Bayes designs could also be used for response surface estimation in case of fitting a lower order model when a higher order model is appropriate. Then the simultaneous minimization of the bias and the variance of the fitted response leads to a criterion which is formally identical with L_B-optimality (see KIEFER (1973) and Section 9.2).

Bayesian experimental design for linear regression models has been discussed in SINHA (1970), BROOKS (1972, 1974, 1976), DUNCAN/DE GROOT (1976), BANDEMER et al. (1977, Chapter 7), BANDEMER/PILZ (1978), O'HAGAN (1978), PILZ (1979 c, d, 1981 b, c, 1983), NÄTHER/PILZ (1980), ALI/SINGH (1981), GLADITZ/PILZ (1982 a, b), CHALONER (1982, 1984), STEINBERG (1985) and BANDEMER/NÄTHER/PILZ (1987). In the context of analysis of variance models, Bayes designs are derived in OWEN (1970), SMITH/VERDINELLI (1980) and GIOVAGNOLI/VERDINELLI (1983). The main results will be reported in the following sections; for a brief review of this and further literature centering around Bayesian experimental design we refer to Section 14.3.

11. Characterization of optimal designs

Now we investigate some properties of the functionals of the optimality criteria introduced in the previous section, especially their convexity. These properties allow to constrain the set Ξ in the optimization and to change to essentially complete classes of designs. Furthermore, they enable us to apply the basic equivalence theorem due to WHITTLE (1973) to get some insight in the structure of optimal designs.

11.1. Properties of the design functionals

The Bayesian optimality criterion and the design criteria introduced in the Definition 10.4 and 10.5 depend on the design only through the Bayesian information matrix. Therefore we next collect some useful analytical properties of this matrix.

Lemma 11.1. $M_B(\xi) \in \mathcal{M}_r^{>}$ *for all* $\xi \in \Xi$.

This is clear from the fact that in $\boldsymbol{M}_B(\boldsymbol{\xi}) = \boldsymbol{M}(\xi) + \frac{1}{n}\boldsymbol{\Phi}^{-1}$ the information matrix

$$\boldsymbol{M}(\boldsymbol{\xi}) = \int_{X_E} \boldsymbol{f}(\boldsymbol{x})\boldsymbol{f}(\boldsymbol{x})^{\top}\xi(\mathrm{d}\boldsymbol{x}) \tag{11.1}$$

is nonnegative definite, whatever the design $\xi \in \Xi$, and the a priori part $\boldsymbol{\Phi}^{-1}$ is positive definite by assumption. Thus, in contrast with the information matrix $\boldsymbol{M}(\boldsymbol{\xi})$ for the LSE, the Bayesian information matrix $\boldsymbol{M}_B(\xi)$ is always of full rank r even if the support supp ξ of the design ξ contains less than r points.
Let

$$\Xi_d = \{\xi \in \Xi : \mathrm{card}(\mathrm{supp}\,\xi) < \infty\}$$

denote the subset of designs with a countable support. The elements of Ξ_d are called *discrete designs*. Note that Ξ_d, as well as Ξ, is convex.

Theorem 11.1. *Let* h *be the dimension of the linear space generated by the products* $f_i \cdot f_j$ $(i, j = 1, \ldots, r)$ *of the components of the regression function* $\boldsymbol{f} = (f_1, \ldots, f_r)^{\top}$. *Then for any* $\xi \in \Xi$ *there exists a discrete design* $\xi_d \in \Xi_d$ *such that*

$$\boldsymbol{M}_B(\xi) = \boldsymbol{M}_B(\xi_d) \quad \textit{and} \quad \mathrm{card}(\mathrm{supp}\,\xi_d) \leqq h + 1 \leqq r(r+1)/2 + 1\,.$$

Proof. According to KIEFER (1959) the above assertion holds for $\boldsymbol{M}(\xi)$ instead of $\boldsymbol{M}_B(\xi)$. But $\boldsymbol{M}_B(\xi) = \boldsymbol{M}(\xi) + \frac{1}{n}\boldsymbol{\Phi}^{-1}$ depends on the design only through the information matrix $M(\xi)$ so that the result holds for $\boldsymbol{M}_B(\xi)$, too. □

Hence, if we are to minimize design functionals of the type

$$Z(\xi) = \bar{Z}(M_B(\xi)), \qquad \bar{Z} \,|\, \mathcal{M}_r^{>} \to \mathbb{R}^1, \tag{11.2}$$

as are all the functionals considered in the previous section, we can restrict our attention to the discrete designs from Ξ_d having a finite support which contains no more than $r(r+1)/2+1$ experimental points. For the solution of design problems with functionals Z of the type (11.2) it is important to note that the Bayesian information matrices form a convex set.

Lemma 11.2. (i) *For all designs $\xi_1, \xi_2 \in \Xi$ and $\alpha \in (0,1)$ it holds*

$$M_B(\alpha\xi_1 + (1-\alpha)\xi_2) = \alpha M_B(\xi_1) + (1-\alpha) M_B(\xi_2).$$

(ii) *If $\Xi_0 \subseteqq \Xi$ is a convex set of designs then the set*

$$M_B(\Xi_0) := \{M_B(\xi) \in \mathcal{M}_r^{>} : \xi \in \Xi_0\} \tag{11.3}$$

of corresponding Bayesian information matrices is convex, too.

Proof. (i) Let $\xi_\alpha = \alpha\xi_1 + (1-\alpha)\xi_2$. Because of $\xi_\alpha(dx) = \alpha\xi_1(dx) + (1-\alpha)\xi_2(dx)$ we then obtain

$$\begin{aligned} M_B(\xi_\alpha) &= \int_{X_E} f(x)f(x)^\top \xi_\alpha(dx) + \frac{1}{n}\Phi^{-1} \\ &= \alpha \int_{X_E} f(x)f(x)^\top \xi_1(dx) + (1-\alpha)\int_{X_E} f(x)f(x)^\top \xi_2(dx) + \frac{1}{n}\Phi^{-1} \\ &= \alpha\left(M(\xi_1) + \frac{1}{n}\Phi^{-1}\right) + (1-\alpha)\left(M(\xi_2) + \frac{1}{n}\Phi^{-1}\right) \\ &= \alpha M_B(\xi_1) + (1-\alpha) M_B(\xi_2). \end{aligned}$$

(ii) follows immediately from (i). □

We now prove the convexity of the functionals defined in the preceding section.

Theorem 11.2. *For any matrix $U_L \in \mathcal{M}_r^{\geqq}$ the functional $L_B(\cdot) = \operatorname{tr} U_L M_B(\cdot)^{-1}$ is convex on Ξ.*

Proof. Let $\xi_1, \xi_2 \in \Xi$ and $\alpha \in (0,1)$. Then, according to Lemma 11.2 (i) and A.30, it holds

$$\begin{aligned} M_B(\alpha\xi_1 + (1-\alpha)\xi_2)^{-1} &= (\alpha M_B(\xi_1) + (1-\alpha) M_B(\xi_2))^{-1} \\ &\leqq \alpha M_B(\xi_1)^{-1} + (1-\alpha) M_B(\xi_2)^{-1}. \end{aligned}$$

From this it follows with A.28:

$$\begin{aligned} \operatorname{tr} U_L M_B(\alpha\xi_1 + (1-\alpha)\xi_2)^{-1} &\leqq \operatorname{tr} U_L(\alpha M_B(\xi_1)^{-1} + (1-\alpha) M_B(\xi_2)^{-1}) \\ &= \alpha \operatorname{tr} U_L M_B(\xi_1)^{-1} + (1-\alpha)\operatorname{tr} U_L M_B(\xi_2)^{-1}. \quad \square \end{aligned}$$

Corollary 11.1. *The functionals $A_B(\cdot)$, $B(\cdot)$, $C_B(\cdot)$ and $I_B(\cdot)$ are convex on Ξ.*

This result is clear from Theorem 11.2 since the Bayesian optimality criterion and the criteria of A_B-, C_B- and I_B-optimality were proved to be special cases of L_B-optimality.

Theorem 11.3. *The functionals* $D_B(\cdot) = \det \boldsymbol{M}_B(\cdot)^{-1}$ *and* $E_B(\cdot) = \lambda_{\max}(\boldsymbol{M}_B(\cdot)^{-1})$ *are convex on* Ξ.

Proof. Let $\xi_1, \xi_2 \in \Xi$, $\alpha \in (0,1)$ and $\xi_\alpha = \alpha\xi_1 + (1-\alpha)\xi_2$. We show that $E'_B(\cdot) = 1/E_B(\cdot)$ and $D'_B(\cdot) = 1/D_B(\cdot)$ are concave, the convexity of $E_B(\cdot)$ and $D_B(\cdot)$ then holds according to A.18.
First, because of $E'_B(\cdot) = 1/\lambda_{\max}(\boldsymbol{M}_B(\cdot)^{-1}) = \lambda_{\min}(\boldsymbol{M}_B(\cdot))$, it follows from A.25 that

$$\begin{aligned} E'_B(\xi_\alpha) &= \lambda_{\min}(\alpha\boldsymbol{M}_B(\xi_1) + (1-\alpha)\boldsymbol{M}_B(\xi_2)) \\ &\geqq \lambda_{\min}(\alpha\boldsymbol{M}_B(\xi_1)) + \lambda_{\min}((1-\alpha)\boldsymbol{M}_B(\xi_2)) \\ &= \alpha\lambda_{\min}(\boldsymbol{M}_B(\xi_1)) + (1-\alpha)\lambda_{\min}(\boldsymbol{M}_B(\xi_2)) \\ &= \alpha E'_B(\xi_1) + (1-\alpha)E'_B(\xi_2). \end{aligned}$$

Further, with $D'_B(\cdot) = 1/\det \boldsymbol{M}_B(\cdot)^{-1} = \det \boldsymbol{M}_B(\cdot)$, the concavity of D'_B results as follows:

$$\begin{aligned} D'_B(\xi_\alpha) &= \det(\alpha\boldsymbol{M}_B(\xi_1) + (1-\alpha)\boldsymbol{M}_B(\xi_2)) \\ &\geqq (\det \boldsymbol{M}_B(\xi_1))^\alpha (\det \boldsymbol{M}_B(\xi_2))^{1-\alpha} \quad \text{(from A.31)} \\ &\geqq \alpha \det \boldsymbol{M}_B(\xi_1) + (1-\alpha)\det \boldsymbol{M}_B(\xi_2) \quad \text{(from A.21)} \\ &= \alpha D'_B(\xi_1) + (1-\alpha)D'_B(\xi_2). \ \square \end{aligned}$$

Next we give some statements concerning the existence and uniqueness of optimal designs and their Bayesian information matrices. It can be shown that, whithin the class Ξ of all approximate designs, there always exists an optimal design and thus also a discrete optimal design with finite support.

Theorem 11.4. *For every functional* $Z \in \{L_B, B, A_B, C_B, I_B, D_B, E_B\}$ *there exists a design* $\xi^* \in \Xi_d$ *such that*

$$Z(\xi^*) = \inf_{\xi \in \Xi} Z(\xi) \quad \textit{and} \quad \operatorname{card}(\operatorname{supp} \xi^*) \leqq r(r+1)/2 + 1.$$

Proof. The above functionals are continuous on $M_B(\Xi)$ with respect to the metric m generated by the usual sum of squares norm for matrices. Furthermore, the set $M_B(\Xi) = \{M_B(\xi) : \xi \in \Xi\}$ is compact with respect to m so that the result follows with Theorem 11.1 from the well-known Weierstrass-theorem guaranteeing the existence of the minimum of a continuous function over a compact set. $\square$

For the criterion of L_B-optimality the upper bound $r(r+1)/2+1$ on the number of points at which to take observations can still be improved.

Theorem 11.5. *Let $U_{\mathrm{L}} \in \mathcal{M}_r^{\geqq}$ have rank $U_{\mathrm{L}} = s$, $1 \leqq s \leqq r$. Then there exists an L_{B}-optimal design in Ξ with respect to U_{L} that includes at most $\frac{s}{2}(2r - s + 1)$ different experimental points.*

(See CHALONER (1984), Theorem 2; cp. also PUKELSHEIM (1980), Theorem 6.)

Particularly, for the case of A_{B}-optimality ($U_{\mathrm{L}} = I_r$, $s = r$) this yields the upper bound $\frac{r}{2}(r+1)$ and for the case of C_{B}-optimality ($U_{\mathrm{L}} = cc^{\top}$, $s = 1$) an optimal design requires at most r observations.

In the following let

$$\Xi_Z = \left\{\xi^* \in \Xi : Z(\xi^*) = \inf_{\xi \in \Xi} Z(\xi)\right\} \tag{11.4}$$

denote the set of all Z-optimal designs, where Z is some continuous functional of the Bayesian information matrix, and

$$M_{\mathrm{B}}(\Xi_Z) = \{M_{\mathrm{B}}(\xi) : \xi \in \Xi_Z\}.$$

Lemma 11.3. *For every $Z \in \{L_{\mathrm{B}}, B, C_{\mathrm{B}}, I_{\mathrm{B}}, D_{\mathrm{B}}, E_{\mathrm{B}}\}$ the sets Ξ_Z and $M_{\mathrm{B}}(\Xi_Z)$ are convex.*

This is clear from the convexity of these functionals. Hence, the convex combination of any two optimal designs again yields an optimal design. It turns out that, for certain functionals, the optimal designs all have the same (Bayesian) information matrix.

Theorem 11.6. *If the matrix U_{L} of the functional $L_{\mathrm{B}}(\cdot) = \operatorname{tr} U_{\mathrm{L}} M_{\mathrm{B}}(\cdot)^{-1}$ is such that*

$$\operatorname{tr} U_{\mathrm{L}} C > 0 \quad \text{for all } C \in \mathcal{M}_r^{\geqq},\ C \neq 0 \tag{11.5}$$

then it holds: $\operatorname{card}(M_{\mathrm{B}}(\Xi_{L_{\mathrm{B}}})) = 1$.

Proof. Under the Condition (11.5) the functional $Z = L_{\mathrm{B}}$, viewed as a function $\bar{Z}(\cdot) = tr\, U_{\mathrm{L}}(\cdot)^{-1}$ on $M_{\mathrm{B}}(\Xi) \subseteqq \mathcal{M}_r^{\geqq}$, is strictly convex (see A.32). The uniqueness of the Bayesian information matrix of the L_{B}-optimal design then comes from A.18. □

Remark 11.1. Especially, the Condition (11.5) is satisfied if U_L is diagonal with positive entries.

Corollary 11.2. *The Bayesian information matrix of A_{B}- and D_{B}-optimal designs is uniquely determined, i.e.,* $\operatorname{card}(M_{\mathrm{B}}(\Xi_Z)) = 1$ *for $Z \in \{A_{\mathrm{B}}, D_{\mathrm{B}}\}$.*

Proof. For $Z(\cdot) = A_{\mathrm{B}}(\cdot) = \operatorname{tr} M_{\mathrm{B}}(\cdot)^{-1}$ this follows immediately from Theorem 11.6, since with $U_{\mathrm{L}} = I_r$ condition (11.5) is satisfied. Further, along the same lines of the proof of Theorem 11.3, it can be shown that $\bar{Z}(\cdot) = \det(\cdot)^{-1}$ is strictly convex on $M_{\mathrm{B}}(\Xi)$, using the fact that for $\xi_1, \xi_2 \in \Xi$ and $\alpha \in (0,1)$ we have strict inequality

$$\det(\alpha \boldsymbol{M}_{\mathrm{B}}(\xi_1) + (1-\alpha)\boldsymbol{M}_{\mathrm{B}}(\xi_2)) > (\det \boldsymbol{M}_{\mathrm{B}}(\xi_1))^{\alpha}(\det \boldsymbol{M}_{\mathrm{B}}(\xi_2))^{1-\alpha}$$

(see A.31). This strict convexity then implies that $\mathrm{card}(M_{\mathrm{B}}(\Xi_{D_{\mathrm{B}}})) = 1$ (see A.18). □

From Theorem 11.6 we can also deduce a sufficient condition for the uniqueness of the information matrix of Bayes designs with respect to some prior measure $P_{\tilde{x}}$ over the prediction region X_{P}.

Corollary 11.3. *If X_{P} contains r points $\boldsymbol{x}_1, \ldots, \boldsymbol{x}_r$ such that $\boldsymbol{f}(\boldsymbol{x}_1), \ldots, \boldsymbol{f}(\boldsymbol{x}_r)$ are linear independent vectors and if supp $P_{\tilde{x}} = X_{\mathrm{P}}$ then it holds*: $\mathrm{card}(M_{\mathrm{B}}(\Xi_{\mathrm{B}})) = 1$.

Proof. We show that under the assumptions of Corollary 11.3 the matrix $\boldsymbol{U}_{\mathrm{L}} = \int_{X_{\mathrm{P}}} \boldsymbol{f}(\boldsymbol{x})\boldsymbol{f}(\boldsymbol{x})^{\top} P_{\tilde{x}}(\mathrm{d}\boldsymbol{x})$, for which we have $\mathrm{tr}\, \boldsymbol{U}_{\mathrm{L}} \boldsymbol{M}_{\mathrm{B}}(\cdot)^{-1} = B(\cdot)$ (cp. Theorem 10.1), satisfies the Condition (11.5).

First we observe that

$$\mathrm{tr}\, \boldsymbol{U}_{\mathrm{L}} \boldsymbol{C} = \int_{X_{\mathrm{P}}} \boldsymbol{f}(\boldsymbol{x})^{\top} \boldsymbol{C} \boldsymbol{f}(\boldsymbol{x})\, P_{\tilde{x}}(\mathrm{d}\boldsymbol{x}) \geqq 0$$

for any $\boldsymbol{C} \in \mathcal{M}_r^{\geqq}$. Now, assume there exists some $\boldsymbol{C}_0 \in \mathcal{M}_r^{\geqq}$, $\boldsymbol{C}_0 \neq \boldsymbol{0}$, for which $\mathrm{tr}\, \boldsymbol{U}_{\mathrm{L}} \boldsymbol{C}_0 = 0$. Let $\boldsymbol{\Lambda} = \mathrm{diag}(\lambda_1, \ldots, \lambda_r)$ and $\boldsymbol{Q} = (\boldsymbol{q}_1, \ldots, \boldsymbol{q}_r)$ be the matrices of eigenvalues and orthonormal eigenvectors of $\boldsymbol{C}_0$, respectively. Since $\boldsymbol{C}_0 \neq \boldsymbol{0}$, at least one of the eigenvalues is positive, say $\lambda_j > 0$. Then, because of our assumption that supp $P_{\tilde{x}} = X_{\mathrm{P}}$, the equality $\mathrm{tr}\, \boldsymbol{U}_{\mathrm{L}} \boldsymbol{C}_0 = 0$ would necessarily imply that

$$\boldsymbol{f}(\boldsymbol{x})^{\top} \boldsymbol{C}_0 \boldsymbol{f}(\boldsymbol{x}) = \boldsymbol{f}(\boldsymbol{x})^{\top} \boldsymbol{Q} \boldsymbol{\Lambda} \boldsymbol{Q} \boldsymbol{f}(\boldsymbol{x}) = \sum_{i=1}^{r} \lambda_i (\boldsymbol{f}(\boldsymbol{x})^{\top} \boldsymbol{q}_i)^2 = 0,$$

for all $\boldsymbol{x} \in X_{\mathrm{P}}$; i.e., especially, $\boldsymbol{f}(\cdot)^{\top} \boldsymbol{q}_j \equiv 0$ everywhere on X_{P}. But, since $\boldsymbol{f}(X_{\mathrm{P}}) = \{\boldsymbol{f}(\boldsymbol{x}) \in \mathbb{R}^r : \boldsymbol{x} \in X_{\mathrm{P}}\}$ contains a system of basis vectors of $\mathbb{R}^r$, this in turn implies that $\boldsymbol{q}_j = \boldsymbol{0}$ which is contradictory to $\boldsymbol{q}_j$ being a nondegenerate eigenvector with $\boldsymbol{q}_j^{\top} \boldsymbol{q}_j = 1$. □

Remark 11.2. From Theorem 10.2 (iii) it is clear that, under the condition on X_{P} given in the above corollary, also the information matrix of I_{B}-optimal designs with respect to p is unique if the weight function p is strictly positive on X_{P}.

We remark, however, that the uniqueness of the optimal information matrix does not mean that the optimal design itself is uniquely determined.

An important feature of all the design criteria considered so far is the following antitonicity property.

Definition 11.1. The functional $Z \mid \Xi \to \mathbb{R}^1$ is called ***M**-antitonic* if it holds

$$\forall(\xi_1, \xi_2) \in \Xi \times \Xi : \boldsymbol{M}(\xi_1) \leqq \boldsymbol{M}(\xi_2) \Rightarrow Z(\xi_1) \geqq Z(\xi_2).$$

Remark 11.3. Since $\boldsymbol{M}(\xi_1) \leqq \boldsymbol{M}(\xi_2)$ is equivalent to $\boldsymbol{M}_{\mathrm{B}}(\xi_2)^{-1} \leqq \boldsymbol{M}_{\mathrm{B}}(\xi_1)^{-1}$, an $\boldsymbol{M}$-antitonic functional Z is isotonic relative to the inverse Bayesian information matrix, i.e.,

$$\boldsymbol{M}_{\mathrm{B}}(\xi_2)^{-1} \leqq \boldsymbol{M}_{\mathrm{B}}(\xi_1)^{-1} \Rightarrow Z(\xi_2) \leqq Z(\xi_1).$$

If we restrict ourselves to exact designs then, in view of Lemma 10.2, $\boldsymbol{M}$-antitonicity means that a design v_n which leads to a smaller expected posterior covariance matrix than another design $\bar{v}_n$, also results in a smaller functional value.

Theorem 11.7. *The functionals $L_{\mathrm{B}}, B, A_{\mathrm{B}}, C_{\mathrm{B}}, I_{\mathrm{B}}, D_{\mathrm{B}}$ and E_{B} are **M**-antitonic.*

Proof. Let $\xi_1, \xi_2 \in \Xi$ and $\boldsymbol{M}(\xi_1) \leqq \boldsymbol{M}(\xi_2)$. Then we have $\boldsymbol{M}_{\mathrm{B}}(\xi_2)^{-1} \leqq \boldsymbol{M}_{\mathrm{B}}(\xi_1)^{-1}$ and with A.28 it follows for any $\boldsymbol{U}_{\mathrm{L}} \in \mathcal{M}_r^{\geqq}$:

$$L_{\mathrm{B}}(\xi_2) = \operatorname{tr} \boldsymbol{U}_{\mathrm{L}} \boldsymbol{M}_{\mathrm{B}}(\xi_2)^{-1} \leqq \operatorname{tr} \boldsymbol{U}_{\mathrm{L}} \boldsymbol{M}_{\mathrm{B}}(\xi_1)^{-1} = L_{\mathrm{B}}(\xi_1).$$

Hence, the statement is true for L_{B} and thus also for its specializations $B, A_{\mathrm{B}}, C_{\mathrm{B}}, I_{\mathrm{B}}$. Furthermore, from A.26 and A.27 we get

$$D_{\mathrm{B}}(\xi_2) = \det \boldsymbol{M}_{\mathrm{B}}(\xi_2)^{-1} \leqq \det \boldsymbol{M}_{\mathrm{B}}(\xi_1)^{-1} = D_{\mathrm{B}}(\xi_1)$$

and

$$E_{\mathrm{B}}(\xi_2) = \lambda_{\max}(\boldsymbol{M}_{\mathrm{B}}(\xi_2)^{-1}) \leqq \lambda_{\max}(\boldsymbol{M}_{\mathrm{B}}(\xi_1)^{-1}) = E_{\mathrm{B}}(\xi_1),$$

so that the statement is true for D_{B} and E_{B}, too. □

11.2. Admissibility and complete classes of designs

The use of the concept of admissibility of designs makes it possible to constrain the set Ξ of all approximate designs competing in the solution of design problems of the type

$$Z(\xi) = \bar{Z}(\boldsymbol{M}_{\mathrm{B}}(\xi)) \stackrel{!}{=} \inf_{\xi \in \Xi}; \qquad \bar{Z} \mid \mathcal{M}_r^{>} \to \mathbb{R}^1$$

and to change to (essentially) complete classes of designs.

Definition 11.2. The class $\Xi_0 \subseteqq \Xi$ is called

(i) *essentially **M**-complete* in Ξ if it holds

$$\forall \xi \in \Xi \setminus \Xi_0 \; \exists \xi_0 \in \Xi_0 : \boldsymbol{M}(\xi) \leqq \boldsymbol{M}(\xi_0),$$

(ii) ***M**-complete* in Ξ if it holds

$$\forall \xi \in \Xi \setminus \Xi_0 \; \exists \xi_0 \in \Xi_0 : \boldsymbol{M}(\xi) \leqq \boldsymbol{M}(\xi_0) \wedge \boldsymbol{M}(\xi_0) \neq \boldsymbol{M}(\xi).$$

Definition 11.3. The design $\xi_0 \in \Xi$ is called ***M-admissible*** in Ξ if there is no design $\xi \in \Xi$ for which $\boldsymbol{M}(\xi_0) \leqq \boldsymbol{M}(\xi)$ and $\boldsymbol{M}(\xi) \neq \boldsymbol{M}(\xi_0)$.

These concepts as they are commonly used in the theory of optimal experimental design are more global and much easier to verify than the corresponding decision theoretic concepts, which are based on the risk function. Some statements concerning the relationships between these different concepts may be found in BANDEMER et al. (1977), PILZ (1979c) and NÄTHER/PILZ (1980). Roughly speaking, every admissible design is $\boldsymbol{M}$-admissible and the (essentially) $\boldsymbol{M}$-complete classes include the (essentially) complete classes.

In a natural way, the possibility of a restriction to an $\boldsymbol{M}$-complete class is closely connected with the $\boldsymbol{M}$-antitonicity of the design functional.

Theorem 11.8. *If Z is an **M**-antitonic functional on Ξ then every essentially **M**-complete class in Ξ contains a Z-optimal design.*

Proof. Let $\Xi_0 \subseteqq \Xi$ be essentially $\boldsymbol{M}$-complete and $\xi^* \in \Xi$ be Z-optimal. If $\xi^* \notin \Xi_0$ then, by Definition 11.2(i), there exists $\xi_0 \in \Xi_0$ such that $\boldsymbol{M}(\xi^*) \leqq \boldsymbol{M}(\xi_0)$. The $\boldsymbol{M}$-antitonicity of Z then implies $Z(\xi_0) \leqq Z(\xi^*)$. But, since ξ^* is Z-optimal, it also holds $Z(\xi^*) \leqq Z(\xi_0)$. Thus, we have $Z(\xi_0) = Z(\xi^*) = \inf_{\xi \in \Xi} Z(\xi)$ so that $\xi_0 \in \Xi_0$ is Z-optimal. □

Corollary 11.4. *Let $Z \in \{L_B, B, A_B, C_B, I_B, D_B, E_B\}$. Then every essentially **M**-complete class in Ξ contains a Z-optimal design.*

We now give an example of an essentially $\boldsymbol{M}$-complete class of designs for the case that the regression function $\boldsymbol{f}$ satisfies a certain convexity property.

Let $\boldsymbol{f}(X_E) = \{\boldsymbol{f}(\boldsymbol{x}) \in \mathbb{R}^r : \boldsymbol{x} \in X_E\}$ be the image of $\boldsymbol{f}$ restricted on X_E. If $\boldsymbol{f}(X_E)$ appears to be a convex set, then let H_f denote the set of all extreme points in $\boldsymbol{f}(X_E)$, i.e., the set of all points from $\boldsymbol{f}(X_E)$ which cannot be written as a proper convex linear combination of any two different points from $\boldsymbol{f}(X_E)$.

Theorem 11.9. *If the regression function **f** is such that $\boldsymbol{f}(X_E)$ is convex then*

$$\Xi_0 := \{\xi \in \Xi : \operatorname{supp} \xi \subseteqq \{\boldsymbol{x} \in X_E : \boldsymbol{f}(\boldsymbol{x}) \in H_f\}\}$$

*is an essentially **M**-complete class of designs in Ξ.*
(See EHRENFELD (1956).)

Remark 11.4. Consider a multiple linear regression

$$\mathbf{E}\tilde{y}(\boldsymbol{x}) = \vartheta_0 + \sum_{i=1}^{k} \vartheta_i x_i, \qquad \boldsymbol{x} = (x_1, \ldots, x_k)^\top \in X_E \subset \mathbb{R}^k,$$

i.e., $\boldsymbol{f}(\boldsymbol{x}) = (1, x_1, \ldots, x_k)^\top$ and $\boldsymbol{\vartheta} = (\vartheta_0, \vartheta_1, \ldots, \vartheta_k)^\top$. If the experimental region X_E is convex then $\boldsymbol{f}(X_E)$ is convex, too, and it suffices to consider the extreme points of X_E. Especially, if X_E is a convex polyhedron or a sphere then the subset of all discrete measures whose supporting points are corner points of the polyhedron or lie on the surface of the sphere, forms an essentially $\boldsymbol{M}$-complete class.

Hence, in case of convexity of $\boldsymbol{f}(X_E)$, the search for an optimal design can be done within the subset Ξ_0 given in Theorem 11.9. A further characterization of the essentially $\boldsymbol{M}$-complete classes of designs may be found in KIEFER (1959).

An essentially $\boldsymbol{M}$-complete class does not necessarily contain all optimal designs. However, it can be shown that, in certain cases, the class of $\boldsymbol{M}$-admissible designs includes all the optimal designs.

Theorem 11.10. (i) *If Z is an $\boldsymbol{M}$-antitonic functional then the class Ξ_{ad} of $\boldsymbol{M}$-admissible designs in Ξ contains at least one design which is Z-optimal in Ξ.*

(ii) *If Z is strictly convex and $\boldsymbol{M}$-antitonic then Ξ_{ad} includes all Z-optimal designs in Ξ.*

Proof. (i) From the Definitions 11.2 and 11.3 it is clear that the class Ξ_{ad} is $\boldsymbol{M}$-complete and thus also essentially $\boldsymbol{M}$-complete. The statement then results from Theorem 11.8.

(ii) Assume there exists a Z-optimal design $\xi^* \in \Xi$ which is not $\boldsymbol{M}$-admissible in Ξ. Then there must exist $\xi_0 \in \Xi$ such that $\boldsymbol{M}(\xi^*) \leqq \boldsymbol{M}(\xi_0)$ and $\boldsymbol{M}(\xi^*) \neq \boldsymbol{M}(\xi_0)$. This yields $Z(\xi_0) \leqq Z(\xi^*)$, since Z is $\boldsymbol{M}$-antitonic. Now, let $\alpha \in (0, 1)$ and $\xi_\alpha = \alpha\xi_0 + (1-\alpha)\,\xi^*$. The strict convexity of Z then yields

$$\begin{aligned} Z(\xi_\alpha) &< \alpha Z(\xi_0) + (1-\alpha)\, Z(\xi^*) \\ &\leqq \alpha Z(\xi^*) + (1-\alpha)\, Z(\xi^*) = Z(\xi^*), \end{aligned}$$

i.e., $Z(\xi_\alpha) < Z(\xi^*)$ in contradiction to Z-optimality of ξ^*. □

Corollary 11.5. *The class Ξ_{ad} contains all A_B- and D_B-optimal designs. If the matrix $\boldsymbol{U}_L \in \mathcal{M}_r^{\geqq}$ satisfies the condition* (11.5) *then Ξ_{ad} contains all L_B-optimal designs with respect to $\boldsymbol{U}_L$, too.*

This follows from Theorem 11.10 (ii), observing that the functionals A_B, D_B and L_B are $\boldsymbol{M}$-antitonic and strictly convex (see Theorem 11.7 and the proofs of Theorem 11.6 and Corollary 11.2).

Similarly, under the conditions on X_P and on the prior distribution $P_{\tilde{x}}$ formulated in Corollary 11.3, Ξ_{ad} includes all Bayes designs.

Concluding, we remark that for the construction of the class of $\boldsymbol{M}$-admissible designs the characterization given in KARLIN/STUDDEN (1966) can be used, which says that $\xi \in \Xi$ is $\boldsymbol{M}$-admissible if there exists a positive definite matrix $\boldsymbol{T}$ such that

$$\boldsymbol{f}(\boldsymbol{x})^\top \boldsymbol{T}\boldsymbol{f}(\boldsymbol{x}) \leqq 1 \quad \text{for all} \quad \boldsymbol{x} \in X_E$$

and

$$\boldsymbol{f}(\boldsymbol{x})^\top \boldsymbol{T}\boldsymbol{f}(\boldsymbol{x}) = 1 \quad \text{for all} \quad \boldsymbol{x} \in \operatorname{supp} \xi.$$

Furthermore, it is shown there that $\boldsymbol{M}$-admissibility only depends on the supporting points of the design but not on the weights given to them.

11.3. Equivalence theorems

We now formulate equivalent conditions for the optimality of designs relative to the criteria introduced in the Definitions 10.4 and 10.5. This will be done by application of the general equivalence theorem due to WHITTLE (1973). Particularly, from this theorem we can also deduce statements on the supporting points of optimal designs and inequalities for the evaluation of the goodness of any design at hand.

Let $\delta_x \in \Xi$ denote the one-point design giving weight $p = 1$ to the single point $\boldsymbol{x} \in X_{\mathrm{E}}$ and zero probability to all the remaining points of the experimental region.

Assumption 11.1. *Let $\bar{\Xi} \subseteqq \Xi$ be a convex subset such that*

$$\bar{\Xi} \supseteqq \{\delta_x \in \Xi : \boldsymbol{x} \in X_{\mathrm{E}}\}.$$

We first formulate Whittle's equivalence theorem for an arbitrary design functional of the type (11.2) which is to be minimized over $\bar{\Xi}$,

$$Z(\xi) = \bar{Z}(\boldsymbol{M}_{\mathrm{B}}(\xi)) \overset{!}{=} \inf_{\xi \in \bar{\Xi}}.$$

Here the subset $\bar{\Xi}$ of Ξ may be, for example, the set Ξ_{d} of all discrete measures or some essentially $\boldsymbol{M}$-complete class, possibly enlarged such that Assumption 11.1 is satisfied.

Assumption 11.2. *Let $\bar{Z} \mid M_{\mathrm{B}}(\bar{\Xi}) \to \mathbb{R}^1$ be convex, lower semicontinuous with respect to the usual sum of squares norm $\| \boldsymbol{M}_{\mathrm{B}}(\xi) \| = (\operatorname{tr} \boldsymbol{M}_{\mathrm{B}}(\xi)^2)^{1/2}$ and such that*

$$\Delta_Z(\xi, \bar{\xi}) := \lim_{\alpha \downarrow 0} \frac{\mathrm{d}}{\mathrm{d}\alpha} Z((1-\alpha)\,\xi + \alpha\bar{\xi}) \tag{11.6}$$

exists for all pairs $(\xi, \bar{\xi}) \in \bar{\Xi} \times \bar{\Xi}$.

The limiting value $\Delta_Z(\xi, \bar{\xi})$ defined by (11.6) is called the *directional derivative* of Z at ξ in the direction of $\bar{\xi}$. It will be shown in the sequel that the directional derivative exists in case of the functionals L_{B}, B, A_{B}, C_{B}, I_{B}, D_{B} and E_{B} so that Assumption 11.2 is satisfied for them.

Definition 11.4. Δ_Z is called *linear* on $\bar{\Xi} \times \bar{\Xi}$ if it holds

$$\Delta_Z(\xi, \bar{\xi}) = \int_{X_{\mathrm{E}}} \Delta_Z(\xi, \delta_x)\, \bar{\xi}(\mathrm{d}\boldsymbol{x}) \quad \text{for all} \quad (\xi, \bar{\xi}) \in \bar{\Xi} \times \bar{\Xi}.$$

Lemma 11.4. *If Δ_Z is linear then it holds*

$$\inf_{\bar{\xi} \in \bar{\Xi}} \Delta_Z(\xi, \bar{\xi}) = \inf_{x \in X_E} \Delta_Z(\xi, \delta_x).$$

Let $\bar{\Xi}_Z$ denote the set of all designs which minimize $Z(\cdot)$ over $\bar{\Xi}$. According to WHITTLE (1973), Z-optimal designs in $\bar{\Xi}$ can be characterized as follows.

Theorem 11.11. *If the Assumptions* 11.1 *and* 11.2 *are satisfied and Δ_Z is linear then it holds:*

(i) $\xi^* \in \bar{\Xi}_Z \Leftrightarrow \inf_{x \in X_E} \Delta_Z(\xi^*, \delta_x) = 0$,

(ii) $\Delta_Z(\xi, \xi^*) \leqq \Delta_Z(\xi^*, \xi^*) = 0 \leqq \Delta_Z(\xi^*, \bar{\xi})$ for all $\xi, \bar{\xi} \in \bar{\Xi}, \xi^* \in \bar{\Xi}_Z$,

(iii) $\operatorname{supp} \xi^* \subseteqq \{x \in X_E : \Delta_Z(\xi^*, \delta_x) = 0\}$ for all $\xi^* \in \bar{\Xi}_Z$.

Note that the lower semicontinuity of $\bar{Z}$ as demanded in our Assumption 11.2 guarantees the applicability of Whittle's equivalence theorem to our Bayes design problems of minimizing $Z(\cdot) = \bar{Z}(\boldsymbol{M}_B(\cdot))$ over $\bar{\Xi}$ (cp. PUKELSHEIM (1980), pp. 354).

The necessary and sufficient condition (i) of Theorem 11.11 allows to prove Z-optimality of a given design and may serve as a basis for the construction of Z-optimal designs. Also Theorem 11.11 (iii), which gives a necessary condition on the supporting points of Z-optimal designs, may be helpful for the construction of such designs. The condition (ii) indicates a saddle-point property of the directional derivative which, together with the following statement, allows to derive bounds for the functional value of a Z-optimal design.

Lemma 11.5. *If Δ_Z is linear then it holds*

$$Z(\xi) + \inf_{x \in X_E} \Delta_Z(\xi, \delta_x) \leqq Z(\xi^*) \leqq Z(\xi)$$

for all $\xi \in \bar{\Xi}$, $\xi^* \in \bar{\Xi}_Z$.

Proof. From the convexity of Z follows that

$$\Delta_Z(\xi, \xi^*) \leqq \lim_{\alpha \downarrow 0} \frac{\mathrm{d}}{\mathrm{d}\alpha}((1-\alpha) Z(\xi) + \alpha Z(\xi^*)) = Z(\xi^*) - Z(\xi).$$

On the other hand, we have from Lemma 11.4 that

$$\inf_{x \in X_E} \Delta_Z(\xi, \delta_x) \leqq \Delta_Z(\xi, \xi^*) \quad \text{and thus}$$

$$\inf_{x \in X_E} \Delta_Z(\xi, \delta_x) \leqq Z(\xi^*) - Z(\xi) = \inf_{\bar{\xi} \in \bar{\Xi}} Z(\bar{\xi}) - Z(\xi) \leqq 0. \square$$

Now we specialize the statements of Theorem 11.11 to our design criteria which are based on the Bayesian information matrix. Before doing so we will state two properties of this matrix which are useful in calculating the corresponding directional derivatives and in proving their linearity.

Lemma 11.6. (i) $\boldsymbol{M}_{\mathrm{B}}(\xi) = \int_{X_{\mathrm{E}}} \boldsymbol{M}_{\mathrm{B}}(\delta_x)\, \xi(\mathrm{d}x)$ *for all* $\xi \in \Xi$,

(ii) $\frac{\mathrm{d}}{\mathrm{d}\alpha} \boldsymbol{M}_{\mathrm{B}}(\xi_\alpha)^{-1} = \boldsymbol{M}_{\mathrm{B}}(\xi_\alpha)^{-1} (\boldsymbol{M}_{\mathrm{B}}(\xi) - \boldsymbol{M}_{\mathrm{B}}(\bar{\xi}))\, \boldsymbol{M}_{\mathrm{B}}(\xi_\alpha)^{-1}$

for all $(\xi, \bar{\xi}) \in \Xi \times \Xi$, $\alpha \in (0,1)$ *and* $\xi_\alpha = (1-\alpha)\,\xi + \alpha\bar{\xi}$.

Proof. (i) For all $\boldsymbol{x} \in X_{\mathrm{E}}$ we have $\boldsymbol{M}_{\mathrm{B}}(\delta_x) = \boldsymbol{M}(\delta_x) + \frac{1}{n}\boldsymbol{\Phi}^{-1} = \boldsymbol{f}(\boldsymbol{x})\boldsymbol{f}(\boldsymbol{x})^\top + \frac{1}{n}\boldsymbol{\Phi}^{-1}$ and thus

$$\int_{X_{\mathrm{E}}} \boldsymbol{M}_{\mathrm{B}}(\delta_x)\, \xi(\mathrm{d}\boldsymbol{x}) = \int_{X_{\mathrm{E}}} \boldsymbol{f}(\boldsymbol{x})\boldsymbol{f}(\boldsymbol{x})^\top \xi(\mathrm{d}\boldsymbol{x}) + \frac{1}{n}\boldsymbol{\Phi}^{-1} = \boldsymbol{M}_{\mathrm{B}}(\xi)\,.$$

$$\text{(ii)}\quad \frac{\mathrm{d}}{\mathrm{d}\alpha} \boldsymbol{M}_{\mathrm{B}}(\xi_\alpha)^{-1} = \boldsymbol{M}_{\mathrm{B}}(\xi_\alpha)^{-1} \left\{ -\frac{\mathrm{d}}{\mathrm{d}\alpha} \boldsymbol{M}_{\mathrm{B}}(\xi_\alpha) \right\} \boldsymbol{M}_{\mathrm{B}}(\xi_\alpha)^{-1} \qquad \text{(see A.38)}$$

$$= \boldsymbol{M}_{\mathrm{B}}(\xi_\alpha)^{-1} \left\{ -\frac{\mathrm{d}}{\mathrm{d}\alpha} [(1-\alpha)\, \boldsymbol{M}_{\mathrm{B}}(\xi) + \alpha \boldsymbol{M}_{\mathrm{B}}(\bar{\xi})] \right\} \boldsymbol{M}_{\mathrm{B}}(\xi_\alpha)^{-1}$$

$$= \boldsymbol{M}_{\mathrm{B}}(\xi_\alpha)^{-1} \{\boldsymbol{M}_{\mathrm{B}}(\xi) - \boldsymbol{M}_{\mathrm{B}}(\bar{\xi})\}\, \boldsymbol{M}_{\mathrm{B}}(\xi_\alpha)^{-1}. \;\square$$

For the functional $Z(\cdot) = L_{\mathrm{B}}(\cdot) = \operatorname{tr} \boldsymbol{U}_{\mathrm{L}} \boldsymbol{M}_{\mathrm{B}}(\cdot)^{-1}$ we obtain the following result on the directional derivative.

Lemma 11.7. (i) $\Delta_{L_{\mathrm{B}}}(\xi, \bar{\xi}) = L_{\mathrm{B}}(\xi) - \operatorname{tr} \boldsymbol{U}_{\mathrm{L}} \boldsymbol{M}_{\mathrm{B}}(\xi)^{-1} \boldsymbol{M}_{\mathrm{B}}(\bar{\xi})\, \boldsymbol{M}_{\mathrm{B}}(\xi)^{-1}$,

(ii) $\Delta_{L_{\mathrm{B}}}$ *is linear.*

Proof. (i) With $\alpha \in (0,1)$ and $\xi_\alpha = (1-\alpha)\,\xi + \alpha\bar{\xi}$ we obtain from (11.6) and Lemma 11.6 (ii):

$$\Delta_{L_{\mathrm{B}}}(\xi, \bar{\xi}) = \lim_{\alpha \downarrow 0} \frac{\mathrm{d}}{\mathrm{d}\alpha} \operatorname{tr} \boldsymbol{U}_{\mathrm{L}} \boldsymbol{M}_{\mathrm{B}}(\xi_\alpha)^{-1}$$

$$= \lim_{\alpha \downarrow 0} \operatorname{tr} \boldsymbol{U}_{\mathrm{L}} \left(\frac{\mathrm{d}}{\mathrm{d}\alpha} \boldsymbol{M}_{\mathrm{B}}(\xi_\alpha)^{-1} \right)$$

$$= \lim_{\alpha \downarrow 0} \operatorname{tr} \boldsymbol{U}_{\mathrm{L}} \boldsymbol{M}_{\mathrm{B}}(\xi_\alpha)^{-1} (\boldsymbol{M}_{\mathrm{B}}(\xi) - \boldsymbol{M}_{\mathrm{B}}(\bar{\xi}))\, \boldsymbol{M}_{\mathrm{B}}(\xi_\alpha)^{-1}$$

$$= \operatorname{tr} \boldsymbol{U}_{\mathrm{L}} \boldsymbol{M}_{\mathrm{B}}(\xi)^{-1} (\boldsymbol{M}_{\mathrm{B}}(\xi) - \boldsymbol{M}_{\mathrm{B}}(\bar{\xi}))\, \boldsymbol{M}_{\mathrm{B}}(\xi)^{-1}.$$

(ii) By help of Lemma 11.6 (i) we get

$$\int_{X_{\mathrm{E}}} \Delta_{L_{\mathrm{B}}}(\xi, \delta_x)\, \bar{\xi}(\mathrm{d}\boldsymbol{x}) = L_{\mathrm{B}}(\xi) - \int_{X_{\mathrm{E}}} \{\operatorname{tr} \boldsymbol{U}_{\mathrm{L}} \boldsymbol{M}_{\mathrm{B}}(\xi)^{-1} \boldsymbol{M}_{\mathrm{B}}(\delta_x)\, \boldsymbol{M}_{\mathrm{B}}(\xi)^{-1}\}\, \bar{\xi}(\mathrm{d}\boldsymbol{x})$$

$$= L_{\mathrm{B}}(\xi) - \operatorname{tr}\left\{ \boldsymbol{U}_{\mathrm{L}} \boldsymbol{M}_{\mathrm{B}}(\xi)^{-1} \left[\int_{X_{\mathrm{E}}} \boldsymbol{M}_{\mathrm{B}}(\delta_x)\, \bar{\xi}(\mathrm{d}\boldsymbol{x}) \right] \boldsymbol{M}_{\mathrm{B}}(\xi)^{-1} \right\}$$

$$= L_{\mathrm{B}}(\xi) - \operatorname{tr} \boldsymbol{U}_{\mathrm{L}} \boldsymbol{M}_{\mathrm{B}}(\xi)^{-1} \boldsymbol{M}_{\mathrm{B}}(\bar{\xi})\, \boldsymbol{M}_{\mathrm{B}}(\xi)^{-1}$$

$$= \Delta_{L_{\mathrm{B}}}(\xi, \bar{\xi}). \;\square$$

From this we can derive the following necessary and sufficient condition for L_B-optimality.

Theorem 11.12. *The design $\xi^* \in \bar{\Xi}$ is L_B-optimal in Ξ if and only if*

$$\sup_{x \in X_E} \| M_B(\xi^*)^{-1} f(x) \|_{U_L}^2 = \operatorname{tr} M_B(\xi^*)^{-1} U_L M_B(\xi^*)^{-1} M(\xi^*) .$$

Proof. For any $x \in X_E$ we have

$$\begin{aligned} \Delta_{L_B}(\xi^*, \delta_x) &= L_B(\xi^*) - \operatorname{tr} U_L M_B(\xi^*)^{-1} M_B(\delta_x) M_B(\xi^*)^{-1} \\ &= \operatorname{tr} U_L M_B(\xi^*)^{-1} (M_B(\xi^*) - M_B(\delta_x)) M_B(\xi^*)^{-1} . \end{aligned}$$

Observing that $M_B(\xi^*) - M_B(\delta_x) = M(\xi^*) - f(x) f(x)^\top$, this yields

$$\Delta_{L_B}(\xi^*, \delta_x) = \operatorname{tr} U_L M_B(\xi^*)^{-1} M(\xi^*) M_B(\xi^*)^{-1} - \| M_B(\xi^*)^{-1} f(x) \|_{U_L}^2 \tag{11.7}$$

and the result follows from Theorem 11.11 (i). □

From Theorem 11.11 (ii) and Lemma 11.5 we obtain bounds for the functional value of L_B-optimal designs.

Theorem 11.13. *Let $(\xi, \bar{\xi}) \in \bar{\Xi} \times \bar{\Xi}$. For all $\xi^* \in \bar{\Xi}_{L_B}$ it holds*

(i) $$\operatorname{tr} U_L M_B(\xi)^{-1} M(\xi) M_B(\xi)^{-1} - \sup_{x \in X_E} \| M_B(\xi)^{-1} f(x) \|_{U_L}^2$$
$$\leqq L_B(\xi^*) - L_B(\xi) ,$$

(ii) $$L_B(\xi) - \operatorname{tr} U_L M_B(\xi)^{-1} M_B(\xi^*) M_B(\xi)^{-1}$$
$$\leqq 0 \leqq L_B(\xi^*) - \operatorname{tr} U_L M_B(\xi^*)^{-1} M_B(\bar{\xi}) M_B(\xi^*)^{-1} .$$

The following theorem gives a necessary condition on the supporting points of L_B-optimal designs.

Theorem 11.14. *For any $\xi^* \in \bar{\Xi}_{L_B}$ it holds:*

$$\operatorname{supp} \xi^* \subseteqq \left\{ x \in X_E : \operatorname{tr} U_L M_B(\xi^*)^{-1} M(\xi^*) M_B(\xi^*)^{-1} = \| M_B(\xi^*)^{-1} f(x) \|_{U_L}^2 \right\} .$$

Observing (11.7), this follows immediately from Theorem 11.11 (iii).

Setting in the above Theorems 11.12–11.14 especially $U_L = I_r$, $U_L = cc^\top$ and $U_L = \int_{X_P} f(x) f(x)^\top P_{\tilde{x}}(dx)$, respectively, we may obtain the corresponding characterizations for A_B-optimal, C_B-optimal and Bayesian designs.

Let us now characterize D_B-optimal designs.

Lemma 11.8. (i) $\Delta_{D_B}(\xi, \bar{\xi}) = D_B(\xi) \cdot \operatorname{tr}\{(M(\xi) - M(\bar{\xi})) M_B(\xi)^{-1}\}$,
(ii) Δ_{D_B} *is linear.*

Proof. Let $\alpha \in (0, 1)$ and $\xi_\alpha = (1 - \alpha) \xi + \alpha \bar{\xi}$.

(i) Applying A.38 and Lemma 11.6 (ii) we get

$$\frac{\mathrm{d}}{\mathrm{d}\alpha}\det \boldsymbol{M}_{\mathrm{B}}(\xi_\alpha)^{-1} = \det \boldsymbol{M}_{\mathrm{B}}(\xi_\alpha)^{-1}\operatorname{tr}\left\{\boldsymbol{M}_{\mathrm{B}}(\xi_\alpha)\frac{\mathrm{d}}{\mathrm{d}\alpha}\boldsymbol{M}_{\mathrm{B}}(\xi_\alpha)^{-1}\right\}$$
$$= D_{\mathrm{B}}(\xi_\alpha)\operatorname{tr}\{(\boldsymbol{M}_{\mathrm{B}}(\xi) - \boldsymbol{M}_{\mathrm{B}}(\bar{\xi}))\,\boldsymbol{M}_{\mathrm{B}}(\xi_\alpha)^{-1}\}.$$

This yields

$$\Delta_{D_{\mathrm{B}}}(\xi, \bar{\xi}) = \lim_{\alpha \downarrow 0}\frac{\mathrm{d}}{\mathrm{d}\alpha}\det \boldsymbol{M}_{\mathrm{B}}(\xi_\alpha)^{-1}$$
$$= D_{\mathrm{B}}(\xi)\operatorname{tr}\{(\boldsymbol{M}_{\mathrm{B}}(\xi) - \boldsymbol{M}_{\mathrm{B}}(\bar{\xi}))\,\boldsymbol{M}_{\mathrm{B}}(\xi)^{-1}\}$$
$$= D_{\mathrm{B}}(\xi)\operatorname{tr}\{(\boldsymbol{M}(\xi) - \boldsymbol{M}(\bar{\xi}))\,\boldsymbol{M}_{\mathrm{B}}(\xi)^{-1}\}.$$

(ii) $$\int_{X_{\mathrm{E}}} \Delta_{D_{\mathrm{B}}}(\xi, \delta_x)\,\bar{\xi}(\mathrm{d}\boldsymbol{x}) = D_{\mathrm{B}}(\xi)\operatorname{tr}\left\{\int_{X_{\mathrm{E}}}(\boldsymbol{M}(\xi) - \boldsymbol{M}(\delta_x))\,\boldsymbol{M}_{\mathrm{B}}(\xi)^{-1}\bar{\xi}(\mathrm{d}\boldsymbol{x})\right\}$$
$$= D_{\mathrm{B}}(\xi)\operatorname{tr}\left\{(\boldsymbol{M}(\xi) - \int_{X_{\mathrm{E}}}\boldsymbol{f}(\boldsymbol{x})\boldsymbol{f}(\boldsymbol{x})^\top\bar{\xi}(\mathrm{d}\boldsymbol{x}))\,\boldsymbol{M}_{\mathrm{B}}(\xi)^{-1}\right\}$$
$$= D_{\mathrm{B}}(\xi)\operatorname{tr}\{(\boldsymbol{M}(\xi) - \boldsymbol{M}(\bar{\xi}))\,\boldsymbol{M}_{\mathrm{B}}(\xi)^{-1}\}$$
$$= \Delta_{D_{\mathrm{B}}}(\xi, \bar{\xi}).\ \square$$

Theorem 11.15. (i) *The design $\xi^* \in \bar{\Xi}$ is D_{B}-optimal in $\bar{\Xi}$ if and only if*

$$\sup_{\boldsymbol{x}\in X_{\mathrm{E}}}\boldsymbol{f}(\boldsymbol{x})^\top\boldsymbol{M}_{\mathrm{B}}(\xi^*)^{-1}\boldsymbol{f}(\boldsymbol{x}) = \operatorname{tr}\boldsymbol{M}(\xi^*)\,\boldsymbol{M}_{\mathrm{B}}(\xi^*)^{-1}.$$

(ii) *For all $\xi \in \bar{\Xi}$ and $\xi^* \in \bar{\Xi}_{D_{\mathrm{B}}}$ it holds*

$$1 \geqq D_{\mathrm{B}}(\xi^*)/D_{\mathrm{B}}(\xi) \geqq 1 + \operatorname{tr}\boldsymbol{M}(\xi)\,\boldsymbol{M}_{\mathrm{B}}(\xi)^{-1} - \sup_{\boldsymbol{x}\in X_{\mathrm{E}}}\boldsymbol{f}(\boldsymbol{x})^\top\boldsymbol{M}_{\mathrm{B}}(\xi)^{-1}\boldsymbol{f}(\boldsymbol{x}).$$

(iii) *The design $\xi^* \in \bar{\Xi}$ can be D_{B}-optimal in $\bar{\Xi}$ only if*

$$\operatorname{supp}\xi^* \subseteqq \{\boldsymbol{x}\in X_{\mathrm{E}} : \boldsymbol{f}(\boldsymbol{x})^\top\boldsymbol{M}_{\mathrm{B}}(\xi^*)^{-1}\boldsymbol{f}(\boldsymbol{x}) = \operatorname{tr}\boldsymbol{M}(\xi^*)\,\boldsymbol{M}_{\mathrm{B}}(\xi^*)^{-1}\}.$$

Proof. According to Lemma 11.8 (i) we first obtain

$$\inf_{\boldsymbol{x}\in X_{\mathrm{E}}}\Delta_{D_{\mathrm{B}}}(\xi, \delta_x) = D_{\mathrm{B}}(\xi)\left\{\inf_{\boldsymbol{x}\in X_{\mathrm{E}}}\operatorname{tr}(\boldsymbol{M}(\xi) - \boldsymbol{f}(\boldsymbol{x})\boldsymbol{f}(\boldsymbol{x})^\top)\,\boldsymbol{M}_{\mathrm{B}}(\xi)^{-1}\right\}$$
$$= D_{\mathrm{B}}(\xi)\left\{\operatorname{tr}\boldsymbol{M}(\xi)\,\boldsymbol{M}_{\mathrm{B}}(\xi)^{-1} - \sup_{\boldsymbol{x}\in X_{\mathrm{E}}}\boldsymbol{f}(\boldsymbol{x})^\top\boldsymbol{M}_{\mathrm{B}}(\xi)^{-1}\boldsymbol{f}(\boldsymbol{x})\right\}.$$

(i) Since $D_{\mathrm{B}}(\xi^*) = \det\boldsymbol{M}_{\mathrm{B}}(\xi^*)^{-1} > 0$, the optimality condition $\inf_{\boldsymbol{x}\in X_{\mathrm{E}}}\Delta_{D_{\mathrm{B}}}(\xi^*, \delta_x) = 0$ (see Theorem 11.11 (i)) is equivalent to

$$\sup_{\boldsymbol{x}\in X_{\mathrm{E}}}\boldsymbol{f}(\boldsymbol{x})^\top\boldsymbol{M}_{\mathrm{B}}(\xi^*)^{-1}\boldsymbol{f}(\boldsymbol{x}) = \operatorname{tr}\boldsymbol{M}(\xi^*)\,\boldsymbol{M}_{\mathrm{B}}(\xi^*)^{-1}.$$

(ii) The result follows from Lemma 11.5 after dividing the inequality

$$D_{\mathrm{B}}(\xi) + \inf_{x \in X_{\mathrm{E}}} \Delta_{D_{\mathrm{B}}}(\xi, \delta_x) \leqq D_{\mathrm{B}}(\xi^*) \leqq D_{\mathrm{B}}(\xi)$$

by $D_{\mathrm{B}}(\xi) > 0$.

(iii) Observing that

$$\Delta_{D_{\mathrm{B}}}(\xi^*, \delta_x) = 0 \Leftrightarrow \operatorname{tr} \boldsymbol{M}(\xi^*)\, \boldsymbol{M}_{\mathrm{B}}(\xi^*)^{-1} = \boldsymbol{f}(\boldsymbol{x})^\top \boldsymbol{M}_{\mathrm{B}}(\xi^*)^{-1} \boldsymbol{f}(\boldsymbol{x}),$$

the result follows from Theorem 11.11 (iii). □

Now we give a characterization of E_{B}-optimal designs. For any $\xi \in \Xi$ let $\lambda_\xi = \lambda_{\min}(\boldsymbol{M}_{\mathrm{B}}(\xi))$ denote the minimum eigenvalue of the Bayesian information matrix and $\boldsymbol{t}_\xi$ the normalized eigenvector corresponding to λ_ξ.

Theorem 11.16. *Let $\xi^* \in \bar{\Xi}$ be such that λ_{ξ^*} is a simple eigenvalue. Then it holds:*

(i) $\xi^* \in \bar{\Xi}_{E_{\mathrm{B}}} \Leftrightarrow \sup_{x \in X_{\mathrm{E}}} (\boldsymbol{f}(\boldsymbol{x})^\top \boldsymbol{t}_{\xi^*})^2 = \lambda_{\xi^*} - \dfrac{1}{n} \boldsymbol{t}_{\xi^*}^\top \boldsymbol{\Phi}^{-1} \boldsymbol{t}_{\xi^*},$

(ii) $\xi^* \in \bar{\Xi}_{E_{\mathrm{B}}} \Rightarrow \operatorname{supp} \xi^* \subseteqq \left\{ x \in X_{\mathrm{E}} \colon \lambda_{\xi^*} - \dfrac{1}{n} \boldsymbol{t}_{\xi^*}^\top \boldsymbol{\Phi}^{-1} \boldsymbol{t}_{\xi^*} = (\boldsymbol{f}(\boldsymbol{x})^\top \boldsymbol{t}_{\xi^*})^2 \right\},$

(iii) $\xi^* \in \bar{\Xi}_{E_{\mathrm{B}}} \Rightarrow \left\{ \dfrac{1}{n} \boldsymbol{t}_\xi^\top \boldsymbol{\Phi}^{-1} \boldsymbol{t}_\xi + \sup_{x \in X_{\mathrm{E}}} (\boldsymbol{f}(\boldsymbol{x})^\top \boldsymbol{t}_\xi)^2 \right\}^{-1} \leqq E_{\mathrm{B}}(\xi^*) \leqq E_{\mathrm{B}}(\xi)$

for all $\xi \in \bar{\Xi}$.

Proof. First we note that, because of $\lambda_{\max}(\boldsymbol{M}_{\mathrm{B}}(\xi)^{-1}) = 1/\lambda_\xi$, the minimization of $E_{\mathrm{B}}(\cdot) = \lambda_{\max}(\boldsymbol{M}_{\mathrm{B}}(\cdot)^{-1})$ is equivalent to minimization of $\bar{E}_{\mathrm{B}}(\cdot) := -\lambda_{\min}(\boldsymbol{M}_{\mathrm{B}}(\cdot))$. According to A.38, the directional derivative of $\bar{E}_{\mathrm{B}}$ takes the form

$$\begin{aligned}
\Delta_{\bar{E}_{\mathrm{B}}}(\xi, \bar{\xi}) &= -\lim_{\alpha \downarrow 0} \frac{\mathrm{d}}{\mathrm{d}\alpha} \lambda_{\min}((1-\alpha)\, \boldsymbol{M}_{\mathrm{B}}(\xi) + \alpha \boldsymbol{M}_{\mathrm{B}}(\bar{\xi})) \\
&= -\boldsymbol{t}_\xi^\top \{\boldsymbol{M}_{\mathrm{B}}(\bar{\xi}) - \boldsymbol{M}_{\mathrm{B}}(\xi)\}\, \boldsymbol{t}_\xi \\
&= \lambda_\xi - \boldsymbol{t}_\xi^\top \boldsymbol{M}_{\mathrm{B}}(\bar{\xi})\, \boldsymbol{t}_\xi .
\end{aligned}$$

Further, the directional derivative is linear, since

$$\begin{aligned}
\int_{X_{\mathrm{E}}} \Delta_{\bar{E}_{\mathrm{B}}}(\xi, \delta_x)\, \bar{\xi}(\mathrm{d}x) &= \lambda_\xi - \boldsymbol{t}_\xi^\top \left\{ \int_{X_{\mathrm{E}}} \boldsymbol{M}_{\mathrm{B}}(\delta_x)\, \bar{\xi}(\mathrm{d}\boldsymbol{x}) \right\} \boldsymbol{t}_\xi \\
&= \lambda_\xi - \boldsymbol{t}_\xi^\top \boldsymbol{M}_{\mathrm{B}}(\bar{\xi})\, \boldsymbol{t}_\xi = \Delta_{\bar{E}_{\mathrm{B}}}(\xi, \bar{\xi}) .
\end{aligned}$$

Observing that

$$\Delta_{\bar{E}_{\mathrm{B}}}(\xi, \delta_x) = \lambda_\xi - \boldsymbol{t}_\xi^\top \boldsymbol{M}_{\mathrm{B}}(\delta_x)\, \boldsymbol{t}_\xi = \lambda_\xi - \frac{1}{n} \boldsymbol{t}_\xi^\top \boldsymbol{\Phi}^{-1} \boldsymbol{t}_\xi - (\boldsymbol{f}(\boldsymbol{x})^\top \boldsymbol{t}_\xi)^2$$

and

$$\inf_{x \in X_E} \Delta_{\bar{E}_B}(\xi, \delta_x) = \lambda_\xi - \frac{1}{n} t_\xi^\top \Phi^{-1} t_\xi - \sup_{x \in X_E} (f(x)^\top t_\xi)^2,$$

the assertions (i) and (ii) follow from the conditions of Theorem 11.11 (i) and (iii), respectively. The third assertion follows from Lemma 11.5 by rewriting the inequality

$$\bar{E}_B(\xi) + \inf_{x \in X_E} \Delta_{\bar{E}_B}(\xi, \delta_x) \leqq \bar{E}_B(\xi^*) \leqq \bar{E}_B(\xi). \ \square$$

Remark 11.5. If, in the above theorem, we drop the assumption of simplicity of λ_{ξ^*} then the directional derivative becomes

$$\Delta_{\bar{E}_B}(\xi, \bar{\xi}) = -\lambda_{\min}(Q^\top [M_B(\bar{\xi}) - M_B(\xi)] Q),$$

where the columns of $Q \in \mathcal{M}_{r \times q}$ are formed by the orthonormal eigenvectors of $M_B(\xi)$ corresponding to the minimum eigenvalue $\lambda_\xi = \lambda_{\min}(M_B(\xi))$ occuring with multiplicity $q \geqq 1$ (cp. KIEFER (1974), Section 4E). In this case, however, $\Delta_{\bar{E}_B}$ is no longer linear and the conditions of Theorem 11.16 (i) and (iii) must be replaced by

(i′) $\xi^* \in \bar{\Xi} \Leftrightarrow \sup_{\bar{\xi} \in \bar{\Xi}} \lambda_{\min}(Q^\top [M_B(\bar{\xi}) - M_B(\xi^*)] Q) = 0$ and

(iii′) $\xi^* \in \bar{\Xi} \Rightarrow \left\{\lambda_\xi - \inf_{\bar{\xi} \in \bar{\Xi}} \Delta_{\bar{E}_B}(\xi, \bar{\xi})\right\}^{-1} \leqq E_B(\xi^*) \leqq E_B(\xi)$ for all $\xi \in \bar{\Xi}$,

respectively, whereas no analogue can be given to Theorem 11.16 (ii).

Concluding, we note that the equivalent characterizations given in this section reduce to well-known conditions for optimal designs with respect to the analogous design criteria for the least squares estimator if we formally set $\Phi^{-1} = \mathbf{0}$ which corresponds to a state of prior ignorance (see Section 8). In this case the Bayesian information matrix coincides with the usual information matrix. The same effect occurs if the number n of observations grows to infinity, which causes the prior knowledge to become more and more dominated by actual knowledge from sampling.

12. Construction of optimal continuous designs

On the basis of the equivalence theorems given in the preceding section we will now develop methods for the construction of optimal designs in $\bar{\Xi}$. Especially, we formulate an algorithm for the computation of the (Bayesian) information matrix of A_B-optimal designs provided that it has the same eigenvectors as the prior covariance matrix. Under a certain condition, this algorithm yields, at the same time, the information matrix of D_B- and E_B-optimal designs and, after some linear transformation, also the information matrix of L_B-optimal designs.

With the optimal information matrix at hand, an optimal design can be obtained as solution of a corresponding nonlinear equation system. For the special case of multiple linear regression we give optimal designs in an explicit form in Section 12.4. Furthermore we set up iterative procedures for the construction of L_B- and D_B-optimal designs. These procedures are well practicable and, in general, they converge very rapidly since the starting design, the choice of which is a crucial point in the application of corresponding iterative design procedures for the least squares estimator, can be chosen objectively within the family of one-point designs.

12.1. Construction of A_B-optimal designs

From Theorem 11.11 it follows that a design $\xi^* \in \Xi$ is A_B-optimal if and only if it satisfies the equation

$$\sup_{x \in X_E} \boldsymbol{f}(\boldsymbol{x})^\top \boldsymbol{M}_B(\xi^*)^{-2} \boldsymbol{f}(\boldsymbol{x}) = \operatorname{tr} \boldsymbol{M}(\xi^*)\, \boldsymbol{M}_B(\xi^*)^{-2}. \tag{12.1}$$

The optimality condition (12.1) depends on the design only through the (Bayesian) information matrix which, in this case, is unique (see Corollary 11.2). This suggests to determine an optimal design ξ^* in such a way that we first try to find the optimal information matrix $\boldsymbol{M}^*$ and then calculate ξ^* as a solution to the integral equation

$$\boldsymbol{M}(\xi^*) = \int_{X_E} \boldsymbol{f}(\boldsymbol{x}) \boldsymbol{f}(\boldsymbol{x})^\top \xi^*(\mathrm{d}\boldsymbol{x}) = \boldsymbol{M}^*. \tag{12.2}$$

Let $\boldsymbol{\Lambda}$ be the diagonal matrix of the eigenvalues of $\boldsymbol{\Phi}^{-1}$, which are all positive, and $\boldsymbol{U}$ be the matrix of the corresponding orthonormal eigenvectors, i.e.,

$$\boldsymbol{\Phi}^{-1} = \boldsymbol{U}\boldsymbol{\Lambda}\boldsymbol{U}^\top,$$

$$\boldsymbol{\Lambda} = \operatorname{diag}(\lambda_1, \ldots, \lambda_r), \qquad \boldsymbol{U}^\top \boldsymbol{U} = \boldsymbol{U}\boldsymbol{U}^\top = \boldsymbol{I}_r.$$

Defining

$$\bar{\boldsymbol{f}}(\boldsymbol{x}) = (\bar{f}_1(\boldsymbol{x}), \ldots, \bar{f}_r(\boldsymbol{x}))^\top := \boldsymbol{U}^\top \boldsymbol{f}(\boldsymbol{x}), \tag{12.3}$$

the transformed information matrix $\bar{\boldsymbol{M}}(\xi) = \int_{X_E} \bar{\boldsymbol{f}}(\boldsymbol{x}) \bar{\boldsymbol{f}}(\boldsymbol{x})^\top \xi(\mathrm{d}\boldsymbol{x})$ takes the form

$$\bar{\boldsymbol{M}}(\xi) = \boldsymbol{U}^\top \boldsymbol{M}(\xi)\, \boldsymbol{U}, \tag{12.4}$$

and for the transformed Bayesian information matrix we get

$$\bar{\boldsymbol{M}}_B(\xi) = \boldsymbol{U}^\top \boldsymbol{M}_B(\xi)\, \boldsymbol{U} = \bar{\boldsymbol{M}}(\xi) + \frac{1}{n}\boldsymbol{\Lambda}. \tag{12.5}$$

Then, for the A_B-functional we have equality

$$\begin{aligned} A_B(\xi) &= \operatorname{tr} \boldsymbol{M}_B(\xi)^{-1} = \operatorname{tr} \boldsymbol{U}^\top \boldsymbol{M}_B(\xi)^{-1} \boldsymbol{U} \\ &= \operatorname{tr} \bar{\boldsymbol{M}}_B(\xi)^{-1}. \end{aligned}$$

Further, let

$$a := \sup_{x\in X_E} \bar{f}(x)^\top \bar{f}(x) = \sup_{x\in X_E} f(x)^\top f(x), \tag{12.6a}$$

$$a_j := \sup_{x\in X_E} \bar{f}_j(x)^2, \qquad j=1,\ldots,r, \tag{12.6b}$$

$$b_j := \inf_{x\in X_E} \bar{f}_j(x)^2, \qquad j=1,\ldots,r, \tag{12.6c}$$

$$\bar{\lambda}_j := \frac{1}{n}\lambda_j + b_j, \qquad j=1,\ldots,r, \tag{12.6d}$$

$$\bar{a} := a - \sum_{j=1}^{r} b_j. \tag{12.6e}$$

Assumption 12.1 *Let exist a design $\xi \in \Xi$ such that $\bar{M}(\xi)$ is diagonal with* $\operatorname{tr}\bar{M}(\xi) = a$.

Remark 12.1. The assumption of diagonality of $\bar{M}(\xi)$ means that there exists a design ξ whose information matrix $M(\xi)$ has the same eigenvectors as the prior covariance matrix Φ and, because of

$$\operatorname{tr}\bar{M}(\xi) = \int_{X_E} \bar{f}(x)^\top \bar{f}(x)\,\xi(dx) = \int_{X_E} f(x)^\top f(x)\,\xi(dx)$$

the additional condition $\operatorname{tr}\bar{M}(\xi) = a$ requires this design to be supported only of experimental points which maximize $f(\cdot)^\top f(\cdot)$ over X_E.

Further, without loss of generality, we assume the quantities $\bar{\lambda}_j$ from (12.6d) to be ordered such that

$$\bar{\lambda}_1 \leqq \bar{\lambda}_2 \leqq \ldots \leqq \bar{\lambda}_r. \tag{12.7}$$

We now calculate quantities $m_1^*, \ldots, m_r^*$ serving as diagonal elements of the transformed information matrix $\bar{M}(\xi^*)$ of an A_B-optimal design $\xi^* \in \Xi$.

Algorithm 12.1. (Calculation of $\bar{M}(\xi^*) = \operatorname{diag}(m_1^*, \ldots, m_r^*)$).

Denotations. For simplicity of representation we introduce a fictive value $\bar{\lambda}_{r+1} := \infty$. Further, let s be an index variable, $J_s \subseteqq \{1, \ldots, r\}$ certain index sets, which are specified successively by the algorithm, and $J_s(i) := J_s \cap \{1, \ldots, i\}$, $i = 1, \ldots, r$. By $N_s(i) := \operatorname{card} J_s(i)$ we denote the number of indices contained in $J_s(i)$, where we set $N_s(i)^{-1} = \infty$ if $N_s(i) = 0$.

Step 0. Put $s = 0$ and define $J_s = J_0 := \{1, \ldots, r\}$ and $\bar{a}_s = \bar{a}_0 := \bar{a}$.

Step 1. Find the smallest integer $i \in J_s$ for which

$$\bar{\lambda}_{i+1} > \frac{1}{N_s(i)}\left(\bar{a}_s + \sum_{j\in J_s(i)} \bar{\lambda}_j\right)$$

and denote it by i_s. For all $i \in J_s(i_s)$ calculate

$$m_i = \frac{1}{N_s(i_s)}\left(\bar{a}_s + \sum_{j \in J_s(i_s)} \bar{\lambda}_j\right) - \bar{\lambda}_i. \tag{12.8}$$

If all these quantities satisfy the inequality

$$m_i \leqq a_i - b_i \tag{12.9}$$

then put

$$m_i^* = \begin{cases} m_i + b_i, & i \in J_s(i_s), \\ b_i, & i = i_s + 1, \ldots, r. \end{cases} \tag{12.10}$$

Otherwise, if (12.9) is violated for at least one index $i \in J_s(i_s)$ then go to Step 2.

Step 2. Define $\bar{J}_s := \{j \in J_s(i_s) \colon m_j > a_j - b_j\}$ and choose

$$m_j^* = a_j \quad \text{for} \quad j \in \bar{J}_s. \tag{12.11}$$

Then calculate

$$\bar{a}_{s+1} = \bar{a}_s - \sum_{j \in \bar{J}_s} (a_j - b_j),$$

$$J_{s+1} = J_s \setminus \bar{J}_s$$

and return to Step 1 replacing s by $s + 1$.

Remark 12.2. If the quantities $\bar{\lambda}_j$ from (12.6d) are not ordered a-priori as indicated in (12.7) then we have first to make a permutation π of $\{1, \ldots, r\}$ such that $\bar{\lambda}_{\pi(1)} \leqq \ldots \leqq \bar{\lambda}_{\pi(r)}$. The corresponding diagonal elements m_i^* then result from Algorithm 12.1 after replacing the integers $i \in \{1, \ldots, r\}$ and i_s by $\pi(i)$ and $\pi(i_s)$, respectively.

The construction mechanism of the algorithm provides a decomposition of the index set $J_0 = \{1, \ldots, r\}$ into three disjoint subsets which, according to (12.10) and (12.11), are given by

$$\begin{aligned} J_{(1)} &= \{j \in J_0 \colon m_j^* = a_j\}, \\ J_{(2)} &= \{j \in J_0 \colon m_j^* = b_j\}, \qquad J_{(3)} = J_0 \setminus J_{(1)} \setminus J_{(2)}. \end{aligned} \tag{12.12}$$

If $J_{(3)}$ is non-empty then for all $i \in J_{(3)}$ we have

$$m_i^* + \frac{1}{n}\lambda_i = \frac{1}{\operatorname{card} J_{(3)}}\left\{a - \sum_{j \in J_{(1)}} a_j - \sum_{j \in J_{(2)}} b_j + \frac{1}{n}\sum_{j \in J_{(3)}} \lambda_j\right\}, \tag{12.13}$$

the right-hand side of (12.13) being independent of i. This means that the diagonal entries of the transformed Bayesian information matrix

$$\bar{M}_{\mathrm{B}}(\xi^*) = \bar{M}(\xi^*) + \frac{1}{n}\Lambda = \operatorname{diag}\left(m_1^* + \frac{1}{n}\lambda_1, \ldots, m_r^* + \frac{1}{n}\lambda_r\right)$$

which correspond to indices $i \in J_{(3)}$ will all have the same value.

Since $a_1 + \ldots + a_r \geqq a$, the Algorithm 12.1 terminates after at most r iterations at Step 1. If Step 2 does not come in use then the algorithm provides a distribution of the "available experimental mass"

$$\bar{a} = \sup_{x \in X_E} \bar{f}(x)^\top \bar{f}(x) - \sum_{j=1}^{r} b_j$$

such that a maximum number of the smallest entries $m_i^* + \frac{1}{n}\lambda_i$ of $\bar{M}_B(\xi^*)$ gets equal. The algorithm generalizes the idea of maximum compactness developed in Covey-Crump/Silvey (1970) for the construction of D-optimal exact designs in a homogeneous multiple linear regression model with previous observations and spherical experimental region. The Algorithm 12.1, however, is not restricted to such a model; it can be applied for arbitrary regression function $f(\cdot)$ and arbitrary experimental region X_E as long as Assumption 12.1 is satisfied.

Remark 12.3. (Heuristical interpretation of Algorithm 12.1)
Let us consider the special case $\Phi^{-1} = \operatorname{diag}(\lambda_1, \ldots, \lambda_r)$ in which the regression coefficients are uncorrelated a-priori, i.e. $\mathbf{Cov}(\tilde{\vartheta} \mid w) = \frac{1}{w}\operatorname{diag}(\lambda_1^{-1}, \ldots, \lambda_r^{-1})$. This implies that $\Lambda = \Phi^{-1}$, $U = I_r$, $\bar{f}(x) = f(x)$ and $\bar{M}(\xi) = M(\xi)$. Then the algorithm determines the optimal information matrix $M(\xi^*) = \operatorname{diag}(m_1^*, \ldots, m_r^*)$ in such a way that the expected posterior variances of the regression coefficients $\tilde{\vartheta}_i$, which are proportional to $\left(m_i^* + \frac{1}{n}\lambda_i\right)^{-1}$ $(i = 1, \ldots, r)$, become as balanced as possible within the given bounds a_i and b_i. Therefore, an optimal design ξ^* provides a certain compensational effect: It puts the greatest experimental mass to those components $\tilde{\vartheta}_i$ which have large a-priori variances λ_i^{-1}, and it pays less attention to components which are relatively well known by previous experience.

Theorem 12.1. *If there exists $\xi^* \in \Xi$ with*

$$\bar{M}(\xi^*) = \operatorname{diag}(m_1^*, \ldots, m_r^*) \tag{12.14}$$

and $m_1^*, \ldots, m_r^*$ obtained according to Algorithm 12.1, *then ξ^* is an A_B-optimal design and it holds*

$$\operatorname{supp} \xi^* \subseteq \left\{ x \in X_E : f(x)^\top f(x) = a \wedge f_j^2(x) = \begin{cases} a_j & \text{for } j \in J_{(1)} \\ b_j & \text{for } j \in J_{(2)} \end{cases} \right\}. \tag{12.15}$$

***Proof*.** We show that ξ^* with transformed information matrix given by (12.14) satisfies the optimality condition (12.1). Observing that $\bar{M}_B(\xi^*)^{-2} = U^\top M_B(\xi^*)^{-2} U$, this condition is equivalent to

$$\operatorname{tr} \bar{M}(\xi^*) \bar{M}_B(\xi^*)^{-2} = \sup_{x \in X_E} \bar{f}(x)^\top \bar{M}_B(\xi^*)^{-2} \bar{f}(x).$$

From (12.14) we have $\bar{M}_B(\xi^*) = \mathrm{diag}(c_1, \ldots, c_r)$ with $c_j = m_j^* + \frac{1}{n}\lambda_j$. Thus, ξ^* is A_B-optimal if and only if

$$\sum_{j=1}^{r} m_j^*/c_j^2 = \sup_{x \in X_E} \sum_{j=1}^{r} \bar{f}_j^2(x)/c_j^2 . \tag{12.16}$$

Now, according to Theorem 11.1 and the presumed existence of a design $\xi^* \in \Xi$ with (12.14), there exists a finite number h of points $x_1, \ldots, x_h \in X_E$ and positive weights $p_1, \ldots, p_h$ such that

$$\sum_{i=1}^{h} p_i f(x_i) f(x_i)^\top = \mathrm{diag}(m_1^*, \ldots, m_r^*) \wedge \sum_{i=1}^{h} p_i = 1 . \tag{12.17}$$

Since $\mathrm{tr}\, \bar{M}(\xi^*) = \sum_{i=1}^{h} p_i \bar{f}(x_i)^\top \bar{f}(x_i) = \sum_{j=1}^{r} m_j^* = a = \sup_{x \in X_E} \bar{f}(x)^\top \bar{f}(x)$, it follows that $\bar{f}(x_i)^\top \bar{f}(x_i) = a$ for all $i = 1, \ldots, h$. Additionally, (12.17) implies that

$$\bar{f}_j^2(x_i) = m_j^* = \begin{cases} a_j, & j \in J_{(1)}, \\ b_j, & j \in J_{(2)}, \end{cases}$$

for $i = 1, \ldots, h$ which proves (12.15). Now let $x^* \in \mathrm{supp}\, \xi^*$, i.e.,

$$\left. \begin{aligned} &\bar{f}_j^2(x^*) = a_j \quad \text{for} \quad j \in J_{(1)}, \qquad \bar{f}_j^2(x^*) = b_j \quad \text{for} \quad j \in J_{(2)}, \\ &\sum_{j \in J_{(3)}} f_j^2(x^*) = a - \sum_{j \in J_{(1)}} a_j - \sum_{j \in J_{(2)}} b_j . \end{aligned} \right\} \tag{12.18}$$

Then we can show that the supremum in the right-hand side expression of (12.16) is attained just for such a point x^*. To this we write

$$\sum_{j=1}^{r} \bar{f}_j^2(x)/c_j^2 = \sum_{j_1 \in J_{(1)}} \bar{f}_{j_1}^2(x)/c_{j_1}^2 + \sum_{j_2 \in J_{(2)}} \bar{f}_{j_2}^2(x)/c_{j_2}^2 + \sum_{j_3 \in J_{(3)}} \bar{f}_{j_3}^2(x)/c_{j_3}^2 \tag{12.19}$$

in accordance with the decomposition (12.12) of J_0. We only consider the case that all the index sets $J_{(1)}$, $J_{(2)}$ and $J_{(3)}$ are non-empty, otherwise the corresponding partial sum in (12.19) would vanish. It is clear from the construction mechanism applied to calculate $m_1^*, \ldots, m_r^*$ that it holds

$$c_{j_1} \leqq c_{j_3} \leqq c_{j_2} \quad \text{for all} \quad j_1 \in J_{(1)}, j_2 \in J_{(2)}, j_3 \in J_{(3)} . \tag{12.20}$$

Moreover, all the quantities c_{j_3} with $j_3 \in J_{(3)}$ are equal, i.e., $c_{j_3} = \text{constant} := c$, say, where c is given by the right-hand side expression from (12.13). Now, the maximum of the weighted sum (12.19) will be attained if the maximum possible experimental mass a is distributed in such a way that the largest possible masses are given successively to those components $\bar{f}_j$ for which the c_j's are smallest and, conversely, the smallest possible masses are given successively to the components with the largest c_j's, until the available mass is exhausted. But this is precisely accomplished by $\boldsymbol{x}^*$ since we obtain from (12.18) and (12.19)

$$\sum_{j=1}^{r} \bar{f}_j^2(\boldsymbol{x}^*)/c_j^2 = \sum_{j \in J_{(1)}} a_j/c_j^2 + \sum_{j \in J_{(2)}} b_j/c_j^2 + c^{-2}\Big(a - \sum_{j \in J_{(1)}} a_j - \sum_{j \in J_{(2)}} b_j\Big),$$

and, as it is clear from (12.20), this value cannot be increased within X_{E}, i.e.,

$$\sup_{\boldsymbol{x} \in X_{\mathrm{E}}} \sum_{j=1}^{r} \bar{f}_j^2(\boldsymbol{x})/c_j^2 = \sum_{j=1}^{r} \bar{f}_j^2(\boldsymbol{x}^*)/c_j^2 .$$

On the other hand, we have from (12.12) and (12.13)

$$\begin{aligned}\sum_{j=1}^{r} m_j^*/c_j^2 &= \sum_{j \in J_{(1)}} m_j^*/c_j^2 + \sum_{j \in J_{(2)}} m_j^*/c_j^2 + \sum_{j \in J_{(3)}} m_j^*/c_j^2 \\ &= \sum_{j \in J_{(1)}} a_j/c_j^2 + \sum_{j \in J_{(2)}} b_j/c_j^2 + c^{-2}\Big(a - \sum_{j \in J_{(1)}} a_j - \sum_{j \in J_{(2)}} b_j\Big)\end{aligned}$$

which proves the required equality (12.16). □

Remark 12.4. We note that Assumption 12.1 is necessary for the existence of a design ξ^* with $\bar{\boldsymbol{M}}(\xi^*) = \mathrm{diag}(m_1^*, \ldots, m_r^*)$, but not sufficient. A further necessary condition for such a design to exist is that there exists at least one point $\boldsymbol{x}^* \in X_{\mathrm{E}}$ which satisfies (12.18). This follows from the proof of Theorem 12.1.

As an immediate consequence of our Algorithm 12.1 we get the following simple specialization of Theorem 12.1.

Corollary 12.1. *If it holds*

$$\bar{\lambda}_{j+1} \leqq \frac{1}{j}\Big(\bar{a} + \sum_{i=1}^{j} \bar{\lambda}_i\Big) \quad \textit{for} \quad j = 1, \ldots, r-1 \tag{12.21}$$

and if, additionally,

$$a_j + \frac{1}{n}\lambda_j \geqq \frac{1}{r}\Big(a + \frac{1}{n}\sum_{i=1}^{r} \lambda_i\Big) \quad \textit{for} \quad j = 1, \ldots, r \tag{12.22}$$

then any design $\xi^* \in \Xi$ having the transformed Bayesian information matrix

$$\bar{\boldsymbol{M}}_{\mathrm{B}}(\xi^*) = c\boldsymbol{I}_r, \quad \textit{where} \quad c = \frac{1}{r}\Big(a + \frac{1}{n}\sum_{i=1}^{r} \lambda_i\Big)$$

is A_{B}-optimal in Ξ.

The situation considered in Corollary 12.1 refers to the completely balanced case in which all diagonal entries of $\bar{\boldsymbol{M}}_{\mathrm{B}}(\xi^*)$ are equal and thus all regression coefficients have the same posterior variance. Observing the ordering (12.7), it is easily seen that the inequalities

$$\frac{1}{n}\lambda_{\max}(\boldsymbol{\Phi}^{-1}) \leqq c - \max(b_1, \ldots, b_r)$$

and

$$\frac{1}{n}\lambda_{\min}(\boldsymbol{\Phi}^{-1}) \geqq c - \min(a_1, \ldots, a_r)$$

are sufficient for (12.21) and (12.22) to hold. The last two inequalities demand that the range between the minimum and maximum eigenvalue of $\boldsymbol{\Phi}$ and thus the range of prior variances of the components of $\tilde{\boldsymbol{\vartheta}}$ is not too wide.

Obviously, a design $\xi^* \in \Xi$ with $\bar{\boldsymbol{M}}(\xi^*) = \operatorname{diag}(m_1^*, \ldots, m_r^*)$ exists if and only if the integral equation

$$\int_{X_E} \boldsymbol{f}(\boldsymbol{x})\boldsymbol{f}(\boldsymbol{x})^\top \xi^*(d\boldsymbol{x}) = \boldsymbol{U}\operatorname{diag}(m_1^*, \ldots, m_r^*)\,\boldsymbol{U}^\top \tag{12.23}$$

has a solution in Ξ, where then $\boldsymbol{M}(\xi^*) = \boldsymbol{U}\operatorname{diag}(m_1^*, \ldots, m_r^*)\,\boldsymbol{U}^\top$ is the uniquely determined information matrix of all A_B-optimal designs. If this system has no solution then there is no A_B-optimal design whose information matrix has such a structure, which means that either $\bar{\boldsymbol{M}}(\xi^*)$ is non-diagonal or $\operatorname{tr}\bar{\boldsymbol{M}}(\xi^*) < a$. In this case a cutting plane algorithm as developed in GRIBIK/KORTANEK (1977) may be used to find the optimal information matrix or approximately optimal designs may be obtained directly by an iterative procedure (see Section 12.5).

Now, if there exists a solution ξ^* to (12.23) then, according to Theorem 11.1, we can assume it to be a discrete measure

$$\xi^* = \{(\boldsymbol{x}_1, p_1), \ldots, (\boldsymbol{x}_m, p_m)\}, \quad m \leqq h \leqq r(r+1)/2$$

with finite support (cp. also Theorem 11.5). Then we can try to find m supporting points $\boldsymbol{x}_1, \ldots, \boldsymbol{x}_m$ and corresponding weights $p_1, \ldots, p_m$ of an A_B-optimal design ξ^* as a solution to the equation system

$$\begin{aligned} \sum_{i=1}^m p_i \boldsymbol{f}(\boldsymbol{x}_i)\boldsymbol{f}(\boldsymbol{x}_i)^\top &= \boldsymbol{U}\operatorname{diag}(m_1^*, \ldots, m_r^*)\,\boldsymbol{U}^\top \\ \sum_{i=1}^m p_i &= 1\,. \end{aligned} \tag{12.24}$$

This is a nonlinear system of $r(r+1)/2 + 1$ equations and $2m$ variables the solution of which will mostly require suitable numerical procedures. In general, one should try to solve (12.24) for sufficiently small m, for example $m = r$ or even smaller if possible. On the one hand, this will keep the numerical effort relatively small and, on the other hand, in case of a true regression setup designs with only few supporting points are preferred to those with larger support.

If the prior covariance matrix is diagonal, which happens if the regression coefficients are uncorrelated a-priori, then it holds $\bar{f}_i(x) = f_i(x)$ for $i = 1, \ldots, r$ and the equation system (12.24) takes the form

$$\sum_{i=1}^{m} p_i f_j(x_i) f_k(x_i) = 0, \qquad k,j = 1, \ldots, r, \quad k \neq j,$$

$$\sum_{i=1}^{m} p_i f_j^2(x_i) = m_j^*, \qquad j = 1, \ldots, r, \tag{12.25}$$

$$\sum_{i=1}^{m} p_i = 1.$$

In a number of cases the numerical effort for the solution of (12.24) and (12.25), respectively, can be reduced significantly. Observing the condition (12.15) of Theorem 12.1, we are only to consider supporting points $x_i \in X_E$ such that

$$f(x_i)^\top f(x_i) = a \quad \text{and} \quad \bar{f}_j^2(x_i) = \begin{cases} a_j, & j \in J_{(1)}, \\ b_j, & j \in J_{(2)}, \end{cases} \tag{12.26}$$

which may restrict the number of possible supporting points drastically. It is then possible to choose a finite number m of points $x_1, \ldots, x_m$ satisfying (12.26) and to insert them into (12.24), which then reduces to a linear algebraic equation system for optimal design weights $p_1, \ldots, p_m$ (see the following example).

Example 12.1. Consider the two-factor model

$$\mathbf{E}\tilde{y}(x) = \vartheta_1 x_1 + \vartheta_2 x_2 + \vartheta_3 x_1 x_2, \quad X_E = \{x = (x_1, x_2)^\top \in \mathbb{R}^2 : x_1^2 + x_2^2 \leq 1\}$$

with prior information from previous observations given by

$$\mathbf{Cov}(\tilde{\vartheta} \mid w) = \frac{1}{w} \boldsymbol{\Phi} = \frac{1}{w} \begin{pmatrix} 0.970 & 0.282 & 0.260 \\ & 1.393 & 0.521 \\ & & 0.492 \end{pmatrix}.$$

The eigenvalues and eigenvectors of $\boldsymbol{\Phi}^{-1}$ are as follows

$$\boldsymbol{\Lambda} = \operatorname{diag}(0.5571, 1.2058, 4.3219)$$

$$\boldsymbol{U} = \begin{pmatrix} 0.4082 & -0.8944 & 0.1826 \\ 0.8165 & 0.4472 & 0.3651 \\ 0.4082 & 0.0000 & -0.9129 \end{pmatrix}.$$

Suppose that $n = 10$ observations have to be made.

The transformation $\bar{f} = U^\top f$ yields

$$\bar{f}(x) = \begin{pmatrix} 0.4082\,[x_1 + 2x_2 + x_1 x_2] \\ 0.4472\,[x_2 - 2x_1] \\ 0.1826\,[2x_2 - x_1 + 5x_1 x_2] \end{pmatrix}, \qquad b_i = \inf_{x \in X_E} \bar{f}_i^2(x) = 0 \quad (i = 1, 2, 3)$$

and thus

$$\bar{\lambda}_1 = 0.0557, \qquad \bar{\lambda}_2 = 0.1206, \qquad \bar{\lambda}_3 = 0.4322\,.$$

Further, we have

$$a = \bar{a} = \sup_{x \in X_E} \boldsymbol{f}(\boldsymbol{x})^\top \boldsymbol{f}(\boldsymbol{x}) = \sup_{x_1^2 + x_2^2 \leqq 1} (x_1^2 + x_2^2 + x_1^2 x_2^2) = 1.25\,.$$

With $J_0 = \{1, 2, 3\}$ and $\bar{a}_0 = \bar{a} = 1.25$ Step 1 of Algorithm 12.1 yields successively

$$i = 1 : \bar{\lambda}_2 = 0.1206 \leqq \bar{a}_0 + \bar{\lambda}_1 = 1.3057, \qquad J_0(i) = \{1\},$$

$$i = 2 : \bar{\lambda}_3 = 0.4322 < \frac{1}{2}(\bar{a}_0 + \bar{\lambda}_1 + \bar{\lambda}_2) = 0.7132, \qquad J_0(i) = \{1, 2\},$$

$$i = 3 : \bar{\lambda}_4 = \infty > \frac{1}{3}(\bar{a}_0 + \bar{\lambda}_1 + \bar{\lambda}_2 + \bar{\lambda}_3) = 0.6195, \qquad J_0(i) = \{1, 2, 3\}\,.$$

Therefore, $i_0 = 3$ and $m_i = \frac{1}{3}(\bar{a}_0 + \bar{\lambda}_1 + \bar{\lambda}_2 + \bar{\lambda}_3) - \bar{\lambda}_i = 0.6195 - \bar{\lambda}_i$ for $i = 1, 2, 3$, i.e.,

$$m_1 = 0.5638, \qquad m_2 = 0.4989, \qquad m_3 = 0.1873\,.$$

It is easily seen that

$$m_i < a_i = \sup_{x \in X_E} \bar{f}_i^2(\boldsymbol{x}), \qquad i = 1, 2, 3\,.$$

This gives

$$\bar{\boldsymbol{M}}(\xi^*) = \operatorname{diag}(m_1^*, m_2^*, m_3^*) = \operatorname{diag}(0.5638, 0.4989, 0.1873)$$

and the information matrix takes the form

$$\boldsymbol{M}(\xi^*) = \boldsymbol{U}\bar{\boldsymbol{M}}(\xi^*)\,\boldsymbol{U}^\top = \begin{pmatrix} 0.4993 & 0.0008 & 0.0627 \\ & 0.5006 & 0.1255 \\ & & 0.2500 \end{pmatrix}.$$

Since the maximum $a = 1.25$ of $\boldsymbol{f}(\cdot)^\top \boldsymbol{f}(\cdot)$ over X_E is attained only for points $\boldsymbol{x}_i = (x_{1i}, x_{2i})^\top \in X_E$ with $|x_{1i}| = |x_{2i}| = \sqrt{2}/2 = 0.7071$, it follows from condition (12.15) of Theorem 12.1 that

$$\operatorname{supp} \xi^* \subseteqq \left\{ \boldsymbol{x}_1 = \begin{pmatrix} -0.7071 \\ -0.7071 \end{pmatrix}, \ \boldsymbol{x}_2 = \begin{pmatrix} 0.7071 \\ -0.7071 \end{pmatrix}, \ \boldsymbol{x}_3 = \begin{pmatrix} -0.7071 \\ 0.7071 \end{pmatrix}, \right.$$

$$\left. \boldsymbol{x}_4 = \begin{pmatrix} 0.7071 \\ 0.7071 \end{pmatrix} \right\}.$$

Therefore, we put $\xi^* = \{(\boldsymbol{x}_1, p_1), (\boldsymbol{x}_2, p_2), (\boldsymbol{x}_3, p_3), (\boldsymbol{x}_4, p_4)\}$ and determine the weights $p_1, \ldots, p_4$ so that

$$M(\xi^*) = \sum_{i=1}^{4} p_i f(x_i) f(x_i)^\top = \begin{pmatrix} \sum p_i x_{1i}^2 & \sum p_i x_{1i} x_{2i} & \sum p_i x_{1i}^2 x_{2i} \\ & \sum p_i x_{2i}^2 & \sum p_i x_{1i} x_{2i}^2 \\ & & \sum p_i x_{1i}^2 x_{2i}^2 \end{pmatrix}$$

and $\sum p_i = 1$.

For the points $x_i = (x_{1i}, x_{2i})^\top$ from above we obtain the values

$$\sum_{i=1}^{4} p_i x_{1i}^2 = \sum_{i=1}^{4} p_i x_{2i}^2 = 0.5 \quad \text{and} \quad \sum_{i=1}^{4} p_i x_{1i}^2 x_{2i}^2 = 0.25$$

which are in good accordance with the main diagonal entries of $M(\xi^*)$. Writing down the remaining equations, we get the system

$$\begin{aligned} \sum p_i x_{1i} x_{2i} &= 0.5(p_1 - p_2 - p_3 + p_4) = 0.0008 \\ \sum p_i x_{1i}^2 x_{2i} &= 0.3535(-p_1 - p_2 + p_3 + p_4) = 0.0627 \\ \sum p_i x_{1i} x_{2i}^2 &= 0.3535(-p_1 + p_2 - p_3 + p_4) = 0.1255 \\ \sum p_i &= p_1 + p_2 + p_3 + p_4 = 1.0000 \end{aligned}$$

which has the solution $p_1 = 0.1173$, $p_2 = 0.2940$, $p_3 = 0.2051$, $p_4 = 0.3835$. Thus

$$\xi^* = \{(x_1, 0.117), (x_2, 0.294), (x_3, 0.205), (x_4, 0.384)\}$$

with $x_1, \ldots, x_4$ as defined above, is an A_B-optimal discrete design. A rounding off-procedure for ξ^* would lead us to the exact design v_{10} requiring $n_1 = 1$ observation at x_1, $n_2 = 3$ observations at x_2, $n_3 = 2$ observations at x_3 and $n_4 = 4$ observations at x_4. Since

$$\frac{1}{n} \Phi^{-1} = \frac{1}{n} U \Lambda U^\top = \begin{pmatrix} 0.1202 & -0.0009 & -0.0628 \\ & 0.1189 & -0.1255 \\ & & 0.3695 \end{pmatrix},$$

the design ξ^* has the Bayesian information matrix

$$M_B(\xi^*) = M(\xi^*) + \frac{1}{n} \Phi^{-1} = 0.6195\, I_3$$

which corresponds to the case of complete balance (see Corollary 12.1). This means, an experiment which is made according to this design assures that, a posteriori, the regression coefficients can be expected to be uncorrelated with the same variance 0.6195.

12.2. L_B-optimal and Bayesian designs

We now turn to the problem of constructing L_B-optimal and Bayesian designs. In the sequel we assume that the matrix U_L occuring in the criterion of L_B-optimality is positive definite.

Assumption 12.2. $U_L \in \mathcal{M}_r^>$.

Then there exists a regular matrix $U_L^{1/2} \in \mathcal{M}_{r\times r}$ such that

$$U_L^{1/2}(U_L^{1/2})^\top = U_L,$$

where $U_L^{1/2}$ may be found, for example, by Cholesky-decomposition of U_L. Now consider the transformations

$$\Phi_L = (U_L^{1/2})^\top \Phi U_L^{1/2}, \qquad f_L(x) = U_L^{-1/2} f(x). \tag{12.27}$$

Then we can show that an L_B-optimal design with respect to U_L can be obtained from an A_B-optimal design based on the transformed regression function f_L and transformed regression parameter $\tilde{\vartheta}_L := (U_L^{1/2})^\top \tilde{\vartheta}$ which has the prior covariance matrix $\mathbf{Cov}(\tilde{\vartheta}_L \mid w) = \frac{1}{w}(U_L^{1/2})^\top \Phi U_L^{1/2} = \frac{1}{w}\Phi_L$. These transformations define a reparameterized model

$$\mathbf{E}\tilde{y}(x) = \eta_L(x, \vartheta_L) := f_L(x)^\top \vartheta_L, \qquad \mathbf{Cov}\,\tilde{\vartheta}_L = \sigma_0^2 \Phi_L. \tag{12.28}$$

Theorem 12.2. *Under the Assumption* 12.2, *any* A_B*-optimal design in the transformed model* (12.28) *is* L_B*-optimal with respect to* U_L *in the original model, and conversely.*

Proof. For any $\xi \in \Xi$, the Bayesian information matrix in the transformed model is given by

$$\begin{aligned} M_B^L(\xi) &= \int_{X_E} f_L(x) f_L(x)^\top \xi(dx) + \frac{1}{n}\Phi_L^{-1} \\ &= U_L^{-1/2}\left\{\int_{X_E} f(x) f(x)^\top \xi(dx) + \frac{1}{n}\Phi^{-1}\right\}(U_L^{-1/2})^\top \\ &= U_L^{-1/2} M_B(\xi)(U_L^{-1/2})^\top \end{aligned}$$

so that it holds

$$\operatorname{tr} M_B^L(\xi)^{-1} = \operatorname{tr}(U_L^{1/2})^\top M_B(\xi)^{-1} U_L^{1/2} = \operatorname{tr} U_L M_B(\xi)^{-1} = L_B(\xi). \quad \square$$

As a consequence, Algorithm 12.1 can be modified to construct an L_B-optimal information matrix and hereafter an L_B-optimal design.

Algorithm 12.2. (Construction of an L_B-optimal design).

Step 1. Determine the diagonal matrix $\Lambda_L = \operatorname{diag}(\lambda_1^L, \ldots, \lambda_r^L)$ of eigenvalues and the corresponding matrix U of orthonormal eigenvectors of Φ_L^{-1}, i.e., find Λ_L and U such that

$$\Phi_L^{-1} = U\Lambda_L U^\top \quad \text{with} \quad UU^\top = U^\top U = I_r.$$

Step 2. Calculate $m_1^*, \ldots, m_r^*$ according to Algorithm 12.1 replacing $f(\cdot)$ and $\lambda_1, \ldots, \lambda_r$ by $f_L(\cdot)$ and $\lambda_1^L, \ldots, \lambda_r^L$, respectively.

Step 3. Find a solution $\xi^* = \{(x_1, p_1), \ldots, (x_m, p_m)\}$, $m \leqq r(r+1)/2$, to the equation system

$$\sum_{i=1}^{m} p_i f(x_i) f(x_i)^\top = U_L^{1/2} U \operatorname{diag}(m_1^*, \ldots, m_r^*) U^\top (U_L^{1/2})^\top$$

$$\sum_{i=1}^{m} p_i = 1 .$$

The assertion of Theorem 12.2 and the above transformations are of special importance for the construction of Bayesian designs which requires minimization of tr $U_L M_B(\cdot)^{-1}$ with U_L given by

$$U_L = \int_{X_P} f(x) f(x)^\top P_{\tilde{x}}(dx)$$

(see Theorem 10.1). This matrix is positive definite if the prediction region X_P and the prior distribution $P_{\tilde{x}}$ over X_P satisfy the following rather weak conditions which also played a role in uniqueness considerations for the information matrix of Bayes designs (see Corollary 11.3).

Lemma 12.1. *Assume that* supp $P_{\tilde{x}} = X_P$ *and* X_P *contains r points* $x_1, \ldots, x_r$ *such that* $f(x_1), \ldots, f(x_r)$ *are linear independent. Then* $U_L = \int_{X_P} f(x) f(x)\, P_{\tilde{x}}(dx)$ *is positive definite.*

Proof. We are to show that $a^\top U_L a \geqq 0$ for any vector $a \in \mathbb{R}^r$, $a \neq 0$. First it is clear that

$$a^\top U_L a = \int_{X_P} (a^\top f(x))^2 P_{\tilde{x}}(dx) \geqq 0$$

for any $a \in \mathbb{R}^r$. Now, assume there exists $a_0 \in \mathbb{R}^r$, $a_0 \neq 0$, such that $a_0^\top U_L a_0 = 0$. Because of supp $P_{\tilde{x}} = X_P$ this would necessarily imply that $a_0^\top f(x) = 0$ for all $x \in X_P$. But, since $f(X_P) = \{f(x) \in \mathbb{R}^r \colon x \in X_P\}$ contains, by assumption, a basis system of vectors from $\mathbb{R}^r$ it follows at once that $a_0^\top c = 0$ for all $c \in \mathbb{R}^r$. This is but only possible if $a_0 = 0$. □

Under the assumptions of Lemma 12.1 we can do the transformation

$$f_L(\cdot) = U_L^{-1/2} f(\cdot) .$$

This transformation implies a change from f to a regression function f_L whose components are orthogonal and normalized with respect to the measure $P_{\tilde{x}}$, for we have

$$\int_{X_P} f_L(x) f_L(x)^\top P_{\tilde{x}}(dx) = U_L^{-1/2} \left\{ \int_{X_P} f(x) f(x)^\top P_{\tilde{x}}(dx) \right\} (U_L^{-1/2})^\top = I_r .$$

Thus, according to Theorem 12.2 it holds

Corollary 12.2. *Let the assumptions of Lemma* 12.1 *be satisfied and* f_L *be the regression function arising by orthogonalization of the components of* f *with respect to the prior distribution* $P_{\tilde{x}}$. *Then* $\xi \in \Xi$ *is a Bayesian design with respect to* $P_{\tilde{x}}$ *if and only if it is* A_B*-optimal in the transformed model* (12.28).

The matrix $U_L^{-1/2}$ and thus the components of the orthogonalized regression function f_L may be obtained by application of Schmidt's orthogonalization procedure in the space of square integrable functions with the scalar product defined as

$$\langle f_i, f_j \rangle := \int_{X_P} f_i(x) f_j(x) \, P_{\tilde{x}}(dx), \qquad i, j = 1, \ldots, r.$$

Remark 12.5. In the special case of one-factor polynomial regression setups this orthogonalization procedure leads us to the well-known orthogonal polynomials: If, for example, $P_{\tilde{x}}$ has a density of the form $p_{\tilde{x}}(x) \propto (a + x)^\alpha (b - x)^\beta$ on $X_P = [a, b] \subset \mathbb{R}^1$, where α and β are arbitrary real numbers, then we get the Jacobian polynomials; if $\alpha = \beta = 0$ then $P_{\tilde{x}}$ is the density of the uniform distribution and the Jacobian polynomials coincide with the Lagrangian polynomials. In case of a normal prior distribution $P_{\tilde{x}}$ on $X_P = \mathbb{R}^1$ the components of f_L are Hermitian polynomials a.s.o. (see, e.g. Szegö (1959)).

We will now give a general result on the construction of L_B-optimal designs in case that the experimental region takes the form

$$X_E = \{x \in \mathbb{R}^k : f(x)^\top H f(x) \leqq 1\}$$

for some given matrix $H \in \mathcal{M}_r^>$. Particularly, when $f(x) = x$, which corresponds to a multiple linear regression setup without an intercept term, X_E describes an ellipsoid. Before turning to the result, we will give an important characterization of the set of Bayesian information matrices associated with the region X_E.

Lemma 12.2. *Let* $X_E = \{x \in \mathbb{R}^k : f(x)^\top H f(x) \leqq 1\}$. *Then it holds*

$$M_B(\Xi) = \left\{ C \in \mathcal{M}_r^> : \operatorname{tr} HC \leqq 1 + \frac{1}{n} \operatorname{tr} H\Phi^{-1} \right\}.$$

Proof. Denote $\mathcal{M}_H = \{M \in \mathcal{M}_r^\geqq : \operatorname{tr} HM \leqq 1\}$. Then we have

$$\operatorname{tr} HM(\xi) = \operatorname{tr} \left\{ H \int_{X_E} f(x) f(x)^\top \xi(dx) \right\} = \int_{X_E} f(x)^\top H f(x) \, \xi(dx) \leqq 1$$

for any $\xi \in \Xi$, i.e., $M(\Xi) = \{M(\xi) : \xi \in \Xi\} \subseteqq \mathcal{M}_H$. We will now show that the converse inclusion $\mathcal{M}_H \subseteqq M(\Xi)$ is also true, which proves the required result, observing that $\operatorname{tr} HM_B(\xi) = \operatorname{tr} HM(\xi) + \frac{1}{n} \operatorname{tr} H\Phi^{-1} \leqq 1 + \frac{1}{n} \operatorname{tr} H\Phi^{-1}$ for any $\xi \in \Xi$.

Let $M \in \mathcal{M}_H$ and denote $M = \lambda_1 m_1 m_1^\top + \ldots + \lambda_s m_s m_s^\top$ the spectral decomposition of $\boldsymbol{M}$, where $\lambda_1, \ldots, \lambda_s$ and $\boldsymbol{m}_1, \ldots, \boldsymbol{m}_s$ are the positive eigenvalues and the corresponding eigenvectors of $\boldsymbol{M}$, respectively, $s \leqq r$. Then define

$$\xi = \{(\boldsymbol{x}_1, p_1), \ldots, (\boldsymbol{x}_s, p_s)\}$$

to be the discrete measure assigning weights $p_i > 0$ to the points $\boldsymbol{x}_i$ such that

$$\boldsymbol{f}(\boldsymbol{x}_i) = a_i^{-1}\boldsymbol{m}_i, \qquad p_i = \lambda_i a_i^2, \quad i = 1, \ldots, s,$$

where $a_i^2 = \boldsymbol{m}_i^\top \boldsymbol{H}\boldsymbol{m}_i$, $i = 1, \ldots, s-1$ and $a_s^2 = \boldsymbol{m}_s^\top \boldsymbol{H}\boldsymbol{m}_s + \lambda_s^{-1}(1 - \operatorname{tr}\boldsymbol{H}\boldsymbol{M})$. Obviously, ξ defines a probability measure over X_{E}, for we have $\boldsymbol{f}(\boldsymbol{x}_i)^\top \boldsymbol{H}\boldsymbol{f}(\boldsymbol{x}_i) = \boldsymbol{m}_i^\top \boldsymbol{H}\boldsymbol{m}_i / a_i^2 = 1$, i.e., $\boldsymbol{x}_i \in X_{\mathrm{E}}$, $i = 1, \ldots, s$, and

$$\sum_{i=1}^{s} p_i = \sum_{i=1}^{s} \lambda_i \boldsymbol{m}_i^\top \boldsymbol{H}\boldsymbol{m}_i + 1 - \operatorname{tr}\boldsymbol{H}\boldsymbol{M} = 1,$$

observing that $\operatorname{tr}\boldsymbol{H}\boldsymbol{M} = \sum_{i=1}^{s} \lambda_i \boldsymbol{m}_i^\top \boldsymbol{H}\boldsymbol{m}_i \leqq 1$. Finally, it holds

$$\boldsymbol{M}(\xi) = \sum_{i=1}^{s} p_i \boldsymbol{f}(\boldsymbol{x}_i)\boldsymbol{f}(\boldsymbol{x}_i)^\top = \sum_{i=1}^{s} \lambda_i \boldsymbol{m}_i \boldsymbol{m}_i^\top = \boldsymbol{M}$$

implying that $\boldsymbol{M} \in \boldsymbol{M}(\Xi)$. □

The preceding proof demonstrates that in case of an ellipsoid-shaped experimental region X_{E} we can restrict ourselves to experimental designs the support of which contains at most r different points from X_{E}. This is an essential reduction compared to the usual bound $r(r+1)/2 + 1$ known from Carathéodory's theorem (cp. Theorem 11.4).

Theorem 12.3. *Let* $X_{\mathrm{E}} = \{\boldsymbol{x} \in \mathbb{R}^k : \boldsymbol{f}(\boldsymbol{x})^\top \boldsymbol{H}\boldsymbol{f}(\boldsymbol{x}) \leqq 1\}$ *and assume that* $\boldsymbol{U}_{\mathrm{L}} \in \mathcal{M}_r^{>}$ *satisfies the inequality*

$$a\boldsymbol{H}^{-1/2}(\boldsymbol{H}^{1/2}\boldsymbol{U}_{\mathrm{L}}\boldsymbol{H}^{1/2})^{1/2}\boldsymbol{H}^{-1/2} \geqq \frac{1}{n}\boldsymbol{\Phi}^{-1},$$

where (12.29)

$$a = \left(1 + \frac{1}{n}\operatorname{tr}\boldsymbol{H}\boldsymbol{\Phi}^{-1}\right) \Big/ \operatorname{tr}(\boldsymbol{H}^{1/2}\boldsymbol{U}_{\mathrm{L}}\boldsymbol{H}^{1/2})^{1/2}.$$

Then any $\xi^* \in \Xi$ *for which*

$$\boldsymbol{M}_{\mathrm{B}}(\xi^*) = \boldsymbol{M}(\xi^*) + \frac{1}{n}\boldsymbol{\Phi}^{-1} = a\boldsymbol{H}^{-1/2}(\boldsymbol{H}^{1/2}\boldsymbol{U}_{\mathrm{L}}\boldsymbol{H}^{1/2})^{1/2}\boldsymbol{H}^{-1/2}$$

forms an L_{B}*-optimal design with respect to* $\boldsymbol{U}_{\mathrm{L}}$.

Proof. First note that our assumptions guarantee the existence of a design $\xi^* \in \Xi$ having the Bayesian information matrix indicated above, for we have

$$\operatorname{tr} \boldsymbol{H}\boldsymbol{M}_{\mathrm{B}}(\xi^*) = a \operatorname{tr} (\boldsymbol{H}^{1/2}\boldsymbol{U}_{\mathrm{L}}\boldsymbol{H}^{1/2})^{1/2} = 1 + \frac{1}{n} \operatorname{tr} \boldsymbol{H}\boldsymbol{\Phi}^{-1},$$

i.e., $\boldsymbol{M}_{\mathrm{B}}(\xi^*) \in M_{\mathrm{B}}(\Xi)$. The L_{B}-optimality of ξ^* then follows from the fact that the necessary and sufficient condition stated in our Theorem 11.12 is satisfied:

$$\operatorname{tr} \boldsymbol{M}_{\mathrm{B}}(\xi^*)^{-1}\boldsymbol{U}_{\mathrm{L}}\boldsymbol{M}_{\mathrm{B}}(\xi^*)^{-1}\boldsymbol{M}(\xi^*) = a^{-2} \operatorname{tr} \boldsymbol{H}\boldsymbol{M}(\xi^*) = a^{-2}$$

and

$$\sup_{x \in X_{\mathrm{E}}} \| \boldsymbol{M}_{\mathrm{B}}(\xi^*)^{-1}\boldsymbol{f}(\boldsymbol{x}) \|^2_{\boldsymbol{U}_{\mathrm{L}}} = \sup_{x \in X_{\mathrm{E}}} a^{-2}\boldsymbol{f}(\boldsymbol{x})^{\top}\boldsymbol{H}\boldsymbol{f}(\boldsymbol{x}) = a^{-2},$$

observing that $\boldsymbol{M}_{\mathrm{B}}(\xi^*)^{-1}\boldsymbol{U}_{\mathrm{L}}\boldsymbol{M}_{\mathrm{B}}(\xi^*)^{-1} = a^{-2}\boldsymbol{H}$. □

Theorem 12.3 represents a generalization of a result due to CHALONER (1984) who considered the special case $\boldsymbol{H} = b^{-2}\boldsymbol{I}_r$ for some $b \geqq 0$. In this case we are led to $\boldsymbol{M}_{\mathrm{B}}(\xi^*) = a_0\boldsymbol{U}_{\mathrm{L}}^{1/2}$ where $a_0 = \left(b^2 + \frac{1}{n} \operatorname{tr} \boldsymbol{\Phi}^{-1}\right) \Big/ \operatorname{tr} \boldsymbol{U}_{\mathrm{L}}^{1/2}$. The definiteness condition (12.29) requires $a_0\boldsymbol{U}_{\mathrm{L}}^{1/2} \geqq \frac{1}{n}\boldsymbol{\Phi}^{-1}$, which will be satisfied provided the radius b of the spherical region $X_{\mathrm{E}} = \{\boldsymbol{x} \in \mathbb{R}^k : \boldsymbol{f}(\boldsymbol{x})^{\top}\boldsymbol{f}(\boldsymbol{x}) \leqq b^2\}$ and/or the number of observations n are sufficiently large. If, additionally, $\boldsymbol{U}_{\mathrm{L}} = \boldsymbol{I}_r$, which corresponds to A_{B}-optimality, then $\boldsymbol{M}_{\mathrm{B}}(\xi^*)$ is required to be a multiple of $\boldsymbol{I}_r$; this is in accordance with the result stated in Corollary 12.1 ('completely balanced case').

A more general result than that one presented in Theorem 12.3, and which does not require a definiteness assumption of the type (12.29), will be given in Section 15.

12.3. D_{B}- and E_{B}-optimality of designs

After we have seen that the problem of the construction of L_{B}-optimal and Bayesian designs can be reduced to that of constructing A_{B}-optimal designs in transformed models, we are now concerned with the criteria of D_{B}- and E_{B}-optimality. It turns out that, in certain cases, A_{B}-optimal designs are also D_{B}- and E_{B}-optimal.

First we give conditions under which A_{B}-optimal designs are D_{B}-optimal, and vice versa.

Let ξ^* be an A_{B}-optimal design, i.e., ξ^* minimizes $\operatorname{tr} \boldsymbol{M}_{\mathrm{B}}(\cdot)^{-1}$ over Ξ, and let the corresponding Bayesian information matrix be represented as

$$\boldsymbol{M}_{\mathrm{B}}(\xi^*) = \boldsymbol{T}\boldsymbol{\Delta}\boldsymbol{T}^{\top}, \quad \text{where} \quad \boldsymbol{T}^{\top}\boldsymbol{T} = \boldsymbol{T}\boldsymbol{T}^{\top} = \boldsymbol{I}_r$$

and (12.30)

$$\boldsymbol{\Delta} = \operatorname{diag}(d_1, \ldots, d_r), \qquad d_i > 0 \quad (i = 1, \ldots, r).$$

Without loss of generality, we assume that $d_1 \leqq d_2 \leqq \ldots \leqq d_r$. In analogy with (12.6) we define

$$\hat{f}(x) := T^\top f(x),$$
$$\alpha := \sup_{x \in X_E} \hat{f}(x)^\top \hat{f}(x) = \sup_{x \in X_E} f(x)^\top f(x),$$
$$\alpha_j := \sup_{x \in X_E} \hat{f}_j^2(x), \qquad \beta_j := \inf_{x \in X_E} \hat{f}_j^2(x), \qquad j = 1, \ldots, r, \tag{12.31}$$
$$\bar{\alpha} := \alpha - \sum_{j=1}^{r} \beta_j.$$

To establish an equivalence relation between A_B- and D_B-optimality we require the following assumption.

Assumption 12.3. *There exist some point $x^* \in X_E$ and a decomposition of the index set $J = \{1, \ldots, r\}$ into three disjoint subsets J_1, J_2 and J_3 depending on $d_1, \ldots, d_r$ such that*

$$d_{j_1} < d_{j_2} < d_{j_3} \quad \textit{for all} \quad j_1 \in J_1, j_2 \in J_2, j_3 \in J_3;$$
$$d_j = \text{constant} := d \quad \textit{for all} \quad j \in J_2$$

and

$$\hat{f}(x^*)^\top \hat{f}(x^*) = \alpha, \qquad \hat{f}_j^2(x^*) = \begin{cases} \alpha_j & \textit{for} \quad j \in J_1, \\ \beta_j & \textit{for} \quad j \in J_3. \end{cases}$$

We add that in Assumption 12.3 up to two of the subsets J_1, J_2 and J_3 are permitted to be empty.

Remark 12.6. If $M_B(\xi^*)$ was generated according to Algorithm 12.1 then Assumption 12.3 is satisfied.

Theorem 12.4. *Let $\xi^* \in \Xi$ have Bayesian information matrix as indicated in* (12.30). *Then, under the Assumption* 12.3, *ξ^* is D_B-optimal if and only if it is A_B-optimal.*

Proof. We show that, under the Assumption 12.3, the condition (12.1), which is necessary and sufficient for A_B-optimality of ξ^*, is equivalent to the condition of D_B-optimality given in Theorem 11.15 (i). To this, assume ξ^* is A_B-optimal. Then, by virtue of (12.1) and (12.30), it holds

$$\sup_{x \in X_E} \hat{f}(x)^\top \Delta^{-2} \hat{f}(x) = \operatorname{tr} T^\top M(\xi^*)\, T \Delta^{-2}.$$

Denoting the diagonal elements of $T^\top M(\xi^*)\, T$ by $\gamma_1, \ldots, \gamma_r$, this equation reads

$$\sup_{x \in X_E} \sum_{j=1}^{r} \hat{f}_j^2(x)/d_j^2 = \sum_{j=1}^{r} \gamma_j/d_j^2. \tag{12.32}$$

Employing our Assumption 12.3, the left-hand sum can be written as

$$\sum_{j=1}^{r} \hat{f}_j^2(x)/d_j^2 = \sum_{j\in J_1} \hat{f}_j^2(x)/d_j^2 + d^{-2}\sum_{j\in J_2} \hat{f}_j^2(x) + \sum_{j\in J_3} \hat{f}_j^2(x)/d_j^2 .$$

Then, by the same arguments as in the proof of Theorem 12.1, it follows from our assumption concerning the point $x^* \in X_E$ that the supremum of the above sum is just attained at x^*. Hence, we have

$$\sum_{j=1}^{r} \gamma_j/d_j^2 = \sum_{j\in J_1} \alpha_j/d_j^2 + d^{-2}\Big(\alpha - \sum_{j\in J_1}\alpha_j - \sum_{j\in J_3}\beta_j\Big) + \sum_{j\in J_3}\beta_j/d_j^2 . \tag{12.33}$$

Furthermore, since $T^\top M(\xi^*)\,T = \int_{X_E} \hat{f}(x)\hat{f}(x)^\top \xi^*(dx)$, we obtain that

$$\beta_j \leqq \gamma_j \leqq \alpha_j, \qquad j = 1, \dots, r$$

and (12.34)

$$\operatorname{tr} T^\top M(\xi^*)\,T = \sum_{j=1}^{r} \gamma_j \leqq \alpha .$$

But then, as an immediate consequence of Assumption 12.3, (12.33) and (12.34), it follows that

$$\gamma_j = \begin{cases} \alpha_j & \text{for } j \in J_1, \\ \beta_j & \text{for } j \in J_3 \end{cases}$$

and (12.35)

$$\sum_{j\in J_2} \gamma_j = \alpha - \sum_{j\in J_1}\alpha_j - \sum_{j\in J_3}\beta_j .$$

In view of the ordering $d_{j_1} < d < d_{j_3}$ $(j_1 \in J_1, j_3 \in J_3)$, it is clear that any choice of $\gamma_1, \dots, \gamma_r$ which is not conformable to (12.35) leads necessarily to a violation of (12.33) or (12.34). On the other hand, owing to monotonicity reasons, the supremum of $\sum_{j=1}^{r} \hat{f}_j^2(x)/d_j$ is also attained at the point x^* specified in Assumption 12.3. Together with (12.35) this implies that

$$\sup_{x\in X_E} \sum_{j=1}^{r} \hat{f}_j^2(x)/d_j = \sum_{j=1}^{r} \hat{f}_j^2(x^*)/d_j = \sum_{j=1}^{r} \gamma_j/d_j . \tag{12.36}$$

By virtue of (12.30) it holds $M_B(\xi^*)^{-1} = T\Delta^{-1}T^\top$ so that (12.36) can be rewritten as

$$\begin{aligned} \sup_{x\in X_E} f(x)^\top M_B(\xi^*)^{-1} f(x) &= \sup_{x\in X_E} \hat{f}(x)^\top \Delta^{-1} \hat{f}(x) \\ &= \operatorname{tr} T^\top M(\xi^*)\,T\Delta^{-1} = \operatorname{tr} M(\xi^*)\,M_B(\xi^*)^{-1} . \end{aligned}$$

But, according to Theorem 11.15 (i), this means that ξ^* is D_B-optimal. Conversely, from the validity of (12.36) follows the equality (12.32), using the samed arguments as before. This completes the proof. □

The following remark is an immediate consequence of the proof of Theorem 12.3.

Remark 12.7. Let Ξ^* be any design with Bayesian information matrix given by (12.30). If there exists a point $\boldsymbol{x}^*$ according to Assumption 12.3 then a structure of the diagonal entries $\gamma_1, \ldots, \gamma_r$ of the transformed information matrix $\boldsymbol{T}^\top \boldsymbol{M}(\xi^*)\,\boldsymbol{T}$ as indicated in (12.35) is necessary for ξ^* to be A_B- or D_B-optimal.
In particular, A_B- and D_B-optimality of a design coincide if its Bayesian information matrix happens to be a multiple of the unity matrix.

Corollary 12.3. *If there exists a design $\xi^* \in \Xi$ such that*

$$\boldsymbol{M}_B(\xi^*) = d\boldsymbol{I}_r \quad \text{for some} \quad d > 0$$

then it holds:

$$\xi^* \in \Xi_{D_B} \Leftrightarrow \xi^* \in \Xi_{A_B}.$$

Proof. With $\boldsymbol{T} = \boldsymbol{I}_r$ and $\Delta = d\boldsymbol{I}_r$ the matrix $\boldsymbol{M}_B(\xi^*) = d\boldsymbol{I}_r = \boldsymbol{T}\Delta\boldsymbol{T}^\top$ is of the form required in (12.30). Further, Assumption 12.3 is satisfied with $J_1 = J_3 = \varnothing$ and $J_2 = \{1, \ldots, r\}$; the existence of a point $\boldsymbol{x}^*$ maximizing $\hat{\boldsymbol{f}}(\cdot)^\top \hat{\boldsymbol{f}}(\cdot) = \boldsymbol{f}(\cdot)^\top \boldsymbol{f}(\cdot)$ over X_E is guaranteed by the continuity of $\boldsymbol{f}(\cdot)$ and the compactness of X_E. The assertion then follows from Theorem 12.3. □

Thus, by Corollary 12.3, the A_B-optimal design constructed in Example 12.1, and for which we had $\boldsymbol{M}_B(\xi^*) = 0.6195\,\boldsymbol{I}_3$, is also D_B-optimal.

We next formulate a sufficient condition for D_B-optimality which is an analogue to a known result from Hoel (1965), and which may be useful in the construction of optimal designs for particular regression functions, e.g., for trigonometric setups.

Theorem 12.5. *If there exists a design $\xi^* \in \Xi$ such that*

$$\boldsymbol{f}(\cdot)^\top \boldsymbol{M}_B(\xi^*)^{-1} \boldsymbol{f}(\cdot) = c = \text{constant}$$

then ξ^* is D_B-optimal in Ξ.

Proof. For every $\bar{\xi} \in \Xi$, the directional derivative as given by Lemma 11.8 (i) takes the form

$$\begin{aligned}\Delta_{D_B}(\xi^*, \bar{\xi}) &= D_B(\xi^*)\,\{\mathrm{tr}\,(\boldsymbol{M}_B(\xi^*) - \boldsymbol{M}_B(\bar{\xi}))\,\boldsymbol{M}_B(\xi^*)^{-1}\} \\ &= D_B(\xi^*)\left\{r - \mathrm{tr}\int_{X_E}\left(\boldsymbol{f}(\boldsymbol{x})\boldsymbol{f}(\boldsymbol{x})^\top + \frac{1}{n}\boldsymbol{\Phi}^{-1}\right)\boldsymbol{M}_B(\xi^*)^{-1}\bar{\xi}(\mathrm{d}\boldsymbol{x})\right\} \\ &= D_B(\xi^*)\left\{r - c - \frac{1}{n}\,\mathrm{tr}\,\boldsymbol{\Phi}^{-1}\boldsymbol{M}_B(\xi^*)^{-1}\right\}.\end{aligned}$$

Since this expression does not depend on $\bar{\xi}$, it holds

$$\Delta_{D_B}(\xi^*, \bar{\xi}) = \Delta_{D_B}(\xi^*, \xi^*) = 0$$

for all $\bar{\xi} \in \Xi$. Because of $D_B(\xi^*) > 0$, it follows that

$$c = r - \frac{1}{n} \operatorname{tr} \boldsymbol{\Phi}^{-1} \boldsymbol{M}_B(\xi^*)^{-1} = \operatorname{tr}\left(\boldsymbol{M}_B(\xi^*) - \frac{1}{n}\boldsymbol{\Phi}^{-1}\right) \boldsymbol{M}_B(\xi^*)^{-1}$$
$$= \operatorname{tr} \boldsymbol{M}(\xi^*)\, \boldsymbol{M}_B(\xi^*)^{-1}.$$

Thus, observing that $c = \sup_{x \in X_E} \boldsymbol{f}(x)^\top \boldsymbol{M}_B(\xi^*)^{-1} \boldsymbol{f}(x)$, ξ^* is D_B-optimal by Theorem 11.15 (i). □

The following small example demonstrates the application of Theorem 12.5.

Example 12.2. Consider the trigonometric setup

$$\mathbf{E}\tilde{y}(x) = \vartheta_1 \sin x + \vartheta_2 \cos x, \qquad x \in X_E = \left[0, \frac{\pi}{2}\right]$$

with a prior covariance matrix of the type $\boldsymbol{\Phi} = a\boldsymbol{I}_2$, where a is some positive real number. It is easily seen that

$$\xi^* = \left\{\left(0, \frac{1}{2}\right), \left(\frac{\pi}{2}, \frac{1}{2}\right)\right\}$$

has the Bayesian information matrix $\boldsymbol{M}_B(\xi^*) = \left(\frac{1}{2} + a\right)\boldsymbol{I}_2$. From this we obtain

$$\boldsymbol{f}(x)^\top \boldsymbol{M}_B(\xi^*)^{-1} \boldsymbol{f}(x) = \left(\frac{1}{2} + a\right)^{-1} \boldsymbol{f}(x)^\top \boldsymbol{f}(x)$$
$$= \left(\frac{1}{2} + a\right)^{-1} (\sin^2 x + \cos^2 x)$$
$$= \left(\frac{1}{2} + a\right)^{-1} = \text{constant}.$$

Hence, by Theorem 12.5, the design ξ^* is D_B-optimal.

Concluding, we turn to the criterion of E_B-optimality. Since an E_B-optimal design maximizes only the smallest eigenvalue of the Bayesian information matrix and does not take account of the remaining eigenvalues, one cannot expect to obtain a general equivalence relation between A_B- or D_B-optimality and E_B-optimality. However, the special structure of the information matrices of the A_B- or D_B-optimal designs as considered before, also leads us to E_B-optimal designs.

Theorem 12.6. *Let $\xi^* \in \Xi$ be any design having a Bayesian information matrix* *as given by* (12.30). *If Assumption* 12.3 *is satisfied, and if* ξ^* *is* A_B*- or* D_B*-optimal, then it is* E_B*-optimal, too.*

(For a proof see GLADITZ/PILZ (1982 a).)

From this result we get the following analogue to Corollary 12.3.

Corollary 12.4. *If there exists* $\xi^* \in \Xi$ *such that* $\boldsymbol{M}_B(\xi^*) = d\boldsymbol{I}_r$ *with some* $d > 0$ *then it holds:*

$$\xi^* \in \Xi_{A_B} \Rightarrow \xi^* \in \Xi_{E_B}.$$

Thus, the designs constructed in the Examples 12.1 and 12.2 are also E_B-optimal.

Now, combining the assertions of the Theorems 12.1, 12.4 and 12.6, we obtain as a final result

Theorem 12.7. *If there exists a design* $\xi^* \in \Xi$ *with an information matrix* $\boldsymbol{M}(\xi^*)$ *constructed according to Algorithm* 12.1 *then* ξ^* *is* A_B*-,* D_B*- and* E_B*-optimal.*

Together with the foregoing results this establishes a remarkable robustness of our design procedure developed in Section 12.1 against the underlying optimality criterion.

12.4. Optimal designs for multiple linear regression

In this section we specialize the Algorithm 12.1 for the construction of an optimal information matrix to the case of multiple linear regression where the experimental region is assumed to be the k-dimensional hypersphere. We construct, both for the homogeneous and the inhomogeneous model, approximate designs having such an information matrix and which then are A_B-, D_B- and E_B-optimal. Hereafter we briefly indicate the construction of optimal designs on the k-dimensional hypercube.

For convenience of representation, we first consider the homogeneous model (regression without an absolute term)

$$\mathbf{E}\tilde{y}(\boldsymbol{x}) = \boldsymbol{\vartheta}^\top \boldsymbol{x} = \sum_{i=1}^{k} \vartheta_i x_i, \qquad \mathbf{Cov}(\tilde{\boldsymbol{\vartheta}} \mid w) = \frac{1}{w}\boldsymbol{\Phi} \tag{12.37}$$

where realizations of $\tilde{y}(\boldsymbol{x})$ can be observed at all points $\boldsymbol{x} = (x_1, \ldots, x_k)^\top$ of the k-dimensional hypersphere

$$X_E = \left\{ \boldsymbol{x} \in \mathbb{R}^k : \boldsymbol{x}^\top \boldsymbol{x} = \sum_{i=1}^{k} x_i^2 \leq 1 \right\}. \tag{12.38}$$

Again, let $\boldsymbol{\Lambda} = \operatorname{diag}(\lambda_1, \ldots, \lambda_k)$ and $\boldsymbol{U} = (\boldsymbol{u}_1, \ldots, \boldsymbol{u}_k)$ be the matrices of eigenvalues and orthonormal eigenvectors of $\boldsymbol{\Phi}^{-1}$, respectively, i.e.,

$$\boldsymbol{\Phi}^{-1} = \boldsymbol{U}\boldsymbol{\Lambda}\boldsymbol{U}^\top, \qquad \boldsymbol{U}^\top\boldsymbol{U} = \boldsymbol{U}\boldsymbol{U}^\top = \boldsymbol{I}_k,$$

where, for notational convenience, we assume that the components of $\boldsymbol{\vartheta} = (\vartheta_1, \ldots, \vartheta_k)^\top$ and $\boldsymbol{x}$ are ordered such that $\lambda_1 \leqq \lambda_2 \leqq \ldots \leqq \lambda_k$. With the information matrix

$$\boldsymbol{M}(\xi) = \int_{X_E} \boldsymbol{x}\boldsymbol{x}^\top \xi(\mathrm{d}\boldsymbol{x}), \qquad \xi \in \Xi \tag{12.39}$$

we obtain the transformed information matrix as

$$\bar{\boldsymbol{M}}(\xi) = \int_{X_E} (\boldsymbol{U}^\top \boldsymbol{x})(\boldsymbol{U}^\top \boldsymbol{x})^\top \xi(\mathrm{d}\boldsymbol{x}) = \boldsymbol{U}^\top \boldsymbol{M}(\xi)\, \boldsymbol{U}$$

so that the Bayesian information matrix can be written in the form

$$\boldsymbol{M}_B(\xi) = \boldsymbol{U}\left(\bar{\boldsymbol{M}}(\xi) + \frac{1}{n}\boldsymbol{\Lambda}\right)\boldsymbol{U}^\top. \tag{12.40}$$

Under the assumption that there exists a design $\xi^* \in \Xi$ for which the transformed information matrix is diagonal with $\operatorname{tr}\bar{\boldsymbol{M}}(\xi^*) = 1 = \sup_{X_E} \boldsymbol{x}^\top \boldsymbol{x}$, the Algorithm 12.1 can be used to generate this (uniquely determined) optimal matrix. Observing that

$$\operatorname{tr}\bar{\boldsymbol{M}}(\xi) = \operatorname{tr}\boldsymbol{M}(\xi) = \int_{X_E} \boldsymbol{x}^\top \boldsymbol{x}\, \xi(\mathrm{d}\boldsymbol{x}) \leqq 1$$

for every $\xi \in \Xi$, the condition $\operatorname{tr}\bar{\boldsymbol{M}}(\xi^*) = 1$ requires that all the supporting points of ξ^* are located on the surface of the sphere X_E. It will be seen in the sequel that such a design with $\bar{\boldsymbol{M}}(\xi^*)$ being diagonal always exists.

We are now going to specialize Algorithm 12.1 to our model (12.37) with the experimental region (12.38). In this case the quantities introduced in (12.6) assume the values

$$a = \sup_{\boldsymbol{x} \in X_E} \boldsymbol{x}^\top \boldsymbol{x} = 1,$$

$$a_j = \sup_{\boldsymbol{x} \in X_E} (\boldsymbol{u}_j^\top \boldsymbol{x})^2 = 1, \qquad b_j = \inf_{\boldsymbol{x} \in X_E} (\boldsymbol{u}_j^\top \boldsymbol{x})^2 = 0 \qquad (j = 1, \ldots, k)$$

$$\bar{a} = a = 1, \qquad \bar{\lambda}_j = \frac{1}{n}\lambda_j \qquad (j = 1, \ldots, k).$$

Therefore, we shall obtain diagonal entries $m_1^*, \ldots, m_k^*$ for the optimal transformed information matrix $\bar{\boldsymbol{M}}(\xi^*) = \operatorname{diag}(m_1^*, \ldots, m_k^*)$ with

$$m_1^* \geqq m_2^* \geqq \ldots \geqq m_k^* \geqq 0 \quad \text{and} \quad \sum_{i=1}^{k} m_i^* = 1$$

in the following way.

Algorithm 12.3.

Step 1. Prove if it holds $\lambda_k \leqq \frac{1}{k}\left(n + \sum_{i=1}^{k} \lambda_i\right)$. If this comes true then put

$$m_j^* = \frac{1}{k}\left(1 + \frac{1}{n}\sum_{i=1}^{k}\lambda_i\right) - \frac{1}{n}\lambda_j, \qquad j = 1, \ldots, k,$$

otherwise goto Step 2.

Step 2. Find the smallest integer $j \in \{1, \ldots, k-1\}$ for which it holds

$$\lambda_{j+1} > \frac{1}{j}\left(n + \sum_{i=1}^{j}\lambda_i\right)$$

and denote it j_0. Then choose

$$m_j^* = \begin{cases} j_0^{-1}\left(1 + \frac{1}{n}\sum_{i=1}^{j_0}\lambda_i\right) - \frac{1}{n}\lambda_j, & j = 1, \ldots, j_0, \\ 0, & j = j_0 + 1, \ldots, k. \end{cases}$$

Theorem 12.8. *If there exists a design* $\xi^* \in \Xi$ *such that* $\bar{\boldsymbol{M}}(\xi^*) = \operatorname{diag}(m_1^*, \ldots, m_k^*)$ *with* $m_1^*, \ldots, m_k^*$ *obtained according to Algorithm* 12.3 *then* ξ^* *is* A_{B}-, D_{B}- *and* E_{B}-*optimal for the homogeneous model with experimental region* (12.38).

Proof. The assertion follows immediately from the Theorems 12.1, 12.7 and Remark 12.6 observing that, by virtue of (12.40), $\boldsymbol{M}_{\mathrm{B}}(\xi^*)$ can be written in the form (12.30) with $\boldsymbol{U} = \boldsymbol{T}$ and $\boldsymbol{\Delta} = \bar{\boldsymbol{M}}(\xi^*) + \frac{1}{n}\boldsymbol{\Lambda}$. □

The following corollary explicitly gives an optimal design.

Corollary 12.5. *For any set of numbers* $\{m_1^*, \ldots, m_k^*\}$ *generated by Algorithm* 12.3 *the design*

$$\xi^* = \{(\boldsymbol{u}_1, m_1^*), \ldots, (\boldsymbol{u}_k, m_k^*)\} \tag{12.41}$$

giving weight m_i^* *to the column* $\boldsymbol{u}_i$ $(i = 1, \ldots, k)$ *of* $\boldsymbol{U}$, *is* A_{B}-, D_{B}- *and* E_{B}-*optimal for homogeneous multiple linear regression on the sphere.*

Proof. For ξ^* from (12.41) it holds

$$\boldsymbol{M}(\xi^*) = \int_{X_{\mathrm{E}}} \boldsymbol{x}\boldsymbol{x}^{\top}\,\xi^*(\mathrm{d}\boldsymbol{x}) = \sum_{i=1}^{k} m_i^*\,\boldsymbol{u}_i\boldsymbol{u}_i^{\top} = \boldsymbol{U}\operatorname{diag}(m_1^*, \ldots, m_k^*)\,\boldsymbol{U}^{\top},$$

which yields $\bar{\boldsymbol{M}}(\xi^*) = \operatorname{diag}(m_1^*, \ldots, m_k^*)$. The assertion then follows from Theorem 12.8 observing that

$$m_i^* \geqq 0, \qquad \sum_{i=1}^{k} m_i^* = 1 \quad \text{and} \quad \boldsymbol{u}_i^{\top}\boldsymbol{u}_i = 1,$$

i.e., $\boldsymbol{u}_i \in X_{\mathrm{E}}$ for each column $\boldsymbol{u}_i$ of $\boldsymbol{U}$. □

Note that ξ^* as defined by (12.41) only requires observations in at most k different experimental points, if Step 2 of Algorithm 12.3 was employed then the number j_0 of supporting points of ξ^* is strictly smaller than the number k of unknown parameters.

Example 12.3. Consider the three-factor model

$$\mathbf{E}\tilde{y}(x) = \vartheta_1 x_1 + \vartheta_2 x_2 + \vartheta_3 x_3, \qquad X_{\mathrm{E}} = \{x = (x_1, x_2, x_3)^\top \in \mathbb{R}^3 \colon x^\top x \leqq 1\}$$

with prior covariance matrix $\mathbf{Cov}(\tilde{\vartheta} \mid w) = \frac{1}{w} \boldsymbol{\Phi}$ and $\boldsymbol{\Phi}$ as given in Example 12.1. Further, assume that $n = 5$ observations have to be made. Then, since $\lambda_3 = 4.3219 > \frac{1}{3}(5 + \lambda_1 + \lambda_2 + \lambda_3) = 3.6949$, Step 2 of Algorithm 12.3 comes in use. Observing that

$$\lambda_2 = 1.2058 < 5 + \lambda_1 = 5.5571$$

and

$$\lambda_3 = 4.3219 > \frac{1}{2}(5 + \lambda_1 + \lambda_2) = 3.3814,$$

we have $j_0 = 2$ which results in choosing

$$m_1^* = \frac{1}{2}\left(1 + \frac{1}{5}(\lambda_1 + \lambda_2)\right) - \frac{1}{5}\lambda_1 = 0.5649,$$

$$m_2^* = \frac{1}{2}\left(1 + \frac{1}{5}(\lambda_1 + \lambda_2)\right) - \frac{1}{5}\lambda_2 = 0.4351, \qquad m_3^* = 0.$$

Thus, by Corollary 12.5,

$$\xi^* = \{(u_1, m_1^*), (u_2, m_2^*)\}, \qquad u_1 = \begin{pmatrix} 0.4082 \\ 0.8165 \\ 0.4082 \end{pmatrix}, \qquad u_2 = \begin{pmatrix} -0.8944 \\ 0.4472 \\ 0 \end{pmatrix}$$

is an A_{B}-, D_{B}- and E_{B}-optimal design.

Remark 12.8. If the experimental region is a hypersphere $X_{\mathrm{E}} = \{x \in \mathbb{R}^k \colon x^\top x \leqq a\}$ with arbitrary radius $a > 0$ then an A_{B}-, D_{B}- and E_{B}-optimal design is obtained by modifying (12.41) according to

$$\xi^* = \{(\sqrt{a}\, u_1, m_1^*/a), \ldots, (\sqrt{a}\, u_k, m_k^*/a)\}.$$

Now we turn to the inhomogeneous model

$$\mathbf{E}\tilde{y}(x) = \vartheta_0 + \boldsymbol{\vartheta}^\top x = \vartheta_0 + \sum_{i=1}^{k} \vartheta_i x_i, \qquad x \in X_{\mathrm{E}} \tag{12.42}$$

with $\boldsymbol{\vartheta}$, x, X_{E} and $\mathbf{Cov}(\tilde{\boldsymbol{\vartheta}} \mid w) = \frac{1}{w} \boldsymbol{\Phi}$ as before. Additionally, let

$$\mathbf{Cov}(\tilde{\vartheta}_0, \tilde{\vartheta}_i \,|\, w) = \frac{1}{w}\varphi_{0i}, \qquad i = 1, \ldots, k$$

so that the covariance matrix of $\tilde{\boldsymbol{\vartheta}}_0 = (\tilde{\vartheta}_0, \tilde{\boldsymbol{\vartheta}}^\top)^\top$ can be subdivided as

$$\mathbf{Cov}(\tilde{\boldsymbol{\vartheta}}_0 \,|\, w) = \frac{1}{w}\boldsymbol{\Phi}_0 := \frac{1}{w}\left(\begin{array}{c|c} \varphi_0 & \boldsymbol{\varphi}^\top \\ \hline \boldsymbol{\varphi} & \boldsymbol{\Phi} \end{array}\right), \qquad \boldsymbol{\varphi} = (\varphi_{01}, \ldots, \varphi_{0k})^\top. \tag{12.43}$$

The information matrix for the above setup takes the form

$$\boldsymbol{M}_0(\xi) = \int_{X_E} \begin{pmatrix} 1 \\ \boldsymbol{x} \end{pmatrix} (1, \boldsymbol{x}^\top)\, \xi(\mathrm{d}\boldsymbol{x}) = \left(\begin{array}{c|c} 1 & \boldsymbol{m}(\xi)^\top \\ \hline \boldsymbol{m}(\xi) & \boldsymbol{M}(\xi) \end{array}\right) \tag{12.44}$$

with $\boldsymbol{M}(\xi)$ given by (12.39) and

$$\boldsymbol{m}(\xi) = \int_{X_E} \boldsymbol{x}\,\xi(\mathrm{d}\boldsymbol{x}), \qquad \xi \in \Xi. \tag{12.45}$$

With these preliminaries, the construction of an optimal information matrix $\boldsymbol{M}_0(\xi^*)$ and an optimal design ξ^* can be accomplished by constructing the optimal $\boldsymbol{M}(\xi^*)$ for the homogeneous case with a redefined covariance matrix and examining an additional condition on $\boldsymbol{m}(\xi^*)$. Here the redefined matrix will be the submatrix $\bar{\boldsymbol{\Phi}}$ of $\boldsymbol{\Phi}_0^{-1}$ subdivided as

$$\boldsymbol{\Phi}_0^{-1} = \left(\begin{array}{c|c} \bar{\varphi}_0 & \bar{\boldsymbol{\varphi}}^\top \\ \hline \bar{\boldsymbol{\varphi}} & \bar{\boldsymbol{\Phi}} \end{array}\right),$$

where, by block inversion (see A.37),

$$\bar{\varphi}_0 = 1/(\varphi_0 - \boldsymbol{\varphi}^\top \boldsymbol{\Phi}^{-1}\boldsymbol{\varphi}), \qquad \bar{\boldsymbol{\varphi}} = -\bar{\varphi}_0 \boldsymbol{\Phi}^{-1}\boldsymbol{\varphi}$$

and

$$\bar{\boldsymbol{\Phi}} = (\boldsymbol{\Phi} - \varphi_0^{-1}\boldsymbol{\varphi}\boldsymbol{\varphi}^\top)^{-1} = \boldsymbol{\Phi}^{-1} + \bar{\varphi}_0 \boldsymbol{\Phi}^{-1}\boldsymbol{\varphi}\boldsymbol{\varphi}^\top \boldsymbol{\Phi}^{-1}.$$

Theorem 12.9: *Let $\lambda_1 \leqq \lambda_2 \leqq \ldots \leqq \lambda_k$ be the eigenvalues and $\boldsymbol{U}$ the matrix of eigenvectors of $\bar{\boldsymbol{\Phi}}$, respectively, and $m_1^*, \ldots, m_k^*$ obtained according to Algorithm* 12.3. *If the design $\xi^* \in \Xi$ is such that*

$$\bar{\boldsymbol{M}}(\xi^*) = \boldsymbol{U}^\top \boldsymbol{M}(\xi^*)\, \boldsymbol{U} = \mathrm{diag}\,(m_1^*, \ldots, m_k^*)$$

and

$$\boldsymbol{m}(\xi^*) = -\frac{1}{n}\bar{\boldsymbol{\varphi}} = \frac{1}{n}\bar{\varphi}_0 \boldsymbol{\Phi}^{-1}\boldsymbol{\varphi} \tag{12.46}$$

then ξ^* is A_B-, D_B- and E_B-optimal for the inhomogeneous model (12.42).

Proof. The Bayesian information matrix for the inhomogeneous model is given by $\boldsymbol{M}_B^0(\xi) = \boldsymbol{M}_0(\xi) + \frac{1}{n}\boldsymbol{\Phi}_0^{-1}$. Let $\boldsymbol{f}_0(\boldsymbol{x}) = (1, \boldsymbol{x}^\top)^\top$,

$$\boldsymbol{U}_0 = \begin{pmatrix} 1 & \boldsymbol{0}^\top \\ \boldsymbol{0} & \boldsymbol{U} \end{pmatrix}, \qquad \bar{\boldsymbol{f}}_0(\boldsymbol{x}) = \boldsymbol{U}_0^\top \boldsymbol{f}_0(\boldsymbol{x}) = \begin{pmatrix} 1 \\ \boldsymbol{U}^\top \boldsymbol{x} \end{pmatrix}$$

and define transformed information matrices as

$$\bar{M}_0(\xi) := \int\limits_{X_E} \bar{f}_0(x)\bar{f}_0(x)^\top \xi(\mathrm{d}x) = U_0^\top M_0(\xi)\, U_0\,,$$

$$\bar{M}_B^0(\xi) := U_0^\top M_B^0(\xi)\, U_0 = \bar{M}_0(\xi) + \frac{1}{n} U_0^\top \Phi_0^{-1} U_0\,.$$

Note that U_0 is orthogonal, i.e., $U_0^\top U_0 = U_0 U_0^\top = I_{k+1}$. Then it follows from Theorem 11.12 that ξ^* is A_B-optimal if and only if

$$\sup_{x \subset X_E} \bar{f}_0(x)^\top \bar{M}_B^0(\xi^*)^{-2} \bar{f}_0(x) = \operatorname{tr} \bar{M}_0(\xi^*)\, \bar{M}_B^0(\xi^*)^{-2}. \tag{12.47}$$

Observing that

$$\bar{M}_0(\xi^*) = \left(\begin{array}{c|c} 1 & m(\xi^*)^\top U \\ \hline u^\top m(\xi^*) & U^\top M(\xi^*)\, U \end{array}\right) = \left(\begin{array}{c|c} 1 & -\frac{1}{n}\bar{\varphi}^\top U \\ \hline -\frac{1}{n} U^\top \bar{\varphi} & \bar{M}(\xi^*) \end{array}\right)$$

and

$$U_0^\top \Phi_0^{-1} U_0 = \left(\begin{array}{c|c} \bar{\varphi}_0 & \bar{\varphi}^\top U \\ \hline U^\top \bar{\varphi} & U^\top \bar{\Phi} U \end{array}\right) = \left(\begin{array}{c|c} \bar{\varphi}_0 & \bar{\varphi}^\top U \\ \hline U^\top \bar{\varphi} & \Lambda \end{array}\right)$$

we have

$$\begin{aligned} \bar{M}_B^0(\xi^*) &= \bar{M}_0(\xi^*) + \frac{1}{n} U_0^\top \Phi_0^{-1} U_0 \\ &= \operatorname{diag}\left(1 + \frac{1}{n}\bar{\varphi}_0,\, m_1^* + \frac{1}{n}\lambda_1, \ldots, m_k^* + \frac{1}{n}\lambda_k\right). \end{aligned}$$

From this it follows that

$$\bar{f}_0(x)^\top \bar{M}_B^0(\xi^*)^{-2} \bar{f}_0(x) = \left(1 + \frac{1}{n}\bar{\varphi}_0\right)^{-2} + x^\top U\left(\bar{M}(\xi^*) + \frac{1}{n}\Lambda\right)^{-2} U^\top x$$

and

$$\operatorname{tr} \bar{M}_0(\xi^*)\, \bar{M}_B^0(\xi^*)^{-2} = \left(1 + \frac{1}{n}\bar{\varphi}_0\right)^{-2} + \operatorname{tr} \bar{M}(\xi^*)\left(\bar{M}(\xi^*) + \frac{1}{n}\Lambda\right)^{-2},$$

i.e., (12.47) holds if and only if

$$\sup_{x \in X_E} x^\top U\left(\bar{M}(\xi^*) + \frac{1}{n}\Lambda\right)^{-2} U^\top x = \operatorname{tr} \bar{M}(\xi^*)\left(\bar{M}(\xi^*) + \frac{1}{n}\Lambda\right)^{-2}.$$

But this is precisely the condition for A_B-optimality of ξ^* in the homogeneous model with redefined covariance matrix $(w\bar{\Phi})^{-1}$. By Theorem 12.8, this condition is satisfied since $\bar{M}(\xi^*)$ is constructed according to Algorithm 12.3 with $\bar{\Phi} = U \Lambda U^\top$.

Similarly, since ξ^* is also D_B-optimal in the homogeneous model it follows that

$$\begin{aligned}\sup_{x\in X_E} \bar{f}_0(x)^\top \bar{M}_B^0(\xi^*)^{-1}\bar{f}_0(x) &= \left(1+\frac{1}{n}\bar{\varphi}_0\right)^{-1} + \sup_{x\in X_E} x^\top U\left(\bar{M}(\xi^*)+\frac{1}{n}\Lambda\right)^{-1} U^\top x \\ &= \left(1+\frac{1}{n}\bar{\varphi}_0\right)^{-1} + \operatorname{tr}\bar{M}(\xi^*)\left(\bar{M}(\xi^*)+\frac{1}{n}\Lambda\right)^{-1} \\ &= \operatorname{tr}\bar{M}_0(\xi^*)\,\bar{M}_B^0(\xi^*)^{-1}\end{aligned}$$

which, by Theorem 11.15 (i), guarantees D_B-optimality of ξ^* in the inhomogeneous model.

To prove E_B-optimality of ξ^* consider the following two cases.

Case 1. $1+\frac{1}{n}\bar{\varphi}_0 < m_1^* + \frac{1}{n}\lambda_1$.

Then $\lambda_{\min}(M_B^0(\xi^*)) = \lambda_{\min}(\bar{M}_B^0(\xi^*)) = 1+\frac{1}{n}\bar{\varphi}_0$; and this eigenvalue is simple. The corresponding eigenvector is given by the first column $u_0 = (1,0,\ldots,0)^\top$ of U_0 so that

$$\sup_{x\in X_E} (f_0(x)^\top u_0)^2 = 1 = \lambda_{\min}(M_B^0(\xi^*)) - \frac{1}{n} u_0^\top \Phi_0^{-1} u_0,$$

which, according to Theorem 11.16 (i), means that ξ^* is E_B-optimal.

Case 2. $1+\frac{1}{n}\bar{\varphi}_0 \geqq m_1^* + \frac{1}{n}\lambda_1$.

Then $\lambda_{\min}(M_B^0(\xi^*)) = m_1^* + \frac{1}{n}\lambda_1$ and, along the same lines as in the proof of Theorem 5 in GLADITZ/PILZ (1982 a), it can be shown that there is no design $\xi \in \Xi$ for which $\lambda_{\min}(M_B^0(\xi)) > m_1^* + \frac{1}{n}\lambda_1$. Thus, ξ^* maximizes the minimal eigenvalue of $M_B^0(\cdot)$ over Ξ. □

From Theorem 12.9 we conclude that an optimal design

$$\xi^* = \{(x_1, p_1), \ldots, (x_m, p_m)\}$$

may be found by solving the system

$$\begin{aligned}\sum_{i=1}^m p_i x_i x_i^\top &= U \operatorname{diag}(m_1^*, \ldots, m_k^*)\, U^\top \\ \sum_{i=1}^m p_i x_i &= -\frac{1}{n}\bar{\varphi} \qquad (12.48) \\ \sum_{i=1}^m p_i &= 1\end{aligned}$$

for some $m \leqq k(k+1)/2$. Here the supporting points x_i $(i = 1, \ldots, m)$ must lie on the surface of the sphere X_E, i.e., $\boldsymbol{x}_i^\top \boldsymbol{x}_i = 1$, to assure that

$$\sum_{i=1}^{m} p_i \boldsymbol{x}_i^\top \boldsymbol{x}_i = \operatorname{tr} \boldsymbol{U} \operatorname{diag}(m_1^*, \ldots, m_k^*) \boldsymbol{U}^\top = \sum_{i=1}^{k} m_i^* = 1$$

which follows necessarily from the first equation of (12.48).

For the special case in which there is no prior correlation between the absolute term $\tilde{\vartheta}_0$ and the remaining components $\tilde{\vartheta}_i$, i.e., $\varphi_{0i} = \mathbf{Cov}(\tilde{\vartheta}_0, \tilde{\vartheta}_i \mid w) = 0$ $(i = 1, \ldots, m)$, an optimal design can be given explicitly as follows.

Corollary 12.6. *Let the prior covariance matrix $\boldsymbol{\Phi}_0$ from* (12.43) *be such that $\boldsymbol{\varphi} = \mathbf{0}$ and $\boldsymbol{\Phi}^{-1} = \boldsymbol{U}\boldsymbol{\Lambda}\boldsymbol{U}^\top$ with $\boldsymbol{\Lambda} = \operatorname{diag}(\lambda_1, \ldots, \lambda_k)$ and orthogonal $\boldsymbol{U} = (\boldsymbol{u}_1, \ldots, \boldsymbol{u}_k)$. Then, with $m_1^*, \ldots, m_k^*$ generated by Algorithm* 12.3, *the design $\xi^* = \{(\boldsymbol{x}_1, p_1), \ldots, (\boldsymbol{x}_{2k}, p_{2k})\}$ giving weights $p_{2s-1} = p_{2s} = m_s^*/2$ to the points $\boldsymbol{x}_{2s-1} = \boldsymbol{u}_s$, $\boldsymbol{x}_{2s} = -\boldsymbol{u}_s$ $(s = 1, \ldots, k)$ is A_B-, D_B- and E_B-optimal.*

Proof. With $\boldsymbol{\varphi} = \mathbf{0}$ the inverted matrix $\boldsymbol{\Phi}_0^{-1}$ simplifies such that

$$\boldsymbol{\Phi}_0^{-1} = \begin{pmatrix} 1/\varphi_0 & \mathbf{0}^\top \\ \mathbf{0} & \boldsymbol{\Phi}^{-1} \end{pmatrix}.$$

Therefore, Theorem 12.9 states that ξ^* is A_B-, D_B- and E_B-optimal if it satisfies (12.46) with $\boldsymbol{m}(\xi^*) = \mathbf{0}$. This is but the case, since

$$\boldsymbol{M}(\xi^*) = \sum_{i=1}^{2k} p_i \boldsymbol{x}_i \boldsymbol{x}_i^\top = \sum_{i=1}^{k} m_i^* \boldsymbol{u}_i \boldsymbol{u}_i^\top = \boldsymbol{U} \operatorname{diag}(m_1^*, \ldots, m_k^*) \boldsymbol{U}^\top$$

and

$$\boldsymbol{m}(\xi^*) = \sum_{i=1}^{2k} p_i \boldsymbol{x}_i = \mathbf{0}. \ \square$$

Let us turn, briefly, to the construction of Bayesian designs for the inhomogeneous model (12.42). The matrix $\boldsymbol{U}_L$ with respect to which these designs are L_B-optimal, takes the form

$$\boldsymbol{U}_L = \int_{X_P} \begin{pmatrix} 1 \\ \boldsymbol{x} \end{pmatrix} (1, \boldsymbol{x}^\top) P_{\tilde{x}}(\mathrm{d}\boldsymbol{x}) = \begin{pmatrix} 1 & \boldsymbol{\mu}_{\tilde{x}}^\top \\ \boldsymbol{\mu}_{\tilde{x}} & \mathbf{E}\tilde{\boldsymbol{x}}\tilde{\boldsymbol{x}}^\top \end{pmatrix}, \tag{12.49}$$

where $\boldsymbol{\mu}_{\tilde{x}} = \mathbf{E}\tilde{\boldsymbol{x}}$ and $\mathbf{E}\tilde{\boldsymbol{x}}\tilde{\boldsymbol{x}}^\top = \mathbf{Cov}\,\tilde{\boldsymbol{x}} + \boldsymbol{\mu}_{\tilde{x}}\boldsymbol{\mu}_{\tilde{x}}^\top$ refer to the first and second order moments of the prior distribution $P_{\tilde{x}}$. This yields

$$\boldsymbol{U}_L^{1/2} = \begin{pmatrix} 1 & \mathbf{0}^\top \\ \boldsymbol{\mu}_{\tilde{x}} & \boldsymbol{C}_{\tilde{x}} \end{pmatrix}, \qquad \boldsymbol{U}_L^{-1/2} = \begin{pmatrix} 1 & \mathbf{0}^\top \\ -\boldsymbol{C}_{\tilde{x}}^{-1}\boldsymbol{\mu}_{\tilde{x}} & \boldsymbol{C}_{\tilde{x}}^{-1} \end{pmatrix},$$

where

$$\boldsymbol{C}_{\tilde{x}} = (\mathbf{Cov}\,\tilde{\boldsymbol{x}})^{1/2}, \qquad \boldsymbol{C}_{\tilde{x}}\boldsymbol{C}_{\tilde{x}}^\top = \mathbf{Cov}\,\tilde{\boldsymbol{x}}.$$

Thus, according to Theorem 12.2, a Bayesian design with respect to $P_{\tilde{x}}$ may be obtained as an A_B-optimal design in the transformed model

$$\mathbf{E}\tilde{y}(\boldsymbol{x}) = \vartheta_0^L + \boldsymbol{\vartheta}_L^\top \boldsymbol{f}_L(\boldsymbol{x}) = \vartheta_0^L + \boldsymbol{\vartheta}_L^\top \boldsymbol{C}_{\tilde{x}}^{-1}(\boldsymbol{x} - \boldsymbol{\mu}_{\tilde{x}})$$

with the covariance matrix

$$\mathbf{Cov}(\widetilde{\vartheta}_0^{\rm L}, \widetilde{\boldsymbol{\vartheta}}_{\rm L}^\top \mid w) = \frac{1}{w}(\boldsymbol{U}_{\rm L}^{1/2})^\top \boldsymbol{\Phi}_0 \boldsymbol{U}_{\rm L}^{1/2}$$

$$= \frac{1}{w}\left(\begin{array}{c|c} \varphi_0 + 2\boldsymbol{\mu}_{\tilde{x}}^\top \boldsymbol{\varphi} + \boldsymbol{\mu}_{\tilde{x}}^\top \boldsymbol{\Phi}\boldsymbol{\mu}_{\tilde{x}} & \boldsymbol{\varphi}^\top \boldsymbol{C}_{\tilde{x}} + \boldsymbol{\mu}_{\tilde{x}}^\top \boldsymbol{\Phi}\boldsymbol{C}_{\tilde{x}} \\ \hline \boldsymbol{C}_{\tilde{x}}^\top \boldsymbol{\varphi} + \boldsymbol{C}_{\tilde{x}}^\top \boldsymbol{\Phi}\boldsymbol{\mu}_{\tilde{x}} & \boldsymbol{C}_{\tilde{x}}^\top \boldsymbol{\Phi}\boldsymbol{C}_{\tilde{x}} \end{array}\right).$$

Up to now we considered designing for a spherical experimental region $X_{\rm E}$. This is an essential assumption for the construction methods developed here. If we specialize Algorithm 12.1 to the case of a cuboidal region then, for arbitrary covariance matrix $\boldsymbol{\Phi}_0$, we cannot guarantee the existence of a design having the required information matrix. However, if the regression coefficients are uncorrelated a priori, then we can give an explicit solution.

Consider the model

$$\begin{aligned} &\mathbf{E}\widetilde{y}(\boldsymbol{x}) = \vartheta_0 + \boldsymbol{\vartheta}^\top \boldsymbol{x}, \qquad \boldsymbol{x} = (x_1, \ldots, x_k)^\top \in X_{\rm E} = [-1, 1]^k \\ &\mathbf{Cov}(\widetilde{\vartheta}_0, \widetilde{\boldsymbol{\vartheta}}^\top \mid w) = \frac{1}{w}\,\mathrm{diag}(\varphi_0, \varphi_1, \ldots, \varphi_k)\,. \end{aligned} \tag{12.50}$$

Then the quantities defined in (12.6) assume the values

$$a = \sup_{x \in X_{\rm E}} \boldsymbol{x}^\top \boldsymbol{x} = k,$$

$$a_j = \sup_{-1 \leqq x_j \leqq 1} x_j^2 = 1, \qquad b_j = \inf_{-1 \leqq x_j \leqq 1} x_j^2 = 0 \qquad (j = 1, \ldots, k)\,.$$

Therefore, Algorithm 12.1 yields $m_1^* = \ldots = m_k^* = 1$ and we are to look for designs ξ^* with $\boldsymbol{M}(\xi^*) = \boldsymbol{I}_k$ and $\boldsymbol{m}(\xi^*) = \boldsymbol{0}$, i.e., $\boldsymbol{M}_0(\xi^*) = \boldsymbol{I}_{k+1}$.

Theorem 12.10. *If the design $\xi^* \in \Xi$ is such that $\boldsymbol{M}_0(\xi^*) = \boldsymbol{I}_{k+1}$ then ξ^* is $A_{\rm B}$-, $D_{\rm B}$- and $E_{\rm B}$-optimal for the model* (12.50).

Especially, the condition $\boldsymbol{M}_0(\xi^*) = \boldsymbol{I}_{k+1}$ demands that the supporting points of an optimal design ξ^* must be corner points of the cube. Hence, an optimal design $\xi^* = \{(\boldsymbol{x}_1, p_1), \ldots, (\boldsymbol{x}_m, p_m)\}$ can be obtained by choosing at most $m = 2^k$ different corner points

$$\boldsymbol{x}_i = (x_{1i}, \ldots, x_{ki})^\top \quad \text{with} \quad |x_{1i}| = \ldots = |x_{ki}| = 1, \; i = 1, \ldots, m$$

and then determining the corresponding weights $p_1, \ldots, p_m \geqq 0$ as a solution to the following system of $k(k+1)/2 + 1$ linear algebraic equations

$$\sum_{i=1}^m p_i x_{si} x_{qi} = 0, \qquad s, q = 1, \ldots, k, \; s < q,$$

$$\sum_{i=1}^m p_i x_{si} = 0, \qquad s = 1, \ldots, k,$$

$$\sum_{i=1}^m p_i = 1\,.$$

We remark that such designs ξ^* with the information matrix $M_0(\xi^*) = I_{k+1}$ are also A-, D- and E-optimal for the least squares estimator of $\boldsymbol{\vartheta}_0 = (\vartheta_0, \boldsymbol{\vartheta}^\top)^\top$ in the inhomogeneous model with $X_E = [-1, 1]^k$ (see, e.g., FEDOROV (1972)).

Clearly, if all corner points are involved, i.e. $m = 2^k$, then the solution ξ^* is not unique since the number of equations $k(k+1)/2 + 1$ is smaller than the number of unknown weights m as long as $k \geqq 3$.

If we drop the assumption of uncorrelatedness of the absolute term $\tilde{\vartheta}_0$ with the remaining coefficients then L_B-optimal designs on the cube, and thus also A_B-, C_B- and I_B-optimal designs, can be obtained with the help of the following result. To this let $\bar{\varphi}_i$ $(i = 0, 1, \ldots, k)$ denote the diagonal elements of $\boldsymbol{\Phi}_0^{-1}$ and $\mathcal{B}$ be the set of matrices

$$\mathcal{B} = \{\boldsymbol{B} = (b_{ij}) \in \mathcal{M}^{>}_{k+1} \colon b_{ii} = 1 + \bar{\varphi}_i/n; \; i = 0, \ldots, k\}.$$

Theorem 12.11. *Let $\boldsymbol{B}^* \in \mathcal{B}$ and $\boldsymbol{D}$ be some diagonal matrix such that*

$$\boldsymbol{B}^*\boldsymbol{D}\boldsymbol{B}^* = \boldsymbol{U}_L \quad \textit{and} \quad \boldsymbol{B}^* - \frac{1}{n}\boldsymbol{\Phi}_0^{-1} \in \mathcal{M}^{\geqq}_{k+1}.$$

If there exists a design $\xi^* \in \Xi$ with $\boldsymbol{M}_0(\xi^*) = \boldsymbol{B}^* - \frac{1}{n}\boldsymbol{\Phi}_0^{-1}$ then ξ^* is L_B-optimal with respect to $\boldsymbol{U}_L$. The solution $\boldsymbol{B}^*$ is unique if and only if $\boldsymbol{U}_L$ is of full rank.

This result is due to CHALONER (1982, Theorem 9). The proof follows by differentiating $\bar{Z}(\boldsymbol{B}) = \operatorname{tr} \boldsymbol{U}_L \boldsymbol{B}^{-1}$ using Lagrange multipliers.

Example 12.4. Let us consider the model

$$\mathbf{E}\tilde{y}(\boldsymbol{x}) = \vartheta_0 + \vartheta_1 x_1 + \vartheta_2 x_2, \qquad \boldsymbol{x} = (x_1, x_2)^\top \in X_E = [-1, 1]^2$$

and assume that $\tilde{\vartheta}_0$ is not correlated with $\tilde{\vartheta}_1$ and $\tilde{\vartheta}_2$, i.e., $\boldsymbol{\Phi}_0^{-1}$ has the form

$$\boldsymbol{\Phi}_0^{-1} = \begin{pmatrix} \bar{\varphi}_0 & 0 & 0 \\ 0 & \bar{\varphi}_1 & \bar{\varphi}_{12} \\ 0 & \bar{\varphi}_{12} & \bar{\varphi}_2 \end{pmatrix}.$$

We wish to determine an A_B-optimal design (i.e. $\boldsymbol{U}_L = \boldsymbol{I}_3$). For arbitrary diagonal matrix $\boldsymbol{D}$, the solution $\boldsymbol{B}^* \in \mathcal{B}$ of $\boldsymbol{B}^*\boldsymbol{D}\boldsymbol{B}^* = \boldsymbol{I}_3$ is uniquely determined by

$$\boldsymbol{B}^* = (b_{ij}) \quad \text{with} \quad b_{ii} = 1 + \frac{1}{n}\bar{\varphi}_i, \; b_{ij} = 0 \quad (i, j = 0, 1, 2; \; i \neq j).$$

Thus, we are to look for a design ξ^* with

$$\boldsymbol{M}(\xi^*) = \boldsymbol{B}^* - \frac{1}{n}\boldsymbol{\Phi}_0^{-1} = \begin{pmatrix} 1 & 0 & 0 \\ 0 & 1 & -\bar{\varphi}_{12}/n \\ 0 & -\bar{\varphi}_{12}/n & 1 \end{pmatrix}.$$

The $2^k = 4$ corner points of X_E at which to take observations are:

$$\boldsymbol{x}_1 = \begin{pmatrix} 1 \\ 1 \end{pmatrix}, \quad \boldsymbol{x}_2 = \begin{pmatrix} -1 \\ -1 \end{pmatrix}, \quad \boldsymbol{x}_3 = \begin{pmatrix} -1 \\ 1 \end{pmatrix}, \quad \boldsymbol{x}_4 = \begin{pmatrix} 1 \\ -1 \end{pmatrix}.$$

Setting $\xi^* = \{(\boldsymbol{x}_1, p_1), \ldots, (\boldsymbol{x}_4, p_4)\}$ and equating

$$\boldsymbol{M}_0(\xi^*) = \begin{pmatrix} 1 & \sum p_i \boldsymbol{x}_i \\ \sum p_i \boldsymbol{x}_i & \sum p_i \boldsymbol{x}_i \boldsymbol{x}_i^\top \end{pmatrix} = \boldsymbol{B}^* - \frac{1}{n} \boldsymbol{\Phi}_0^{-1}$$

yields the system

$$p_1 + p_2 - (p_3 + p_4) = -\bar{\varphi}_{12}/n \qquad p_1 - p_2 - p_3 + p_4 = 0$$

$$p_1 + p_2 + (p_3 + p_4) = 1 \qquad p_1 - p_2 + p_3 - p_4 = 0 .$$

The solution is

$$p_1 = p_2 = \frac{1}{4}(1 - \bar{\varphi}_{12}/n)$$

$$p_3 = p_4 = \frac{1}{4}(1 + \bar{\varphi}_{12}/n)$$

and ξ^* is A_B-optimal provided that $|\bar{\varphi}_{12}| \leqq n$.

We note that for this covariance matrix the design ξ^* is also L_B-optimal with respect to arbitrary diagonal matrix $\boldsymbol{U}_L \in \mathcal{M}_3^>$. This is due to the fact that $\boldsymbol{B}^*$ from above also solves the equation

$$\boldsymbol{B}^* \boldsymbol{D} \boldsymbol{B}^* = \operatorname{diag}(u_{11}, u_{22}, u_{33})$$

for any choice of $u_{11}, u_{22}, u_{33} > 0$.

Concluding, we apply the preceding results to obtain optimal designs for the simple linear regression model

$$\mathbf{E}\tilde{y}(x) = \vartheta_0 + \vartheta_1 x, \qquad x \in X_E = [-1, 1]$$

$$\mathbf{Cov}((\tilde{\vartheta}_0, \tilde{\vartheta}_1)^\top | w) = \frac{1}{w} \boldsymbol{\Phi}_0 = \frac{1}{w} \begin{pmatrix} \varphi_0 & \varphi_{01} \\ \varphi_{01} & \varphi_1 \end{pmatrix}. \tag{12.51}$$

In this case Theorem 12.9 tells us that if we can find a design ξ^* such that

$$\boldsymbol{M}_0(\xi^*) = \begin{pmatrix} 1 & m(\xi^*) \\ m(\xi^*) & m_1^* \end{pmatrix} = \begin{pmatrix} 1 & -\dfrac{\varphi_{01}}{n \det \boldsymbol{\Phi}_0} \\ -\dfrac{\varphi_{01}}{n \det \boldsymbol{\Phi}_0} & 1 \end{pmatrix} \tag{12.52}$$

then ξ^* is A_B-, D_B- and E_B-optimal. The condition

$$\int_{-1}^{1} x^2 \xi^*(\mathrm{d}x) = m_1^* = 1$$

necessarily implies that ξ^* must be of the form $\xi^* = \{(-1, p_1), (1, p_2)\}$. The optimal weights then result from the conditions

$$m(\xi^*) = \int_{-1}^{1} x\xi^*(\mathrm{d}x) = -p_1 + p_2 = -\varphi_{01}/n \det \boldsymbol{\Phi}_0$$

and

$$p_1 + p_2 = 1, p_1 \geqq 0, p_2 \geqq 0,$$

which yields

$$p_1 = \frac{1}{2}(1 - \varphi_{01}/n \det \boldsymbol{\Phi}_0), \; p_2 = \frac{1}{2}(1 + \varphi_{01}/n \det \boldsymbol{\Phi}_0) \tag{12.53}$$

provided that $|\varphi_{01}| \leqq n \det \boldsymbol{\Phi}_0 = n(\varphi_0\varphi_1 - \varphi_{01}^2)$. Thus we have proved

Theorem 12.12. *If it holds* $|\varphi_{01}| \leqq n \det \boldsymbol{\Phi}_0$ *then* $\xi^* = \{(-1, p_1), (1, p_2)\}$ *with* p_1 *and* p_2 *given by* (12.53) *is* A_{B}-, D_{B}- *and* E_{B}-*optimal for the simple linear regression model* (12.51).

Alternatively, if $|\varphi_{01}| > n \det \boldsymbol{\Phi}_0$, which may happen in case of high prior correlation between $\tilde{\vartheta}_0$ and $\tilde{\vartheta}_1$, then one-point designs turn out to be optimal: $\xi^* = \delta_{-1}$ in case that $\varphi_{01} < -n \det \boldsymbol{\Phi}_0$ and $\xi^* = \delta_{+1}$ in case that $\varphi_{01} > n \det \boldsymbol{\Phi}_0$ (see Section 14.1 and PILZ (1981b)).

The designs just obtained are also Bayesian designs with respect to arbitrary symmetric prior distributions over the prediction region.

Theorem 12.13. *Let* $X_{\mathrm{P}} = X_{\mathrm{E}} = [-1, 1]$ *and* $P_{\tilde{x}}$ *be a symmetric prior over* X_{P}, *i.e.,* $\mathbf{E}\tilde{x} = 0$. *Then* ξ^* *as indicated in Theorem* 12.11 *is a Bayesian design with respect to* $P_{\tilde{x}}$ *in the model* (12.51).

Proof. According to Theorem 10.1, the assertion comes true if and only if ξ^* is L_{B}-optimal with respect to

$$\boldsymbol{U}_{\mathrm{L}} = \int_{X_{\mathrm{P}}} \begin{pmatrix} 1 \\ x \end{pmatrix} (1, x)\, P_{\tilde{x}}(\mathrm{d}x) = \begin{pmatrix} 1 & \mathbf{E}\tilde{x} \\ \mathbf{E}\tilde{x} & \mathbf{E}\tilde{x}^2 \end{pmatrix} = \begin{pmatrix} 1 & 0 \\ 0 & \sigma_{\tilde{x}}^2 \end{pmatrix},$$

where $\sigma_{\tilde{x}}^2 = \operatorname{Var} \tilde{x} = \mathbf{E}\tilde{x}^2$. By Theorem 11.12 this is but the case, since with $\boldsymbol{M}_0(\xi^*)$ from (12.52) and

$$\boldsymbol{M}_{\mathrm{B}}^0(\xi^*) = \boldsymbol{M}_0(\xi^*) + \frac{1}{n}\boldsymbol{\Phi}_0^{-1} = \operatorname{diag}(1 + \varphi_1/n \det \boldsymbol{\Phi}_0, 1 + \varphi_0/n \det \boldsymbol{\Phi}_0)$$

we have

$$\begin{aligned} \sup_{x \in X_{\mathrm{E}}} \| \boldsymbol{M}_{\mathrm{B}}^0(\xi^*)^{-1} \boldsymbol{f}_0(x) \|_{\boldsymbol{U}_{\mathrm{L}}}^2 &= (1 + \varphi_1/n \det \boldsymbol{\Phi}_0)^{-1} + \sigma_{\tilde{x}}^2(1 + \varphi_0/n \det \boldsymbol{\Phi}_0)^{-2} \\ &= \operatorname{tr} \boldsymbol{M}_0(\xi^*)\, \boldsymbol{M}_{\mathrm{B}}^0(\xi^*)^{-1} \boldsymbol{U}_{\mathrm{L}} \boldsymbol{M}_{\mathrm{B}}^0(\xi^*)^{-1}. \; \square \end{aligned}$$

Again, if it holds $|\varphi_{01}| > n \det \boldsymbol{\Phi}_0$ then the one-point designs mentioned above are Bayesian with respect to symmetric $P_{\tilde{x}}$.

From the proof of Theorem 12.13 it is clear that $\xi^* = \{(-1, p_1), (1, p_2)\}$ with p_1 and p_2 defined by (12.53) is also L_B-optimal with respect to every positive definite diagonal matrix $\boldsymbol{U}_L$.

Bayesian designs with respect to arbitrary, not necessarily symmetric, priors over X_P and L_B-optimal designs with respect to arbitrary positive definite matrices $\boldsymbol{U}_L$ can be obtained by constructing A_B-optimal designs in the transformed model (12.28) (see Theorem 12.2).

12.5. Iteration procedures for the construction of approximately optimal designs

In the preceding sections we presented methods which, under certain assumptions, allowed the construction of the information matrix of an optimal design. The experimental points and weights of an optimal design then could be obtained as solutions to a corresponding system of (nonlinear) equations. But in case of complicated regression setups and increasing numbers of parameters and controllable factors it will be difficult to solve this system. In this case iterative methods can be employed to construct approximately optimal designs.

First we describe a general iteration procedure for convex design functionals Z having a linear directional derivative Δ_Z (in the sense of Definition 11.4). Starting with an initial design ξ_0, at every stage $s = 1, 2, \ldots$ of the iteration we generate a design

$$\xi_s = (1 - \alpha_s)\, \xi_{s-1} + \alpha_s \delta_{x_s} \tag{12.54}$$

by adding a point $\boldsymbol{x}_s \in X_E$ to the support of the design ξ_{s-1} obtained at the preceding stage and define the weights of the supporting points of the new design ξ_s such that it yields a maximal decrease of the functional value. That is, we have to find ξ_s such that

$$Z(\xi_{s-1}) - Z(\xi_s) = \sup_{\xi \in \Xi_s} (Z(\xi_{s-1}) - Z(\xi)),$$

where

$$\Xi_s = \{\xi \in \Xi : \xi = (1 - \alpha)\, \xi_{s-1} + \alpha \delta_x,\ \alpha \in [0, 1],\ \boldsymbol{x} \in X_E\}. \tag{12.55}$$

The following modification explicitly yields a criterion for the choice of the new supporting point and the choice of the weighting factor (cp. Fedorov (1972), Whittle (1973)): We choose $\boldsymbol{x}_s \in X_E$ such that it minimizes the directional derivative of ξ_{s-1} in the direction of all possible one-point designs, i.e.,

$$\Delta_Z(\xi_{s-1}, \delta_{x_s}) = \inf_{x \in X_E} \Delta_Z(\xi_{s-1}, \delta_x) \tag{12.56}$$

and hereafter determine $\alpha_s \in [0, 1]$ such that

$$Z((1 - \alpha_s)\, \xi_{s-1} + \alpha_s \delta_{x_s}) = \inf_{\alpha \in [0,1]} Z((1 - \alpha)\, \xi_{s-1} + \alpha \delta_{x_s}). \tag{12.57}$$

Then the sequence $\{Z(\xi_s)\}_{s \in \mathbb{N}}$ of functional values converges and it holds

$$\lim_{s \to \infty} Z(\xi_s) = \inf_{\xi \in \Xi} Z(\xi).$$

Remark 12.9. In general, the choice of $\boldsymbol{x}_s$ according to (12.56) is not equivalent to that indicated in (12.55). In (12.55) the new supporting point $\boldsymbol{x}_s$ is chosen so as to maximize the decrease of the functional value itself whereas, by definition of Δ_Z (see (11.6)), the choice of $\boldsymbol{x}_s$ according to (12.56) is such that it lies in the direction of the steepest descent of Z at ξ_{s-1}. As a rule, the latter choice requires less numerical effort.

The main difficulty with the above iteration procedure is the minimization problem (12.57) of finding optimal weights α_s at every stage of the iteration. To avoid this, several authors propose to choose a fixed sequence $\{\alpha_s\}_{s \in \mathbb{N}}$ of preassigned weights (see, for example, FEDOROV (1972), WYNN (1970), (1975), ATWOOD (1976)). Here, any sequence $\{\alpha_s\}_{s \in \mathbb{N}}$ such that

$$\lim_{s \to \infty} \alpha_s = 0, \qquad \sum_{s \in \mathbb{N}} \alpha_s = \infty, \qquad \alpha_s \in [0, 1), \qquad s = 1, 2, \ldots \tag{12.58}$$

leads us to a decreasing sequence $\{Z(\xi_s)\}_{s \in \mathbb{N}}$ of functional values and thus to an approximately Z-optimal design if $\boldsymbol{x}_s$ and α_s are chosen according to (12.56) and (12.54), respectively. Some theoretical and numerical investigations on the choice of appropriate sequences $\{\alpha_s\}_{s \in \mathbb{N}}$ may be found, for example, in WU/WYNN (1978), WU (1978) and BÖHNING (1981), (1985).

A very simple choice, which works but well, would be

$$\alpha_s = 1/(n_0 + s), \qquad s = 1, 2, \ldots,$$

where $n_0 = \text{card}\,(\text{supp}\,\xi_0)$ stands for the number of supporting points of the initial design, a choice first proposed by WYNN (1970). It is clear, however, that the convergence of iteration with weights chosen optimally at every stage will be faster than with fixed weights.

If the functional Z depends on the design only through the Bayesian information matrix, as it is the case with all the functionals $Z \in \{L_B, B, A_B, C_B, D_B, E_B, I_B\}$ considered so far, then the choice of an initial design rises no problems. Since $\boldsymbol{M}_B(\xi)$ is regular whatever the design $\xi \in \Xi$, the starting design may be taken as one-point design $\xi_0 = \delta_{x_0}$ where $\boldsymbol{x}_0$ should be such that

$$Z(\xi_0) = \inf_{x \in X_E} Z(\delta_x). \tag{12.59}$$

Then the weights α_s needed for the construction of the sth iteration design ξ_s may be taken as

$$\alpha_s = 1/(1 + s), \qquad s = 1, 2, \ldots \tag{12.60}$$

Moreover, if there exists a Z-optimal design for which the weights of the supporting points are rational numbers then, with α_s chosen according to

(12.60), the sequence $\{\xi_s\}_{s \in \mathbb{N}}$ may terminate in such an optimal design after a finite number of iterations.

The stage s_0 at which the iteration procedure should be stopped depends on the goodness with which we wish to approximate the Z-optimal designs. If, for example, we demand that the approximation ξ_{s_0} comes within $\varepsilon > 0$ to minimize $Z(\cdot)$ over Ξ, i.e.,

$$Z(\xi_{s_0}) - \inf_{\xi \in \Xi} Z(\xi) \leqq \varepsilon \tag{12.61}$$

then the iteration should be stopped at that stage s_0 for which

$$\inf_{x \in X_E} \Delta_Z(\xi_{s_0}, \delta_x) \geqq -\varepsilon. \tag{12.62}$$

In fact, it follows from Lemma 11.5 that (12.61) is guaranteed if (12.62) comes true.

Another, more realistic, way of evaluating the goodness of an approximation ξ for an optimal ξ^* would be to consider the ratio of the functional values instead of their difference. This leads us to the concept of Z-efficiency of a design.

Definition 12.1. Let $Z^* = \inf_{\xi \in \Xi} Z(\xi)$ and $\xi \in \Xi$ be such that $Z(\xi) \neq 0$. The ratio

$$e_Z(\xi) := Z^*/Z(\xi)$$

is called the *Z-efficiency* of the design ξ.

Clearly, we have $e_Z(\xi) = 1$ if and only if ξ is Z-optimal. We note, however, that the exact computation of the efficiency of a given design requires the knowledge of the functional value Z^* of a Z-optimal design which will but rarely be available. But at least we can find lower bounds for the efficiency.

Lemma 12.2. *Let* $Z(\cdot) = \bar{Z}(\boldsymbol{M}_B(\cdot))$ *be a convex functional according to Assumption* 11.2. *Further, let* $Z(\xi) > 0$ *for every* $\xi \in \Xi$ *and define*

$$d_Z(\xi) = \inf_{x \in X_E} \Delta_Z(\xi, \delta_x). \tag{12.63}$$

Then it holds

$$1 + d_Z(\xi)/Z(\xi) \leqq e_Z(\xi) \leqq 1$$

for all $\xi \in \Xi$.

This result follows immediately from Lemma 11.5.

This way, an iteration procedure should be stopped at the stage s_0 if

$$1 + d_Z(\xi_{s_0})/Z(\xi_{s_0}) \geqq e_0, \tag{12.64}$$

where $e_0 \in (0, 1)$ is some predetermined efficiency that is to be guaranteed. Any design $\xi_{s_0} \in \Xi$ satisfying (12.64) we call *Z-e_0-optimal*.

As a specialization of Lemma 12.2, we can give the following lower bounds for the efficiency associated with the criteria of L_B- and D_B-optimality.

Corollary 12.7. *For every design $\xi \in \Xi$ it holds*

(i) $1 \geqq e_{L_B}(\xi) \geqq 2 - \left\{\frac{1}{n} \operatorname{tr} \boldsymbol{Q}(\xi)\, \boldsymbol{\Phi}^{-1} + \sup_{x \in X_E} \boldsymbol{f}(\boldsymbol{x})^\top \boldsymbol{Q}(\xi) \boldsymbol{f}(\boldsymbol{x})\right\} \Big/ L_B(\xi)$,

where $\boldsymbol{Q}(\xi) := \boldsymbol{M}_B(\xi)^{-1} \boldsymbol{U}_L \boldsymbol{M}_B(\xi)^{-1}$,

(ii) $1 \geqq e_{D_B}(\xi) \geqq 1 + \operatorname{tr} \boldsymbol{M}(\xi)\, \boldsymbol{M}_B(\xi)^{-1} - \sup_{x \in X_E} \boldsymbol{f}(\boldsymbol{x})^\top \boldsymbol{M}_B(\xi)^{-1} \boldsymbol{f}(\boldsymbol{x})$.

With Lemma 12.2, these results follow from the Theorems 11.13 (i) and 11.15 (ii), respectively.

Now we are going to formulate iteration procedures for the construction of approximately L_B- and D_B-optimal designs. To this let $e_0 \in (0, 1)$ be a predetermined value for the efficiency which we wish to achieve with the iterated designs.

Algorithm 12.4. Construction of approximately L_B-optimal designs

Step 1. Choose an initial design $\xi_0 = \delta_{x_0}$ with

$$\boldsymbol{x}_0 = \arg \sup_{x \in X_E} \frac{\boldsymbol{f}(\boldsymbol{x})^\top \boldsymbol{\Phi} \boldsymbol{U}_L \boldsymbol{\Phi} \boldsymbol{f}(\boldsymbol{x})}{1 + n \boldsymbol{f}(\boldsymbol{x})^\top \boldsymbol{\Phi} \boldsymbol{f}(\boldsymbol{x})}.$$

Step 2. Beginning with $s = 1$, form the design $\xi_B = (1 - \alpha_s)\, \xi_{s-1} + \alpha_s \delta_{x_s}$, where $\boldsymbol{x}_s$ and α_s are to be chosen such that

$$\boldsymbol{x}_s = \arg \sup_{x \in X_E} \boldsymbol{f}(\boldsymbol{x})^\top \boldsymbol{M}_B(\xi_{s-1})^{-1} \boldsymbol{U}_L \boldsymbol{M}_B(\xi_{s-1})^{-1} \boldsymbol{f}(\boldsymbol{x})$$

and

$$\alpha_s = \arg \inf_{\alpha \in [0,1]} \operatorname{tr} \boldsymbol{U}_L((1 - \alpha)\, \boldsymbol{M}_B(\xi_{s-1}) + \alpha \boldsymbol{M}_B(\delta_{x_s}))^{-1}.$$

Step 3. Calculate

$$e_{0,s} := 2 - \left\{\frac{1}{n} \operatorname{tr} \boldsymbol{Q}(\xi_s)\, \boldsymbol{\Phi}^{-1} + \sup_{x \in X_E} \boldsymbol{f}(\boldsymbol{x})^\top \boldsymbol{Q}(\xi_s) \boldsymbol{f}(\boldsymbol{x})\right\} \Big/ L_B(\xi_s)$$

with $Q(\cdot)$ as defined in Corollary 12.7 (i).

If $e_{0,s} \geqq e_0$ then ξ_s is L_B-e_0-optimal, otherwise continue the iteration according to Step 2 with s replaced by $s + 1$.

The choice of the initial design ξ_0 in Step 1 proceeds according to (12.59), for we have

$$\begin{aligned} L_B(\delta_x) &= \operatorname{tr} \boldsymbol{U}_L \left(\frac{1}{n} \boldsymbol{\Phi}^{-1} + \boldsymbol{f}(\boldsymbol{x}) \boldsymbol{f}(\boldsymbol{x})^\top\right)^{-1} \\ &= n \operatorname{tr} \boldsymbol{U}_L \boldsymbol{\Phi} - \frac{n^2 \boldsymbol{f}(\boldsymbol{x})^\top \boldsymbol{\Phi} \boldsymbol{U}_L \boldsymbol{\Phi} \boldsymbol{f}(\boldsymbol{x})}{1 + n \boldsymbol{f}(\boldsymbol{x})^\top \boldsymbol{\Phi} \boldsymbol{f}(\boldsymbol{x})} \qquad \text{(by A.34)}. \end{aligned}$$

Furthermore, with $\boldsymbol{Q}(\xi_{s-1}) = \boldsymbol{M}_{\mathrm{B}}(\xi_{s-1})^{-1}\boldsymbol{U}_{\mathrm{L}}\boldsymbol{M}_{\mathrm{B}}(\xi_{s-1})^{-1}$ and Lemma 11.7 we have

$$\begin{aligned}\Delta_{L_{\mathrm{B}}}(\xi_{s-1}, \delta_x) &= L_{\mathrm{B}}(\xi_{s-1}) - \operatorname{tr}\boldsymbol{Q}(\xi_{s-1})\,\boldsymbol{M}_{\mathrm{B}}(\delta_x)\\ &= L_{\mathrm{B}}(\xi_{s-1}) - \frac{1}{n}\operatorname{tr}\boldsymbol{Q}(\xi_{s-1})\,\boldsymbol{\Phi}^{-1} - \|\boldsymbol{f}(x)\|^2_{\boldsymbol{Q}(\xi_{s-1})}\end{aligned}$$

so that the choice of x_s in Step 2 is in accordance with (12.56).

If the minimization problem for the optimal weight α_s, as formulated in Step 2, becomes numerically unfeasible then we recommend to choose fixed weights $\alpha_s = 1/(1+s)$; $s = 1, 2, \dots$ In general, the Algorithm 12.4 converges very rapidly, even if the weights are chosen fixed.

Using the theorem given in Böhning (1981) we can prove

Theorem 12.14. *For any sequence* $\{\xi_s\}_{s\in\mathbb{N}}$ *generated according to* (12.54), *for which* $\{\alpha_s\}_{s\in\mathbb{N}}$ *and* $\{x_s\}_{s\in\mathbb{N}}$ *have been chosen according to* (12.58) *and Step* 2 *of Algorithm* 12.4, *respectively, it holds either one of the following assertions:*

a) $\{\xi_s\}_{s\in\mathbb{N}}$ *is finite with the last element being* L_{B}*-optimal*

b) $\{\xi_s\}_{s\in\mathbb{N}}$ *is infinite with* $\lim\limits_{s\to\infty} L_{\mathrm{B}}(\xi_s) = \inf\limits_{\xi\in\Xi} L_{\mathrm{B}}(\xi)$.

(For a proof see Pilz (1987b).)

Example 12.5. Consider the quadratic regression setup

$$\mathsf{E}\tilde{y}(x) = \vartheta_0 + \vartheta_1 x + \vartheta_2 x^2, \qquad x \in X_{\mathrm{E}} = [-1, 1]$$

with

$$\mathsf{Cov}((\tilde{\vartheta}_0, \tilde{\vartheta}_1, \tilde{\vartheta}_2)^{\top} \mid w) = \frac{1}{w}\begin{pmatrix}0.1 & 0 & 0\\ 0 & 0.1 & 0\\ 0 & 0 & 0.2\end{pmatrix},$$

i.e., $\boldsymbol{\Phi} = \operatorname{diag}(0.1, 0.1, 0.2)$. Suppose that $n = 5$ observations have to be made and we wish to find an approximately L_{B}-optimal design with respect to

$$\boldsymbol{U}_{\mathrm{L}} = \begin{pmatrix}1 & 0 & 1/5\\ 0 & 1/5 & 0\\ 1/5 & 0 & 3/35\end{pmatrix}$$

which at least has an efficiency $e_0 = 0.95$. We remark that this matrix $\boldsymbol{U}_{\mathrm{L}}$ occurs, for example, when searching for a Bayesian design with respect to the prior distribution $P_{\tilde{x}}$ having density $p_{\tilde{x}}(x) \propto (1 - x^2)$.

The application of Algorithm 12.4 then yields the following scheme.

$$s = 0\colon\ x_0 = \arg\sup_{-1\leq x\leq 1} \frac{0.01(1 + x^2 + 4x^4)}{0.5(3 + x^2 + 2x^4)} = 1, \qquad \xi_0 = \delta_1.$$

With $\frac{1}{n}\boldsymbol{\Phi}^{-1} = \operatorname{diag}(2, 2, 1)$ we obtain

$$M_B(\xi_0) = \begin{pmatrix} 3 & 1 & 1 \\ 1 & 3 & 1 \\ 1 & 1 & 2 \end{pmatrix}, \qquad M_B(\xi_0)^{-1} = \frac{1}{12}\begin{pmatrix} 5 & -1 & -2 \\ -1 & 5 & -2 \\ -2 & -2 & 8 \end{pmatrix},$$

$$Q(\xi_0) = M_B(\xi_0)^{-1} U_L M_B(\xi_0)^{-1}$$

$$= \begin{pmatrix} 0.1496 & -0.0504 & -0.0151 \\ -0.0504 & 0.0496 & -0.0151 \\ -0.0151 & -0.0151 & 0.0270 \end{pmatrix},$$

$$\frac{1}{n} \operatorname{tr} Q(\xi_0) \Phi^{-1} = 0.4254, \qquad L_B(\xi_0) = \operatorname{tr} U_L M_B(\xi_0)^{-1} = 0.4905,$$

$$f(x)^\top Q(\xi_0) f(x) = 0.1496 - 0.1008x + 0.0194x^2 - 0.0302x^3 + 0.0270x^4$$

$$\sup_{-1 \leqq x \leqq 1} f(x)^\top Q(\xi_0) f(x) = 0.3270.$$

Thus, $e_{L_B}(\xi_0) \geqq 2 - (0.4254 + 0.3270)/0.4905 = 0.466 < e_0$.

$s = 1$: $x_1 = \arg \sup_{-1 \leqq x \leqq 1} f(x)^\top Q(\xi_0) f(x) = -1$,

$$(1-\alpha) M_B(\xi_0) + \alpha M_B(\delta_{-1}) = \begin{pmatrix} 3 & 1-2\alpha & 1 \\ 1-2\alpha & 3 & 1-2\alpha \\ 1 & 1-2\alpha & 2 \end{pmatrix},$$

$$\alpha_1 = \arg \inf_{0 \leqq \alpha \leqq 1} \frac{206 + 96\alpha - 96\alpha^2}{12(1+\alpha-\alpha^2)} = 0.5,$$

$$\xi_1 = 0.5\,\xi_0 + 0.5\,\delta_{-1} = \{(-1, 0.5), (1, 0.5)\},$$

$$M_B(\xi_1) = \begin{pmatrix} 3 & 0 & 1 \\ 0 & 3 & 0 \\ 1 & 0 & 2 \end{pmatrix}, \qquad M_B(\xi_1)^{-1} = \frac{1}{15}\begin{pmatrix} 6 & 0 & -3 \\ 0 & 5 & 0 \\ -3 & 0 & 9 \end{pmatrix},$$

$$Q(\xi_1) = M_B(\xi_1)^{-1} U_L M_B(\xi_1)^{-1} = \begin{pmatrix} 0.1314 & 0 & -0.0343 \\ 0 & 0.0222 & 0 \\ -0.0343 & 0 & 0.0229 \end{pmatrix},$$

$$\frac{1}{n} \operatorname{tr} Q(\xi_1) \Phi^{-1} = 0.3302, \qquad L_B(\xi_1) = 0.4381,$$

$$f(x)^\top Q(\xi_1) f(x) = 0.1314 - 0.0464x^2 + 0.0229x^4,$$

$$e_{0,1} = 2 - (0.3302 + 0.1314)/0.4381 = 0.9464 < e_0.$$

$s = 2$: $x_2 = \arg \sup_{-1 \leqq x \leqq 1} f(x)^\top Q(\xi_1) f(x) = 0$,

$$(1-\alpha)\,\boldsymbol{M}_{\mathrm{B}}(\xi_1)+\alpha\boldsymbol{M}_{\mathrm{B}}(\delta_0)=\begin{pmatrix}3 & 0 & 1-\alpha\\ 0 & 3-\alpha & 0\\ 1-\alpha & 0 & 2-\alpha\end{pmatrix},$$

$$\alpha_2=\arg\inf_{0\leqq\alpha\leqq 1}\frac{230-135\alpha+7\alpha^2}{15-8\alpha-2\alpha^2+\alpha^3}\approx 0.2\,,$$

$$\xi_2=0.8\,\xi_1+0.2\,\delta_0=\{(-1,0.4),(0,0.2),(1,0.4)\}\,,$$

$$\boldsymbol{M}_{\mathrm{B}}(\xi_2)=\begin{pmatrix}3 & 0 & 0.8\\ 0 & 2.8 & 0\\ 0.8 & 0 & 1.8\end{pmatrix},$$

$$\boldsymbol{Q}(\xi_2)=\begin{pmatrix}0.1200 & 0 & -0.0193\\ 0 & 0.0255 & 0\\ -0.0193 & 0 & 0.0199\end{pmatrix},$$

$$\frac{1}{n}\operatorname{tr}\boldsymbol{Q}(\xi_2)\,\boldsymbol{\Phi}^{-1}=0.3109,\qquad L_{\mathrm{B}}(\xi_2)=0.4364\,,$$

$$\boldsymbol{f}(x)^{\top}\boldsymbol{Q}(\xi_2)\boldsymbol{f}(x)=0.12-0.0131x^2+0.0199x^4\,,$$

$$\sup_{-1\leqq x\leqq 1}\boldsymbol{f}(x)^{\top}\boldsymbol{Q}(\xi_2)\boldsymbol{f}(x)=0.1268\,,$$

$$e_{0,2}=2-(0.3109+0.1268)/0.4364=0.997>e_0\,.$$

Thus, ξ_2 guarantees the required efficiency.

If we had chosen fixed weights $\alpha_s=1/(1+s)$ then the iteration would have stopped at stage $s=2$, too, yielding but an iteration design with a slightly smaller efficiency than ξ_2 from above. Briefly, the iteration steps are the following.

$s=0$: as above.

$s=1$: x_1 as above, $\alpha_1=1/(1+s)=1/2$,

ξ_1, $\boldsymbol{Q}(\xi_1)$ and $e_{0,1}$ as above.

$s=2$: x_2 as above, $\alpha_2=1/3$,

$$\xi_2=\frac{2}{3}\,\xi_1+\frac{1}{3}\,\delta_0=\left\{\left(-1,\frac{1}{3}\right),\left(0,\frac{1}{3}\right),\left(1,\frac{1}{3}\right)\right\},$$

$$\boldsymbol{M}_{\mathrm{B}}(\xi_2)=\begin{pmatrix}3 & 0 & 2/3\\ 0 & 8/3 & 0\\ 2/3 & 0 & 5/3\end{pmatrix},$$

$$\boldsymbol{Q}(\xi_2)=\begin{pmatrix}0.1143 & 0 & -0.0093\\ 0 & 0.0281 & 0\\ -0.0093 & 0 & 0.0200\end{pmatrix},$$

$$\frac{1}{n}\operatorname{tr} \boldsymbol{Q}(\xi_2)\,\boldsymbol{\Phi}^{-1} = 0.3048, \qquad L_{\mathrm{B}}(\xi_2) = 0.4387,$$

$$\boldsymbol{f}(x)^{\top}\boldsymbol{Q}(\xi_2)\boldsymbol{f}(x) = 0.1143 + 0.0095x^2 + 0.02x^4,$$

$$\sup_{-1 \leq x \leq 1} \boldsymbol{f}(x)^{\top}\boldsymbol{Q}(\xi_2)\boldsymbol{f}(x) = 0.1438,$$

$$e_{0,2} = 2 - (0.3048 + 0.1438)/0.4387 = 0.977 > e_0.$$

We remark that two further iterations would lead us to identically the same iteration design as ξ_2 in the former iteration with optimally chosen weights, i.e., with $x_3 = 1$, $\alpha_3 = 1/4$, $\xi_3 = \left\{\left(-1, \frac{1}{4}\right), \left(0, \frac{1}{4}\right), \left(1, \frac{1}{2}\right)\right\}$ at the third stage we would obtain $x_4 = -1$, $\alpha_4 = 1/5$, and

$$\xi_4 = \left\{\left(-1, \frac{2}{5}\right), \left(0, \frac{1}{5}\right), \left(+1, \frac{2}{5}\right)\right\}.$$

Particularly, if we set $\boldsymbol{U}_{\mathrm{L}} = \boldsymbol{I}_r$, $\boldsymbol{U}_{\mathrm{L}} = \boldsymbol{c}\boldsymbol{c}^{\top}$, $\boldsymbol{U}_{\mathrm{L}} = \int_{X_{\mathrm{P}}} \boldsymbol{f}(\boldsymbol{x})\boldsymbol{f}(\boldsymbol{x})^{\top} P_{\tilde{x}}(\mathrm{d}\boldsymbol{x})$ then the Algorithm 12.4 yields approximately A_{B}-optimal, C_{B} optimal and Bayesian designs, respectively.

For the determinant criterion $Z = D_{\mathrm{B}}$ we have

$$\begin{aligned} Z(\delta_x) &= \det\left(\boldsymbol{f}(\boldsymbol{x})\boldsymbol{f}(\boldsymbol{x})^{\top} + \frac{1}{n}\boldsymbol{\Phi}^{-1}\right)^{-1} \\ &= (n^r \det \boldsymbol{\Phi})\,(1 + n\boldsymbol{f}(\boldsymbol{x})^{\top}\boldsymbol{\Phi}\boldsymbol{f}(\boldsymbol{x}))^{-1} \end{aligned}$$

so that δ_{x_0} with $\boldsymbol{x}_0$ maximizing $\boldsymbol{f}(\cdot)^{\top}\boldsymbol{\Phi}\boldsymbol{f}(\cdot)$ over X_{E} can serve as an initial design for an iteration procedure. Further, by Lemma 11.8 (i) it holds

$$\Delta_Z(\xi_{s-1}, \delta_x) = D_{\mathrm{B}}(\xi_{s-1}) \cdot \operatorname{tr} \boldsymbol{M}(\xi_{s-1})\,\boldsymbol{M}_{\mathrm{B}}(\xi_{s-1})^{-1} - D_{\mathrm{B}}(\xi_{s-1}) \cdot \boldsymbol{f}(\boldsymbol{x})^{\top}\boldsymbol{M}_{\mathrm{B}}(\xi_{s-1})^{-1}\boldsymbol{f}(\boldsymbol{x})$$

so that the new supporting point $\boldsymbol{x}_s$ to be added to ξ_{s-1} is to maximize $\boldsymbol{f}(\cdot)^{\top}\boldsymbol{M}_{\mathrm{B}}(\xi_{s-1})^{-1}\boldsymbol{f}(\cdot)$ over X_{E}.

Algorithm 12.5. Construction of approximately D_{B}-optimal designs

Step 1. Choose an initial design $\xi_0 = \delta_{x_0}$ with

$$\boldsymbol{x}_0 = \arg \sup_{x \in X_{\mathrm{E}}} \boldsymbol{f}(\boldsymbol{x})^{\top}\boldsymbol{\Phi}\boldsymbol{f}(\boldsymbol{x}).$$

Step 2. Beginning with $s = 1$, form the design

$$\xi_s = (1 - \alpha_s)\,\xi_{s-1} + \alpha_s \delta_{x_s},$$

where $\boldsymbol{x}_s$ and α_s are to be chosen such that

$$\boldsymbol{x}_s = \arg \sup_{x \in X_{\mathrm{E}}} \boldsymbol{f}(\boldsymbol{x})^{\top}\boldsymbol{M}_{\mathrm{B}}(\xi_{s-1})^{-1}\boldsymbol{f}(\boldsymbol{x})$$

and

$$\alpha_s = \arg \sup_{0 \leq \alpha \leq 1} \det((1 - \alpha)\,\boldsymbol{M}_{\mathrm{B}}(\xi_{s-1}) + \alpha\boldsymbol{M}_{\mathrm{B}}(\delta_{x_s})).$$

Step 3. Calculate

$$e_{0,s} := 1 + \operatorname{tr} \boldsymbol{M}(\xi_s)\, \boldsymbol{M}_{\mathrm{B}}(\xi_s)^{-1} - \sup_{x \in X_{\mathrm{E}}} \boldsymbol{f}(\boldsymbol{x})^{\top} \boldsymbol{M}_{\mathrm{B}}(\xi_s)^{-1} \boldsymbol{f}(\boldsymbol{x}).$$

If $e_{0,s} \geqq e_0$ then ξ_s is D_{B}-e_0-optimal, otherwise continue the iteration according to Step 2 with s replaced by $s+1$.

Example 12.6. Consider again the quadratic regression problem described in the preceding example where we now wish to find a D_{B}-optimal design with efficiency not smaller than $e_0 = 0.95$. The application of Algorithm 12.5 then yields the following iteration steps.

$s=0$: $x_0 = \arg \sup_{-1 \leqq x \leqq 1} (0.1 + 0.1x^2 + 0.2x^4) = 1,$

$$\xi_0 = \delta_1, \quad \boldsymbol{M}_{\mathrm{B}}(\xi_0) \quad \text{as in Example 12.5},$$

$$\boldsymbol{f}(x)^{\top} \boldsymbol{M}_{\mathrm{B}}(\xi_0)^{-1} \boldsymbol{f}(x) = \frac{1}{12}(5 - 2x - x^2 - 4x^3 + 8x^4).$$

$s=1$: $x_1 = \arg \sup_{-1 \leqq x \leqq 1} \boldsymbol{f}(x)^{\top} \boldsymbol{M}_{\mathrm{B}}(\xi_0)^{-1} \boldsymbol{f}(x) = -1,$

$$(1-\alpha)\, \boldsymbol{M}_{\mathrm{B}}(\xi_0) + \alpha \boldsymbol{M}_{\mathrm{B}}(\delta_{-1}) = \begin{pmatrix} 3 & 1-2\alpha & 1 \\ 1-2\alpha & 3 & 1-2\alpha \\ 1 & 1-2\alpha & 2 \end{pmatrix}$$

$$\alpha_1 = \arg \sup_{0 \leqq \alpha \leqq 1} 12(1 + \alpha - \alpha^2) = 0.5,$$

$$\xi_1 = 0.5\, \delta_1 + 0.5\, \delta_{-1} = \{(-1, 0.5), 1, 0.5)\},$$

$$\boldsymbol{M}(\xi_1) = \begin{pmatrix} 1 & 0 & 1 \\ 0 & 1 & 0 \\ 1 & 0 & 1 \end{pmatrix}, \qquad \boldsymbol{M}_{\mathrm{B}}(\xi_1)^{-1} = \frac{1}{15}\begin{pmatrix} 6 & 0 & -3 \\ 0 & 5 & 0 \\ -3 & 0 & 9 \end{pmatrix},$$

$$\operatorname{tr} \boldsymbol{M}(\xi_1)\, \boldsymbol{M}_{\mathrm{B}}(\xi_1)^{-1} = \frac{14}{15},$$

$$\boldsymbol{f}(x)^{\top} \boldsymbol{M}_{\mathrm{B}}(\xi_1)^{-1} \boldsymbol{f}(x) = \frac{1}{15}(6 - x^2 + 9x^4),$$

$$\sup_{-1 \leqq x \leqq 1} \boldsymbol{f}(x)^{\top} \boldsymbol{M}_{\mathrm{B}}(\xi_1)^{-1} \boldsymbol{f}(x) = \frac{14}{15}.$$

Thus, $e_{0,1} = 1$, which means that ξ_1 is exactly D_{B}-optimal.

The same design ξ_1 we would have obtained with the weight α_1 at stage $s=1$ chosen fixed according to $\alpha_s = 1/(1+s) = 0.5$. Note that the D_{B}-optimal design ξ_1 has only two supporting points whereas the regression setup contains three parameters.

13. Construction of exact optimal designs

13.1. Rounding of discrete optimal designs

As we had pointed out already in Section 10.1, the change from the original exact design problem (10.9) to the extended design problem (10.10) in terms of continuous designs offered the possibility to handle this problem within the framework of convex optimization. This way we were able to derive necessary and sufficient optimality conditions and to use them for the construction of optimal approximate designs. Particularly, we could restrict attention to discrete designs with finite support and, under certain conditions, we could obtain such an optimal design as a solution of a corresponding equation system guaranteeing an information matrix as specified by Algorithm 12.1 and its modifications.

As before, let $n \in \mathbb{N}$ be a fixed integer standing for the number of observations to be performed. If a discrete optimal design can be written as an exact design, i.e., if its weights are integer multiples of $1/n$, then it is also optimum within the class V_n of exact designs of size n. But, in general, this is not the case and a discrete design cannot be realized by observation.

Nevertheless, a discrete optimal design or some approximation of it can be taken as a starting point for the construction of an exact (approximately) optimal design. The change from discrete to exact designs may be accomplished by help of some rounding procedure. A rather simple, but powerful method would be as follows: Assume $\xi^* = \{(x_1, p_1), \dots, (x_m, p_m)\}$, $m \geqq 1$, to be a discrete Z-optimal design where $Z(\cdot) = \bar{Z}(M_B(\cdot))$ is some functional of the Bayesian information matrix. Then we form the set

$$\Xi_n = \left\{ \xi_{v_n} = \left\{ \left(x_1, \frac{n_1}{n}\right), \dots, \left(x_m, \frac{n_m}{n}\right) \right\} : n_i \geqq [np_i],\ n_i \in \{0, \dots, n\},\ \sum_{i=1}^{m} n_i = n \right\} \tag{13.1}$$

which is a subset of V_n, since any $\xi_{v_n} \in \Xi_n$ can be identified with an exact design

$$v_n = (\underbrace{x_1, \dots, x_1}_{n_1 \text{ times}},\ \underbrace{x_2, \dots, x_2}_{n_2 \text{ times}},\ \dots,\ \underbrace{x_m, \dots, x_m}_{n_m \text{ times}})$$

and hereafter we choose

$$\xi^*_{v_n} = \arg \inf_{\xi_{v_n} \in \Xi_n} \bar{Z}(M_B(\xi_{v_n})) \tag{13.2}$$

as an approximation for an exact Z-optimal design. (In (13.1), $[np_i]$ stands for the greatest integer less or equal to np_i.) If the above set Ξ_n, which is finite for every $n \in \mathbb{N}$, becomes too large then as a rough rounding procedure we recommend to choose $\xi^*_{v_n} = \{(x_1, n_1/n), \dots, (x_m, n_m/n)\}$ such that $n_i = [np_i] + 1$ for those $n_0 = n - \sum [np_i]$ indices $i \in \{1, \dots, m\}$ which yield the largest differences $h_i = np_i - [np_i]$ and $n_i = [np_i]$ for the remaining $n - n_0$ indices i with the smallest differences h_i.

Example 13.1. We look for a rounding of the discrete design $\xi^* = \{(-1, 0.5), (1, 0.5)\}$ which was proved to be D_B-optimal for the quadratic regression problem considered in the Examples 12.5, 12.6 and where $n = 5$ observations had to be made.

Here we have $[np_1] = [np_2] = [5/2] = 2$ and, according to (13.1),

$$\Xi_5 = \left\{ \xi_{v_5}^{(1)} = \left\{ \left(-1, \frac{3}{5}\right), \left(1, \frac{2}{5}\right) \right\}, \xi_{v_5}^{(2)} = \left\{ \left(-1, \frac{2}{5}\right), \left(1, \frac{3}{5}\right) \right\} \right\}.$$

This yields

$$M_B(\xi_{v_5}^{(1)}) = \begin{pmatrix} 1 & -1/5 & 1 \\ -1/5 & 1 & -1/5 \\ 1 & -1/5 & 1 \end{pmatrix} + \frac{1}{5} \begin{pmatrix} 10 & 0 & 0 \\ 0 & 10 & 0 \\ 0 & 0 & 5 \end{pmatrix} = \frac{1}{5} \begin{pmatrix} 15 & -1 & 5 \\ -5 & 15 & -1 \\ 5 & -1 & 10 \end{pmatrix},$$

$$M_B(\xi_{v_5}^{(2)}) = \begin{pmatrix} 1 & 1/5 & 1 \\ 1/5 & 1 & 1/5 \\ 1 & 1/5 & 1 \end{pmatrix} + \frac{1}{5} \begin{pmatrix} 10 & 0 & 0 \\ 0 & 10 & 0 \\ 0 & 0 & 5 \end{pmatrix} = \frac{1}{5} \begin{pmatrix} 15 & 1 & 5 \\ 1 & 15 & 1 \\ 5 & 1 & 10 \end{pmatrix}$$

and

$$\bar{Z}(M_B(\xi_{v_5}^{(1)})) = 1/\det M_B(\xi_{v_5}^{(1)}) = \frac{25}{372} = 1/\det M_B(\xi_{v_5}^{(2)}) = \bar{Z}(M_B(\xi_{v_5}^{(2)})).$$

Therefore, both $v_5^{(1)} = (-1, -1, -1, 1, 1)$ and $v_5^{(2)} = (-1, -1, 1, 1, 1)$ could be taken as approximately exact D_B-optimal designs. The increase in the functional value due to rounding is negligibly small: $\det M_B(\xi^*)^{-1} = 0.0667$, $\det M_B(v_5^{(1)})^{-1} = 0.0672$.

Having obtained an exact design by some rounding procedure, then it is important to evaluate its efficiency within the class of exact designs. Some proposals for such an evaluation will be given in the sequel. If the efficiency of a given exact design is not yet satisfactory then it can be improved by help of some iteration procedure as will be demonstrated in Section 13.3.

Let Z be a design functional defined on the set of positive definite matrices $\mathcal{M}_r^>$, as it is the case with all the functionals of the Bayesian information matrix considered so far. If the functional value of Z-optimal discrete designs is known then it can be used to derive bounds for the Z-efficiency of exact designs.

Theorem 13.1. *Let $\bar{Z} \mid \mathcal{M}_r^> \to \mathbb{R}^1$ be such that*

$$\bar{Z}(\alpha M_B(\xi)) = g(\alpha)\, \bar{Z}(M_B(\xi)) \quad \textit{for all} \quad \xi \in \Xi,\ \alpha \in \mathbb{R}^+, \qquad (13.3)$$

where g is some real-valued function defined on $\mathbb{R}^+$. Further, let $\xi^* \in \Xi$ be Z-optimal in Ξ with card(supp ξ^*) $= m < n$. *Then it holds*

$$\bar{Z}(\boldsymbol{M}_{\mathrm{B}}(\xi^*)) \leqq \inf_{v_n \in V_n} \bar{Z}(\boldsymbol{M}_{\mathrm{B}}(v_n)) \leqq g\left(\frac{n-m}{n}\right) \bar{Z}(\boldsymbol{M}_{\mathrm{B}}(\xi^*))$$

(cp. FEDOROV (1972)).

According to Definition 12.1, the Z-efficiency of exact designs $v_n \in V_n$ is given by

$$e_Z(v_n) = \inf_{\bar{v}_n \in V_n} \bar{Z}(\boldsymbol{M}_{\mathrm{B}}(\bar{v}_n))/\bar{Z}(\boldsymbol{M}_{\mathrm{B}}(v_n)) \tag{13.4}$$

provided that $\bar{Z}(\boldsymbol{M}_{\mathrm{B}}(v_n)) > 0$. In this case the Z-efficiency of exact designs $v_n \in V_n$ relative to functionals Z satisfying (13.3) is constrained by

$$\frac{\bar{Z}(\boldsymbol{M}_{\mathrm{B}}(\xi^*))}{\bar{Z}(\boldsymbol{M}_{\mathrm{B}}(v_n))} \leqq e_Z(v_n) \leqq g\left(\frac{n-m}{n}\right) \frac{\bar{Z}(\boldsymbol{M}_{\mathrm{B}}(\xi^*))}{\bar{Z}(\boldsymbol{M}_{\mathrm{B}}(v_n))}. \tag{13.5}$$

Particularly, for the functionals $Z = L_{\mathrm{B}}$ and $Z = D_{\mathrm{B}}$ we can give the following bounds.

Corollary 13.1. *Let $\xi^* \in \Xi$ be L_{B}-optimal and D_{B}-optimal, respectively, with* card(supp ξ^*) $= m < n$. *Then it holds*

(i) $L_{\mathrm{B}}(\xi^*)/L_{\mathrm{B}}(v_n) \leqq e_{L_{\mathrm{B}}}(v_n) \leqq nL_{\mathrm{B}}(\xi^*)/(n-m)\,L_{\mathrm{B}}(v_n)$,

(ii) $D_{\mathrm{B}}(\xi^*)/D_{\mathrm{B}}(v_n) \leqq e_{D_{\mathrm{B}}}(v_n) \leqq n^r D_{\mathrm{B}}(\xi^*)/(n-m)^r D_{\mathrm{B}}(v_n)$.

***Proof*.** Observing that $\operatorname{tr} \boldsymbol{U}_{\mathrm{L}}(\alpha \boldsymbol{M}_{\mathrm{D}}(\xi))^{-1} = \alpha^{-1} L_{\mathrm{B}}(\xi)$ and $\det(\alpha \boldsymbol{M}_{\mathrm{B}}(\xi))^{-1} = \alpha^{-r} D_{\mathrm{B}}(\xi)$, the functionals $Z = L_{\mathrm{B}}$ and $Z = D_{\mathrm{B}}$ satisfy Condition (13.3) with $g(\alpha) = \alpha^{-1}$ and $g(\alpha) = \alpha^{-r}$, respectively. The above bounds then result from Theorem 13.1 and (13.5). □

Especially, if we apply some rounding procedure to a discrete L_{B}-optimal design as described before then a lower bound for the L_{B}-efficiency of a rounded design can be conveniently evaluated as follows.

Theorem 13.2. *Let $\xi^* = \{(\boldsymbol{x}_1, p_1), \ldots, (\boldsymbol{x}_m, p_m)\}$ be an L_{B}-optimal design in Ξ and v_n an exact design taking $n_i \geqq 1$ observations at $\boldsymbol{x}_i$ $(i = 1, \ldots, m)$ where $n_1 + \ldots + n_m = n$. Then it holds*

$$e_{L_{\mathrm{B}}}(v_n) \geqq L_{\mathrm{B}}(\xi^*)/L_{\mathrm{B}}(v_n) \geqq 1 - (1 + q_n)^{-1},$$

where

$$q_n = n \min(n_i) \Big/ \left\{\sum_{i=1}^{m} |np_i - n_i|\right\}^2$$

(see CHALONER (1984), Theorem 5).

The rounding procedure and the evaluation of efficiency as described above is promising if an optimal discrete design or some approximation of it is known. If we have not such knowledge then we can evaluate the Z-efficiency of a given design $v_n \in V_n$ as in the discrete case by the following lower bound:

$$e_Z(v_n) \geqq 1 + d_Z(v_n)/Z(v_n),$$

where (13.6)

$$d_Z(v_n) = \inf_{x \in X_E} \Delta_Z(\xi_{v_n}, \delta_x)$$

and ξ_{v_n} coincides with v_n written as a discrete design (see Lemma 12.2). For the special case of L_B-optimality with respect to U_L this reads

$$e_{L_B}(v_n) \geqq 2 - \frac{1}{L_B(v_n)} \left\{ \frac{1}{n} \operatorname{tr} Q(v_n)\, \Phi^{-1} + \sup_{x \in X_E} f(x)^\top Q(v_n) f(x) \right\},$$

where (13.7)

$$Q(v_n) = M_B(v_n)^{-1} U_L M_B(v_n)^{-1}$$

and for the case of D_B-optimality we have

$$e_{D_B}(v_n) \geqq 1 + \operatorname{tr} M(v_n)\, M_B(v_n)^{-1} - \sup_{x \in X_E} f(x)^\top M_B(v_n)^{-1} f(x) \quad (13.8)$$

(see Corollary 12.7). Clearly, $v_n \in V_n$ is Z-optimal in V_n if the lower bound on the right-hand side of (13.6) is equal to one, which is the case if and only if $d_Z(v_n) = 0$. We remark, however, that this condition is only sufficient for Z-optimality in V_n, but it is not necessary. This means that $v_n \in V_n$ may be Z-optimal in V_n even if $d_Z(v_n) \neq 0$, whereas in the extended design problem the condition $d_Z(\xi^*) = 0$ is both sufficient and necessary for ξ^* to be Z-optimal in Ξ (see Theorem 11.11 (i)).

In the preceding chapter we demonstrated that, under certain conditions, a discrete optimal design ξ^* can be obtained by solving an equation system of the type $M(\xi^*) = M^*$ with a predetermined optimal information matrix M^*. Since the solution $\xi^* \in \Xi$, provided there exists one, need not be unique, it makes also sense to look for exact designs $v_n^* \in V_n$ having such an information matrix $M(v_n^*) = M^*$, even if a particular discrete solution ξ^* does not represent an exact design. From Theorem 12.6 follows immediately

Corollary 13.2. *Let U and Λ be the matrices of eigenvectors and eigenvalues of Φ^{-1}, respectively, and $m_1^*, \ldots, m_r^*$ nonnegative numbers obtained according to Algorithm 12.1. If there exists an exact design $v_n^* = (x_1, \ldots, x_n) \in V_n$ such that*

$$\frac{1}{n} \sum_{i=1}^{n} f(x_i) f(x_i)^\top = U \operatorname{diag}(m_1^*, \ldots, m_r^*)\, U^\top \quad (13.9)$$

then it is A_B-, D_B- and E_B-optimal in V_n.

In the sequel we will apply this idea to obtain exact optimal designs for the multiple linear regression case.

13.2. Exact optimal designs for multiple linear regression

As in our previous considerations on the approximate design problem for multiple linear regression, we will investigate the existence of exact designs with an information matrix as indicated in (13.9) separately for the homogeneous and the inhomogeneous model. Again, the experimental region will be assumed to be the hypersphere

$$X_{\mathrm{E}} = \{\boldsymbol{x} \in \mathbb{R}^k : \boldsymbol{x}^\top \boldsymbol{x} \leqq 1\}, \tag{13.10}$$

the case of a cuboidal region will be considered briefly at the end of this section. In any case, the existence of an exact design satisfying (13.9) depends on the number n of observations to be made.

The homogeneous model

In the homogeneous model case

$$\mathbf{E}\tilde{y}(\boldsymbol{x}) = \boldsymbol{\vartheta}^\top \boldsymbol{x},\ \boldsymbol{x} \in X_{\mathrm{E}}, \qquad \mathbf{Cov}(\tilde{\boldsymbol{\vartheta}} \mid w) = \frac{1}{w}\boldsymbol{\Phi} \tag{13.11}$$

we have $\boldsymbol{f}(\boldsymbol{x}) = \boldsymbol{x}$, $r = k$ and the columns of the transposed design matrix are identical with the experimental points $\boldsymbol{x}_i = (x_{i1}, \ldots, x_{ik})^\top \in X_{\mathrm{E}}$ $(i = 1, \ldots, n)$ of the design $v_n = (\boldsymbol{x}_1, \ldots, \boldsymbol{x}_n)$, i.e.,

$$\boldsymbol{F}(v_n)^\top = \begin{pmatrix} x_{11} & x_{21} \ldots x_{n1} \\ x_{12} & x_{22} \ldots x_{n2} \\ \vdots & \vdots \quad \vdots \\ x_{1k} & x_{2k} \ldots x_{nk} \end{pmatrix}.$$

Again, let $\lambda_1 \leqq \lambda_2 \leqq \ldots \leqq \lambda_k$ be the ordered sequence of eigenvalues of $\boldsymbol{\Phi}^{-1}$ and

$$j_0 = \min\left\{ j \in \{1, \ldots, k\} : \frac{1}{n}\lambda_{j+1} > \frac{1}{j}\left(1 + \frac{1}{n}\sum_{i=1}^{j} \lambda_i\right)\right\}, \tag{13.12}$$

where we set $\lambda_{k+1} = \infty$. Then, applying Algorithm 12.3, we obtain nonnegative numbers

$$m_1^* \geqq m_2^* \geqq \ldots \geqq m_{j_0}^* > m_{j_0+1}^* = \ldots = m_k^* = 0.$$

Now, if $n \geqq j_0$ then there always exists an exact design $v_n = (\boldsymbol{x}_1, \ldots, \boldsymbol{x}_n)$, even with $\boldsymbol{x}_i^\top \boldsymbol{x}_i = 1$ $(i = 1, \ldots, n)$, for which it holds

$$\boldsymbol{M}(v_n) = \frac{1}{n}\boldsymbol{F}(v_n)^\top \boldsymbol{F}(v_n) = \boldsymbol{U}\,\mathrm{diag}(m_1^*, \ldots, m_k^*)\,\boldsymbol{U}^\top \tag{13.13}$$

and which then is A_{B}-, D_{B}- and E_{B}-optimal by Corollary 13.2. To this let $\boldsymbol{U}$ be the matrix of eigenvectors of $\boldsymbol{\Phi}^{-1}$ and

$$\boldsymbol{U}_{j_0} = (\boldsymbol{u}_1 \ldots \boldsymbol{u}_{j_0}), \qquad \boldsymbol{U}_{j_0}^\top \boldsymbol{U}_{j_0} = \boldsymbol{I}_{j_0}$$

the $(k \times j_0)$-matrix which arises from $\boldsymbol{U}$ by deleting the columns $\boldsymbol{u}_{j_0+1}, \ldots, \boldsymbol{u}_k$ for which we have $m^*_{j_0+1} = \ldots = m^*_k = 0$.

Theorem 13.3. *Let $n \geqq j_0$ and $\boldsymbol{\Omega} = (\omega_{ij}) \in \mathcal{M}_{n \times j_0}$ with $\boldsymbol{\Omega}^\top \boldsymbol{\Omega} = \boldsymbol{I}_{j_0}$ and Hadamard product $\boldsymbol{\Omega} \circ \boldsymbol{\Omega} = (\omega^2_{ij})$ be such that*

$$n(\boldsymbol{\Omega} \circ \boldsymbol{\Omega})(m^*_1, \ldots, m^*_{j_0})^\top = \mathbf{1}_n, \tag{13.14}$$

where $\mathbf{1}_n = (1, \ldots, 1)^\top \in \mathbb{R}^n$. Then the columns of

$$\boldsymbol{F}(v_n)^\top = \sqrt{n}\, \boldsymbol{U}_{j_0} \operatorname{diag}\left(\sqrt{m^*_1}, \ldots, \sqrt{m^*_{j_0}}\right) \boldsymbol{\Omega}^\top \tag{13.15}$$

form an exact A_B-, D_B- and E_B-optimal design v_n.

Proof. First, it follows from a result due to Horn (1954) that, for $n \geqq j_0$, there always exists a matrix $\boldsymbol{\Omega}$ as required above (see Gladitz/Pilz (1982b)). Secondly, it follows from (13.14) that $\boldsymbol{F}(v_n)$ as given by (13.15) in fact represents the design matrix of an exact design $v_n = (\boldsymbol{x}_1, \ldots, \boldsymbol{x}_n)$: The matrix $\boldsymbol{F}(v_n)\boldsymbol{F}(v_n)^\top = n\boldsymbol{\Omega} \operatorname{diag}(m^*_1, \ldots, m^*_{j_0}) \boldsymbol{\Omega}^\top$ has main diagonal elements

$$\boldsymbol{x}_i^\top \boldsymbol{x}_i = n \sum_{j=1}^{j_0} \omega^2_{ij} m^*_j = 1 \qquad (i = 1, \ldots, n)$$

since $(\boldsymbol{x}_1^\top \boldsymbol{x}_1, \ldots, \boldsymbol{x}_n^\top \boldsymbol{x}_n) = n(\boldsymbol{\Omega} \circ \boldsymbol{\Omega})(m^*_1, \ldots, m^*_{j_0})^\top = \mathbf{1}_n$. Therefore, the columns $\boldsymbol{x}_i$ of $\boldsymbol{F}(v_n)^\top$ have unity length, thus belonging to the experimental region X_E. Finally, observing that $\boldsymbol{\Omega}$ is orthogonal and $m^*_{j_0+1} = \ldots = m^*_k = 0$, (13.15) yields

$$\begin{aligned}\frac{1}{n}\boldsymbol{F}(v_n)^\top \boldsymbol{F}(v_n) &= \boldsymbol{U}_{j_0} \operatorname{diag}(m^*_1, \ldots, m^*_{j_0})\, \boldsymbol{U}_{j_0}^\top \\ &= \boldsymbol{U} \operatorname{diag}(m^*_1, \ldots, m^*_k)\, \boldsymbol{U}^\top .\end{aligned}$$

Thus, (13.9) is satisfied and the result follows from Corollary 13.2. □

We will now give a method for the construction of a matrix $\boldsymbol{\Omega}$ as required in the above theorem. The starting point for the construction are Hadamard matrices. These are quadratic matrices $\boldsymbol{H}_n = (h_{ij}) \in \mathcal{M}_{n \times n}$ with components $h_{ij} \in \{-1, +1\}$ such that $\boldsymbol{H}_n^\top \boldsymbol{H}_n = n\boldsymbol{I}_n$. Their existence has been proved for $n = 2^r$ $(r = 0, 1, \ldots)$, the existence for all integer multiples $n = 4s$ $(s = 1, 2, \ldots)$ is an open problem. For $n = 2$ and all integers $n \leqq 200$ divisible by 4, such matrices may be found, for example, in Spence (1967) and Hedayat/Wallis (1978).

Now let n be such that there exists a Hadamard matrix $\boldsymbol{H}_n$. Then

$$\boldsymbol{\Omega}_n = n^{-1/2}\boldsymbol{H}_n =: (\boldsymbol{\omega}_1 \ldots \boldsymbol{\omega}_n) \tag{13.16}$$

is an orthogonal matrix and all elements of $\boldsymbol{\Omega}_n \circ \boldsymbol{\Omega}_n$ are equal to n^{-1}. Hence, the matrix

$$\boldsymbol{\Omega} = (\boldsymbol{\omega}_1 \ldots \boldsymbol{\omega}_{j_0}) \tag{13.17}$$

containing the first $j_0 \leqq n$ columns of $\boldsymbol{\Omega}_n$ yields $\boldsymbol{\Omega}^\top \boldsymbol{\Omega} = \boldsymbol{I}_{j_0}$ and $n(\boldsymbol{\Omega} \circ \boldsymbol{\Omega})(m_1^*, \ldots, m_{j_0}^*)^\top = \mathbf{1}_n \mathbf{1}_{j_0}^\top (m_1^*, \ldots, m_{j_0}^*)^\top = \mathbf{1}_n$ since

$$(m_1^*, \ldots, m_{j_0}^*)\, \mathbf{1}_{j_0} = \sum_{j=1}^{j_0} m_j^* = 1 .$$

Therefore, (13.14) is satisfied and $\boldsymbol{\Omega}$ has all the properties required in Theorem 13.3. An orthogonal matrix $\boldsymbol{\Omega}_{n+1}$ of order $n+1$ can be computed from such of order n as given in (13.16) simply by setting

$$\boldsymbol{\Omega}_{n+1} = \begin{pmatrix} a \cdot \omega_1 & \omega_2 & \ldots & \omega_n & b \cdot \omega_1 \\ b & 0 & \ldots & 0 & -a \end{pmatrix},$$

where (13.18)

$$a^2 = \frac{m_1^* - 1/(n+1)}{m_1^* - m_{n+1}^*}, \qquad b^2 = 1 - a^2$$

and a, b chosen arbitrarily such that $a^2 + b^2 = 1$ in case of $m_1^* = m_{n+1}^*$. (Note that $m_n^* = 0$ whenever $n > j_0$.) Again, deleting the last $(n+1) - j_0$ columns of $\boldsymbol{\Omega}_{n+1}$ we obtain an $((n+1) \times j_0)$-matrix $\boldsymbol{\Omega}$ which, according to (13.15), leads us to an A_{B}-, D_{B}- and E_{B}-optimal design of size $n+1$. This way we can construct exact optimal designs of any size $n = 4s$ and, by help of the above procedure, also for the intermediate sizes $n = 4s+1$, $4s+2$, $4s+3$, starting with a Hadamard matrix of order $4s$.

Covey-Crump/Silvey (1970) have given an alternative method for the construction of an exact optimal design v_n^* starting with an orthogonal matrix $\boldsymbol{\Omega}_{j_0}$ of order j_0, adjoining some $((n-j_0) \times j_0)$ matrix $\boldsymbol{Z}$ to form $\boldsymbol{Z}_0^\top = (\boldsymbol{Z}^\top \mid \boldsymbol{\Omega}_{j_0}^\top)$ and computing $\boldsymbol{F}(v_n^*)^\top = \boldsymbol{U}_{j_0} \boldsymbol{Z}_0^\top$. They proved the D_{B}-optimality of v_n^* for a design problem with previous observations, where the matrix $\boldsymbol{\Phi}^{-1}$ plays the role of the information matrix of the 'prior' design used to obtain these observations. However, if n is large compared to j_0, the construction of the matrix Z_0 becomes complicated.

We will now give an example demonstrating the application of Theorem 13.3 to the construction of an A_{B}-, D_{B}- and E_{B}-optimal design.

Example 13.2. Consider the three-factor model

$$\mathbf{E}\tilde{y}(\boldsymbol{x}) = \vartheta_1 x_1 + \vartheta_2 x_2 + \vartheta_3 x_3, \qquad \sum_{i=1}^{3} x_i^2 \leqq 1$$

with prior covariance matrix $\mathbf{Cov}(\tilde{\boldsymbol{\vartheta}} \mid w) = \frac{1}{w} \boldsymbol{\Phi}$ given by

$$\boldsymbol{\Phi} = \begin{pmatrix} 1.015 & 0.468 & 0.344 \\ & 1.717 & 0.689 \\ & & 0.573 \end{pmatrix}.$$

The eigenvalues and eigenvectors of Φ^{-1} are as follows:

$$\lambda_1 = 0.4357, \qquad \lambda_2 = 1.2804, \qquad \lambda_3 = 4.3693,$$

$$u_1 = \begin{pmatrix} 0.4082 \\ 0.8165 \\ 0.4082 \end{pmatrix}, \qquad u_2 = \begin{pmatrix} -0.8944 \\ 0.4472 \\ 0 \end{pmatrix}, \qquad u_3 = \begin{pmatrix} 0.1826 \\ 0.3651 \\ -0.9129 \end{pmatrix}.$$

Suppose that $n = 5$ observations have to be made. Then, since $\frac{1}{n}\lambda_2 < 1 + \frac{1}{n}\lambda_1$ and $\frac{1}{n}\lambda_3 > \frac{1}{2}\left(1 + \frac{1}{n}(\lambda_1 + \lambda_2)\right)$, we have $j_0 = 2$ and step 2 of Algorithm 12.3 yields

$$m_1^* = 0.5 + (\lambda_2 - \lambda_1)/10 = 0.5845,$$

$$m_2^* = 0.5 - (\lambda_2 - \lambda_1)/10 = 0.4155, \quad m_3^* = 0.$$

Starting with a Hadamard-matrix of order 4 we first obtain

$$\Omega_4 = \frac{1}{2} H_4 = \begin{pmatrix} 1 & 1 & 1 & 1 \\ 1 & 1 & -1 & -1 \\ 1 & -1 & 1 & -1 \\ 1 & -1 & -1 & 1 \end{pmatrix}.$$

Then, according to (13.18), we have

$$a = \sqrt{1 - 1/5m_1^*} = 0.8111, \qquad b = 1/\sqrt{5m_1^*} = 0.5850,$$

$$\Omega_5 = \frac{1}{2}\begin{pmatrix} a & 1 & 1 & 1 & b \\ a & 1 & -1 & -1 & b \\ a & -1 & 1 & -1 & b \\ a & -1 & -1 & 1 & b \\ 2b & 0 & 0 & 0 & -2a \end{pmatrix}, \qquad \Omega = \begin{pmatrix} 0.4056 & 0.5 \\ 0.4056 & 0.5 \\ 0.4056 & -0.5 \\ 0.4056 & -0.5 \\ 0.5850 & 0 \end{pmatrix}.$$

This yields the A_B-, D_B- and E_B-optimal design

$$F(v_n)^\top = \sqrt{n}\,(u_1 \vdots u_2)\,\mathrm{diag}\left(\sqrt{m_1^*}, \sqrt{m_2^*}\right)\Omega^\top$$

$$= \begin{pmatrix} -0.3616 & -0.3616 & 0.9276 & 0.9276 & 0.4082 \\ 0.8883 & 0.8883 & 0.2437 & 0.2437 & 0.8165 \\ 0.2830 & 0.2830 & 0.2830 & 0.2830 & 0.4082 \end{pmatrix}.$$

For comparison, the procedure of Covey-Crump and Silvey leads to

$$Z_0^\top = \begin{pmatrix} 1 & 0 & 1 & \sqrt{\sigma_1/2} & \sqrt{\sigma_1/2} \\ 0 & 1 & 0 & \sqrt{\sigma_2/2} & -\sqrt{\sigma_2/2} \end{pmatrix}, \qquad \sigma_i = \frac{1}{2}(\lambda_1 + \lambda_2) + 1 - \lambda_i \quad (i = 1, 2)$$

which, according to $F(v_n^*)^\top = (u_1 \vdots u_2) Z_0^\top$, yields

$$F(v_n^*)^\top = \begin{pmatrix} 0.4082 & -0.8944 & 0.4082 & -0.3793 & 0.9338 \\ 0.8165 & 0.4472 & 0.8165 & 0.8827 & 0.2262 \\ 0.4082 & 0 & 0.4082 & 0.2772 & 0.2772 \end{pmatrix}.$$

Though v_n and v_n^* are different, they have the same information matrix .

$$M(v_n) = M(v_n^*) = \begin{pmatrix} 0.4298 & 0.0286 & 0.0974 \\ & 0.4727 & 0.1948 \\ & & 0.0974 \end{pmatrix} = \sum_{i=1}^{2} m_i^* u_i u_i^\top .$$

Thus, v_n^* is A_B-, D_B- and E_B-optimal, too.

For the case that less than j_0 observations have to be made, there does not exist a solution $v_n \in V_n$ to the equations (13.9) and (13.13), respectively, since then it holds rank $M(v_n) = n < j_0$ and $U_{j_0}\,\mathrm{diag}(m_1^*, \ldots, m_{j_0}^*)\,U_{j_0}^\top$ has rank j_0. In this case we propose to determine an exact experimental design $v_n = (x_1, \ldots, x_n)$ as a solution to the minimum problem

$$\mathrm{tr}\left\{\frac{1}{n}\sum_{i=1}^{n} x_i x_i^\top - U_{j_0}\,\mathrm{diag}(m_1^*, \ldots, m_{j_0}^*)\,U_{j_0}^\top\right\}^2 \overset{!}{=} \inf_{x_i \in \bar{X}_E} , \tag{13.19}$$

where

$$\bar{X}_E = \{x \in X_E : \|x\| = 1\} .$$

This way, v_n is determined such that its information matrix $M(v_n)$ has minimum (Euclidean) distance from the optimum (discrete) information matrix. The following statement, which is proved in GLADITZ/PILZ (1982b), can be used to find a solution to (13.19).

Lemma 13.1. *Let* $A \in \mathcal{M}_k^{\geqq}$ *be such that* $\mathrm{rank}\,A = s \leqq k$, $\mathrm{tr}\,A = 1$ *and* $A = \sum_{i=1}^{s} \alpha_i e_i e_i^\top$ *where* $\alpha_1 \geqq \alpha_2 \geqq \ldots \geqq \alpha_s > 0$ *and* $e_1, \ldots, e_s$ *are the positive eigenvalues and the corresponding orthonormal eigenvectors of* A, *respectively. Further let*

$$\mathcal{M} = \{M \in \mathcal{M}_k^{\geqq} : \mathrm{rank}\,M = n < s,\ \mathrm{tr}\,M = 1\} .$$

Then, with M^* *defined as*

$$M^* = \sum_{i=1}^{n}\left(\alpha_i + \frac{1}{n}\left(1 - \sum_{j=1}^{n} \alpha_j\right)\right) e_i e_i^\top$$

it holds

$$\mathrm{tr}\,(A - M^*)^2 = \inf_{M \in \mathcal{M}} \mathrm{tr}\,(A - M)^2 .$$

From Lemma 13.1 it follows immediately that in the case $n < j_0$ any design $v_n = (x_1, \ldots, x_n)$ satisfying the equation system

$$\frac{1}{n}\sum_{i=1}^{n} x_i x_i^\top = \sum_{i=1}^{n} \bar{m}_i^* u_i u_i^\top ,$$

where

$$\bar{m}_i^* = m_i^* + \frac{1}{n}\left(1 - \sum_{j=1}^{n} m_j^*\right), \qquad (i = 1, \ldots, n),$$

yields a solution to the minimum problem (13.19). Obviously, this is accomplished by choosing the experimental points as columns of

$$F(v_n)^\top = \sqrt{n}\, U_n \operatorname{diag}\left(\sqrt{\bar{m}_1^*}, \ldots, \sqrt{\bar{m}_n^*}\right) \Omega_n^\top ,$$

where $U_n = (u_1, \ldots, u_n)$ contains the first n columns of U and Ω_n is an orthogonal $(n \times n)$-matrix such that

$$n(\Omega_n \circ \Omega_n)(\bar{m}_1^*, \ldots, \bar{m}_n^*)^\top = 1_n .$$

Again, since $(\bar{m}_1^*, \ldots, \bar{m}_n^*)\, 1_n = 1$, such a matrix Ω_n can easily be obtained on the basis of Hadamard matrices as described before (see formulae (13.16) and (13.18)).

The inhomogeneous model

For the inhomogeneous model case

$$\mathsf{E}\tilde{y}(x) = \vartheta_0 + \boldsymbol{\vartheta}^\top x, \qquad x \in X_{\mathrm{E}} = \{x \in \mathbb{R}^k : x^\top x \leqq 1\}$$

$$\mathsf{Cov}((\tilde{\vartheta}_0, \tilde{\boldsymbol{\vartheta}}^\top)^\top \mid w) = \frac{1}{w}\Phi_0 = \frac{1}{w}\left(\begin{array}{c|c} \varphi_0 & \boldsymbol{\varphi}^\top \\ \hline \boldsymbol{\varphi} & \Phi \end{array}\right),$$

where $\boldsymbol{\varphi} = (\varphi_{01}, \ldots, \varphi_{0k})^\top = (\mathsf{Cov}(\tilde{\vartheta}_0, \tilde{\vartheta}_1), \ldots, \mathsf{Cov}(\tilde{\vartheta}_0, \tilde{\vartheta}_k))^\top$ and $\varphi_0 = \operatorname{Var}\tilde{\vartheta}_0$, the design matrix of the exact design $v_n = (x_1, \ldots, x_n) \in X_{\mathrm{E}}^n$ takes the form

$$F_0(v_n) = (1_n \,\vdots\, F(v_n))$$

with $F(v_n)$ as before. Then, according to the result formulated in Theorem 12.9, v_n is A_{B}-, D_{B}- and E_{B}-optimal if, additionally to (13.13), it holds

$m(v_n) = \frac{1}{n}\sum_{i=1}^{n} x_i = -\frac{1}{n}\bar{\boldsymbol{\varphi}}$ where $\bar{\boldsymbol{\varphi}}$ is the subvector of Φ_0^{-1} subdivided as

$$\Phi_0^{-1} = \left(\begin{array}{c|c} \bar{\varphi}_0 & \bar{\boldsymbol{\varphi}}^\top \\ \hline \bar{\boldsymbol{\varphi}} & \bar{\Phi} \end{array}\right).$$

Thus, if v_n satisfies the system

$$\begin{aligned} \frac{1}{n}\sum_{i=1}^{n} x_i x_i^\top &= \sum_{i=1}^{k} m_i^* u_i u_i^\top \\ \sum_{i=1}^{n} x_i &= -\bar{\boldsymbol{\varphi}}, \end{aligned} \qquad (13.20)$$

where now $m_1^*, \ldots, m_k^*$ must be determined according to Algorithm 12.3 with $\boldsymbol{\Phi}^{-1}$ replaced by $\bar{\boldsymbol{\Phi}}$, then v_n is A_B-, D_B- and E_B-optimal in $V_n = X_E^n$.

Let us consider the special case in which there is no prior correlation between $\tilde{\vartheta}_0$ and $\tilde{\vartheta}_i$ $(i = 1, \ldots, k)$, i.e., $\boldsymbol{\varphi} = \mathbf{0}$. Then we have $\bar{\boldsymbol{\Phi}} = \boldsymbol{\Phi}^{-1}$ and the quantities $m_1^*, \ldots, m_k^*$ assume the same values as in the homogeneous case. Whereas optimal discrete designs exist whatever the values of $m_1^*, \ldots, m_k^*$ and n, the existence of an exact design $v_n = (\boldsymbol{x}_1, \ldots, \boldsymbol{x}_n)$ satisfying

$$\frac{1}{n} \sum_{i=1}^{n} \boldsymbol{x}_i \boldsymbol{x}_i^\top = \sum_{i=1}^{k} m_i^* \boldsymbol{u}_i \boldsymbol{u}_i^\top$$
$$\sum_{i=1}^{n} \boldsymbol{x}_i = \mathbf{0} \tag{13.21}$$

depends on its size n. For certain values of n, such designs can be constructed by help of Hadamard matrices.

Theorem 13.4. *Let $\boldsymbol{\varphi} = \mathbf{0}$ and $n > j_0$ such that there exists a Hadamard matrix $\boldsymbol{H}_n$ of order n. Further, let $m_1^* \geqq \ldots \geqq m_{j_0}^* > 0$ be obtained according to Algorithm 12.3 and $\boldsymbol{U}_{j_0} = (\boldsymbol{u}_1 \ldots \boldsymbol{u}_{j_0})$ be the matrix of eigenvectors corresponding to the smallest eigenvalues $\lambda_1 \leqq \ldots \leqq \lambda_{j_0} \leqq \ldots \leqq \lambda_k$ of $\boldsymbol{\Phi}^{-1}$. Then the columns of*

$$\boldsymbol{F}(v_n)^\top = \boldsymbol{U}_{j_0} \operatorname{diag}\left(\sqrt{m_1^*}, \ldots, \sqrt{m_{j_0}^*}\right) \boldsymbol{H}_{n,j_0}^\top \tag{13.22}$$

form an A_B-, D_B- and E_B-optimal design v_n, where $\boldsymbol{H}_{n,j_0}$ arises from $\boldsymbol{H}_n$ by deleting $(n - j_0)$ columns such that $\boldsymbol{H}_{n,j_0}^\top \mathbf{1}_n = \mathbf{0}$.

Proof. First we observe that $v_n = (\boldsymbol{x}_1, \ldots, \boldsymbol{x}_n)$ according to (13.22) represents an exact design, i.e., $\boldsymbol{x}_i \in X_E$ $(i = 1, \ldots, n)$, for we can write

$$\boldsymbol{F}(v_n)^\top = \sqrt{n}\, \boldsymbol{U}_{j_0} \operatorname{diag}\left(\sqrt{m_1^*}, \ldots, \sqrt{m_{j_0}^*}\right) \boldsymbol{\Omega}^\top$$

with $\boldsymbol{\Omega} = n^{-1/2} \boldsymbol{H}_{n,j_0}$ satisfying $n(\boldsymbol{\Omega} \circ \boldsymbol{\Omega})(m_1^*, \ldots, m_{j_0}^*)^\top = \mathbf{1}_n$ so that $\boldsymbol{x}_1^\top \boldsymbol{x}_1 = \ldots = \boldsymbol{x}_n^\top \boldsymbol{x}_n = 1$ (see the proof of Theorem 13.3).

Further, from (13.22) it follows that $\frac{1}{n} \boldsymbol{F}(v_n)^\top \boldsymbol{F}(v_n) = \boldsymbol{U}_{j_0} \boldsymbol{M}^* \boldsymbol{U}_{j_0}^\top$ with $\boldsymbol{M}^* = \operatorname{diag}(m_1^*, \ldots, m_{j_0}^*)$, i.e., the first equation in (13.21) is satisfied. Since we can always find a Hadamard matrix $\boldsymbol{H}_n = (\boldsymbol{h}_1 \ldots \boldsymbol{h}_n)$ containing the column $\boldsymbol{h}_1 = \mathbf{1}_n$ and for which the components of the remaining columns sum up to zero, i.e., $\boldsymbol{h}_2^\top \mathbf{1}_n = \ldots = \boldsymbol{h}_n^\top \mathbf{1}_n = 0$, any matrix $\boldsymbol{H}_{n,j_0}$ obtained by deleting $\boldsymbol{h}_1$ and $(n - j_0 - 1)$ arbitrary further columns from $\boldsymbol{H}_n$ assures that $\boldsymbol{H}_{n,j_0}^\top \mathbf{1}_n = \mathbf{0}$. Then it holds

$$\sum_{i=1}^{n} \boldsymbol{x}_i = \boldsymbol{F}(v_n)^\top \mathbf{1}_n = \boldsymbol{U}_{j_0} \operatorname{diag}\left(\sqrt{m_1^*}, \ldots, \sqrt{m_{j_0}^*}\right) \boldsymbol{H}_{n,j_0}^\top \mathbf{1}_n = \mathbf{0},$$

i.e., the second equation in (13.21) is satisfied too, which concludes the proof. □

Example 13.3. Consider again the three-factor model from Example 12.2, but now with an additional absolute term,

$$\mathbf{E}\tilde{y}(x) = \vartheta_0 + \vartheta_1 x_1 + \vartheta_2 x_2 + \vartheta_3 x_3, \qquad \sum_{i=1}^{3} x_i^2 \leqq 1,$$

$\mathbf{Cov}((\tilde{\vartheta}_1, \tilde{\vartheta}_2, \tilde{\vartheta}_3)^\top | w) = \dfrac{1}{w}\boldsymbol{\Phi}$ as before and

$$\mathbf{Cov}(\tilde{\vartheta}_0, \tilde{\vartheta}_i | w) = 0, \qquad i = 1, 2, 3.$$

Assume that $n = 4$ observations have to be made.

Since the absolute term is not correlated with the remaining coefficients, we have $\bar{\boldsymbol{\Phi}} = \boldsymbol{\Phi}^{-1}$ and λ_i, $\boldsymbol{u}_i$ $(i = 1, 2, 3)$ as before so that Algorithm 12.3 yields the following scheme

$$\frac{1}{n}\lambda_2 = 0.3201 < 1 + \frac{1}{n}\lambda_1,$$

$$\frac{1}{n}\lambda_3 = 1.0923 > \frac{1}{2}\left(1 + \frac{1}{n}(\lambda_1 + \lambda_2)\right) = 0.7145, \qquad j_0 = 2,$$

$$m_1^* = 0.7145 - \frac{1}{n}\lambda_1 = 0.6056, \qquad m_2^* = 0.7145 - \frac{1}{n}\lambda_2 = 0.3944, \qquad m_3^* = 0.$$

Thus, with $\boldsymbol{H}_4$ and the truncated version $\boldsymbol{H}_{4,2}$ chosen as

$$\boldsymbol{H}_4 = \begin{pmatrix} 1 & 1 & 1 & 1 \\ 1 & 1 & -1 & -1 \\ 1 & -1 & 1 & -1 \\ 1 & -1 & -1 & 1 \end{pmatrix}, \qquad \boldsymbol{H}_{4,2} = \begin{pmatrix} 1 & 1 \\ -1 & -1 \\ 1 & -1 \\ -1 & 1 \end{pmatrix}$$

the A_{B}-, D_{B}- and E_{B}-optimal design takes the form

$$\boldsymbol{F}(v_n)^\top = (\boldsymbol{u}_1 \vdots \boldsymbol{u}_2)\,\mathrm{diag}\left(\sqrt{m_1^*}, \sqrt{m_2^*}\right)\boldsymbol{H}_{4,2}^\top$$

$$= \begin{pmatrix} 0.4082 & -0.8944 \\ 0.8165 & 0.4472 \\ 0.4082 & 0 \end{pmatrix} \begin{pmatrix} 0.7782 & 0 \\ 0 & 0.6280 \end{pmatrix} \begin{pmatrix} 1 & -1 & 1 & -1 \\ 1 & -1 & -1 & 1 \end{pmatrix}$$

$$= \begin{pmatrix} -0.2440 & 0.2440 & 0.8793 & -0.8793 \\ 0.9162 & -0.9162 & 0.3546 & -0.3546 \\ 0.3177 & -0.3177 & 0.3177 & -0.3177 \end{pmatrix}.$$

As already mentioned, the construction methods applied here essentially make use of the invariance of $X_{\mathrm{E}} = \{\boldsymbol{x} \in \mathbb{R}^k \colon \boldsymbol{x}^\top \boldsymbol{x} \leqq 1\}$ with respect to orthogonal transformations, and a transfer of the methods to other experimental regions, for example, polyhedra, seems not to be obvious. However, for the case that all the coefficients in our inhomogeneous model are uncorrelated with each other, we can obtain exact optimal designs on the hypercube directly from Hadamard matrices.

Theorem 13.5. *Let $n > k$, $\boldsymbol{H}_n$ some Hadamard matrix of order n and $\boldsymbol{H}_{n,k}$ a truncated matrix obtained from $\boldsymbol{H}_n$ by deleting $n-k$ columns such that $\boldsymbol{H}_{n,k}^\top \mathbf{1}_n = \mathbf{0}$. Then the columns of $\boldsymbol{F}(v_n)^\top = \boldsymbol{H}_{n,k}^\top$ form an exact A_B-, D_B- and E_B-optimal design of size n for the model*

$$\begin{aligned} &\mathbf{E}\tilde{y}(\boldsymbol{x}) = \vartheta_0 + \boldsymbol{\vartheta}^\top \boldsymbol{x}, \qquad \boldsymbol{x} \in X_E = [-1,1]^k, \\ &\mathbf{Cov}(\tilde{\vartheta}_i, \tilde{\vartheta}_j) = 0, \qquad i,j = 0,\ldots,k, \qquad i \neq j. \end{aligned} \tag{13.23}$$

Proof. With $\boldsymbol{F}(v_n) = \boldsymbol{H}_{n,k}$ the experimental points of v_n have components which are either $+1$ or -1. Observing that $\boldsymbol{H}_{n,k}^\top = \mathbf{0}$, v_n has the information matrix

$$\boldsymbol{M}_0(v_n) = \frac{1}{n} \boldsymbol{F}_0(v_n)^\top \boldsymbol{F}_0(v_n) = (\mathbf{1}_n \vdots \boldsymbol{F}(v_n))^\top (\mathbf{1}_n \vdots \boldsymbol{F}(v_n))$$

$$= \frac{1}{n} \left(\begin{array}{c|c} n & \mathbf{1}_n^\top \boldsymbol{H}_{n,k} \\ \hline \boldsymbol{H}_{n,k}^\top \mathbf{1}_n & \boldsymbol{H}_{n,k}^\top \boldsymbol{H}_{n,k} \end{array} \right) - \left(\begin{array}{c|c} 1 & \mathbf{0}^\top \\ \hline \mathbf{0} & \boldsymbol{I}_k \end{array} \right) - \boldsymbol{I}_{k+1}.$$

The assertion then follows from Theorem 12.10. □

Thus, if n is of the form $n = 2^{k-p}$ with some $p \in \{0,\ldots,k\}$ then the well-known (fractional) factorial designs are A_B-, D_B- and E_B-optimal for the model (13.23). These designs are also A-, D- and E-optimal for the least squares estimator. In other words, referring to the determinant, trace and eigenvalue criteria and a cuboidal experimental region, the optimal designs for the least squares and Bayes estimator of a multiple linear regression equation coincide if, in the Bayesian model, the coefficients can be considered uncorrelated.

Finally, let us consider again the special case of simple linear regression

$$\mathbf{E}\tilde{y}(x) = \vartheta_0 + \vartheta_1 x, \qquad x \in X_E = [-1,1]$$

$$\mathbf{Cov}((\tilde{\vartheta}_0, \tilde{\vartheta}_1)^\top | w) = \frac{1}{w} \boldsymbol{\Phi}_0, \qquad \boldsymbol{\Phi}_0^{-1} = \begin{pmatrix} \bar{\varphi}_0 & \bar{\varphi}_{01} \\ \bar{\varphi}_{01} & \bar{\varphi}_1 \end{pmatrix}.$$

Whereas in the discrete case we had coincidence of A_B-, D_B- and E_B-optimal designs (see Theorem 12.12), this must not generally be true for the corresponding exact optimal designs.

Theorem 13.6. *Assume that $\bar{\varphi}_{01} = 0$.*

(i) *If n is even then the design v_n requiring $\frac{n}{2}$ observations both at -1 and at $+1$ is A_B-, D_B- and E_B-optimal.*

(ii) *Let n be odd. Then the design v_n requiring $\frac{n-1}{2}$ observations at -1* (or

+1) and $\frac{n+1}{2}$ *observations at* $+1$ *(or* -1*) is* D_B*-optimal and, provided that* $\bar{\varphi}_1 \leqq \bar{\varphi}_0(\bar{\varphi}_0 + 2n - 1) + (n-1)^2 =: c_n$*, also* A_B*-optimal. If* $\bar{\varphi}_1 > c_n$ *then the* A_B*-optimal design is to take* $\frac{n-1}{2}$ *observations both at* -1 *and at* $+1$ *and one observation at the origin.*

(See NÄTHER/PILZ (1980), GLADITZ/PILZ (1982b).)

In other words, if there is no prior correlation between $\tilde{\vartheta}_0$ and $\tilde{\vartheta}_1$ then the D_B-optimal design coincides with the classical D-optimal design for the regression line whereas the A_B- and A-optimal designs coincide only in case that $\bar{\varphi}_1 \leqq c_n$ which will but mostly be satisfied except when the prior variances of $\tilde{\vartheta}_0$ and $\tilde{\vartheta}_1$ are extremely unbalanced.

If $\tilde{\vartheta}_0$ and $\tilde{\vartheta}_1$ are correlated, i.e., $\bar{\varphi}_{01} \neq 0$, then it can easily be verified that the D_B-optimal design is to take n_0 observations at -1 and $n - n_0$ observations at $+1$ where n_0 is to be chosen such that

$$n_0(n - n_0) + \bar{\varphi}_{01} n_0 \overset{!}{=} \max_{n_0 \in \{0, \ldots, n\}} .$$

Clearly, if $\bar{\varphi}_{01} = 0$ then this is accomplished by choosing the design as indicated in the above theorem. If $\bar{\varphi}_{01} \neq 0$ then the optimal choice of n_0 will be either $[(n + \bar{\varphi}_{01})/2]$ or $[(n + \bar{\varphi}_{01})/2] + 1$, provided that $|\bar{\varphi}_{01}| < n$. The A_B-optimal design, however, is to take n_1 observations at -1, n_2 observations at $+1$ and $n_0 \leqq 1$ additional observation at some point $x_0 \in X_E = [-1, 1]$ such that

$$\operatorname{tr} M_B(v_n)^{-1} = \frac{n + \bar{\varphi}_0 + m_2 + \bar{\varphi}_1}{(n + \bar{\varphi}_0)(m_2 + \bar{\varphi}_1) - (m_1 + \bar{\varphi}_{01})^2}, \text{ where } \quad \begin{matrix} m_1 = n_0 x_0 + n_2 - n_1, \\ m_2 = n_0 x_0^2 + n - n_0 \end{matrix}$$

becomes a minimum under all designs for which

$$|m_1 + \bar{\varphi}_{01}| \leqq 1 \quad \text{and} \quad n_0 \in \{0, 1\}, \quad n_1, n_2 \in \{0, \ldots, n\}, \quad n_0 + n_1 + n_2 = n.$$

In any case, if $|\bar{\varphi}_{01}| \geqq n$ then the A_B-, D_B- and E_B-optimal design requires to take all of the n observations at the same experimental point, namely at -1 if $\bar{\varphi}_{01} \geqq n$ and at $+1$ if $\bar{\varphi}_{01} \leqq -n$ (see PILZ (1981b)).

13.3. Iteration procedures

In this section we formulate iteration procedures for the construction of approximately optimal exact designs which are based mainly on results by FEDOROV (1972) and WYNN (1972). Contrary to the iteration procedures for the construction of optimal discrete designs, now, in general, we cannot prove convergence to the functional value of an exact optimal design. However, the iteration assures a stepwise improvement of a given starting design, i.e., the sequence of functional values decreases monotonically. The iteration then can be stopped if, at a certain stage $s_0 \geqq 1$, a preassigned value of efficiency $e_0 \in (0,1)$ is attained.

First we formulate the iteration procedure for a general design functional Z of the Bayesian information matrix and then we indicate corresponding specializations to the case of L_B- and D_B-optimality.

In the following denote

$$v_{n,s} + (x_{n+1,s}) = (x_{1,s}, \ldots, x_{n,s}, x_{n+1,s})$$

the design which arises by adding some point $x_{n+1,s} \in X_E$ to an exact design $v_{n,s} = (x_{1,s}, \ldots, x_{n,s})$ at some stage $s \geqq 1$ of the iteration and

$$v^j_{n,s} = v_{n+1,s} - (x_{j,s}) := (x_{1,s}, \ldots, x_{j-1,s}, x_{j+1,s}, \ldots, x_{n+1,s})$$

the design obtained from $v_{n+1,s}$ by deleting the point $x_{j,s}$.

Algorithm 13.1. Construction of approximately Z-optimal exact designs

Step 1. Choose some initial design $v_{n,1} = (x_{1,1}, \ldots, x_{n,1}) \in X^n_E$ of size n.

Step 2. Beginning with $s = 1$, form the design $v_{n+1,s} = v_{n,s} + (x_{n+1,s})$ by adding the point

$$x_{n+1,s} = \arg \inf_{x \in X_E} Z(M_B(v_{n,s} + (x))) \tag{13.24}$$

to $v_{n,s}$. Then form $v^{j^*}_{n,s} = v_{n+1,s} - (x_{j^*,s})$ by deleting that point $x_{j^*,s}$ from $v_{n+1,s}$ for which

$$Z(M_B(v^{j^*}_{n,s})) = \min_{j \in \{1,\ldots,n+1\}} Z(M_B(v^j_{n,s})). \tag{13.25}$$

Step 3. Calculate $e_{0,s} := 1 + d_Z(v^{j^*}_{n,s})/Z(M_B(v^{j^*}_{n,s}))$. If $e_{0,s} \geqq e_0$ then $v^{j^*}_{n,s}$ is Z-e_0-optimal, otherwise continue the iteration according to Step 2 with s and $v_{n,s}$ replaced by $s+1$ and $v_{n,s+1} := v^{j^*}_{n,s}$, respectively.

We note that the point $x_{n+1,s}$ to be added to $v_{n,s}$ at stage $s = 1, 2, \ldots$ is chosen such that it effects a maximal improvement of the functional value,

$$Z(M_B(v_{n,s})) - Z(M_B(v_{n+1,s})) = \sup_{x \in X_E} \{Z(M_B(v_{n,s})) - Z(M_B(v_{n,s} + (x)))\},$$

whereas the choice of the point $x_{j^*,s}$ to be deleted is such that the increase in the functional value, which necessarily results from deleting an experimental point, will be minimal.

Theorem 13.7. *For any sequence $\{v_{n,s}\}_{s \in \mathbf{N}}$ of exact designs generated according to Algorithm 13.1, the sequence $\{Z(M_B(v_{n,s}))\}_{s \in \mathbf{N}}$ of functional values decreases monotonically and it holds*

$$\lim_{s \to \infty} Z(M_B(v_{n,s})) \geqq \inf_{v_n \in V_n} Z(M_B(v_n))$$

(cp. Fedorov (1972)).

For the important special cases of L_B- and D_B-optimality the optimization problems to be solved in Step 2 of Algorithm 13.1 can be conveniently reformulated as follows.

L_B-optimality: Observing that

$$\boldsymbol{M}_B(v_{n,s}+(\boldsymbol{x}))^{-1}=\left\{\frac{n}{n+1}\boldsymbol{M}_B(v_{n,s})+\frac{1}{n+1}\boldsymbol{f}(\boldsymbol{x})\boldsymbol{f}(\boldsymbol{x})^\top\right\}^{-1}$$
$$=\frac{n+1}{n}\left\{\boldsymbol{M}_B(v_{n,s})^{-1}-\frac{\boldsymbol{M}_B(v_{n,s})^{-1}\boldsymbol{f}(\boldsymbol{x})\boldsymbol{f}(\boldsymbol{x})^\top\boldsymbol{M}_B(v_{n,s})^{-1}}{n+\boldsymbol{f}(\boldsymbol{x})^\top\boldsymbol{M}_B(v_{n,s})^{-1}\boldsymbol{f}(\boldsymbol{x})}\right\} \tag{13.26}$$

(see A.34), choose the point to be added according to

$$\boldsymbol{x}_{n+1,s}=\arg\sup_{x\in X_E}\frac{\boldsymbol{f}(\boldsymbol{x})^\top\boldsymbol{M}_B(v_{n,s})^{-1}\boldsymbol{U}_L\boldsymbol{M}_B(v_{n,s})^{-1}\boldsymbol{f}(\boldsymbol{x})}{n+\boldsymbol{f}(\boldsymbol{x})^\top\boldsymbol{M}_B(v_{n,s})^{-1}\boldsymbol{f}(\boldsymbol{x})}. \tag{13.27}$$

Furthermore, it holds

$$\boldsymbol{M}_B(v^j_{n,s})^{-1}=\left\{\frac{n+1}{n}\boldsymbol{M}_B(v_{n+1,s})-\frac{1}{n}\boldsymbol{f}(\boldsymbol{x}_{j,s})\boldsymbol{f}(\boldsymbol{x}_{j,s})^\top\right\}^{-1}$$
$$=\frac{n}{n+1}\left\{\boldsymbol{M}_B(v_{n+1,s})^{-1}\right. \tag{13.28}$$
$$\left.+\frac{\boldsymbol{M}_B(v_{n+1,s})^{-1}\boldsymbol{f}(\boldsymbol{x}_{j,s})\boldsymbol{f}(\boldsymbol{x}_{j,s})^\top\boldsymbol{M}_B(v_{n+1,s})^{-1}}{n+1-\boldsymbol{f}(\boldsymbol{x}_{j,s})^\top\boldsymbol{M}_B(v_{n+1,s})^{-1}\boldsymbol{f}(\boldsymbol{x}_{j,s})}\right\}$$

and therefore we have to delete the point $\boldsymbol{x}_{j^*,s}$ for which

$$j^*=\arg\min_{1\leq j\leq n+1}\frac{\boldsymbol{f}(\boldsymbol{x}_{j,s})^\top\boldsymbol{M}_B(v_{n+1,s})^{-1}\boldsymbol{U}_L\boldsymbol{M}_B(v_{n+1,s})^{-1}\boldsymbol{f}(\boldsymbol{x}_{j,s})}{n+1-\boldsymbol{f}(\boldsymbol{x}_{j,s})^\top\boldsymbol{M}_B(v_{n+1,s})^{-1}\boldsymbol{f}(\boldsymbol{x}_{j,s})}.$$

D_B-optimality: By virtue of A.35, it follows from (13.26) and (13.28) that

$$\det\boldsymbol{M}_B(v_{n,s}+(\boldsymbol{x}))=\left(\frac{n}{n+1}\right)^r\left(1+\frac{1}{n}\boldsymbol{f}(\boldsymbol{x})^\top\boldsymbol{M}_B(v_{n,s})^{-1}\boldsymbol{f}(\boldsymbol{x})\right)\det\boldsymbol{M}_B(v_{n,s}),$$

$$\det\boldsymbol{M}_B(v^j_{n,s})=\left(1-\frac{1}{n+1}\boldsymbol{f}(\boldsymbol{x}_{j,s})^\top\boldsymbol{M}_B(v_{n+1,s})^{-1}\boldsymbol{f}(\boldsymbol{x}_{j,s})\right)\det\left(\frac{n+1}{n}\boldsymbol{M}_B(v_{n+1,s})\right).$$

Therefore, the points $\boldsymbol{x}_{n+1,s}$ and $\boldsymbol{x}_{j^*,s}$ to be added and deleted, respectively, have to be chosen such that

$$\boldsymbol{x}_{n+1,s}=\arg\sup_{x\in X_E}\boldsymbol{f}(\boldsymbol{x})^\top\boldsymbol{M}_B(v_{n,s})^{-1}\boldsymbol{f}(\boldsymbol{x}) \tag{13.29}$$

and

$$j^*=\arg\min_{1\leq j\leq n+1}\boldsymbol{f}(\boldsymbol{x}_{j,s})^\top\boldsymbol{M}_B(v_{n+1,s})^{-1}\boldsymbol{f}(\boldsymbol{x}_{j,s}).$$

As in the case of the iteration procedures for the construction of discrete optimal designs, the choice of an initial design does not cause any difficulties. It need not be regular and we can proceed such that, starting with some v_{n_0} of size $n_0 < n$ (possibly Z-optimal within V_{n_0}), we choose $(n - n_0)$ additional points by stepwise application of (13.24) and take the resulting sum of these points and v_{n_0} as an initial design for the iteration. Particularly, we can start with a single-point design $v_1 = (\boldsymbol{x}_1)$ taken to be Z-optimal within the class of exact designs $v = (\boldsymbol{x})$, $\boldsymbol{x} \in X_{\mathrm{E}}$, of size $n_0 = 1$. We formalize this approach in the following algorithm, where

$$\boldsymbol{M}_{\mathrm{B}}(\boldsymbol{x}) = \boldsymbol{f}(\boldsymbol{x})\boldsymbol{f}(\boldsymbol{x})^{\top} + \boldsymbol{\Phi}^{-1}, \qquad \boldsymbol{x} \in X_{\mathrm{E}},$$

denotes the Bayesian information matrix of the single-point design $v = (\boldsymbol{x})$. (Note the difference between $v = (\boldsymbol{x})$ and the discrete one-point design $\delta_{\boldsymbol{x}}$ requiring n observations at the point $\boldsymbol{x}$.)

Algorithm 13.2. Generation of an initial design

Step 1. Choose $\boldsymbol{x}_1 \in X_{\mathrm{E}}$ such that

$$\boldsymbol{x}_1 = \arg \inf_{\boldsymbol{x} \in X_{\mathrm{E}}} Z(\boldsymbol{M}_{\mathrm{B}}(\boldsymbol{x}))$$

and set $v_1 = (\boldsymbol{x})$.

Step 2. Beginning with $i = 1$, find $\boldsymbol{x}_{i+1}$ such that

$$\boldsymbol{x}_{i+1} = \arg \inf_{\boldsymbol{x} \in X_{\mathrm{E}}} Z(\boldsymbol{M}_{\mathrm{B}}(v_i + (\boldsymbol{x})))$$

and form $v_{i+1} = v_i + (\boldsymbol{x}_{i+1})$. Continue with i replaced by $i+1$ until $i + 1 = n$.

Step 3. If $i + 1 = n$ then stop and take $v_{n,1} = (\boldsymbol{x}_1, \ldots, \boldsymbol{x}_n)$ as an initial design.

Particularly, for the criteria of L_{B}- and D_{B}-optimality the determination of the experimental points of the starting design $v_{n,1}$ proceeds as follows.

L_{B}-optimality. Observing that, by virtue of A.34,

$$\operatorname{tr} \boldsymbol{U}_{\mathrm{L}} \boldsymbol{M}_{\mathrm{B}}(\boldsymbol{x})^{-1} = \operatorname{tr} \boldsymbol{U}_{\mathrm{L}} \boldsymbol{\Phi} - \frac{\boldsymbol{f}(\boldsymbol{x})^{\top} \boldsymbol{\Phi} \boldsymbol{U}_{\mathrm{L}} \boldsymbol{\Phi} \boldsymbol{f}(\boldsymbol{x})}{1 + \boldsymbol{f}(\boldsymbol{x})^{\top} \boldsymbol{\Phi} \boldsymbol{f}(\boldsymbol{x})}$$

we have to choose

$$\boldsymbol{x}_1 = \arg \sup_{\boldsymbol{x} \in X_{\mathrm{E}}} \frac{\boldsymbol{f}(\boldsymbol{x})^{\top} \boldsymbol{\Phi} \boldsymbol{U}_{\mathrm{L}} \boldsymbol{\Phi} \boldsymbol{f}(\boldsymbol{x})}{1 + \boldsymbol{f}(\boldsymbol{x})^{\top} \boldsymbol{\Phi} \boldsymbol{f}(\boldsymbol{x})}.$$

The remaining points, in analogy with (13.27), have to be chosen according to

$$\boldsymbol{x}_{i+1} = \arg \sup_{\boldsymbol{x} \in X_{\mathrm{E}}} \frac{\boldsymbol{f}(\boldsymbol{x})^{\top} \boldsymbol{M}_{\mathrm{B}}(v_i)^{-1} \boldsymbol{U}_{\mathrm{L}} \boldsymbol{M}_{\mathrm{B}}(v_i)^{-1} \boldsymbol{f}(\boldsymbol{x})}{i + \boldsymbol{f}(\boldsymbol{x})^{\top} \boldsymbol{M}_{\mathrm{B}}(v_i)^{-1} \boldsymbol{f}(\boldsymbol{x})} \qquad (i = 1, \ldots, n-1).$$

D_{B}-optimality. Observing that, by virtue of A.35,

$$\det \boldsymbol{M}_{\mathrm{B}}(\boldsymbol{x})^{-1} = (1 + \boldsymbol{f}(\boldsymbol{x})^{\top} \boldsymbol{\Phi} \boldsymbol{f}(\boldsymbol{x}))^{-1} \det \boldsymbol{\Phi},$$

we have to choose

$$x_1 = \arg \sup_{x \in X_E} f(x)^\top \Phi f(x).$$

The remaining points, in analogy with (13.29), have to be chosen according to

$$x_{i+1} = \arg \sup_{x \in X_E} f(x)^\top M_B(v_i)^{-1} f(x), \qquad i = 1, \ldots, n-1.$$

Example 13.4. Let us consider a simple linear regression model

$$\mathbf{E}\tilde{y}(x) = \vartheta_0 + \vartheta_1 x, \qquad x \in X_E = [-1, 1],$$

$$\mathbf{Cov}((\tilde{\vartheta}_0, \tilde{\vartheta}_1) \mid w) = \frac{1}{w}\begin{pmatrix} 2 & -1 \\ -1 & 1 \end{pmatrix}.$$

We wish to determine an initial design of size $n = 3$ which can serve as a basis for the construction of an exact A_B-optimal design. The application of Algorithm 13.2 and the above specialization to L_B-optimality with respect to $U_L = I_2$ and $f(x) = (1, x)^\top$ then yields the following scheme:

Step 1. $$\frac{f(x)^\top \Phi^2 f(x)}{1 + f(x)^\top \Phi f(x)} = \frac{5 - 6x + 2x^2}{3 - 2x + x^2},$$

$$x_1 = \arg \sup_{-1 \leq x \leq 1} \frac{5 - 6x + 2x^2}{3 - 2x + x^2} = -1, \qquad v_1 = (-1),$$

$$M_B(v_1) = \begin{pmatrix} 1 & -1 \\ -1 & 1 \end{pmatrix} + \begin{pmatrix} 1 & 1 \\ 1 & 2 \end{pmatrix} = \begin{pmatrix} 2 & 0 \\ 0 & 3 \end{pmatrix}.$$

Step 2. $$\frac{f(x)^\top M_B(v_1)^{-2} f(x)}{1 + f(x)^\top M_B(v_1)^{-1} f(x)} = \frac{9 + 4x^2}{9 + 2x^2},$$

$$x_2 = \arg \sup_{-1 \leq x \leq 1} \frac{9 + 4x^2}{9 + 2x^2} = 1, \qquad v_2 = (-1, 1),$$

$$M_B(v_2) = \frac{1}{2}\begin{pmatrix} 3 & 1 \\ 1 & 4 \end{pmatrix}, \qquad M_B(v_2)^{-1} = \frac{2}{11}\begin{pmatrix} 4 & -1 \\ -1 & 3 \end{pmatrix},$$

$$M_B(v_2)^{-2} = \frac{4}{121}\begin{pmatrix} 17 & -7 \\ -7 & 10 \end{pmatrix},$$

$$\frac{f(x)^\top M_B(v_2)^{-2} f(x)}{2 + f(x)^\top M_B(v_2)^{-1} f(x)} = \frac{17 - 14x + 10x^2}{15 - 2x + 3x^2},$$

$$x_3 = \arg \sup_{-1 \leq x \leq 1} \frac{17 - 14x + 10x^2}{15 - 2x + 3x^2} = -1.$$

Step 3. Initial design $v_{3,1} = (-1, -1, 1)$.

We point out that in most cases the Algorithm 13.2 already yields a very good approximation of a Z-optimal design. Thus, for example, the design $v_3 = (-1, -1, 1)$ just obtained is A_B-optimal since, by (13.7),

$$e_{A_B}(v_3) \geqq 2 - \left\{\frac{1}{3} \operatorname{tr} \boldsymbol{M}_B(v_3)^{-2} \boldsymbol{\Phi}^{-1} + \sup_{-1 \leqq x \leqq 1} \boldsymbol{f}(x)^\top \boldsymbol{M}_B(v_3)^{-2} \boldsymbol{f}(x)\right\} \Big/ \operatorname{tr} \boldsymbol{M}_B(v_3)^{-1}$$

$$= 2 - \left\{\frac{171}{400} + \frac{369}{400}\right\} \Big/ \frac{27}{20} = 1\,.$$

Therefore, it will often suffice to perform an iteration according to Algorithm 13.2 and the original iteration procedure described by Algorithm 13.1 has to be examined only if the efficiency of the initial design is not yet satisfactory.

It is also possible to obtain an initial design $v_{n,1}$ from a given design v_N of size $N > n$ (possibly Z-optimal within V_N). Then $v_{n,1}$ can be taken that design which arises from v_N by deleting those $N - n$ points which keep the resulting increase in the functional value as small as possible. This is just the case if these points are chosen stepwise analogously to (13.25).

14. Further problems

14.1. Bayesian one-point designs

A striking feature of Bayesian experimental design is the existence of optimal one-point designs, i.e., designs whose supports contain only a single point. Such designs have no correspondence in the design problem for the least squares estimator where, except in case of particular extrapolation problems (see, e.g. Elfving (1952)) or truncated optimality criteria (D_A-optimality a.s.o.), the support of the design must include at least as many distinct points as indicated by the number of regression coefficients. The one-point designs are of particular importance since they are exact designs which are easy to implement and which keep the experimental effort minimal. Their application is but crucial in cases where the adequacy of the regression setup is doubtful.

In any case, an optimization within the crude but highly tractable subclass of one-point designs will always provide a benchmark against which to compare competing designs. This way, one-point designing suggests itself for the construction of efficient sequential Bayes estimation and design strategies.

The possibility of the existence of optimal one-point designs arises from the fact that the Bayesian information matrix is positive definite whatever the design. In particular, there is good hope for the optimality of such designs if the prior precision matrix $\boldsymbol{\Phi}^{-1}$ has a convenient structure, for example such that the prior knowledge arises from previous observations with a suitable (almost optimal) "prior" design. In the following we briefly formulate necessary and sufficient conditions for the L_B- and D_B-optimality of one-point designs, for details and proofs we refer to PILZ (1981b).

Let h and φ be functions on $X_E \times X_E$ and X_E, respectively, defined by

$$h(x, v) := f(x)^\top M_B(\delta_v)^{-1} U_L M_B(\delta_v)^{-1} f(x)$$

and (14.1)

$$\varphi(x) := f(x)^\top \Phi f(x).$$

A necessary condition for the L_B-optimality of a design δ_v is that it minimizes $L_B(\delta_x) = \operatorname{tr} U_L M_B(\delta_x)^{-1}$ among all one-point designs δ_x, $x \in X_E$. Applying A.34 to rearrange $M_B(\delta_x)^{-1}$, we get

$$\operatorname{tr} U_L M_B(\delta_x)^{-1} = n \operatorname{tr} U_L \Phi - \frac{n^2}{1 + n\varphi(x)} f(x)^\top \Phi U_L \Phi f(x). \quad (14.2)$$

Then, by virtue of Theorem 11.12, it holds

Theorem 14.1. (i) *The design δ_v can be L_B-optimal in Ξ only if*

$$v = \arg \sup_{x \in X_E} (1 + n\varphi(x))^{-1} f(x)^\top \Phi U_L \Phi f(x). \quad (14.3)$$

(ii) *The design δ_v is L_B-optimal in Ξ if and only if*

$$\sup_{x \in X_E} h(x, v) = h(v, v). \quad (14.4)$$

Thus, a practicable way of verifying the existence of L_B-optimal one-point designs would be to determine experimental points $v \in X_E$ satisfying (14.3) and then to prove if any of these points comes to maximize $h(\cdot, v)$ over X_E. If there is no point $v \in X_E$ for which (14.3) and (14.4) come true simultaneously then there does not exist an L_B-optimal one-point design.

Again, by particular choice of U_L, we may obtain corresponding characterizations of A_B-, B-, C_B- and I_B-optimal one-point designs.

Example 14.1. Let us prove if there exists an A_B-optimal one-point design for the two-factor model

$$\mathsf{E}\tilde{y}(x) = \vartheta_0 + \vartheta_1 x_1 + \vartheta_2 x_2, \qquad x = (x_1, x_2)^\top \in X_E = [-1, 1]^2$$

with prior covariance matrix $\mathsf{Cov}(\tilde{\vartheta} \mid w) = \frac{1}{w} \Phi$ given by

$$\Phi = \begin{pmatrix} 1.00 & 1.25 & -1{,}75 \\ & 2.00 & -2.75 \\ & & 4.00 \end{pmatrix}.$$

Suppose that $n = 4$ observations have to be made. Then, with $U_L = I_3$,

$$\frac{f(x)^\top \Phi^2 f(x)}{1 + n\varphi(x)}$$

$$= \frac{5.625 + 13.125x_1^2 + 26.625x_2^2 + 17.125x_1 - 24.375x_2 - 37.375x_1x_2}{17 + 32x_1^2 + 64x_2^2 + 40x_1 - 56x_2 - 88x_1x_2}$$

is maximized (uniquely) by the point $v = (1, -1)^\top \in X_E$. Observing that $M_B(\delta_v)^{-2} = n^2(1 + n\varphi(v))^{-2}\{(1 + n\varphi(v))\,\Phi - n\Phi f(v) f(v)^\top \Phi\}^2$, we obtain

$$M_B(\delta_v)^{-2} = \begin{pmatrix} 50.284 & 76.725 & -109.350 \\ & 118.537 & -169.036 \\ & & 241.331 \end{pmatrix}.$$

Obviously, $v = (1, -1)^\top$ is also a maximum point of $h(\cdot, v) = f(\cdot)^\top \times M_B(\delta_v)^{-2} f(\cdot)$, i.e., δ_v is A_B-optimal. (δ_v is also D_B-optimal.)

For the special case of C_B-optimality, Chaloner (1982, 1984) has given a characterization of optimal one-point designs. In particular, it is shown there that δ_x with

$$x = \frac{1}{\lambda}(nI_r + \Phi^{-1})^{-1}c, \qquad \lambda = \|(nI_r + \Phi^{-1})^{-1}c\|$$

is C_B-optimal which respect to $c \in \mathbb{R}^r$ if the experimental region X_E is taken to be the unit sphere. This means, especially, that in case of diagonality of Φ the design δ_c is C_B-optimal in Ξ with respect to unit vectors $c = (0, \ldots, 0, 1, 0, \ldots, 0)^\top$ with 1 in the ith co-ordinate ($i = 1, \ldots, r$).

In any case, the existence of a C_B-optimal one-point design can be assured provided the symmetric image of the experimental region satisfies the following convexity condition.

Corollary 14.1. *Let $c \in \mathbb{R}^r$, $c \neq 0$. Provided that*

$$f(X) := \{f(x)\colon x \in X\} \cup \{-f(x)\colon x \in X\}$$

forms a convex set, there always exists $x^* \in X$ such that δ_{x^*} is C_B-optimal in Ξ with respect to c. Moreover, δ_{x^*} is C_B-optimal in Ξ with respect to c if and only if

$$\sup_{x \in X} [f(x)^\top(\Phi + M^*)^{-1}c]^2 = [f(x^*)^\top(\Phi + M^*)^{-1}c]^2,$$

where $M^* = f(x^*) f(x^*)^\top$.

Remark upon the proof. The existence of a C_B-optimal one-point design is guaranteed according to the corollary to Theorem 3 in Chaloner (1984). The above necessary and sufficient optimality condition follows immediately from Theorem 14.1 (ii).

Turning to the D_B-optimality criterion, let us define a function g on $X_E \times X_E$ by

$$\begin{aligned} g(x, v) &:= f(x)^\top M_B(\delta_v)^{-1} f(x) \\ &= n\varphi(x) - n^2(1 + n\varphi(v))^{-1}(f(x)^\top \Phi^{-1} f(v))^2. \end{aligned} \tag{14.5}$$

Then we get from Theorem 11.15 and the fact that $D_B(\delta_x) = \det M_B(\delta_x)^{-1} = (n^r \det \Phi)(1 + n\varphi(x))^{-1}$ the following analogue to Theorem 14.1.

Theorem 14.2. (i) *The design δ_v with $v \in X_E$ can be D_B-optimal only if v is a maximum point of $\varphi(\cdot)$.*

(ii) *The design δ_v with $v \in X_E$ is D_B-optimal in Ξ if and only if*

$$\sup_{x \in X_E} g(x, v) = g(v, v) = 1 - (1 + n\varphi(v))^{-1}.$$

Moreover, it can be shown that the function g satisfies the inequality

$$\sup_{x \in X_E} \inf_{v \in X_E} g(x, v) \leqq \sup_{x \in X_E} g(x, x) \leqq \inf_{v \in X_E} \sup_{x \in X_E} g(x, v)$$

from which it follows that if g has a saddle point $(v, v) \in X_E \times X_E$ then the necessary condition from Theorem 14.2 (i) is also sufficient for D_B-optimality of δ_v (see Pilz (1981 b), Theorem 2).

Now let us consider the multiple linear regression equation

$$\mathbf{E}\tilde{y}(x) = \vartheta_0 x_0 + \sum_{i=1}^{k} \vartheta_i x_i, \qquad x = (x_1, \ldots, x_k)^\top \in X_E \tag{14.6}$$

with the dummy variable $x_0 \in \{0, 1\}$ chosen according to the homogeneous and inhomogeneous case, respectively, and X_E being a compact and convex subregion of $\mathbb{R}^k$. For this setup the D_B-optimality of one-point designs can be proved by help of the following simpler and more tractable condition.

Corollary 14.2. *Denote G_φ and G_v, respectively, the subsets of boundary points of X_E which maximize the functions φ and $g(\cdot, v)$ over X_E. A necessary and sufficient condition for the existence of a D_B-optimal one-point design for the setup* (14.6) *is that for at least one point $v \in G_v$. The design δ_v with $v \in X_E$ is D_B-optimal in Ξ if and only if*

$$v \in G_\varphi \cap G_v.$$

This follows from Theorem 14.2 (ii), observing that with $f(x) = (1, x^\top)^\top$ the functions φ and $g(\cdot, v)$ are convex over X_E.

An example of an one-point design which is D_B-optimal for a three-factor model, whatever the number n of observations to be made, is given in Pilz (1981 b), Section 4.

Concludingly, we state the following result on the existence of optimal one-point designs for the simple linear regression case.

Corollary 14.3. *For the simple linear regression model* (12.51) *there exists an A_B-, D_B- and E_B-optimal one-point design if and only if $|\varphi_{01}| \geqq n \det \Phi_0$. In this case*

$$\delta_v = \begin{cases} \delta_{-1} & \text{if} \quad \varphi_{01} \leqq -n \det \Phi_0 \\ \delta_{+1} & \text{if} \quad \varphi_{01} \geqq n \det \Phi_0 \end{cases}$$

is A_B-, D_B- and E_B-optimal (see Pilz (1981 b), Section 5).

In terms of the prior correlation coefficient $\varrho = \varphi_{01}/(\varphi_0\varphi_1)^{1/2}$ between $\tilde{\vartheta}_0$ and $\tilde{\vartheta}_1$, the existence condition $|\varphi_{01}| \geqq n \det \Phi_0$ can be rewritten as

$|\varrho| \geqq n \det \Phi_0/(\varphi_0\varphi_1)^{1/2}$, or, equivalently,

$$|\varrho|/(1-\varrho^2) \geqq n(\varphi_0\varphi_1)^{1/2}.$$

This means that the existence of an optimal one-point design requires the prior correlation to exceed a certain lower bound depending on the size of the design and the prior variances of the coefficients.

14.2. Cost-optimal designs

The design problem considered in the preceding sections can be modified to include the cost of experimentation. On the one hand, this can be done by restricting the class of designs in such a way that it includes only designs satisfying a certain cost constraint. For example, let $k \mid X_E \to \mathbb{R}^+$ be a function evaluating the costs of observing $\tilde{y}(\boldsymbol{x})$ at the various levels $\boldsymbol{x} \in X_E$ and $c \mid V \to \mathbb{R}^+$ defined by

$$\forall v_n = (\boldsymbol{x}_1, \ldots, \boldsymbol{x}_n) \in V_n : c(v_n) = \sum_{i=1}^{n} k(\boldsymbol{x}_i)$$

the function measuring the total costs of observation at the points of an exact experimental design. Setting forth this function to some class $\bar{\Xi} \subseteqq \Xi$ of continuous designs, we have

$$c(\xi) = \int_{X_E} k(\boldsymbol{x})\, \xi(\mathrm{d}\boldsymbol{x}), \qquad \xi \in \bar{\Xi}. \tag{14.7}$$

Given that $\bar{\Xi}$ is a convex set of designs, it is clear that the class

$$\bar{\Xi}_c = \{\xi \in \bar{\Xi} : c(\xi) \leqq c_0\}, \qquad c_0 \in \mathbb{R}^+$$

of cost-constrained designs is convex, too. Hence, all the results on the construction of optimal designs obtained before can be transferred to include the cost of experimentation when replacing $\bar{\Xi}$ by $\bar{\Xi}_c$. On the other hand, the costs could also be incorporated directly into the design procedure by adding the cost function c to the loss function L_η (cp. Section 2.1 and Remark 5.2). This leads us to the minimization of the functional

$$Z_c(\xi) = Z(\xi) + c(\xi)$$

over $\bar{\Xi}$, where Z is some general design functional as considered before. It follows immediately that, given the convexity of $Z(\cdot)$ and c chosen according to (14.7), the functional Z_c is also convex. Thus, from Theorem 11.11 we can derive a necessary and sufficient condition for a design to minimize this functional.

Theorem 14.3. *Let $\bar{\Xi}$ be a convex set of designs, c as defined by* (14.7) *and $Z(\cdot)$ a convex functional on $\bar{\Xi}$. Further, let the Assumptions* 11.1 *and* 11.2 *be satisfied and assume the directional derivative Δ_Z to be linear in the sense of De-*

finition 11.4. *Then it holds*

$$Z_c(\xi^*) = \inf_{\xi \in \bar{\Xi}} Z_c(\xi) \Leftrightarrow \inf_{x \in X_E} \{\Delta_Z(\xi^*, \delta_x) + k(x)\} = c(\xi^*)\,.$$

Proof. Let be $\xi, \bar{\xi} \in \bar{\Xi}$, $\alpha \in (0,1)$ and $\xi_\alpha = (1-\alpha)\,\xi + \alpha\bar{\xi}$. Then, according to (11.6), we have

$$\begin{aligned}\Delta_{Z_c}(\xi, \bar{\xi}) &= \lim_{\alpha \downarrow 0} \frac{1}{\alpha} \{Z(\xi_\alpha) + c(\xi_\alpha) - (Z(\xi) + c(\xi))\} \\ &= \Delta_Z(\xi, \bar{\xi}) + \lim_{\alpha \downarrow 0} \frac{1}{\alpha} \{c(\xi_\alpha) - c(\xi)\}\,.\end{aligned}$$

Applying L'Hospital's differentiation rule, we get

$$\lim_{\alpha \downarrow 0} \frac{1}{\alpha} \{c(\xi_\alpha) - c(\xi)\} = c(\bar{\xi}) - c(\xi)$$

and thus

$$\Delta_{Z_c}(\xi, \delta_x) = \Delta_Z(\xi, \delta_x) + k(x) - c(\xi)$$

for arbitrary $x \in X_E$. Since Δ_Z is linear, so is Δ_{Z_c} and the assertion follows from Theorem 11.11 (i). □

Inserting the special functionals L_B, B, A_B a.s.o., Theorem 14.3 together with the equivalence theorems formulated in Section 11.3 yields the corresponding conditions on cost-optimal designs. This has been detailed to some extent in Tuchscherer (1983). Moreover, for the simple linear regression model, Brooks (1972) has given optimal designs under various cost functions, where he proceeds with an ignorance prior distribution and his approach is different from the above and does not refer to equivalence theorems.

14.3. Further literature and research

1. In the literature there have also been developed design criteria which do not arise from the minimization of the risk function of some estimator. In particular, interest has been focused on criteria which are based on some measure of information about the regression coefficients which comes available after observation at the points of an experimental design (see Lindley (1956), Stone (1959), Mallows (1959)). These authors propose to design the experiment such that it maximizes the expected change in entropy of the posterior distribution compared to the prior distribution. This is equivalent to minimization of the expected entropy of the posterior distribution and is usually referred to as the criterion of S-optimality. Lindley (1956) gives a comparison of this criterion with Blackwell's information criteria (see Blackwell (1951), (1953)).

For the special case of normal error distributions with known variance and a normal prior distribution for the regression parameter the criterion of S-optimality is equivalent to D_B-optimality (see Stone (1959), Bandemer et al. (1977, Section 7.4)). Moreover, a decision theoretic interpretation of D_B-optimality in terms of expected information has been given by Bernardo (1979).

2. Lindley (1968) and Brooks (1972), (1974) developed a Bayesian approach to the prediction problem for the response surface in case of vague prior knowledge and consider the choice of which explanatory variables to include in the model. The resulting design problems have a similar structure as those considered here. SINHA (1970) deals with the problem of finding optimal weights for a fixed set of experimental points (finite cardinality) in a homogeneous multiple linear regression model. Optimal weights are given with respect to the usual ignorance prior distribution and the design criteria of A_B- and D_B-optimality.

3. COVEY-CRUMP and SILVEY (1970) construct D_B- and E_B-optimal designs for homogeneous multiple linear regression with spherical experimental region and prior knowledge given by previous observations. They also discuss the efficiency of designs generated according to Wynn's iteration procedure relative to their exact D_B-optimal designs. For the same model, SILVEY (1969) discusses C_B-optimality when only $n = 1$ additional observation is to be made. GUTTMAN (1971) has shown that Covey-Crump and Silvey's design criterion is equivalent to minimization of the determinant of the preposterior covariance matrix of the regression coefficients based on the ignorance prior distribution for $(\tilde{\vartheta}, \tilde{\sigma})$.

4. In the context of analysis of variance, the criterion of A_B-optimality has first been considered by OWEN (1970) who is concerned with inference about the treatment parameters in a simple no interaction model. SMITH and VERDINELLI (1980) derive D_B-optimal designs for one-way analysis of variance models and polynomial regression under a three-stage hierarchical prior structure. Bayes designs for a two-way analysis of variance model are considered by GIOVAGNOLI and VERDINELLI (1983), in particular they construct D_B- and E_B-optimal designs.

A feasible directions method for computing Bayes E-optimal block designs is given in SIMEONE/VERDINELLI (1988).

5. The criterion of L_B-optimality has been discussed in DUNCAN/DE GROOT (1976), BROOKS (1976), BANDEMER/PILZ (1978), PILZ (1979c, d), ALI/SINGH (1981), GLADITZ/PILZ (1982a, b) and CHALONER (1982, 1984).

DUNCAN and DE GROOT (1976) derived L_B-optimal designs for simple one-way analysis of variance and simple linear regression. BROOKS (1976) is concerned with the construction of optimal designs for inhomogeneous multiple linear regression with spherical experimental region. BANDEMER and PILZ (1978) derived some L_B-optimal designs for arbitrary regression setups and arbitrary experimental region assuming a full rank matrix U_L. In GLADITZ/PILZ (1982a, b) there are developed construction methods for L_B-optimal designs which, under certain conditions, also lead to D_B and E_B-optimal designs. CHALONER (1982, 1984) derives L_B-optimal designs under various specifications of the regression function and the experimental region, in particular she gives C_B-optimal interpolation and extrapolation designs for polynomial regression.

6. The experimental design considerations in Part III are based on the assumptions of a true linear regression setup, homoscedastic and independent errors and quadratic loss. There is clearly a need for design of experiments to investigate the adequacy of these assumptions as well as to estimate the parameters or to predict the expected response. The problem of designing an experiment when the exact form of the model is unknown has been considered by O'HAGAN (1978), in a non-Bayesian context this has been studied e.g. by ATKINSON/FEDOROV (1975), MARCUS/SACKS (1977), COOK/NACHTSHEIM (1982) and SACKS/YLVISAKER (1978), (1984). O'HAGAN introduced a Bayesian localized model in which ϑ is replaced by $\vartheta(x)$ and where $\vartheta(x)$ is assumed to follow a second-order stationary Gaussian process that reflects beliefs about the local stability of the true response function. For such a model, he found that a design criterion based on the (pre-) posterior variance favoured placing more points near the centre of the experimental region when compared with our Bayes design criteria for a fixed model. However, the design problems to be solved have a complicated structure. SMITH and VERDINELLI (1980) adopted a hierarchical Bayesian model incorporating a low-degree polynomial but also reflecting the degree of confidence in the adequacy of the model and studied the effects on the choice of design using the determinant criterion. In PILZ (1981a, Section 6) it is shown that the design procedures developed for the Bayes estimator show a certain robustness when changing from quadratic to arbitrary convex loss functions.

Bayes Theory for finding model robust designs for the estimation of linear functionals of a regression function subject to a stochastic process prior is developed in SACKS/YLVISAKER (1985), where also a connection is made with the minimax theory studied in SACKS/YLVISAKER (1984).

7. Throughout the book, it has been assumed that all the regressor variables are non-stochastic and can be chosen by the experimenter. There may but be situations in which only some of these variables can be controlled whereas the remaining are stochastic. Moreover, it may be possible that the distributions can be chosen (e.g. expectations and certain other moments). Experimental design considerations in this direction may be found in BROOKS (1974) and BANDEMER et al. (1977, Chapter 12).

8. Little is known about experimental design in non-linear situations. Any reasonable approach to handle this problem requires some a priori assumption about the location of the regression parameter. Experimental design for non-linear models using a prior distribution for $\tilde{\vartheta}$ is considered e.g. in BOX/LUCAS (1959), DRAPER/HUNTER (1967), ZACKS (1977), DENISOV (1977), DUBOV (1981), CHALONER (1986), CHALONER/LARNTZ (1986) and BUONACCORSI (1986). But, in any case, further research is needed in this field.

9. Further attention should also deserve the following fields of experimental design, for which, in our opinion, Bayes methods seem to be predestinate or even indispensable:

(i) Experimental design for empirical Bayes estimators and sequential decision problems. Some ideas and first results on designing in the empirical Bayes case may be found in GLADITZ (1981), for results in sequential Bayes design we refer to HECKENDORFF (1982) and CHALONER (1986).

(ii) Experimental design for optimum control situations and for seeking the optimum of a response function. For results in optimum control see BROOKS (1977) and CHANG (1979), for application of Bayes methods in the latter field we refer to MOCKUS (1983), (1989).

(iii) Experimental design for regression problems with correlated errors and for random processes. For a review of results on such problems we refer to NÄTHER (1985), BANDEMER/NÄTHER/PILZ (1987) and YLVISAKER (1987).

IV. Minimax and admissible linear regression estimation with restricted parameter regions

In the preceding parts we have considered various ways and the effects of incorporating prior knowledge about the regression parameters using a Bayesian approach. The most simple and most widespread form of incorporating prior knowledge consists in a restriction of the parameter space. Since the early sixties, regression estimation using additional prior information in form of a restricted parameter space has received increasing interest. The first papers in this area studied linear equality constraints (see, e.g. THEIL/GOLDBERGER (1961), CHIPMAN/RAO (1964)). This development was followed by the study of linear inequality constraints which permits more flexibility in the specification of prior information. The constrained least squares estimators that have been proposed require the numerical solution of quadratic programming problems and cannot be given in a simple closed form analytical representation unless the number of restrictions is very small (see, e.g. ESCOBAR/SCARPNESS (1984) and the references cited there). Moreover, only little is known about the sampling and risk properties of these estimators. There is also a small portion of papers dealing with a Bayesian analysis of parameter constraints, we refer, for example, to O'HAGAN (1973), DAVIS (1978), BUNKE/BUNKE (1986, Section 7.1.4) and BANDEMER/PILZ/FELLENBERG (1986). However, the corresponding Bayes estimates and risks rarely can be given explicitly, their computation requires efficient numerical methods of multidimensional integration. OMAN (1983) derived a maximum likelihood estimator for the regression parameter when there are implicit parameter restrictions due to a bounded response.

In the recent literature, increasing attention was paid to the principle of minimax linear estimation as an important alternative to the above mentioned approaches to incorporating prior knowledge, let us mention, for example, KUKS/OL'MAN (1971), (1972), KUKS (1972), BUNKE (1975 a, b), LÄUTER (1975), HOFFMANN (1979), PEELE/RYAN (1982), BUNKE/MÖHNER (1982), TOUTENBURG (1982), OL'MAN (1983), STAHLECKER (1985), KOZÁK (1985), KRAFFT (1986), PILZ (1984), (1985), (1986 a, b), (1987), (1988 a, b), STAHLECKER/TRENKLER (1987), HERING et al. (1987), ALSON (1988), GAFFKE/HEILIGERS (1989), LAUTERBACH (1989). In the following we deal with the situation that the regression parameter is known to belong to an arbitrary, compact subset $\Theta \subset \mathbb{R}^r$, and determine minimax linear and minimax affine estimators. Such estimators are attractive for several reasons: First, they represent globally robust Bayes estimators (cp. Section 6.3), second, they provide admissible and even optimal linear (affine) improvements, respectively, over the LSE and, finally, they do not require further specifications with regard to prior knowledge and the underlying distribution of the observations. The second point yields a strong impetus for developing minimax linear estimators, since the LSE and a number of well-known and widely used alternatives to the LSE are inadmissible, even within the class of linear estimators, when we are faced with a restricted compact parameter region (see MARQUARDT (1970), HOFFMANN (1977), (1980) and TERÄSVIRTA (1983)).

Up to now, explicit forms for minimax linear estimators were only available in some particular cases, essentially for ellipsoidal regions and special types of quadratic loss and experimental design matrices, see KUKS/OL'MAN (1972), LÄUTER (1975), HOFFMANN (1979), GAFFKE/HEILIGERS (1987) and ALSON (1988), who also gives minimax solutions for special convex polyhedral regions. Approximate solutions for the special case of ellipsoid restrictions, partly in combination with additional equality restrictions were given by TOUTENBURG (1982), OL'MAN (1983), STAHLECKER (1985), SCHIPP et al. (1985), KOZÁK (1985), KRAFFT (1986), STAHLECKER/TRENKLER (1987), HERING et al. (1987), GAFFKE/MATHAR (1988) and LAUTERBACH (1989). For symmetric compact parameter regions, KUKS (1972) developed an iterative minimax estimation procedure.

In the following chapter we will give explicit form representations for minimax linear and affine estimators in case of an arbitrary compact parameter region. Our approach is based on the well-known relationship between minimax estimation and Bayes estimation with respect to least favourable prior distributions. We establish a fundamental duality between the problems of finding a least favourable prior distribution over a restricted parameter set and of optimum experimental design for Bayes (affine-) linear estimators which allows the successfull application of ideas and methods of Bayesian experimental design layed out in Part III to the construction of minimax (affine-) linear estimators. An essential element of this duality relation is the interpretation of the second order moment matrix of a prior distribution on the parameter region as the information matrix of an experimental design measure on this region.

In this chapter we give characterizations of minimax linear and minimax affine estimators as well as methods for their construction both for the case of symmetric compact and arbitrary compact parameter regions, using the theory of Bayes (affine-) linear estimation and experimental design as a key methodological tool. For the important special cases of ellipsoid and linear inequality constraints we develop estimators in an explicit form. In Chapter 16 we give characterizations of admissible linear and affine-linear estimators under restricted compact parameter regions, sharpening thereby RAO's (1976) results for unconstrained Θ, and we indicate conditions under which admissible (affine-) linear estimators represent universal improvements over the LSE. Finally, in Chapter 17, we extend our results to the problem of minimax estimation in inadequate models and determine minimax (affine-) linear approgression estimators for the unknown response surface.

Throughout the following chapters, we assume the error covariance matrix $\mathbf{Cov}\,\tilde{e} = \Sigma$ to be known and write shortly $L(\vartheta, \hat{\vartheta}(y))$ and $R(\vartheta, \hat{\vartheta})$ instead of $L(\vartheta, w; \hat{\vartheta}(y))$ and $R(\vartheta, w; \hat{\vartheta})$, respectively. Our results can be easily generalized to the case of $\mathbf{Cov}\,\tilde{e} \leqq \Sigma$ with some known upper bound Σ (see Section 6.3), for some initial findings on using estimated variances in Bayes linear estimators we also refer to GOLDSTEIN (1983).

15. Minimax linear and affine estimation in true models

15.1. Minimax linear estimators

15.1.1. Characterization of minimax linear estimators

Consider the linear regression model

$$\mathbf{E}\tilde{y} = F\vartheta, \quad \mathbf{Cov}\,\tilde{e} = \Sigma \in \mathcal{M}_n^{>} \tag{15.1}$$

with unknown parameter $\vartheta \in \Theta$ and let be satisfied

Assumption 15.1. *The parameter region $\Theta \subset \mathbb{R}^r$ is compact.*

This assumption covers a lot of interesting special cases of practical importance. The scope of applications of linear models with the above type of parameter restrictions visualize for, example, DAVIS (1978), ITO (1980), SCHMIDT (1981), OMAN (1983) and STAHLECKER (1985), also a number of models of isotonic regression is covered (see, e.g. BARLOW/BRUNK (1972)).

Our goal is to determine a minimax linear estimator for ϑ with respect to quadratic loss

$$L(\vartheta, \hat{\vartheta}(y)) = \|\vartheta - \hat{\vartheta}(y)\|_U^2, \qquad U \in \mathcal{M}_r^{\geqq}. \tag{15.2}$$

Definition 15.1. Let $D_l = \{\hat{\vartheta}(\tilde{y}) = Z\tilde{y} : Z \in \mathcal{M}_{r \times n}\}$ the class of linear estimators. Then $\hat{\vartheta}_M \in D_l$ is called a *minimax linear estimator* if it holds

$$\sup_{\vartheta \in \Theta} R(\vartheta, \hat{\vartheta}_M) = \inf_{\hat{\vartheta} \in D_l} \sup_{\vartheta \in \Theta} R(\vartheta, \hat{\vartheta}).$$

The risk $R(\vartheta, \hat{\vartheta}) = \mathsf{E}_{\tilde{y}|\vartheta}\|\vartheta - \hat{\vartheta}(\tilde{y})\|_U^2$ of an arbitrary linear estimator $\hat{\vartheta} = Z\tilde{y}$ takes the form

$$R(\vartheta, \hat{\vartheta}) = \operatorname{tr} UZ\Sigma Z^\top + \vartheta^\top (I_r - ZF)^\top U(I_r - ZF)\vartheta. \tag{15.3}$$

We approach the minimax estimation problem by changing it to an equivalent Bayesian estimation problem with respect to some least favourable prior distribution on the parameter region.

Since Θ is compact and $R(\cdot, \hat{\vartheta})$ is continuous and bounded on Θ, it holds

$$\sup_{\vartheta \in \Theta} R(\vartheta, \hat{\vartheta}) = \sup_{P \in \Theta^*} \int_\Theta R(\vartheta, \hat{\vartheta}) P(\mathrm{d}\vartheta) < \infty, \qquad \forall \hat{\vartheta} \in D_1 \tag{15.4}$$

cp. Lemma 6.5. (Again, Θ^* denotes the set of all prior distributions $P = P_{\tilde{\vartheta}}$ for $\tilde{\vartheta}$ over the region Θ). Denoting

$$\mathcal{M} = \left\{ M_P = \int_\Theta \vartheta\vartheta^\top P(\mathrm{d}\vartheta) : P \in \Theta^* \right\} \tag{15.5}$$

the class of second order moment matrices associated with the priors from Θ^*, the Bayes risk reads

$$\int_\Theta R(\vartheta, \hat{\vartheta}) P(\mathrm{d}\vartheta) = \operatorname{tr} U(Z\Sigma Z^\top + Z_F M_P Z_F^\top) =: \varrho(M_P, Z), \tag{15.6}$$

where

$$Z_F = I_r - ZF.$$

The minimax problem

$$\sup_{P \in \Theta^*} \int_\Theta R(\vartheta, \hat{\vartheta}_{\mathrm{M}}) P(\mathrm{d}\vartheta) = \inf_{\hat{\vartheta} \in D_1} \sup_{P \in \Theta^*} \int_\Theta R(\vartheta, \hat{\vartheta}) P(\mathrm{d}\vartheta)$$

which, by virtue of (15.4), is equivalent to the original problem formulated in Definition 15.1, can be solved by interchanging the order of minimization and maximization.

Lemma 15.1. *Under the Assumption* 15.1 *the set of second order moment matrices* $\mathcal{M}$ *defined by* (15.5) *is convex and compact (with respect to* $\|M_P\| = (\operatorname{tr} M_P^2)^{1/2}$) *and it holds*

$$\inf_{Z \in \mathcal{M}_{r\times n}} \sup_{M_P \in \mathcal{M}} \varrho(M_P, Z) = \sup_{M_P \in \mathcal{M}} \inf_{Z \in \mathcal{M}_{r\times n}} \varrho(M_P, Z).$$

Proof. The assertion on $\mathcal{M}$ follows immediately from Lemma 5.1.8 in Bandemer et al. (1977), cp. also A.19 (i). The interchange of infimum and supremum is justified by a minimax theorem due to Sion (1958), see A.15, since $\varrho(\cdot, Z)$ is concave (actually it is linear) and continuous on the compact set $\mathcal{M}$ (with respect to $\|M_P\| = (\operatorname{tr} M_P^2)^{1/2}$) for any $Z \in \mathcal{M}_{r\times n}$ and, moreover, $\varrho(M_P, \cdot)$ is convex on $\mathcal{M}_{r\times n}$ for any $M_P \in \mathcal{M}$.

Together with (15.4) and (15.6) we thus get

$$\inf_{\hat{\vartheta} \in D_1} \sup_{\vartheta \in \Theta} R(\vartheta, \hat{\vartheta}) = \sup_{M_P \in \mathcal{M}} \inf_{Z \in \mathcal{M}_{r\times n}} \varrho(M_P, Z). \quad \square \tag{15.7}$$

Lemma 15.1 states that the statistical two-persons-zero-sums game $[\Theta^*, D_1, \varrho]$ has a value. We now determine this value and prove the existence of a least favourable prior distribution.

Lemma 15.2. *For any* $P \in \Theta^*$ *define* $Z_P = M_P F^\top (F M_P F^\top + \Sigma)^{-1}$. *Then there exists a prior* $P_0 \in \Theta^*$ *such that*

$$\varrho(M_{P_0}, Z_{P_0}) = \sup_{P \in \Theta^*} \varrho(M_P, Z_P) \tag{15.8}$$

and any such P_0 *is least favourable within* Θ^*, i.e.,

$$\varrho(M_{P_0}, Z_{P_0}) = \sup_{P \in \Theta^*} \inf_{Z \in \mathcal{M}_{r \times n}} \varrho(M_P, Z). \tag{15.9}$$

Proof. The existence of a least favourable prior, which is completely characterized by a least favourable moment matrix $M_0 = M_{P_0} \in \mathcal{M}$, follows from the fact that $\varrho(M_P, Z_P)$ is continuous with respect to M_P on the compact set $\mathcal{M}$. Further, $\hat{\vartheta}_P = Z_P \tilde{y}$ is Bayesian in D_1 with respect to $P \in \Theta^*$ (cp. Theorem 5.3 and (5.31)), i.e.,

$$\varrho(M_P, Z_P) = \inf_{Z \in \mathcal{M}_{r \times n}} \varrho(M_P, Z)$$

and thus

$$\sup_{P \in \Theta^*} \inf_{Z \in \mathcal{M}_{r \times n}} \varrho(M_P, Z) = \sup_{P \in \Theta^*} \varrho(M_P, Z_P) = \varrho(M_{P_0}, Z_{P_0}). \quad \square$$

The following result is basic for the characterization of minimax linear estimators.

Lemma 15.3. *The estimator* $\hat{\vartheta}_0 \in D_1$ *is minimax in* D_1 *if and only if it is Bayesian in* D_1 *with respect to some least favourable prior distribution defined by* (15.8).

Proof. Let $\hat{\vartheta}_0$ a Bayes linear estimator with respect to P_0. Then $\hat{\vartheta}_0$ can be represented as

$$\hat{\vartheta}_0 = [U^+ U Z_{P_0} + (I_r - U^+ U) G] \tilde{y} \tag{15.10}$$

with Z_{P_0} defined as in Lemma 15.2 and $G \in \mathcal{M}_{r \times n}$ chosen arbitrarily (cp. Theorem 5.3). The Bayes risk is given by $\varrho(M_{P_0}, Z_{P_0})$ independent of the choice of G which, together with (15.7) and (15.9), implies the minimax optimality.

Now, conversely, assume $\hat{\vartheta}_0$ to be minimax in D_1. Then, by virtue of A.11, the Bayes optimality of $\hat{\vartheta}_0$ in D_1 with respect to P_0 follows immediately from Lemmas 15.1 and 15.2. $\square$

Remark 15.1. The existence of a least favourable prior distribution, which results from the compactness of Θ, is essential for the preceding results. Without this existence being guaranteed, we could only prove the almost Bayes optimality of the minimax solutions. By virtue of Lemma 15.3, minimax linear estimators are characterized as Bayes linear estimators of the form (15.10).

Theorem 15.1. *Under the Assumption* 15.1, *an estimator* $\hat{\boldsymbol{\vartheta}}_{\mathrm{M}}$ *is minimax linear if and only if it holds*

$$\hat{\boldsymbol{\vartheta}}_{\mathrm{M}} = [\boldsymbol{U}^{+}\boldsymbol{U}\boldsymbol{M}_0\boldsymbol{F}^{\top}(\boldsymbol{F}\boldsymbol{M}_0\boldsymbol{F}^{\top}+\boldsymbol{\Sigma})^{-1} + (\boldsymbol{I}_r - \boldsymbol{U}^{+}\boldsymbol{U})\boldsymbol{G}]\tilde{\boldsymbol{y}} \tag{15.11}$$

with some $\boldsymbol{G} \in \mathcal{M}_{r\times n}$ *and* $\boldsymbol{M}_0 = \boldsymbol{M}_{P_0}$ *being the matrix of second order moments of a least favourable prior* $P_0 \in \Theta^*$. *The minimax risk, which is independent of the choice of* $\boldsymbol{G}$, *is given by*

$$\sup_{\vartheta\in\Theta} R(\boldsymbol{\vartheta}, \hat{\boldsymbol{\vartheta}}_{\mathrm{M}}) = \mathrm{tr}\,\{\boldsymbol{U}\boldsymbol{M}_0 - \boldsymbol{U}\boldsymbol{M}_0\boldsymbol{F}^{\top}(\boldsymbol{F}\boldsymbol{M}_0\boldsymbol{F}^{\top}+\boldsymbol{\Sigma})^{-1}\boldsymbol{F}\boldsymbol{M}_0\}\,. \tag{15.12}$$

Proof. It only remains to show that the minimax risk has the form indicated in (15.12). But, with the Bayes optimality of $\hat{\boldsymbol{\vartheta}}_{\mathrm{M}}$ with respect to P_0 this follows immediately from Theorem 5.3 observing that

$$\sup_{\vartheta\in\Theta} R(\boldsymbol{\vartheta}, \hat{\boldsymbol{\vartheta}}_{\mathrm{M}}) = \sup_{P\in\Theta^*} \varrho(\boldsymbol{M}_P, \boldsymbol{Z}_P) = \varrho(\boldsymbol{M}_{P_0}, \boldsymbol{Z}_{P_0})\,. \quad \square \tag{15.13}$$

In case of a full rank weight matrix $\boldsymbol{U}$ (15.11) reduces to

$$\hat{\boldsymbol{\vartheta}}_{\mathrm{M}} = \boldsymbol{M}_0\boldsymbol{F}^{\top}(\boldsymbol{F}\boldsymbol{M}_0\boldsymbol{F}^{\top}+\boldsymbol{\Sigma})^{-1}\tilde{\boldsymbol{y}}, \tag{15.14}$$

where it is to be noted that the least favourable moment matrix $\boldsymbol{M}_0$ is not necessarily uniquely determined. We shall see, however, that under the additional assumption of a full rank design matrix $\boldsymbol{F}$ the matrix $\boldsymbol{M}_0$ and thus also the minimax linear estimator are unique. We note, however, that the solution $\boldsymbol{M}_0 = \boldsymbol{M}_{P_0}$ to the maximization problem (15.8) heavily depends on the concrete form of the weight matrix $\boldsymbol{U}$, since

$$\varrho(\boldsymbol{M}_P, \boldsymbol{Z}_P) = \mathrm{tr}\,\{\boldsymbol{U}\boldsymbol{M}_P - \boldsymbol{U}\boldsymbol{M}_P\boldsymbol{F}^{\top}(\boldsymbol{F}\boldsymbol{M}_P\boldsymbol{F}^{\top}+\boldsymbol{\Sigma})^{-1}\boldsymbol{F}\boldsymbol{M}_P\} \tag{15.15}$$

(see Theorem 5.3). But, on the other hand, we shall see that the solution $\boldsymbol{M}_0$ corresponding to a given positive definite $\boldsymbol{U}$ leads us to a minimax linear estimator which is universally Θ-admissible, i.e., $(\boldsymbol{U}, \Theta)$-admissible for any $\boldsymbol{U} \in \mathcal{M}_{r\times n}$ (see Section 16.1).

Under the assumption that $\boldsymbol{F}^{\top}\boldsymbol{\Sigma}^{-1}\boldsymbol{F}$ has full rank we can show that the minimax linear estimators effect an improvement in risk over the (generalized) LSE. But before doing so we provide a simplified representation for the Bayes risk $\varrho(\boldsymbol{M}_P, \boldsymbol{Z}_P)$ defined by (15.15).

Lemma 15.4. *Assume that* rank $\boldsymbol{F}^{\top}\boldsymbol{\Sigma}^{-1}\boldsymbol{F} = r$ *and denote* $\boldsymbol{V} = (\boldsymbol{F}^{\top}\boldsymbol{\Sigma}^{-1}\boldsymbol{F})^{-1}$. *Further, let* $P \in \Theta^*$ *and* $\boldsymbol{Z}_P$ *as defined in Lemma* 15.2. *Then the Bayes risk of* $\hat{\boldsymbol{\vartheta}}_P = \boldsymbol{Z}_P\tilde{\boldsymbol{y}}$ *with respect to* P *can be written as*

$$\varrho(\boldsymbol{M}_P, \boldsymbol{Z}_P) = \mathrm{tr}\,\boldsymbol{U}\boldsymbol{V} - \mathrm{tr}\,\boldsymbol{V}\boldsymbol{U}\boldsymbol{V}(\boldsymbol{M}_P + \boldsymbol{V})^{-1}\,. \tag{15.16}$$

Proof. Setting $\boldsymbol{C}_1 = \boldsymbol{M}_P\boldsymbol{F}^{\top}$, $\boldsymbol{C}_2 = \boldsymbol{\Sigma}^{-1}\boldsymbol{F}$ and applying the matrix identity $(\boldsymbol{I}_r + \boldsymbol{C}_1\boldsymbol{C}_2^{\top})^{-1} = \boldsymbol{I}_r - \boldsymbol{C}_1(\boldsymbol{I}_n + \boldsymbol{C}_2^{\top}\boldsymbol{C}_1)^{-1}\boldsymbol{C}_2^{\top}$ (see A.33) we obtain

$$\boldsymbol{I}_r - \boldsymbol{Z}_P\boldsymbol{F} = \boldsymbol{I}_r - \boldsymbol{C}_1(\boldsymbol{I}_n + \boldsymbol{C}_2^{\top}\boldsymbol{C}_1)^{-1}\boldsymbol{C}_2^{\top} = (\boldsymbol{I}_r + \boldsymbol{M}_P\boldsymbol{V}^{-1})^{-1} = \boldsymbol{V}(\boldsymbol{M}_P + \boldsymbol{V})^{-1}$$

and thus

$$U(I_r - Z_P F)M_P = UV[I_r - (M_P + V)^{-1}V].$$

This leads to the representation (15.16) observing that $\varrho(M_P, Z_P) = \operatorname{tr} U(I_r - Z_P F)M_P$ (see (15.16)). □

Moreover, we note that for a full rank matrix $F^\top \Sigma^{-1} F$ the minimax estimators given by (15.11) can be written equivalently as

$$\hat{\vartheta}_M = [U^+ UM_0(M_0 + V)^{-1} + (I_r - U^+ U)G]\hat{\vartheta}_\Sigma, \quad G \in \mathcal{M}_{r\times n} \quad (15.17)$$

with $\hat{\vartheta}_\Sigma = VF^\top \Sigma^{-1}\tilde{y}$ (cp. Remark 5.3, Remark 5.4).

Corollary 15.1. *Assume quadratic loss* (15.2) *and* rank $F^\top \Sigma^{-1} F = r$. *Then it holds*

$$R(\vartheta, \hat{\vartheta}_M) \leqq R(\vartheta, \hat{\vartheta}_\Sigma), \qquad \forall \in \Theta$$

for any weight matrix $U \in \mathcal{M}_r^{\geqq}$ *and any* $\hat{\vartheta}_M$ *of the form* (15.17). *In case of* rank $U = r$ *this inequality holds strictly, i.e., the minimax linear estimators* $\hat{\vartheta}_M$ *yield strict* (U, Θ)*-improvements* over $\hat{\vartheta}_\Sigma$ *in the sense of Definition* 7.1 (i).

Proof. By virtue of (15.13) and (15.16) we have

$$\sup_{\vartheta \in \Theta} R(\vartheta, \hat{\vartheta}_M) = \operatorname{tr} UV - \operatorname{tr} UV(M_0 + V)^{-1}V \quad (15.18)$$

Further, $\operatorname{tr} UV(M_0 + V)^{-1}V \geqq 0$ due to the fact that $V(M_0 + V)^{-1}V$ is positive definite, in case of rank $U = r$ we have strict inequality. From this the assertion follows observing that $R(\vartheta, \hat{\vartheta}_\Sigma) \equiv \operatorname{tr} UV$. □

In addition to Corollary 15.1 it can be shown that the minimax linear estimators effect an optimal improvement over $\hat{\vartheta}_\Sigma$ within the class of linear estimators.

Corollary 15.2. *For given* $\hat{\vartheta} \in D_1$ *denote*

$$\bar{R}(\hat{\vartheta}) = \inf_{\vartheta \in \Theta} \{R(\vartheta, \hat{\vartheta}_\Sigma) - R(\vartheta, \hat{\vartheta})\}.$$

Then any $\hat{\vartheta}_M$ *maximizes the minimum guaranteed improvement in risk over* $\hat{\vartheta}_\Sigma$ *within the class* D_1, *i.e.,*

$$\bar{R}(\hat{\vartheta}_M) = \sup_{\hat{\vartheta} \in D_1} \bar{R}(\hat{\vartheta}) = \inf_{M_P \in \mathcal{M}} \operatorname{tr} VUV(M_P + V)^{-1}.$$

(cp. Bunke (1975 a)).

This follows together with (15.18) from the fact that

$$\sup_{\hat{\vartheta} \in D_1} \bar{R}(\hat{\vartheta}) = \operatorname{tr} UV - \inf_{\hat{\vartheta} \in D_1} \sup_{\vartheta \in \Theta} R(\vartheta, \hat{\vartheta}) = \operatorname{tr} UV - \sup_{\vartheta \in \Theta} R(\vartheta, \hat{\vartheta}_M).$$

15.1.2. Analogies with Bayesian experimental design

For the numerical computation of a minimax linear estimator we need to know a least favourable moment matrix $\boldsymbol{M}_0$. We will now show that such a matrix can be obtained as solution to a dual Bayesian experimental design problem.

From Lemma 15.4 we see that the maximization of the Bayes risk of $\hat{\boldsymbol{\vartheta}}_P$ is equivalent to minimizing the functional

$$L_{\mathrm{B}}(P) := \operatorname{tr} \bar{\boldsymbol{U}}(\boldsymbol{M}_P + \boldsymbol{V})^{-1}, \qquad \bar{\boldsymbol{U}} = \boldsymbol{VUV} \tag{15.19}$$

over the convex and compact set Θ^* of all (prior) distributions for $\tilde{\boldsymbol{\vartheta}}$. This means that a least favourable moment matrix $\boldsymbol{M}_0$ satisfies

$$\boldsymbol{M}_0 = \boldsymbol{M}_{P_0} \quad \text{with} \quad P_0 = \arg \sup_{P \in \Theta^*} \varrho(\boldsymbol{M}_P, \boldsymbol{Z}_P) = \arg \inf_{P \in \Theta^*} L_{\mathrm{B}}(P)$$

and the minimax risk is given by

$$\sup_{\boldsymbol{\vartheta} \in \Theta} R(\boldsymbol{\vartheta}, \hat{\boldsymbol{\vartheta}}_{\mathrm{M}}) = \operatorname{tr} \boldsymbol{UV} - \inf_{P \in \Theta^*} L_{\mathrm{B}}(P). \tag{15.20}$$

The matrices $\boldsymbol{M}_P \in \mathcal{M}$ can be interpreted as information matrices

$$\boldsymbol{M}_P = \int_{\Theta} \boldsymbol{t}\boldsymbol{t}^{\top} P(\mathrm{d}\boldsymbol{t})$$

of experimental design measures P for a (fictitious) multiple linear regression setup $\mathbf{E}\tilde{y}(\boldsymbol{t}) = \boldsymbol{t}^{\top}\boldsymbol{\beta}$ with regression parameter $\boldsymbol{\beta} \in \mathbb{R}^r$ and experimental region $\Theta \subset \mathbb{R}^r$. The minimization of the functional $L_{\mathrm{B}}(\cdot)$ defined by (15.19) is equivalent to the problem of determining an L_{B}-optimal design minimizing the Bayes risk of the Bayes (affine-) linear estimator for $\boldsymbol{\beta}$ with respect to the regression setup

$$\mathbf{E}\tilde{y}(\boldsymbol{t}) = \boldsymbol{t}^{\top}\boldsymbol{\beta}, \qquad \boldsymbol{t} \in \Theta, \qquad \mathbf{Cov}\,\tilde{\boldsymbol{\beta}} \propto \boldsymbol{V}^{-1} \tag{15.21}$$

and quadratic loss $\|\boldsymbol{\beta} - \hat{\boldsymbol{\beta}}(\boldsymbol{y})\|_{\boldsymbol{U}}^2$ (cp. Definition 10.4).

However, compared to our original model

$$\mathbf{E}\tilde{\boldsymbol{y}} = \boldsymbol{F}\boldsymbol{\vartheta}, \qquad \boldsymbol{\vartheta} \in \Theta, \qquad \mathbf{E}\,\tilde{\boldsymbol{\vartheta}}\tilde{\boldsymbol{\vartheta}}^{\top} = \boldsymbol{M}_P \tag{15.22}$$

we have changed the roles of the information matrix and of the prior second order moment matrix: Whereas in the model (15.22) $\boldsymbol{M}_P$ plays the role of the prior moment matrix $\boldsymbol{M}_P$ represents the information matrix of the experimental design measure $P \in \Theta^*$ in the above model (15.21).

Corollary 15.3. *The following problems are equivalent:*

(i) *Find a least favourable prior distribution on the compact parameter region Θ for the model* (15.22).

(ii) *Find an L_{B}-optimal design with respect to $\bar{\boldsymbol{U}} = \boldsymbol{VUV}$ for the model* (15.21).

In this sense, minimax linear estimation with a restricted parameter space and optimal experimental design for Bayes linear estimators can be thought of as being dual problems. This offers the possibility to apply all the powerful results of Bayesian experimental design theory to provide characterizations and methods for the computation of a least favourable moment matrix.

Corollary 15.4. *Let* rank $\bar{U} = s$. *Then there exists a least favourable prior distribution* $P_0 \in \Theta^*$ *minimizing the functional* L_B *defined by* (15.19) *and the support of which includes at most* $s(2r - s + 1)/2$ *different boundary points from* Θ. *For any supporting point of* P_0 *it holds:*

$$\vartheta_0 \in \operatorname{supp} P_0 \Rightarrow \vartheta_0 = \arg\sup_{\vartheta \in \Theta} \vartheta^\top (M_0 + V)^{-1} \bar{U} (M_0 + V)^{-1} \vartheta, \quad (15.23)$$

where $M_0 = M_{P_0}$.

This follows with Corollary 15.3 from Theorem 11.5 and Theorem 11.14.

Remark 15.2. Geometrically, the condition stated in (15.23) means that the supporting points of P_0 lie on a surface $\vartheta^\top A \vartheta = \text{constant}$, $A = (M_0 + V)^{-1} \bar{U} (M_0 + V)^{-1}$, centered at the origin and containing the region Θ. Thus, a least favourable distribution will only include points which lie in the intersection of this surface and the boundary of Θ. If, additionally, the parameter region Θ is convex then a least favourable prior distribution need only include points which are extreme points of Θ (cp. Theorem 11.9).

On the basis of the necessary and sufficient optimality condition for L_B-optimal experimental designs we can supplement the characterization of minimax linear estimators given in Theorem 15.1 as follows.

Theorem 15.2. *Under the assumptions* 15.1 *and* rank $F^\top \Sigma^{-1} F = r$ *the estimator* $\hat{\vartheta}_M$ *is minimax linear with respect to quadratic loss* (15.2) *if it can be written in the form* (15.17) *with a matrix* M_0 *satisfying the following conditions*

(i) $M_0 \in \mathcal{M} = \{M_P \in \mathcal{M}_r^{\geqq} : P \in \Theta^*\}$.

(ii) $\sup_{\vartheta \in \Theta} \vartheta^\top (M_0 + V)^{-1} \bar{U} (M_0 + V)^{-1} \vartheta = \operatorname{tr} (M_0 + V)^{-1} \bar{U} (M_0 + V)^{-1} M_0$.

The proof follows with Theorem 6.1 and Corollary 15.3 from Theorem 11.12 replacing there $M_B(\xi^*) = M(\xi^*) + \frac{1}{n} \Phi^{-1}$ by $M_0 + V$.

Remark 15.3. a) If $\bar{U} = VUV$ has full rank then the functional L_B defined by (15.19) is strictly convex on $\mathcal{M}$ which in turn implies that M_0 is uniquely determined and $\hat{\vartheta}_M = M_0(M_0 + V)^{-1} \hat{\vartheta}_\Sigma$ is the unique minimax estimator in D_1 (cp. Theorem 11.6).

b) Any least favourable moment matrix M_0 is a boundary point of $\mathcal{M}$ with respect to the metric m defined by $m(M_1, M_2) = (\operatorname{tr}(M_1 - M_2)^2)^{1/2}$, $M_1, M_2 \in \mathcal{M}$.

With the help of Corollary 15.3, the latter assertion can be proved as Theorem 5.3.1 in BANDEMER et al. (1977).

In most cases we may need much less supporting points to form a least favourable prior P_0 than indicated by the upper bound in Corollary 15.4. In case of ellipsoidal parameter restrictions we require at most r supporting points (cp. the remark preceding Theorem 12.3). There are situations in which a least favourable prior can even be concentrated at a single point, especially for low rank matrices $\bar{U} = VUV$. Let us consider, for example, the important special case of estimating a linear combination $c^\top \vartheta$ of the regression coefficients, $c \in \mathbb{R}^r$. Then, in case of the convexity of the parameter region Θ, such a concentration to a least favourable one-point distribution is always possible. The following corollary shows how to find such a point.

Corollary 15.5. *Let $U = cc^\top$ for some $c \in \mathbb{R}^r$, $c \neq 0$, and assume that $\Theta \subset \mathbb{R}^r$ is convex. Define $\Theta_0 \subset \Theta$ to be the set of all points ϑ_0 maximizing*

$$g(\vartheta) = (c^\top \vartheta)^2/(1 + \vartheta^\top V^{-1}\vartheta). \tag{15.24}$$

If $\vartheta_0 \in \Theta_0$ is chosen such that

$$\sup_{\vartheta \in \Theta} [c^\top V(V + \vartheta_0\vartheta_0^\top)^{-1}\vartheta]^2 = [c^\top V(V + \vartheta_0\vartheta_0^\top)^{-1}\vartheta_0]^2 \tag{15.25}$$

then $P_0 = \delta_{\vartheta_0}$ is a least favourable prior distribution in Θ^* and, correspondingly, $M_0 = \vartheta_0\vartheta_0^\top$ is a least favourable moment matrix in $\mathcal{M}$.

This follows immediately from Corollary 14.1, see also CHALONER (1984) and PILZ (1986 a), Corollary 3. Interestingly, the minimax linear estimator formed with $M_0 = \vartheta_0\vartheta_0^\top$ also acts as a maximin estimator in D_1 (see Section 6.1.3).

The analogy between minimax linear estimation and Bayesian experimental design also suggests the application of iteration procedures as described in Section 12.5 for finding an approximately (L_B-optimal) least favourable moment matrix M_0.

To this, one starts with an initial measure $P_0 \in \Theta^*$ having finite support and then, at every stage of the iteration one forms a measure

$$P_s = (1 - \alpha_s)P_{s-1} + \alpha_s \delta_{\vartheta_s}, \qquad s = 1, 2, \ldots \tag{15.26}$$

The sequence of weights $\{\alpha_s\}_{s \in \mathbb{N}}$ and the supporting points $\vartheta_s \in \Theta$ have to be chosen such that

$$\lim_{s \to \infty} \alpha_s = 0, \quad \sum_{s=1}^{\infty} \alpha_s = \infty, \quad \alpha_s \in [0, 1), \quad s = 1, 2, \ldots \tag{15.27}$$

and

$$\vartheta_s = \arg \inf_{\vartheta \in \Theta} \Delta_{L_B}(P_{s-1}, \delta_\vartheta),$$

where Δ_{L_B} is the Fréchet directional derivative of the functional L_B. According to Lemma 11.7 (i), the latter condition amounts to

$$\vartheta_s = \arg\sup_{\vartheta \in \Theta} \vartheta^\top (M_{P_{s-1}} + V)^{-1} \bar{U} (M_{P_{s-1}} + V)^{-1} \vartheta \tag{15.28}$$

(cp. Algorithm 12.4).

The initial measure P_0 may be taken as a one-point measure $P_0 = \delta_{\vartheta_0}$ with ϑ_0 chosen such that it minimizes $L_B(\delta_\vartheta)$ among all one-point measures δ_ϑ, $\vartheta \in \Theta$, which implies that

$$\vartheta_0 = \arg\sup_{\vartheta \in \Theta} \vartheta^\top U \vartheta / (1 + \vartheta^\top V^{-1} \vartheta), \qquad P_0 = \delta_{\vartheta_0}. \tag{15.29}$$

Accordingly, the weights could be chosen as $\alpha_s = 1/(1+s)$, $s = 1, 2, \ldots$ Clearly, in view of a faster convergence, the weights can also be chosen optimally as indicated in (12.57) and Algorithm 12.4, respectively, for example with the help of some line search algorithm as proposed in BÖHNING (1985). However, to avoid numerical complexities, we recommend to choose a fixed sequence of weights according to (15.27).

Theorem 15.3. *Let Assumption* 15.1 *be satisfied and assume that* rank $F^\top \Sigma^{-1} F = r$. *Then, for any sequence of Bayes linear estimators*

$$\hat{\vartheta}_{P_s} = M_{P_s}(M_{P_s} + V)^{-1} V F^\top \Sigma^{-1} \tilde{y} =: Z_{P_s} \tilde{y}, \qquad s = 1, 2, \ldots$$

associated with $\{P_s\}_{s \in \mathbb{N}}$ *generated according to* (15.26) – (15.29), *it holds either one of the following assertions:*

a) $\{\hat{\vartheta}_{P_s}\}_{s \in \mathbb{N}}$ *is finite with the last element being minimax in* D_1 *or*

b) $\{\hat{\vartheta}_{P_s}\}_{s \in \mathbb{N}}$ *is infinite with the sequence of Bayes risks converging to the minimax risk:* $\lim_{s \to \infty} \varrho(M_{P_s}, Z_{P_s}) = \inf_{\hat{\vartheta} \in D_1} \sup_{\vartheta \in \Theta} R(\vartheta, \hat{\vartheta})$.

This follows immediately from Theorem 12.14, the convergence of the Bayes risks can be concluded from (15.20).

As a stopping criterion we can use, again, the notion of $L_B - e_0$-optimality, where $e_0 \in (0, 1)$ is some predetermined value for the efficiency $e_{L_B}(P_s) = \inf_{P \in \Theta^*} L_B(P)/L_B(P_s)$ that is to be guaranteed at the final stage of the iteration (see Section 12.5). A sufficient condition for $e_{L_B}(P_{s_0}) \geqq e_0$ to hold is that

$$(2 - e_0) L_B(P_s) \geqq \operatorname{tr} V G_s + \sup_{\vartheta \in \Theta} \vartheta^\top G_s \vartheta,$$

where

$$G_s = (M_{P_s} + V)^{-1} \bar{U} (M_{P_s} + V)^{-1} \tag{15.30}$$

(cp. Corollary 12.7 (i)). The following inequalities for the absolute differences in risk can also be used as a basis for an appropriate stopping criterion.

Lemma 15.5. *Let* $\varepsilon > 0$ *and* $s \in \mathbb{N}$ *such that*

$$\operatorname{tr} V G_s + \sup_{\vartheta \in \Theta} \vartheta^\top G_s \vartheta \leqq L_B(P_s) + \varepsilon. \tag{15.31}$$

Then it holds: $0 \leqq \sup_{P \in \Theta^*} \varrho(M_P, Z_P) - \varrho(M_{P_s}, Z_{P_s}) \leqq \varepsilon$.

Proof. The lower bound is trivial. On the other hand, (15.31) is equivalent to

$$\inf_{\vartheta \in \Theta} \Delta_{L_B}(P_s, \delta_\vartheta) \geqq -\varepsilon \tag{15.32}$$

(see Lemma 11.5 and Lemma 11.7). Observing that

$$L_B(P_s) + \inf_{\vartheta \in \Theta} \Delta_{L_B}(P_s, \delta_\vartheta) \leqq \inf_{P \in \Theta^*} L_B(P)$$

(see Lemma 12.2) we then have from (15.16) and (15.32):

$$\sup_{P \in \Theta^*} \varrho(\boldsymbol{M}_P, \boldsymbol{Z}_P) - \varrho(\boldsymbol{M}_{P_s}, \boldsymbol{Z}_{P_s}) = L_B(P_s) - \inf_{P \in \Theta^*} L_B(P) \leqq \varepsilon. \quad \square$$

Thus, the iteration procedure can be stopped at that stage s at which (15.31) is fulfilled for the first time, for a sufficiently small $\varepsilon > 0$.

We remark that in case of the convexity of Θ the search for the supporting points according to (15.28) and (15.29) can be restricted to the set of extreme points of Θ (cp. Theorem 11.9 and Remark 11.4). In particular, if prior knowledge suggests linear inequality constraints for the regression coefficients then the search can be done within the finite set of corner points of Θ which substantially simplifies the numerical computation.

15.1.3. Maximin linear estimators

The counterpart to the minimax estimator is the maximin estimator. Contrary to the minimax estimator, this estimator can be computed relatively easy. Krafft (1986), who considered the special case of ellipsoidal parameter restrictions, proposed to use this estimator as an approximation for the minimax estimator in D_l.

Definition 15.2. The estimator $\hat{\boldsymbol{\vartheta}} \in D_l$ is called a *maximin linear estimator* if it holds

$$\sup_{\vartheta \in \Theta} \inf_{\hat{\vartheta} \in D_l} R(\boldsymbol{\vartheta}, \hat{\boldsymbol{\vartheta}}) = \sup_{\vartheta \in \Theta} R(\boldsymbol{\vartheta}, \hat{\boldsymbol{\vartheta}}).$$

It is well-known that

$$\sup_{\vartheta \in \Theta} \inf_{\hat{\vartheta} \in D_l} R(\boldsymbol{\vartheta}, \hat{\boldsymbol{\vartheta}}) \leqq \inf_{\hat{\vartheta} \in D_l} \sup_{\vartheta \in \Theta} R(\boldsymbol{\vartheta}, \hat{\boldsymbol{\vartheta}}) \tag{15.33}$$

and the question arises whether there can occur equality in (15.33), which would guarantee the minimax optimality of the maximin linear estimator.

Lemma 15.6. *Let* $\boldsymbol{\vartheta}_0 \in \Theta$ *a parameter value chosen according to* (15.29). *Then*

$$\tilde{\boldsymbol{\vartheta}} = (1 + \boldsymbol{\vartheta}_0^\top \boldsymbol{V}^{-1} \boldsymbol{\vartheta}_0)^{-1} \boldsymbol{\vartheta}_0 \boldsymbol{\vartheta}_0^\top \boldsymbol{F}^\top \boldsymbol{\Sigma}^{-1} \tilde{\boldsymbol{y}} \tag{15.34}$$

is a maximin linear estimator for $\boldsymbol{\vartheta}$.

Proof. We consider $R(\boldsymbol{\vartheta}, \hat{\boldsymbol{\vartheta}})$, $\hat{\boldsymbol{\vartheta}} = \boldsymbol{Z}\tilde{\boldsymbol{y}} \in D_l$, for fixed $\boldsymbol{\vartheta} \in \Theta$. Then, observing that with $P = \delta_\vartheta$ we have equality $R(\vartheta, \hat{\vartheta}) = \varrho(M_P, Z)$, the minimum of $R(\vartheta, \cdot)$ in D_1 is attained by choosing

$$Z_P = \vartheta\vartheta^\top F^\top (F\vartheta\vartheta^\top F^\top + \Sigma)^{-1} = (1 + \vartheta^\top V^{-1}\vartheta)^{-1}\vartheta\vartheta^\top F^\top \Sigma^{-1}$$

(see Theorem 5.3 and (5.31)), the latter rearrangement follows from A.34. This implies that

$$\inf_{\hat{\vartheta} \in D_1} R(\vartheta, \hat{\vartheta}) = \varrho(M_P, Z_P) = \operatorname{tr} UV - L_B(P) = \vartheta^\top U\vartheta/(1 + \vartheta^\top V^{-1}\vartheta). \quad (15.35)$$

Thus, we have: $\sup_{\vartheta \in \Theta} \inf_{\hat{\vartheta} \in D_1} R(\vartheta, \hat{\vartheta}) = R(\vartheta_0, \tilde{\vartheta}) = \sup_{\vartheta \in \Theta} R(\vartheta, \tilde{\vartheta})$. □

We remark that for singular matrices U, again, we obtain a whole class of maximin estimators

$$\tilde{\vartheta}_A = [U^+ U\vartheta_0\vartheta_0^\top F^\top \Sigma^{-1}/(1 + \vartheta_0^\top V^{-1}\vartheta_0) + (I_r - U^+ U)A]\tilde{y} \quad (15.36)$$

with arbitrary $A \in \mathcal{M}_{r\times n}$, for it holds $R(\vartheta_0, \tilde{\vartheta}) = R(\vartheta_0, \tilde{\vartheta}_A)$.

From the characterization theorem 15.1 we see that the maximin estimators $\tilde{\vartheta}$ from (15.34) or $\tilde{\vartheta}_A$ from (15.36), respectively, are minimax optimal in D_1 if and only if it holds $M_0 = \vartheta_0\vartheta_0^\top$, i.e., if $P_0 = \delta_{\vartheta_0}$ is a least favourable prior distribution in Θ^*. According to Theorem 15.2, this is the case if and only if ϑ_0 satisfies

$$\sup_{\vartheta \in \Theta} \|(\vartheta_0\vartheta_0^\top + V)^{-1}\vartheta\|_U^2 = \|(\vartheta_0\vartheta_0^\top + V)^{-1}\vartheta_0\|_U^2. \quad (15.37)$$

Summarizing, we can state

Corollary 15.6. *Assume Θ to be compact,* rank $F = r$ *and let ϑ_0 as defined in* (15.29). *A maximin linear estimator of the form* (15.36) *is minimax linear if and only if the one-point-measure δ_{ϑ_0} is least favourable in Θ^*. This is the case if and only if* (15.37) *comes true.*

Remark 15.4. The maximin estimators $\tilde{\vartheta}_A$ represent locally best linear estimators at the point $\vartheta = \vartheta_0$, since

$$R(\vartheta_0, \tilde{\vartheta}_A) = \inf_{\hat{\vartheta} \in D_1} R(\vartheta_0, \hat{\vartheta})$$

(see the proof of Lemma 15.6). In the literature, estimators of this type are known as MCMSE *(minimum conditional mean square error) estimators*, see, e.g. Farebrother (1975), Swamy/Mehta (1976).

The above results also lead us to an interesting interpretation of the iteration procedure described before. According to Lemma 15.6, at the initial stage $s = 0$ we generate a maximin linear estimator. If this estimator already turns out to be sufficiently efficient in the sense of (15.30) and (15.31), respectively, then the procedure stops and the maximin estimator provides a fairly good approximation of the minimax estimator. Given that (15.30) and (15.31) even hold with $e_0 = 1$ and $\varepsilon = 0$, respectively, then the maximin estimator is exactly minimax. This situation is reflected in the iteration procedure by the fact that for the point $\boldsymbol{\vartheta}_1$ to be added according to (15.28) at the stage $s=1$ we have

$$\boldsymbol{\vartheta}_1 \in \Theta_0 = \left\{\boldsymbol{\vartheta}_0 \in \Theta : \boldsymbol{\vartheta}_0 = \arg\sup_{\boldsymbol{\vartheta}\in\Theta} \boldsymbol{\vartheta}^\top \boldsymbol{U}\boldsymbol{\vartheta}/(1+\boldsymbol{\vartheta}^\top \boldsymbol{V}^{-1}\boldsymbol{\vartheta})\right\}$$

and (15.28) coincides with the necessary and sufficient optimality condition (15.37). If the maximin estimator is not yet efficient enough, then it will be improved step by step in the course of the iteration procedure into the "direction" of the minimax linear estimator.

In general, the least favourable moment matrix will have rank greater than one and the maximin estimator will not be minimax. The existence of a least favourable one-point distribution heavily depends on the structure of the matrices $\boldsymbol{U}$ and $\boldsymbol{V}$. In the special case rank $\boldsymbol{U}=1$ the existence of such a distribution is assured provided that Θ is convex (see Corollary 15.5). An example of the existence of a least favourable one-point distribution in case of ellipsoidal restrictions and rank $\boldsymbol{U}>1$ may be found in Pilz (1986 b), Section 4.

15.1.4. Projection of the minimax estimator

The realizations of a minimax linear estimator do not necessarily lie within the restricted parameter region Θ. This can be considered a disadvantage of this estimator if the parameter restrictions are sharp and result for example from a physical theory. However, if the restrictions are not based on a firm theory and are more or less vague then it seems to be unrealistic to put probability one on the restricted region Θ and to constrain the estimates to lie in Θ. Viewed in this light, minimax linear estimators can be considered as special types of contraction estimators which shrink towards a region that the parameter is believed to lie either within or close to. Thus, the situation is similiar to that considered by Berger (1980 b), (1982 a) and Bock (1982) who derive (nonlinear) shrinkage estimators which shrink least squares estimates towards ellipsoids and closed convex polyhedra, respectively.

On the other hand, we can easily construct a projection of the minimax linear estimator into Θ and the resulting nonlinear estimator possibly effects a uniform improvement in risk. We now show that such an improvement can in fact be attained under the additional assumption of convexity of Θ.

Theorem 15.4. *Assume $\Theta \subset \mathbb{R}^r$ to be compact and convex. Further, assume quadratic loss with a full rank weight matrix $\boldsymbol{U}$ and let $\hat{\boldsymbol{\vartheta}}_{\mathrm{M}}$ a minimax linear estimator. Then the projection estimator $\hat{\boldsymbol{\vartheta}}^*$ defined by*

$$\hat{\boldsymbol{\vartheta}}^* = \arg\min_{\boldsymbol{\vartheta}\in\Theta} \|\boldsymbol{\vartheta} - \hat{\boldsymbol{\vartheta}}_{\mathrm{M}}\|_U^2 \tag{15.38}$$

represents a $(\boldsymbol{U}, \Theta)$-improvement over $\hat{\boldsymbol{\vartheta}}_{\mathrm{M}}$, i.e.,

$$R(\boldsymbol{\vartheta}, \hat{\boldsymbol{\vartheta}}^*) \leqq R(\boldsymbol{\vartheta}, \hat{\boldsymbol{\vartheta}}_{\mathrm{M}}), \qquad \forall \boldsymbol{\vartheta} \in \Theta.$$

Proof. The assertion is proved if we can show for any $\vartheta \in \Theta$ that

$$L(\vartheta, \hat{\vartheta}^*(\tilde{y})) \leqq L(\vartheta, \hat{\vartheta}_M(\tilde{y})), \qquad P_{\tilde{y}/\vartheta} - \text{a.e.} \tag{15.39}$$

Observing that

$$\|\vartheta - \hat{\vartheta}_M\|_U^2 = \|\vartheta - \hat{\vartheta}^*\|_U^2 + \|\hat{\vartheta}^* - \hat{\vartheta}_M\|_U^2 + 2(\vartheta - \hat{\vartheta}^*)^\top U(\hat{\vartheta}^* - \hat{\vartheta}_M),$$

we see that (15.39) comes true if it holds $P_{\tilde{y}/\vartheta} - \text{a.e.}$

$$(\vartheta - \hat{\vartheta}^*)^\top U(\hat{\vartheta}^* - \hat{\vartheta}_M) \geqq 0, \qquad \forall \vartheta \in \Theta. \tag{15.40}$$

Now, let $y \in \mathbb{R}^n$ an arbitrary realization of $\tilde{y}$. In case that $\hat{\vartheta}_M(y) \in \Theta$ we have equality $\hat{\vartheta}^*(y) = \hat{\vartheta}_M(y)$ and (15.40) holds trivially. Therefore, assume $\hat{\vartheta}_M(y) \notin \Theta$. Then, for arbitrary $\vartheta \neq \hat{\vartheta}^*(y)$ consider the distance of $\hat{\vartheta}_M(y)$ to the point $a\vartheta + (1-a)\hat{\vartheta}^*(y)$ as a function of $a \in \mathbb{R}^1$:

$$\begin{aligned} g(a) &= \|a\vartheta + (1-a)\hat{\vartheta}^*(y) - \hat{\vartheta}_M(y)\|_U^2 \\ &= a^2\|\vartheta - \hat{\vartheta}^*(y)\|_U^2 + 2a(\vartheta - \hat{\vartheta}^*(y))^\top U(\hat{\vartheta}^*(y) - \hat{\vartheta}_M(y)) \\ &\quad + \|\hat{\vartheta}_M(y) - \hat{\vartheta}^*(y)\|_U^2. \end{aligned}$$

Since U is positive definite and thus $\|\vartheta - \hat{\vartheta}^*(y)\|_U^2 > 0$, the function g has a unique minimum

$$a_0 = -(\vartheta - \hat{\vartheta}^*(y))^\top U(\hat{\vartheta}^*(y) - \hat{\vartheta}_M(y))/\|\vartheta - \hat{\vartheta}^*(y)\|_U^2.$$

Since $g(1) = \|\vartheta - \hat{\vartheta}_M(y)\|_U^2 > \|\hat{\vartheta}^*(y) - \hat{\vartheta}_M(y)\|_U^2 = g(0)$, it follows that $a_0 < 1$. Further, for $0 < a < 1$ we have a $\vartheta + (1-a)\hat{\vartheta}^*(y) \in \Theta$ due to the convexity of Θ, which implies that $a_0 \notin (0, 1)$, otherwise we would have a contradiction to the fact that there exists no point in Θ having a smaller distance to $\hat{\vartheta}_M(y)$ than $\hat{\vartheta}^*(y)$ has. Hence, it follows $a_0 \leqq 0$ which proves (15.40). □

We remark that $\hat{\vartheta}^*$ is uniformly better than $\hat{\vartheta}_M$, i.e., $R(\vartheta, \hat{\vartheta}^*) < R(\vartheta, \hat{\vartheta}_M)$ for any $\vartheta \in \Theta$, provided that

$$P_{\tilde{y}/\vartheta}(\hat{\vartheta}_M(\tilde{y})) \notin \Theta > 0 \quad \text{and } \operatorname{supp} P_{\tilde{y}/\vartheta} = \mathbb{R}^n.$$

The improvement will be the more significant the more Θ is restricted, i.e., the smaller the diameter of Θ is. However, if Θ is not too "small" and there is some evidence for the compatibility of prior knowledge and actual data then the probability that the minimax linear estimator lies outside Θ and thus the improvement over this estimator should be small.

15.2. Minimax affine estimators

Now we consider the problem of finding minimax estimators in the class D_a of affine-linear estimators. Under the additional assumption that the parameter region is symmetric around some center point the results of the preceding section can be immediately transferred to give solutions to the present problem.

Definition 15.3. (i) The parameter region Θ is said to be *symmetric around* $\vartheta_0 \in \mathbb{R}^r$ if $\bar{\Theta} := \{\bar{\vartheta} = \vartheta - \vartheta_0 : \vartheta \in \Theta\}$ is symmetric around the origin, i.e., if $\bar{\vartheta} \in \bar{\Theta}$ implies that $-\bar{\vartheta} \in \bar{\Theta}$.

(ii) $\hat{\vartheta}^{\mathrm{a}}_{\mathrm{M}} \in D_{\mathrm{a}}$ is called a *minimax affine estimator* if

$$\sup_{\vartheta \in \Theta} R(\vartheta, \hat{\vartheta}^{\mathrm{a}}_{\mathrm{M}}) = \inf_{\hat{\vartheta} \in D_{\mathrm{a}}} \sup_{\vartheta \in \Theta} R(\vartheta, \hat{\vartheta}).$$

The risk of an affine estimator $\hat{\vartheta} = Z\tilde{y} + z$, $Z \in \mathcal{M}_{r \times n}$, $z \in \mathbb{R}^r$, with respect to quadratic loss reads

$$R(\vartheta, \hat{\vartheta}) = \operatorname{tr} UZ\Sigma Z^{\top} + (Z_F\vartheta - z)^{\top} U (Z_F\vartheta - z), \qquad Z_F = I_r - ZF. \tag{15.41}$$

In case of a symmetric parameter region we can restrict attention to the subclass of estimators

$$D^0_{\mathrm{a}} := \{\hat{\vartheta} = Z(\tilde{y} - F\vartheta_0) + \vartheta_0 : Z \in \mathcal{M}_{r \times n}\} \subset D_{\mathrm{a}} \tag{15.42}$$

Lemma 15.7. *Let Θ compact and symmetric around ϑ_0. Then any minimax estimator in D^0_{a} is a minimax affine estimator.*

Proof. Writing $z = Z_F\vartheta_0 + g$ with arbitrary $g \in \mathbb{R}^r$, we have

$$\|Z_F\vartheta - z\|^2_U = \|Z_F(\vartheta - \vartheta_0)\|^2_U - 2(\vartheta - \vartheta_0)^{\top} Z_F^{\top} U g + \|g\|^2_U.$$

Now, let $\vartheta^* = \vartheta^*(Z) \in \Theta$ a parameter such that

$$\|Z_F(\vartheta^* - \vartheta_0)\|^2_U = \sup_{\vartheta \in \Theta} \|Z_F(\vartheta - \vartheta_0)\|^2_U.$$

Then, if $(\vartheta^* - \vartheta_0)^{\top} Z_F^{\top} U g \geqq 0$, it follows that

$$\sup_{\vartheta \in \Theta} \|Z_F\vartheta - z\|^2_U \geqq \sup_{\vartheta \in \Theta} \|Z_F(\vartheta - \vartheta_0)\|^2_U, \qquad \forall z \in \mathbb{R}^r. \tag{15.43}$$

If, on the other hand, $(\vartheta^* - \vartheta_0)^{\top} Z_F^{\top} U g < 0$ then we change to the symmetric counterpart $\vartheta_* = -(\vartheta^* - \vartheta_0)$ and obtain:

$$\begin{aligned} \|Z_F\vartheta^* - z\|^2_U &= \|Z_F\vartheta_*\|^2_U + 2\vartheta_*^{\top} Z_F^{\top} U g + \|g\|^2_U \\ &> \|Z_F\vartheta_*\|^2_U = \sup_{\vartheta \in \Theta} \|Z_F(\vartheta - \vartheta_0)\|^2_U, \end{aligned}$$

observing that $\|Z_F(\vartheta^* - \vartheta_0)\|^2_U = \|Z_F\vartheta_*\|^2_U$. Thus, (15.43) holds irrespective of the sign of $(\vartheta^* - \vartheta_0)^{\top} Z_F^{\top} U g$, from which the assertion follows. □

The risk of an estimator $\hat{\vartheta} = Z(\tilde{y} - F\vartheta_0) + \vartheta_0 \in D^0_{\mathrm{a}}$ is given by

$$R(\vartheta, \hat{\vartheta}) = \operatorname{tr} UZ\Sigma Z^{\top} + \|Z_F(\vartheta - \vartheta_0)\|^2_U. \tag{15.44}$$

Denoting

$$\bar{\Theta} = \{\bar{\vartheta} \in \mathbb{R}^r : \bar{\vartheta} = \vartheta - \vartheta_0, \ \vartheta \in \Theta\} \tag{15.45}$$

the translation of Θ to the origin, the matrices of centered second order moments of the prior distributions from Θ^* coincide with the moment matrices with respect to the corresponding distributions $\bar{P}$ from $\bar{\Theta}^*$:

$$\int_{\Theta} (\boldsymbol{\vartheta} - \boldsymbol{\vartheta}_0)(\boldsymbol{\vartheta} - \boldsymbol{\vartheta}_0)^\top P(\mathrm{d}\boldsymbol{\vartheta}) = \int_{\bar{\Theta}} \bar{\boldsymbol{\vartheta}}\bar{\boldsymbol{\vartheta}}^\top \bar{P}(\mathrm{d}\bar{\boldsymbol{\vartheta}}) = \boldsymbol{M}_{\bar{P}}, \qquad \bar{P} \in \bar{\Theta}^*. \tag{15.46}$$

According to (15.44), the Bayes risk of $\hat{\boldsymbol{\vartheta}} \in D_{\mathrm{a}}^0$ with respect to $P \in \Theta^*$ takes the form

$$\int_{\Theta} R(\boldsymbol{\vartheta}, \hat{\boldsymbol{\vartheta}}) P(\mathrm{d}\boldsymbol{\vartheta}) = \operatorname{tr} \boldsymbol{U}(\boldsymbol{Z\Sigma Z}^\top + \boldsymbol{Z}_F \boldsymbol{M}_{\bar{P}} \boldsymbol{Z}_F^\top) =: \varrho(\boldsymbol{M}_{\bar{P}}, \boldsymbol{Z}) \tag{15.47}$$

It has exactly the same structure as the Bayes risk of a linear estimator $\hat{\boldsymbol{\vartheta}} = \boldsymbol{Z}\tilde{\boldsymbol{y}}$ for the parameter $\bar{\boldsymbol{\vartheta}} = \boldsymbol{\vartheta} - \boldsymbol{\vartheta}_0 \in \bar{\Theta}$ (see (15.6)).

Since the translation of Θ to $\bar{\Theta}$ does not affect the compactness of the parameter region it follows that in case of a symmetric compact parameter region the corresponding minimax estimation problems in D_{l} and D_{a}^0, respectively, are identical. Hence, all the preceding assertions on the characterization and construction of minimax linear estimators hold analogously for minimax estimators in D_{a}^0 (see PILZ (1986 a)).

Theorem 15.5. *Under the assumptions of Lemma* 15.7 $\hat{\boldsymbol{\vartheta}}_{\mathrm{M}}^{\mathrm{a}}$ *is minimax in* D_{a}^0 *with respect to quadratic loss if and only if it can be represented as*

$$\hat{\boldsymbol{\vartheta}}_{\mathrm{M}}^{\mathrm{a}} = [\boldsymbol{U}^+\boldsymbol{U}\boldsymbol{M}_0\boldsymbol{F}^\top(\boldsymbol{F}\boldsymbol{M}_0\boldsymbol{F}^\top + \boldsymbol{\Sigma})^{-1} + (\boldsymbol{I}_r - \boldsymbol{U}^+\boldsymbol{U})\boldsymbol{A}](\tilde{\boldsymbol{y}} - \boldsymbol{F}\boldsymbol{\vartheta}_0) + \boldsymbol{\vartheta}_0$$

with some $\boldsymbol{A} \in \mathcal{M}_{r\times n}$, *where* $\boldsymbol{M}_0$ *is the matrix of second order moments of a least favourable distribution* $\bar{P}_0 \in \bar{\Theta}^*$ *determined by*

$$\bar{P}_0 = \arg \sup_{\bar{P} \in \bar{\Theta}^*} \varrho(\boldsymbol{M}_{\bar{P}}, \boldsymbol{Z}_{\bar{P}}), \qquad \boldsymbol{Z}_{\bar{P}} = \boldsymbol{M}_{\bar{P}}\boldsymbol{F}^\top(\boldsymbol{F}\boldsymbol{M}_{\bar{P}}\boldsymbol{F}^\top + \boldsymbol{\Sigma})^{-1}.$$

If, additionally, rank $\boldsymbol{F}^\top\boldsymbol{\Sigma}^{-1}\boldsymbol{F} = r$ *then the Conditions* (i) *and* (ii) *given in Theorem* 15.2 *with* Θ *replaced by* $\bar{\Theta}$ *are necessary and sufficient for the optimality of* $\boldsymbol{M}_0$ (cp. PILZ (1986 a), Theorems 1 and 2).

Now we drop the symmetry assumption on the parameter region and consider the minimax estimation problem in D_{a} for arbitrary compact Θ. Also in this case we can proceed to an equivalent Bayesian estimation problem with respect to some least favourable prior distribution and hereafter provide characterizations and methods for the construction of minimax affine estimators, where then, however, besides the second order moments of the components of $\tilde{\boldsymbol{\vartheta}}$ the first order moments become of importance, too.

For a given prior distribution $P \in \Theta^*$ denote

$$\boldsymbol{\mu}_P = \int_{\Theta} \boldsymbol{\vartheta} P(\mathrm{d}\boldsymbol{\vartheta})$$

and (15.48)

$$\boldsymbol{\Phi}_P = \int_{\Theta} (\boldsymbol{\vartheta} - \boldsymbol{\mu}_P)(\boldsymbol{\vartheta} - \boldsymbol{\mu}_P)^\top P(\mathrm{d}\boldsymbol{\vartheta}) = \boldsymbol{M}_P - \boldsymbol{\mu}_P\boldsymbol{\mu}_P^\top$$

the expectation vector and covariance matrix associated with P, respectively. The Bayes risk of an estimator $\hat{\vartheta} = Z\tilde{y} + z \in D_a$ can then be rewritten as

$$\int_{\Theta} R(\vartheta, \hat{\vartheta})P(\mathrm{d}\vartheta) = \operatorname{tr} U(\bar{Z}\bar{A}\bar{Z}^{\top} - \bar{Z}\bar{B} - \bar{B}^{\top}\bar{Z}^{\top} + M_P)$$

with (15.49)

$$\bar{Z} = [z \mid Z], \quad \bar{A} = \left[\begin{array}{c|c} 1 & \mu_P^{\top}F^{\top} \\ \hline F\mu_P & FM_PF^{\top} + \Sigma \end{array}\right], \quad \bar{B} = \left[\begin{array}{c} \mu_P^{\top} \\ \hline FM_P \end{array}\right]$$

(see (5.32) and (5.33)). Further, any Bayes affine estimator of ϑ with respect to $P \in \Theta^*$ admits a representation

$$\begin{aligned} \hat{\vartheta}_P^{a} = {} & [U^+U\Phi_PF^{\top}(F\Phi_PF^{\top} + \Sigma)^{-1} + (I_r - U^+U)G](\tilde{y} - F\mu_P) \\ & + U^+U(\mu_P - g) + g, \; G \in \mathcal{M}_{r\times n}, \; g \in \mathbb{R}^r \end{aligned} \tag{15.50}$$

(see Theorem 5.4). Defining

$$\bar{Z}_P = [z_P \mid Z_P], \quad Z_P = \Phi_PF^{\top}(F\Phi_PF^{\top} + \Sigma)^{-1}, \quad z_P = (I_r - Z_PF)\mu_P, \tag{15.51}$$

$\hat{\vartheta}_P^{a}$ from (15.50) can be written in a "quasi-linear" form

$$\hat{\vartheta}_P^{a} = [U^+U\bar{Z}_P + (I_r - U^+U)\bar{G}]\tilde{y}_0 \quad \text{with} \quad \tilde{y}_0 = \left[\begin{array}{c} 1 \\ \hline \tilde{y} \end{array}\right] \tag{15.52}$$

and arbitrary $\bar{G} = [g \mid G] \in \mathcal{M}_{r\times(n+1)}$. In case of a full rank weight matrix U the Bayes affine estimator with respect to P is uniquely determined by $\hat{\vartheta}_P^{a} = \bar{Z}_P\tilde{y}_0 = Z_P\tilde{y} + z_P$.

In order to find a minimax affine estimator we proceed to the equivalent problem

$$\sup_{P\in\Theta^*} \int_{\Theta} R(\vartheta, \hat{\vartheta})P(\mathrm{d}\vartheta) \to \inf_{\bar{Z}\in\mathcal{M}_{r\times(n+1)}}. \tag{15.53}$$

Introducing the modified moment matrix (of first and second order moments)

$$\bar{M}_P = \left[\begin{array}{c|c} 1 & \mu_P^{\top} \\ \hline \mu_P & M_P \end{array}\right], \qquad P \in \Theta^* \tag{15.54}$$

the minimax problem (15.53) can be solved using the framework developed in the preceding section.

Lemma 15.8. *Let $\hat{\vartheta} = Z\tilde{y} + z \in D_a$ and $P \in \Theta^*$. Then it holds for the Bayes risk of $\hat{\vartheta}$ with respect to P*

$$\begin{aligned} \int_{\Theta} R(\vartheta, \hat{\vartheta})P(\mathrm{d}\vartheta) &= \operatorname{tr} U[\bar{Z}\Sigma_0\bar{Z}^{\top} + (\bar{Z}F_0 - J)\bar{M}_P(\bar{Z}F_0 - J)^{\top}] \\ &=: \varrho(\bar{M}_P, \bar{Z}) \end{aligned} \tag{15.55}$$

with

$$\Sigma_0 = \left[\begin{array}{c|c} 0 & \mathbf{0}_n^\top \\ \hline \mathbf{0}_n & \Sigma \end{array}\right], \qquad F_0 = \left[\begin{array}{c|c} 1 & \mathbf{0}_r^\top \\ \hline \mathbf{0}_n & F \end{array}\right], \qquad J = [\mathbf{0}_r \mid I_r].$$

Proof. One easily verifies the following identities:

$$\bar{A} = \Sigma_0 + F_0 \bar{M}_P F_0^\top, \quad \bar{B} = F_0 \bar{M}_P J^\top, \quad M_P = J\bar{M}_P J^\top. \tag{15.56}$$

Then we obtain

$$\bar{\varrho}(\bar{M}_P, \bar{Z}) = \operatorname{tr} U(\bar{Z}\bar{A}\bar{Z}^\top - \bar{Z}\bar{B} - \bar{B}^\top \bar{Z}^\top + M_P),$$

i.e., (15.49) and (15.55) are identical representations for the Bayes risk of $\hat{\vartheta}$. □

Lemma 15.9. *Let Θ be compact and $\bar{Z}_P$ as defined in* (15.51). *Then there exist a least favourable prior distribution and a minimax affine estimator and it holds:*

(i) *P_0 is least favourable within Θ^* if and only if*

$$\bar{\varrho}(\bar{M}_{P_0}, \bar{Z}_{P_0}) = \sup_{P \in \Theta^*} \bar{\varrho}(\bar{M}_P, \bar{Z}_P). \tag{15.57}$$

(ii) *$\hat{\vartheta}^{\mathrm{a}}_{\mathrm{M}}$ is minimax in D_{a} if and only if $\hat{\vartheta}^{\mathrm{a}}_{\mathrm{M}}$ is Bayesian in D_{a} with respect to some least favourable prior distribution $P_0 \in \Theta^*$.*

(iii) $\inf\limits_{\hat{\vartheta} \in D_{\mathrm{a}}} \sup\limits_{\vartheta \in \Theta} R(\vartheta, \hat{\vartheta}) = \sup\limits_{P \in \Theta^*} \bar{\varrho}(\bar{M}_P, \bar{Z}_P)$.

Proof. Denote $\bar{\mathcal{M}}$ the set of all moment matrices generated by Θ^*,

$$\bar{\mathcal{M}} = \{\bar{M}_P \in \mathcal{M}^{\geq}_{r+1} : P \in \Theta^*\}. \tag{15.58}$$

This set is convex and compact with respect to the metric generated by $\|\bar{M}_P\| = (\operatorname{tr} \bar{M}_P^2)^{1/2}$. The convexity of $\bar{\mathcal{M}}$ follows from the convexity of Θ^* and the compactness follows immediately from A.19 (i) observing that

$$\bar{M}_P = \int_\Theta f(\vartheta) f(\vartheta)^\top P(\mathrm{d}\vartheta) \quad \text{with} \quad f(\vartheta) := \begin{pmatrix} 1 \\ \vartheta \end{pmatrix}, \quad \vartheta \in \Theta \tag{15.59}$$

and $f(\cdot)$ being a continuous function on the compact set Θ. Then, by virtue of Lemma 6.5 and Lemma 15.8, we have

$$\inf_{\hat{\vartheta} \in D_{\mathrm{a}}} \sup_{\vartheta \in \Theta} R(\vartheta, \hat{\vartheta}) = \inf_{\bar{Z} \in \mathcal{M}_{r \times (n+1)}} \sup_{\bar{M}_P \in \bar{\mathcal{M}}} \bar{\varrho}(\bar{M}_P, \bar{Z}). \tag{15.60}$$

The function $\bar{\varrho}$ satisfies the assumptions of Sion's theorem (see A.20) so that we can interchange the order of minimization and maximization in the right-hand side expression of (15.60). From this the Assertions (i) and (iii) follow using the fact that $\hat{\vartheta}^{\mathrm{a}}_P = \bar{Z}_P \bar{y}_0$ is Bayesian in D_{a} with respect to P and thus $\bar{\varrho}(\bar{M}_P, \cdot)$ is minimized by $\bar{Z}_P = \bar{Z}_P(\bar{M}_P)$. Since $\bar{Z}_P$ minimizes the trace functional given in (15.49) with respect to $\bar{Z}$, it follows from Lemma 5.3 that $\bar{Z}_P = \bar{B}^\top \bar{A}^{-1}$, which by virtue of (15.56), implies that

$$\bar{Z}_P = J\bar{M}_P F_0^\top (F_0 \bar{M}_P F_0^\top + \Sigma_0)^{-1}. \tag{15.61}$$

Further, according to Lemma 5.3 and (15.56), the minimum value of the trace functional is given by

$$\bar{\varrho}(\bar{M}_P, \bar{Z}_P) = \operatorname{tr} U(M_P - \bar{B}^\top \bar{A}^{-1}\bar{B}) \\ = \operatorname{tr} UJ[\bar{M}_P - \bar{M}_P F_0^\top (F_0 \bar{M}_P F_0^\top + \Sigma_0)^{-1} F_0 \bar{M}_P] J^\top . \tag{15.62}$$

The existence of a least favourable prior distribution P_0 satisfying (15.57), and which is completely determined by its moment matrix $\bar{M}_P$, follows from the fact that $\bar{\varrho}(\bar{M}_P, \bar{Z}_P)$ is a continuous functional on the compact set $\bar{\mathcal{M}}$ with respect to the metric generated by $\|\bar{M}_P\| = (\operatorname{tr} \bar{M}_P^2)^{1/2}$. The Assertion (ii) can be proved along the same lines as in the proof of Lemma 15.3 replacing there (15.10) by (15.52) with $P = P_0$ and $\varrho(M_{P_0}, Z_{P_0})$ by $\bar{\varrho}(\bar{M}_{P_0}, \bar{Z}_{P_0})$. Finally, the existence of a minimax affine estimator follows from Assertion (ii) and the existence of a least favourable prior. □

With the results of Lemma 15.9 the minimax affine estimators are characterized as Bayes affine estimators of the type (15.52), formed with a least favourable prior $P_0 \in \Theta^*$.

Theorem 15.6. *In case of a compact parameter region Θ the estimator $\hat{\vartheta}_{\mathrm{M}}^{\mathrm{a}}$ is minimax affine with respect to quadratic loss if and only if it has the form*

$$\hat{\vartheta}_{\mathrm{M}}^{\mathrm{a}} = [U^+ UJ\bar{M}_0 F_0^\top (F_0 \bar{M}_0 F_0^\top + \Sigma_0)^{-1} + (I_r - U^+ U)\bar{G}]\tilde{y}_0 \tag{15.63}$$

with some $\bar{G} \in \bar{\mathcal{M}}_{r \times (n+1)}$, where $\bar{M}_0 = \bar{M}_{P_0}$ is the moment matrix associated with a least favourable $P_0 \in \Theta^*$. The minimax risk, which is independent of the choice of $\bar{G}$, is given by

$$\sup_{\vartheta \in \Theta} R(\vartheta, \hat{\vartheta}_{\mathrm{M}}^{\mathrm{a}}) = \operatorname{tr} J^\top UJ[\bar{M}_0 - \bar{M}_0 F_0^\top (F_0 \bar{M}_0 F_0^\top + \Sigma_0)^{-1} F_0 \bar{M}_0] . \tag{15.64}$$

Under the assumption that the design matrix has full rank we can show, again, that the least favourable moment matrix can be obtained as solution to a Bayes-L-optimal design problem.

Lemma 15.10. *Assume* rank $F^\top \Sigma^{-1} F = r$ *and denote*

$$V_0 = \left[\begin{array}{c|c} 0 & 0_r^\top \\ \hline 0_r & V \end{array}\right], \qquad U_0 = V_0 J^\top UJV_0 . \tag{15.65}$$

Then P_0 is least favourable within Θ^* if and only if it minimizes the functional

$$\bar{L}_{\mathrm{B}}(P) := \operatorname{tr} U_0(\bar{M}_P + V_0)^{-1}, \qquad P \in \Theta^* \tag{15.66}$$

defined over the set Θ^*.

Proof. With Z_P as defined in (15.51) the Bayes risk can be alternatively represented as

$$\bar{\varrho}(\bar{M}_P, \bar{Z}_P) = \operatorname{tr} U(I_r - Z_P F)\Phi_P ,$$

cp. Theorem 5.4. With the same rearrangements as in the proof of

Lemma 15.4, except that M_P must be replaced by Φ_P, this implies

$$\bar{\varrho}(\bar{M}_P, \bar{Z}_P) = \operatorname{tr} UV - \operatorname{tr} VUV(\Phi_P + V)^{-1}. \tag{15.67}$$

On the other hand, applying the block inversion formula A.37 we obtain:

$$(\bar{M}_P + V_0)^{-1} = \left[\begin{array}{c|c} (1 - \mu_P^\top (M_P + V)^{-1} \mu_P)^{-1} & -\mu_P^\top (\Phi_P + V)^{-1} \\ \hline -(\Phi_P + V)^{-1} \mu_P & (\Phi_P + V)^{-1} \end{array}\right]. \tag{15.68}$$

This yields $\bar{L}_B(P) = \operatorname{tr} UJV_0(\Phi_P + V)^{-1} V_0 J^\top = \operatorname{tr} UV(\Phi_P + V)^{-1} V$ which, by virtue of (15.67), gives

$$\bar{\varrho}(\bar{M}_P, \bar{Z}_P) = \operatorname{tr} UV - \bar{L}_B(P). \tag{15.69}$$

The assertion then follows from Lemma 15.9 (i). □

Remark 15.5. Note that the matrix V_0 occuring in the functional $\bar{L}_B$ is singular. The sum $\bar{M}_P + V_0$, however, is regular and thus positive definite for any $P \in \Theta^*$, since with $M_P = \Phi_P + \mu_P \mu_P^\top$ and $M_P + V \in \mathcal{M}_r^>$ it follows from A.37 and A.34 that

$$\det(\bar{M}_P + V_0) = \det(M_P + V)/(1 + \mu_P^\top (\Phi_P + V)^{-1} \mu_P) > 0.$$

Also, note that the matrix U_0 defined in (15.65) is singular so that a least favourable moment matrix must not necessarily be uniquely determined.

With (15.69) we immediately get from Lemma 15.9 (iii) the following identity for the minimax risk in D_a:

$$\inf_{\hat{\vartheta} \in D_a} \sup_{\vartheta \in \Theta} R(\vartheta, \hat{\vartheta}) = \operatorname{tr} UV - \inf_{P \in \Theta^*} \bar{L}_B(P). \tag{15.70}$$

Thus, in complete analogy with Corollaries 15.1 and 15.2, we conclude that minimax affine estimators effect (U, Θ)-improvements over the generalized LSE and these improvements are optimal within the class D_a. Comparing the two functionals L_B and $\bar{L}_B$ defined by (15.19) and (15.66), respectively, we obtain

$$\bar{L}_B(P) = \operatorname{tr} VUV(\Phi_P + V)^{-1} \geqq \operatorname{tr} VUV(M_P + V)^{-1} = L_B(P), \qquad \forall P \in \Theta^*$$

due to the fact that $\Phi_P \leqq M_P = \Phi_P + \mu_P \mu_P^\top$. Referring to (15.20) and (15.70) we get the following relationship between the corresponding minimax risks in D_l and D_a, respectively:

$$\sup_{\vartheta \in \Theta} R(\vartheta, \hat{\vartheta}_M) - \sup_{\vartheta \in \Theta} R(\vartheta, \hat{\vartheta}_M^a) = \inf_{P \in \Theta^*} \bar{L}_B(P) - \inf_{P \in \Theta^*} L_B(P) \geqq 0.$$

Now we draw a link between the problem of minimax affine estimation and a related problem in Bayesian experimental design. From the representation (15.59) we infer that the moment matrices $\bar{M}_P$ can be interpreted as information matrices of experimental design measures P for the multiple linear regression setup

$$\mathbf{E}\,\tilde{y}(\boldsymbol{t}) = \boldsymbol{f}(\boldsymbol{t})^\top \boldsymbol{\beta} = \beta_0 + \beta_1 t_1 + \ldots + \beta_r t_r,$$
$$\boldsymbol{t} = (t_1, \ldots, t_r)^\top \in \Theta \subset \mathbb{R}^r, \qquad \boldsymbol{\beta} = (\beta_0, \ldots, \beta_r)^\top \in \mathbb{R}^{r+1}, \tag{15.71}$$

where $\boldsymbol{\beta}$ is an unknown regression parameter including an intercept term β_0 and Θ plays the role of the experimental region. The minimization of the functional $\bar{L}_B$ is then equivalent to determining a Bayes-L-optimal design for estimating $\boldsymbol{\beta}$ assuming that $\mathbf{Cov}\,\tilde{\boldsymbol{\beta}} \propto \boldsymbol{V}_0^+$, i.e.,

$$\mathbf{Cov}\,(\tilde{\beta}_1, \ldots, \tilde{\beta}_r)^\top \propto \boldsymbol{V}^{-1}, \qquad \mathrm{Var}\,\tilde{\beta}_0 = \infty, \qquad \mathbf{Cov}\,(\tilde{\beta}_0, \tilde{\beta}_i) = 0, \qquad i = 1, \ldots, r,$$

and assuming quadratic loss

$$\|\boldsymbol{\beta} - \hat{\boldsymbol{\beta}}\|^2_{U_0} = \|\boldsymbol{J}\boldsymbol{V}_0\boldsymbol{\beta} - \boldsymbol{J}\boldsymbol{V}_0\hat{\boldsymbol{\beta}}\|^2_U = \|\boldsymbol{V}\boldsymbol{\beta}_{(0)} - \boldsymbol{V}\hat{\boldsymbol{\beta}}_{(0)}\|^2_U,$$

where $\boldsymbol{\beta}_{(0)} = (\beta_1, \ldots, \beta_r)^\top$. Due to the particular structure of

$$\boldsymbol{U}_0 = \left[\begin{array}{c|c} 0 & \boldsymbol{0}_r^\top \\ \hline \boldsymbol{0}_r & \boldsymbol{V}\boldsymbol{U}\boldsymbol{V} \end{array}\right] \tag{15.72}$$

this corresponds to experimental designing based on a "truncated" Bayes-L-optimality criterion for estimating the parameter $\boldsymbol{\gamma} = \boldsymbol{J}\boldsymbol{V}_0\boldsymbol{\beta} = \boldsymbol{V}\boldsymbol{\beta}_{(0)}$, i.e., the intercept term β_0 is discarded. With the preceding interpretation we can now supplement the characterization of minimax affine estimators given in Theorem 15.6 as follows.

Theorem 15.7. *Under the assumptions of compactness of* Θ *and* $\operatorname{rank} \boldsymbol{F}^\top \boldsymbol{\Sigma}^{-1} \boldsymbol{F} = r$ *the estimator* $\hat{\boldsymbol{\vartheta}}^{\mathrm{a}}_{\mathrm{M}}$ *is minimax in* D_{a} *if and only if it can be written in the form* (15.63) *with a matrix* $\bar{\boldsymbol{M}}_0$ *satisfying the following conditions:*

(i) $\bar{\boldsymbol{M}}_0 \in \bar{\mathcal{M}} = \{\bar{\boldsymbol{M}}_P : P \in \Theta^*\}$.

(ii) $\sup\limits_{\boldsymbol{\vartheta} \in \Theta} \boldsymbol{f}(\boldsymbol{\vartheta})^\top (\bar{\boldsymbol{M}}_0 + \boldsymbol{V}_0)^{-1} \boldsymbol{U}_0 (\bar{\boldsymbol{M}}_0 + \boldsymbol{V}_0)^{-1} \boldsymbol{f}(\boldsymbol{\vartheta})$
$= \operatorname{tr}(\bar{\boldsymbol{M}}_0 + \boldsymbol{V}_0)^{-1} \boldsymbol{U}_0 (\bar{\boldsymbol{M}}_0 + \boldsymbol{V}_0)^{-1} \bar{\boldsymbol{M}}_0$, *where* $\boldsymbol{f}(\boldsymbol{\vartheta}) = (1, \boldsymbol{\vartheta}^\top)^\top$.

We remark that the second condition follows from Theorem 11.12 replacing there $\boldsymbol{M}_B(\xi^*)$ by $\bar{\boldsymbol{M}}_0 + \boldsymbol{V}_0$ and observing Lemma 15.10 and the representation (15.59).

Remark 15.6. In case that the design matrix $\boldsymbol{F}$ has full rank, using the matrix identity $\boldsymbol{\Phi}_0 \boldsymbol{F}^\top (\boldsymbol{F} \boldsymbol{\Phi}_0 \boldsymbol{F}^\top + \boldsymbol{\Sigma})^{-1} \boldsymbol{F} = \boldsymbol{\Phi}_0 (\boldsymbol{\Phi}_0 + \boldsymbol{V})^{-1}$, the minimax affine estimators can be written more conveniently as

$$\hat{\boldsymbol{\vartheta}}^{\mathrm{a}}_{\mathrm{M}} = [\boldsymbol{U}^+ \boldsymbol{U} \boldsymbol{\Phi}_0 (\boldsymbol{\Phi}_0 + \boldsymbol{V})^{-1} + (\boldsymbol{I}_r - \boldsymbol{U}^+ \boldsymbol{U}) \boldsymbol{C}] (\hat{\boldsymbol{\vartheta}}_{\Sigma} - \boldsymbol{\mu}_0)$$
$$+ \boldsymbol{U}^+ \boldsymbol{U} (\boldsymbol{\mu}_0 - \boldsymbol{g}) + \boldsymbol{g}, \; \boldsymbol{C} \in \mathcal{M}_{r \times r}, \; \boldsymbol{g} \in \mathbb{R}^r, \tag{15.73}$$

where $\boldsymbol{\mu}_0$ and $\boldsymbol{\Phi}_0$, respectively, denote the expection vector and covariance matrix associated with a least favourable prior distribution $P_0 \in \Theta^*$. Also, using (15.68) and the special structure of $\boldsymbol{U}_0$ indicated in (15.72), Condition (ii) of Theorem 15.7 can be replaced by the following simpler and equivalent condition:

$$\sup_{\vartheta \in \Theta} \| V(\Phi_0 + V)^{-1}(\vartheta - \mu_0)\|_U^2 = \operatorname{tr}(\Phi_0 + V)^{-1} VUV(\Phi_0 + V)^{-1}\Phi_0 . \qquad (15.74)$$

Finally, the result of Lemma 15.10 allows, again, the application of the iteration procedure described in Section 12.5 (especially Algorithm 12.4) for an approximate computation of a least favourable moment matrix $\bar{M}_0$. To this one constructs a sequence of measures $\{P_s\}_{s \in \mathbb{N}} \subset \Theta^*$ according to (15.26) and (15.27) and then determines the supporting point ϑ_s to be chosen at stage s such that it maximizes the quadratic form $\|(\bar{M}_{P_{s-1}} + V_0)^{-1} f(\vartheta)\|_{U_0}^2$ with respect to $\vartheta \in \Theta$ (cp. Algorithm 12.4). This amounts to choosing

$$\vartheta_s = \arg \sup_{\vartheta \in \Theta} \| V(\Phi_{P_{s-1}} + V)^{-1}(\vartheta - \mu_{P_{s-1}})\|_U^2, \quad s = 1, 2, \ldots \qquad (15.75)$$

Again, the initial measure may be taken as a one-point-measure $P_0 = \delta_{\vartheta_0}$ according to

$$\vartheta_0 = \arg \inf_{\vartheta \in \Theta} \bar{L}_B(\delta_\vartheta) = \arg \sup_{\vartheta \in \Theta} \operatorname{tr} VUV(\Phi_{\delta_\vartheta} + V)^{-1}$$

(see (15.67) and (15.69)). Observing, however, that $\Phi_{\delta_\vartheta} = \mathbf{0}_{r,r}$ for any $\vartheta \in \Theta$, we arrive at the surprising fact that the iteration procedure may be started with an arbitrary one-point measure δ_ϑ, $\vartheta \in \Theta$.

15.3. Minimax estimation with ellipsoid constraints

In this section we will find minimax affine estimators for the special case of ellipsoidal restrictions for the regression parameter. Under an additional condition we will give the explicit solution for quadratic loss with arbitrary weight matrix $U \in \mathcal{M}_r^{\geqq}$, thus unifying and generalizing previous results of Kuks/Ol'man (1972), Läuter (1975 a), Hoffmann (1979), Toutenburg (1982), Pilz (1986 a, b), (1987), Stahlecker (1987), Stahlecker/Trenkler (1987) and Gaffke/Heiligers (1989). In order to obtain characterizations and methods for the construction of the minimax affine estimators we make essential use of the established analogies with Bayesian experimental design theory. Throughout this section let be satisfied.

Assumption 15.2. rank $F^\top \Sigma^{-1} F = r$.

We consider an ellipsoidal parameter region

$$\Theta = \mathscr{E}(H, \vartheta_0) := \{\vartheta \in \mathbb{R}^r : (\vartheta - \vartheta_0)^\top H(\vartheta - \vartheta_0) \leqq 1\} \qquad (15.76)$$

with some given shape matrix $H \in \mathcal{M}_r^{>}$ and center point $\vartheta_0 \in \Theta$. Because of the symmetry of Θ with respect to ϑ_0 we can restrict the search for a minimax affine estimator to the subclass $D_a^0 = \{\hat{\vartheta} = Z(\tilde{y} - F\vartheta_0) + \vartheta_0 : Z \in \mathcal{M}_{r \times n}\}$, see Lemma 15.7, and, moreover, the minimax estimation problem in D_a^0 is equivalent to that in D_l when changing to the translated parameter region

$$\bar{\Theta} = \mathscr{E}(\boldsymbol{H}) := \{\bar{\boldsymbol{\vartheta}} \in \mathbb{R}^r : \bar{\boldsymbol{\vartheta}}^\top \boldsymbol{H} \bar{\boldsymbol{\vartheta}} \leqq 1\}. \tag{15.77}$$

According to Theorem 15.2 and Theorem 15.5, every minimax estimator in D_{a}^0 can be written in the form

$$\begin{aligned} \hat{\boldsymbol{\vartheta}}_{\mathrm{M}}^{\mathrm{a}} = [&\boldsymbol{U}^+ \boldsymbol{U} \boldsymbol{M}_0 (\boldsymbol{M}_0 + \boldsymbol{V})^{-1} \\ &+ (\boldsymbol{I}_r - \boldsymbol{U}^+ \boldsymbol{U}) \boldsymbol{C}] (\hat{\boldsymbol{\vartheta}}_\Sigma - \boldsymbol{\vartheta}_0) + \boldsymbol{\vartheta}_0, \quad \boldsymbol{C} \in \mathscr{M}_{r \times r}, \end{aligned} \tag{15.78}$$

where $\boldsymbol{M}_0$ is the moment matrix of a least favourable prior distribution from $\bar{\Theta}^* = \mathscr{E}(\boldsymbol{H})^*$, the set of all probability measures defined on $\mathscr{E}(\boldsymbol{H})$. From the proof of Lemma 12.2 we know that the set of moment matrices generated by $\bar{\Theta}^*$ is given by

$$\mathscr{M} = \{\boldsymbol{M}_P : P \in \bar{\Theta}^*\} = \{\boldsymbol{M} \in \mathscr{M}_r^{\geqq} : \operatorname{tr} \boldsymbol{H}\boldsymbol{M} \leqq 1\}. \tag{15.79}$$

According to Corollary 15.4, the supporting points of a least favourable distribution P_0 necessarily maximize the convex quadratic function $\boldsymbol{g}(\bar{\boldsymbol{\vartheta}}) = \bar{\boldsymbol{\vartheta}}^\top (\boldsymbol{M}_0 + \boldsymbol{V})^{-1} \boldsymbol{V}\boldsymbol{U}\boldsymbol{V} (\boldsymbol{M}_0 + \boldsymbol{V})^{-1} \bar{\boldsymbol{\vartheta}}$ with respect to $\bar{\boldsymbol{\vartheta}} \in \bar{\Theta}$ and thus must be boundary points of the convex set $\bar{\Theta}$, i.e., it holds

$$\operatorname{supp} P_0 \subset T := \{\boldsymbol{t} \in \mathbb{R}^r : \boldsymbol{t}^\top \boldsymbol{H} \boldsymbol{t} = 1\}. \tag{15.80}$$

Corollary 15.7. *Denote* $\mathscr{M}_{\boldsymbol{H}} = \{\boldsymbol{M} \in \mathscr{M}_r^{\geqq} : \operatorname{tr} \boldsymbol{H}\boldsymbol{M} = 1\}$. *Under the Assumption* 15.2 $\boldsymbol{M}_0$ *is a least favourable moment matrix in* $\mathscr{M}$ *if and only if*

$$\boldsymbol{M}_0 = \arg \inf_{\boldsymbol{M} \in \mathscr{M}_{\boldsymbol{H}}} \operatorname{tr} \boldsymbol{V}\boldsymbol{U}\boldsymbol{V} (\boldsymbol{M} + \boldsymbol{V})^{-1}. \tag{15.81}$$

Proof. From (15.80) it follows necessarily that

$$\operatorname{tr} \boldsymbol{H}\boldsymbol{M}_0 = \operatorname{tr} \left\{ \boldsymbol{H} \int_T \boldsymbol{t}\boldsymbol{t}^\top P_0(\mathrm{d}\boldsymbol{t}) \right\} = \int_T \boldsymbol{t}^\top \boldsymbol{H} \boldsymbol{t} P_0(\mathrm{d}\boldsymbol{t}) = 1.$$

This, together with Theorem 15.5 and Lemma 15.4, yields the assertion. (The necessity of the condition $\operatorname{tr} \boldsymbol{H}\boldsymbol{M}_0 = 1$ also follows from the fact that $\boldsymbol{M}_0$ is a boundary point of $\mathscr{M}$, see Remark 15.3 b). □

Summarizing, we get the following characterization of minimax affine estimators under ellipsoidal restrictions.

Theorem 15.8. *Let* $\Theta = \mathscr{E}(\boldsymbol{H}, \boldsymbol{\vartheta}_0)$ *and Assumption* 15.2 *be satisfied. The estimator* $\hat{\boldsymbol{\vartheta}}_{\mathrm{M}}^{\mathrm{a}}$ *is minimax in* D_{a}^0 *if and only if it can be written in the form* (15.78) *with a matrix* $\boldsymbol{M}_0$ *satisfying the following conditions:*

(i) $\boldsymbol{M}_0 \in \mathscr{M}_{\boldsymbol{H}}$.

(ii) $\lambda_{\max}(\boldsymbol{H}^{-1/2} \boldsymbol{M}_0^* \boldsymbol{H}^{-1/2}) = \operatorname{tr} \boldsymbol{M}_0^* \boldsymbol{M}_0$, *where* $\boldsymbol{M}_0^* = (\boldsymbol{M}_0 + \boldsymbol{V})^{-1} \boldsymbol{V}\boldsymbol{U}\boldsymbol{V} (\boldsymbol{M}_0 + \boldsymbol{V})^{-1}$.

This follows with Corollary 15.7 immediately from Theorem 15.2 and Theorem 15.5, respectively, observing that

$$\sup_{\bar{\vartheta} \in \bar{\Theta}} \bar{\vartheta}^\top (M_0 + V)^{-1} VUV (M_0 + V)^{-1} \bar{\vartheta} = \sup_{z^\top z \leqq 1} z^\top H^{-1/2} M_0^* H^{-1/2} z$$
$$= \lambda_{\max}(H^{-1/2} M_0^* H^{-1/2}).$$

The following lemma indicates a matrix equation for the least favourable moment matrix, which may be helpful for the construction of such a matrix.

Lemma 15.11. *Let $M_0 \in \mathcal{M}_H$. Then the following condition is equivalent to that given in Theorem* 15.8 (ii):

$$(\lambda H - M_0^*) M_0 = 0_{r,r}, \quad \textit{where} \quad \lambda = \lambda_{\max}(H^{-1/2} M_0^* H^{-1/2}). \quad (15.82)$$

Proof. First, by definition of λ, it is clear that

$$\lambda H - M_0^* = H^{-1/2}(\lambda I_r - H^{-1/2} M_0^* H^{-1/2}) H^{1/2} \in \mathcal{M}_r^{\geqq}. \quad (15.83)$$

Observing that $M_0 \in \mathcal{M}_r^{\geqq}$ and $\operatorname{tr} HM_0 = 1$, this implies

$$\operatorname{tr}(\lambda H - M_0^*) M_0 = \lambda - \operatorname{tr} M_0^* M_0 \geqq 0$$

and the equality $\lambda = \operatorname{tr} M_0^* M_0$, i.e., Condition (ii) of Theorem 15.8, is fulfilled if and only if (15.82) holds. □

Together with Lemma 15.11, the result of Theorem 15.8 generalizes a characterization theorem due to Hoffmann (1979) who proved the following assertion for the special case $\vartheta_0 = 0_r$ (which entails $D_a^0 = D_1$).

Theorem 15.9. *Let $\Theta = \mathscr{E}(H, \vartheta_0)$ and assume* $\operatorname{rank} F^\top \Sigma^{-1} F = \operatorname{rank} U = r$. *Then the estimator $\hat{\vartheta}_M^a = Z(\tilde{y} - F\vartheta_0) + \vartheta_0$ is minimax in D_a^0 if and only if there exists a matrix $A \in \mathcal{M}_{r \times r}$ such that $Z = AF^\top \Sigma^{-1}$ and the following conditions are satisfied:*

(1) $V^{-1/2} A V^{-1/2} \in \mathscr{A}_{[0,1)}$.

(2) $\operatorname{tr} H[(I_r - ZF)^{-1} - I_r] V = 1$.

(3) $[\lambda H - (I_r - ZF)^\top U (I_r - ZF)][(I_r - ZF)^{-1} - I_r] V = 0_{r,r}$,

where $\lambda = \lambda_{\max}\{H^{-1/2}(I_r - ZF)^\top U (I_r - ZF) H^{-1/2}\}$.

(See Hoffmann (1979), Theorem 1.)

Replacing Condition (ii) of Theorem 15.8 by (15.82), the characterization given in that theorem exactly yields Hoffmann's result for the special case $\operatorname{rank} U = r$: In this case we have

$$\hat{\vartheta}_M^a = M_0(M_0 + V)^{-1}(\hat{\vartheta}_\Sigma - \mu) + \mu = Z(\tilde{y} - F\mu) + \mu \quad (15.84)$$

with $Z = M_0(M_0 + V)^{-1} VF^\top \Sigma^{-1}$ and $\hat{\vartheta}_M^a$ is the unique minimax estimator in D_a^0. Setting $A = M_0(M_0 + V)^{-1} V = V - V(M_0 + V)^{-1} V$, the condition $M_0 \in \mathcal{M}_r^{\geqq}$ is equivalent to Condition (1) of Theorem 15.9. Finally, observing that $I_r - ZF = V(M_0 + V)^{-1}$, one easily verifies that the conditions $\operatorname{tr} HM_0 = 1$ and (15.82), respectively, are equivalent to the Equations (2) and (3) of Theorem 15.9.

We will now deal with the problem of how to determine explicitly a least favourable moment matrix. First, by virtue of Corollary 15.7, we can immediately apply the result of Theorem 12.3 on Bayes L-optimal designs to our present problem.

Corollary 15.8. *Let the assumptions of Theorem* 15.9 *be satisfied and, additionally, assume that*

$$aH^{-1/2}(H^{1/2}VUVH^{1/2})^{1/2}H^{-1/2} \geqq V \quad \text{with} \quad a = (1 + \operatorname{tr} HV)/\operatorname{tr}(H^{1/2}VUVH^{1/2})^{1/2}. \tag{15.85}$$

Then the unique least favourable moment matrix is given by

$$M_0 = aH^{-1/2}(H^{1/2}VUVH^{1/2})^{1/2}H^{-1/2} - V.$$

Proof. The result follows after replacing in Theorem 12.3 U_L, $\frac{1}{n}\Phi^{-1}$ and $M(\xi^*)$ by VUV, V and M_0, respectively. The uniqueness of M_0 results from the assumption that U has full rank. □

The above corollary is a slight generalization of a result due to GAFFKE/HEILIGERS (1989) who considered the special case $U = I_r$. Also, the result gives a corrected version of Theorem 4 of PILZ (1986 a), for this theorem to hold true generally one requires the additional assumption that $VHU = UHV$.

Example 15.1. Let the assumptions of Theorem 15.9 be satisfied and assume $H = h^{-2}I_r$ for some $h > 0$, i.e., the parameter region is assumed to be a hypersphere with radius h. Then, provided that

$$h_0(VUV)^{1/2} \geqq V \quad \text{with} \quad h_0 = (h^2 + \operatorname{tr} V)/\operatorname{tr}(VUV)^{1/2} \tag{15.86}$$

holds true, Corollary 15.8 says that $M_0 = h(VUV)^{1/2} - V$ is a least favourable moment matrix and the corresponding minimax estimator reads

$$\hat{\vartheta}^a_M = \hat{\vartheta}_\Sigma - (1/h_0)V(VUV)^{-1/2}(\hat{\vartheta}_\Sigma - \mu).$$

The above condition of nonnegative definiteness of $h_0(VUV)^{1/2} - V$ is satisfied provided that the radius h of the sphere is sufficiently large. Particularly, in case of $U = I_r$ we have $h_0 = 1 + h^2/\operatorname{tr} V > 1$ and (15.86) holds automatically. Then $M_0 = (h^2/\operatorname{tr} V)V$ is least favourable and the minimax estimator takes the form of a simple convex combination of the generalized LSE and the midpoint ϑ_0 of the sphere:

$$\hat{\vartheta}^a_M = (1 - 1/h_0)\hat{\vartheta}_\Sigma + (1/h_0)\mu.$$

For $\mu = 0_k$ this reduces to a shrunken estimator $g\hat{\vartheta}_\Sigma$ with shrinkage factor $g = h^2/(h^2 + \operatorname{tr} V) \in (0, 1)$. This estimator is already known from LÄUTER (1975 a) to be minimax in the special case of $\Theta = \{\vartheta \in \mathbb{R}^r : \vartheta'\vartheta \leqq h^2\}$, $U = I_r$.

In the sequel we will give a further generalization of Corollary 15.8

which for some special cases of practical importance leads to an explicit solution. In particular, note that the weight matrix $\boldsymbol{U}$ is not required to have full rank.

To this let $\boldsymbol{S}$ be the matrix which simultaneously diagonalizes $(\boldsymbol{H}^{1/2}\boldsymbol{VUVH}^{1/2})^{1/2}$ and $\boldsymbol{H}^{1/2}\boldsymbol{VH}^{1/2}$, and denote $\boldsymbol{B} = \operatorname{diag}(b_1, \ldots, b_r)$ the diagonal matrix of the corresponding generalized eigenvalue problem, i.e.,

$$\boldsymbol{H}^{1/2}\boldsymbol{VH}^{1/2} = \boldsymbol{SS}^{\top}, \qquad (\boldsymbol{H}^{1/2}\boldsymbol{VUVH}^{1/2})^{1/2} = \boldsymbol{SBS}^{\top}, \tag{15.87}$$

$$\begin{aligned} &\det((\boldsymbol{H}^{1/2}\boldsymbol{VUVH}^{1/2})^{1/2} - b_i\boldsymbol{H}^{1/2}\boldsymbol{VH}^{1/2}) = 0, \\ &b_1 \geqq \ldots \geqq b_r \geqq 0. \end{aligned} \tag{15.88}$$

We remark that $\boldsymbol{S}$ is a regular matrix and every column $\boldsymbol{s}_i$ of $\boldsymbol{S}$ which corresponds to a simple eigenvalue b_i, $1 \leqq i \leqq r$, is uniquely determined (apart from a reversal of sign). Appropriately, for any information matrix out of $\mathcal{M}_H = \{\boldsymbol{M} \in \mathcal{M}_r^{\geqq} : \operatorname{tr} \boldsymbol{HM} = 1\}$ define $\boldsymbol{D_M} = \operatorname{diag}(d_1, \ldots, d_r)$ to be the diagonal matrix of the ordered eigenvalues of

$$\det(\boldsymbol{H}^{1/2}\boldsymbol{MH}^{1/2} - d_i\boldsymbol{H}^{1/2}\boldsymbol{VH}^{1/2}) = 0, \quad d_1 \geqq d_2 \geqq \ldots \geqq d_r \geqq 0 \tag{15.89}$$

and $\boldsymbol{S_M}$ the corresponding matrix of eigenvectors, i.e.,

$$\boldsymbol{H}^{1/2}\boldsymbol{VH}^{1/2} = \boldsymbol{S_M S_M^{\top}}, \qquad \boldsymbol{H}^{1/2}\boldsymbol{MH}^{1/2} = \boldsymbol{S_M D_M S_M^{\top}}. \tag{15.90}$$

The functional to be minimized then reads

$$\begin{aligned} \operatorname{tr} \boldsymbol{VUV}(\boldsymbol{M} + \boldsymbol{V})^{-1} &= \operatorname{tr}(\boldsymbol{H}^{1/2}\boldsymbol{VUVH}^{1/2})(\boldsymbol{H}^{1/2}\boldsymbol{MH}^{1/2} + \boldsymbol{H}^{1/2}\boldsymbol{VH}^{1/2})^{-1} \\ &= \operatorname{tr}(\boldsymbol{SBS}^{\top})^2(\boldsymbol{S_M D_M S_M^{\top}} + \boldsymbol{SS}^{\top})^{-1} \end{aligned} \tag{15.91}$$

and the side condition takes the form

$$\operatorname{tr} \boldsymbol{HM} = \operatorname{tr} \boldsymbol{S_M^{\top} S_M D_M} = 1. \tag{15.92}$$

Now, provided there exists a least favourable moment matrix $\boldsymbol{M} = \boldsymbol{M}_0$ for which (15.87) and (15.90) have a common solution, i.e. $\boldsymbol{S_M} = \boldsymbol{S}$, then we infer from Corollary 15.7, (15.91) and (15.92) that this matrix admits the following representation:

$$\begin{aligned} &\boldsymbol{M}_0 = \boldsymbol{H}^{-1/2}\boldsymbol{SD}_0\boldsymbol{S}^{\top}\boldsymbol{H}^{-1/2}, \quad \text{where} \\ &\boldsymbol{D}_0 = \arg\inf_{\boldsymbol{D} \in \mathcal{D}} \operatorname{tr} \boldsymbol{S}^{\top}\boldsymbol{SB}^2(\boldsymbol{D} + \boldsymbol{I}_r)^{-1} \\ \text{and} \quad & \\ &\mathcal{D} = \{\boldsymbol{D} = \operatorname{diag}(d_1, \ldots, d_r) \in \mathcal{M}_r^{\geqq} : \operatorname{tr} \boldsymbol{S}^{\top}\boldsymbol{SD} = 1\}. \end{aligned} \tag{15.93}$$

Lemma 15.12. *Let i_0 the smallest integer from $\{1, \ldots, r\}$ such that*

$$b_{i_0+1} \leqq \frac{\sum_{i=1}^{i_0} b_i \|\boldsymbol{s}_i\|^2}{1 + \sum_{i=1}^{i_0} \|\boldsymbol{s}_i\|^2} =: a < b_{i_0}, \tag{15.94}$$

where we put $b_{r+1} = 0$, and $\boldsymbol{S} = (\boldsymbol{s}_1, \ldots, \boldsymbol{s}_r)$ and $\boldsymbol{B} = \operatorname{diag}(b_1, \ldots, b_r)$ are as de-

fined in (15.87) *and* (15.88), *respectively. Then*

$$\boldsymbol{D}_0 = \operatorname{diag}(d_1^*, \ldots, d_r^*) \text{ with } d_i^* = \begin{cases} b_i/a - 1 & \text{for} \quad i = 1, \ldots, i_0 \\ 0 & \text{for} \quad i = i_0 + 1, \ldots, r \end{cases} \tag{15.95}$$

is a solution of the optimization problem (15.93).

Proof. According to (15.93), we have to minimize the function

$$h(d_1, \ldots, d_r) = \sum_{i=1}^{r} b_i^2 \|\boldsymbol{s}_i\|^2 (1 + d_i)^{-1}$$

over the convex region of all vectors $(d_1, \ldots, d_r)^\top \in \mathbb{R}^r$ with $d_1 \geqq d_2 \geqq \ldots \geqq d_r \geqq 0$ satisfying the constraint

$$\operatorname{tr} \boldsymbol{S}^\top \boldsymbol{S} \boldsymbol{D} = \sum_{i=1}^{r} \|\boldsymbol{s}_i\|^2 d_i = 1 . \tag{15.96}$$

The function h is (strictly) convex over this region, which, together with the linearity of the constraint (15.96), implies that the local Kuhn-Tucker conditions

$$\partial G / \partial d_i \geqq 0 , \quad d_i \partial G / \partial d_i = 0 , \quad d_i \geqq 0 , \quad i = 1, \ldots, r \tag{15.97}$$

and $\partial G / \partial \lambda = 0$ are necessary and sufficient for the optimality of a solution $\boldsymbol{D}_{M_0} = \operatorname{diag}(d_1^*, \ldots, d_r^*)$ of (15.93), see, e.g. BAZARAA/SHETTY (1979), Theorem 4.3.7. Here

$$G(d_1, \ldots, d_r, \lambda) = h(d_1, \ldots, d_r) + \lambda \left(\sum_{i=1}^{r} \|\boldsymbol{s}_i\|^2 d_i - 1 \right)$$

denotes the Lagrange function and λ stands for the Lagrange multiplier. Observing that $\partial G / \partial d_i = \lambda \|\boldsymbol{s}_i\|^2 - b_i^2 \|\boldsymbol{s}_i\|^2 (1 + d_i)^{-2}$, (15.97) implies that

$$\lambda \geqq b_i^2 (1 + d_i)^{-2} , \qquad \lambda d_i = b_i^2 d_i (1 + d_i)^{-2} , \qquad i = 1, \ldots, r .$$

Introducing two disjoint index sets I_0 and I_+ such that

$$d_i > 0 \text{ for } i \in I_+ , \; d_i = 0 \text{ for } i \in I_0 , \; I_+ \cup I_0 = \{1, \ldots, r\} \tag{15.98}$$

we thus have

$$d_i = b_i \lambda^{-1/2} - 1 \quad \text{for} \quad i \in I_+ \tag{15.99}$$

and

$$\lambda \geqq b_i^2 \quad \text{for} \quad i \in I_0 . \tag{15.100}$$

Further, from the equation $\partial G / \partial \lambda = 0$ we get

$$\lambda^{-1/2} = \left(1 + \sum_{i \in I_+} \|\boldsymbol{s}_i\|^2 \right) \Big/ \sum_{i \in I_+} b_i \|\boldsymbol{s}_i\|^2 . \tag{15.101}$$

By virtue of (15.99) and (15.100) we have

$$\lambda^{1/2} < b_i \quad \text{for} \quad i \in I_+ , \qquad \lambda^{1/2} \geqq b_i \quad \text{for} \quad i \in I_0 , \tag{15.102}$$

from which I_+ and I_0 arise in a natural way by ordering $b_1 \geqq b_2 \geqq \dots \geqq b_r \geqq b_{r+1} = 0$ and setting $I_+ = \{1, 2, \dots, i_0\}$, $I_0 = \{i_0 + 1, \dots, r\}$. From (15.101) and (15.102) it then follows that $b_{i_0+1} \leqq \lambda^{1/2} = a < b_{i_0}$, in accordance with (15.94). With (15.99) and $d_i = 0$ for $i \in I_0$ we thus obtain the solution indicated in (15.94). □

Summarizing, we can state the following result.

Theorem 15.10. *Let* $\Theta = \mathscr{E}(\boldsymbol{H}, \boldsymbol{\mu})$, $\operatorname{rank} \boldsymbol{F} = r$ *and assume quadratic loss* (15.2) *with arbitrary weight matrix* $\boldsymbol{U} \in \mathscr{M}_r^{\geqq}$. *Further, let* $\boldsymbol{S}$, $\boldsymbol{B}$, a *and* $\boldsymbol{D}_0$ *as defined in Lemma* 15.12. *Then*

$$\boldsymbol{M}_0 = \boldsymbol{H}^{-1/2} \boldsymbol{S} \boldsymbol{D}_0 \boldsymbol{S}^\top \boldsymbol{H}^{-1/2} \tag{15.103}$$

is a least favourable moment matrix if and only if

$$\lambda_{\max}((\boldsymbol{SBS}^\top)^2, (\boldsymbol{SD}_0\boldsymbol{S}^\top + \boldsymbol{SS}^\top)^2) \leqq a^2 . \tag{15.104}$$

(Here $\lambda_{\max}(\boldsymbol{A}, \boldsymbol{B})$ *stands for the maximum eigenvalue of* $\det(\boldsymbol{A} - \lambda \boldsymbol{B}) = 0$.)

Proof. The assertion follows from Corollary 15.7 and Theorem 15.8 observing that $\boldsymbol{M}_0$ satisfies the following equations:

$$\begin{aligned}
\operatorname{tr} \boldsymbol{M}_0^* \boldsymbol{M}_0 &= \operatorname{tr}(\boldsymbol{H}^{-1/2} \boldsymbol{M}_0^* \boldsymbol{H}^{-1/2})(\boldsymbol{H}^{1/2} \boldsymbol{M}_0 \boldsymbol{H}^{1/2}) \\
&= \operatorname{tr}[\boldsymbol{S}(\boldsymbol{D}_0 + \boldsymbol{I}_r)\boldsymbol{S}^\top]^{-1} (\boldsymbol{SBS}^\top)^2 [\boldsymbol{S}(\boldsymbol{D}_0 + \boldsymbol{I}_r)\boldsymbol{S}^\top]^{-1} \boldsymbol{SD}_0\boldsymbol{S}^\top \\
&= \operatorname{tr} \boldsymbol{SB}^2 (\boldsymbol{D}_0 + \boldsymbol{I}_r)^{-2} \boldsymbol{D}_0 \boldsymbol{S}^\top = \sum_{i=1}^{i_0} b_i^2 \|s_i\|^2 d_i^* (1 + d_i^*)^{-2} = a^2 ,
\end{aligned}$$

and $\det(\boldsymbol{H}^{-1/2} \boldsymbol{M}_0^* \boldsymbol{H}^{-1/2} - \lambda \boldsymbol{I}_r) = \det((\boldsymbol{SBS}^\top)^2 - \lambda[\boldsymbol{S}(\boldsymbol{D}_0 + \boldsymbol{I}_r)\boldsymbol{S}^\top]^2)$. □

Remark 15.7. Given that the ordered eigenvalues $b_1 \geqq \dots \geqq b_r$ fulfill the inequalities

$$b_{j+1}\left(1 + \sum_{i=1}^{j} \|s_i\|^2\right) > \sum_{i=1}^{j} b_i \|s_i\|^2 , \qquad j = 1, \dots, r-1 \tag{15.105}$$

then it holds $i_0 = r$ and $\boldsymbol{D}_0 = \mathrm{a}^{-1}\boldsymbol{B} - \boldsymbol{I}_r$ with

$$a^{-1} = (1 + \operatorname{tr} \boldsymbol{S}^\top \boldsymbol{S}) / \operatorname{tr} \boldsymbol{SBS}^\top = (1 + \operatorname{tr} \boldsymbol{HV}) / \operatorname{tr}(\boldsymbol{H}^{1/2} \boldsymbol{VUVH}^{1/2})^{1/2} .$$

Moreover, since $\boldsymbol{SD}_0\boldsymbol{S}^\top + \boldsymbol{SS}^\top = a^{-1}\boldsymbol{SBS}^\top$, Condition (15.104) is satisfied, hence

$$\boldsymbol{M}_0 = \boldsymbol{H}^{-1/2} \boldsymbol{S}(a^{-1}\boldsymbol{B} - \boldsymbol{I}_r)\boldsymbol{S}^\top \boldsymbol{H}^{-1/2}$$

is a least favourable moment matrix. Observing the factorization (15.87), this is exactly the solution given in Corollary 15.8.

If the matrices $\boldsymbol{H}^{1/2}\boldsymbol{VH}^{1/2}$ and $\boldsymbol{H}^{1/2}\boldsymbol{VUVH}^{1/2}$ have the same eigenvectors then the system (15.87), (15.88) reduces to a simple eigenvalue problem, i.e., the matrix $\boldsymbol{S}$ takes the form $\boldsymbol{S} = \boldsymbol{S}_0 \boldsymbol{C}^{1/2}$, where $\boldsymbol{S}_0$ and

$C = \operatorname{diag}(c_1, \dots, c_r)$ are the matrices of eigenvectors and eigenvalues of $H^{1/2}VH^{1/2}$, respectively, and it holds $\|s_i\|^2 = c_i$, $i = 1, \dots, r$. In this case we have

$$\lambda_{\max}((SBS^\top)^2, (SD_0S^\top + SS^\top)^2) = \lambda_{\max}(B^2, (D_0 + I_r)^2)$$
$$= \max(a^2, b_{i_0+1}^2, \dots, b_r^2) = a^2$$

(cp. (15.94)), i.e., Condition (15.104) is satisfied. Thus, we have proved

Corollary 15.9. *Let the assumptions of Theorem* 15.10 *be satisfied and assume that*

$$H^{1/2}VH^{1/2} = S_0CS_0^\top \quad \text{and} \quad H^{1/2}VUVH^{1/2} = S_0AS_0^\top$$

with an orthogonal matrix S_0 and C and A being the diagonal matrices of the corresponding eigenvalues of $H^{1/2}VH^{1/2}$ and $H^{1/2}VUVH^{1/2}$, respectively. Further, let $B = A^{1/2}C^{-1} = \operatorname{diag}(b_1, \dots, b_r)$ and, without loss of generality, assume that $b_1 \geqq b_2 \geqq \dots \geqq b_r \geqq b_{r+1} := 0$. Then, with $\|s_i\|^2 = c_i$, $i = 1, \dots, r$, and i_0, a and D_0 as defined in Lemma 15.12,

$$M_0 = H^{-1/2}S_0CD_0S_0^\top H^{-1/2}$$

is least favourable.

Remark 15.8. The result presented in Theorem 2.2 of ALSON (1988) does not hold in its stated generality, since $A_1 \geqq A_2$ does not necessarily imply that $A_1A_3A_1 \geqq A_2A_3A_2$ for given matrices $A_1, A_2, A_3 \in \mathcal{M}_r^{\geqq}$. This is only true under the additional symmetry assumption that $A_1A_3A_2 = A_2A_3A_1$.

We now consider two important special cases for which the application of Corollary 15.9 leads to an explicit solution.

Example 15.2. Let $U = V^{-1} = F^\top\Sigma^{-1}F$, i.e., an estimator of ϑ is evaluated on the basis of the "design norm" (prediction risk). Then we have $H^{1/2}VUVH^{1/2} = H^{1/2}VH^{1/2}$ and the basic assumption of Corollary 15.9 is satisfied. Now, let $S_0CS_0^\top$ the spectral decomposition of $H^{1/2}VH^{1/2}$, i.e.,

$$H^{1/2}VH^{1/2} = S_0CS_0^\top, \qquad S_0^\top S_0 = S_0S_0^\top = I_r$$
$$C = \operatorname{diag}(c_1, \dots, c_r), \qquad 0 < c_1 \leqq c_2 \leqq \dots \leqq c_r.$$

Then, with $B = C^{-1/2}$ the integer i_0 is determined by

$$i_0 = \min_{1 \leqq i \leqq r}\left\{i : \sqrt{c_{i+1}} \geqq \left(1 + \sum_{j=1}^{i} c_j\right) \Big/ \sum_{j=1}^{i} \sqrt{c_j} > \sqrt{c_i}\right\},$$

where we have to put $c_{r+1} = \infty$. From this we obtain

$$a^{-1} = \left(1 + \sum_{j=1}^{i_0} c_j\right) \Big/ \sum_{j=1}^{i_0} \sqrt{c_j},$$
$$D_0 = \operatorname{diag}(1/a\sqrt{c_1} - 1, \dots, 1/a\sqrt{c_{i_0}} - 1, 0, \dots, 0),$$

and Corollary 15.9 tells us that $M_0 = H^{-1/2} S_0 C D_0 S_0^\top H^{-1/2}$ is a least favourable moment matrix.

Example 15.3. Consider the case that $U \in \mathcal{M}_r^{\geqq}$, V and H have the same eigenvectors, i.e., these matrices admit spectral decompositions of the form

$$U = \sum_{i=1}^{r} u_i t_i t_i^\top, \qquad V = \sum_{i=1}^{r} v_i t_i t_i^\top, \qquad H = \sum_{i=1}^{r} h_i t_i t_i^\top$$

with $u_i \geqq 0$, $v_i > 0$, $h_i > 0$, $i = 1, \ldots, r$, and an orthonormal system of eigenvectors $t_1, \ldots, t_r \in \mathbb{R}^r$. Here the situation required in Corollary 15.9 applies with $S_0 = (t_1, \ldots, t_r)$ and

$$B = \operatorname{diag}\left(\sqrt{u_1/h_1}, \ldots, \sqrt{u_r/h_r}\right),$$

where it is assumed that

$$u_1/h_1 \geqq u_2/h_2 \geqq \ldots \geqq u_r/h_r \geqq 0 =: u_{r+1}/h_{r+1}.$$

Thus, observing that $c_i = h_i v_i$, $i = 1, \ldots, r$, we obtain

$$i_0 = \min_{i \leqq i \leqq r} \left\{ i : \frac{u_{i+1}}{h_{i+1}} \leqq \sum_{j=1}^{i} \sqrt{u_j h_j}\, v_j \Big/ \left(1 + \sum_{j=1}^{i} h_j v_j\right) < \frac{u_i}{h_i} \right\},$$

$$a = \sum_{j=1}^{i_0} \sqrt{u_j h_j}\, v_j \Big/ \left(1 + \sum_{j=1}^{i_0} h_j v_j\right),$$

$$d_i^* = \begin{cases} a^{-1} \sqrt{u_i/h_i} - 1 & \text{for} \quad i = 1, \ldots, i_0 \\ 0 & \text{for} \quad i = i_0 + 1, \ldots, r \end{cases}.$$

Corollary 15.9 then tells us that $M_0 = H^{-1/2} S_0 D_1 S_0^\top H^{-1/2}$, where $D_1 = \operatorname{diag}(h_1 v_1 d_1^*, \ldots, h_r v_r d_r^*)$, is least favourable. The resulting estimator $\hat{\vartheta}_{\mathrm{M}}^{\mathrm{a}} = M_0 (M_0 + V)^{-1} (\hat{\vartheta}_\Sigma - \mu) + \mu$ is already known from KUKS/OL'MAN (1971) and HOFFMANN (1979) to be a minimax-affine estimator for the special case under consideration.

If, for a given configuration of the matrices U, V and H, the condition (15.104) is not satisfied then the matrix M_0 defined by (15.103) may still serve as an approximation of a least favourable moment matrix. The goodness of the approximation can be evaluated with the help of the inequality

$$0 \leqq \operatorname{tr} VUV(M_0 + V)^{-1} - \inf_{M \in \mathcal{M}_H} \operatorname{tr} VUV(M + V)^{-1} \leqq \lambda - \operatorname{tr} M_0^* M_0, \qquad (15.106)$$

where M_0^* and λ are defined as in Theorem 15.8 and Lemma 15.11, respectively. This follows immediately from Theorem 11.13 (i) replacing there U_{L} and $M_{\mathrm{B}}(\xi)$ by VUV and $M_0 + V$, respectively, and observing that

$$\sup_{\vartheta \in \Theta} (\vartheta - \mu)^\top (M_0 + V)^{-1} VUV (M_0 + V)^{-1} (\vartheta - \mu)$$

$$= \lambda_{\max}(H^{-1/2} M_0^* H^{-1/2}). \qquad (15.107)$$

In case that M_0 is least favourable we have equality in (15.106), cp. Theorem 15.8 (ii).

From our considerations in Section 15.1.3 we already know that even in case of a full rank weight matrix U there may exist a least favourable moment matrix M_0 having rank $M_0 = 1$. This is the case if and only if there exists a least favourable one-point distribution $P_0 = \delta_{\vartheta^*}$ with some $\vartheta^* \in \Theta = \mathscr{E}(H, \mu)$. Then the minimax-affine estimator $\hat{\vartheta}^{a}_{M} = M_0(M_0 + V)^{-1}(\hat{\vartheta}_{\Sigma} - \mu) + \mu$ formed with $M_0 = (\vartheta^* - \mu)(\vartheta^* - \mu)^{\top}$ coincides with the maximin affine estimator, i.e., it holds

$$\sup_{\vartheta \in \Theta} R(\vartheta, \hat{\vartheta}^{a}_{M}) = \sup_{\vartheta \in \Theta} \inf_{\hat{\vartheta} \in D^0_a} R(\vartheta, \hat{\vartheta}) = R(\vartheta^*, \hat{\vartheta}^{a}_{M}).$$

Lemma 15.13. *Let F arbitrary and denote $G = H^{1/2}(H + V^{-1})^{-1}H^{1/2}$. Further, let $\lambda_* = \lambda_{\max}(G^{1/2}H^{-1/2}UH^{-1/2}G^{1/2})$, t_* the corresponding normed eigenvector and*

$$M_* = H^{-1/2}G^{1/2}t_*t_*^{\top}G^{1/2}H^{-1/2}/t_*^{\top}Gt_*. \tag{15.108}$$

Then any estimator of the form

$$\hat{\vartheta}^*_A = [(t_*^{\top}Gt_*)U^+UM_*F^{\top}\Sigma^{-1} + (I_r - U^+U)A](\tilde{y} - F\mu) + \mu \tag{15.109}$$

with arbitrary $A \in \mathscr{M}_{r\times n}$ is a maximin-affine estimator, and it holds

$$\sup_{\vartheta \in \Theta} \inf_{\hat{\vartheta} \in D^0_a} R(\vartheta, \hat{\vartheta}) = \lambda_*. \tag{15.110}$$

Proof. Reasoning as in the proof of Lemma 15.6, one easily verifies that

$$\inf_{\hat{\vartheta} \in D^0_a} R(\vartheta, \hat{\vartheta}) = (\vartheta - \mu)^{\top}U(\vartheta - \mu)/[1 + (\vartheta - \mu)^{\top}V^{-1}(\vartheta - \mu)], \text{ for all } \vartheta \in \Theta.$$

Thus, it holds

$$\sup_{\vartheta \in \Theta} \inf_{\hat{\vartheta} \in D^0_a} R(\vartheta, \hat{\vartheta}) = \sup (\bar{\vartheta}^{\top}U\bar{\vartheta}/\bar{\vartheta}^{\top}(H + V^{-1})\bar{\vartheta} : \bar{\vartheta}^{\top}H\bar{\vartheta} = 1\}$$

$$= \sup \{t^{\top}H^{-1/2}UH^{-1/2}t/t^{\top}G^{-1}t : t^{\top}t = 1\} = \lambda_*$$

and the maximum λ_* is attained at $t = t_0 := H^{-1/2}G^{1/2}t_*/(t_*^{\top}Gt_*)^{1/2}$. Recalling that for any one-point distribution $\bar{P} = \delta_{\bar{\vartheta}}$ with $\bar{\vartheta} = \vartheta - \mu$, $\vartheta \in \Theta$, and any estimator $\hat{\vartheta} = Z(\tilde{y} - F\mu) + \mu \in D^0_a$ we have identity $R(\vartheta, \hat{\vartheta}) = \varrho(M_{\bar{P}}, Z)$, it is clear that the maximin-affine estimator coincides with the Bayes-affine estimator with respect to $\bar{P}_0 = \delta_{t_0}$. This estimator, in turn, is given by

$$\hat{\vartheta}^{a}_{B} = [U^+UZ_* + (I_r - U^+U)A](\tilde{y} - F\mu) + \mu$$

with

$$\begin{aligned} Z_* &= t_0t_0^{\top}F^{\top}(Ft_0t_0^{\top}F^{\top} + \Sigma)^{-1} = t_0t_0^{\top}F^{\top}\Sigma^{-1}/(1 + t_0^{\top}V^{-1}t_0) \\ &= M_*F^{\top}\Sigma^{-1}/t_0^{\top}(H + V^{-1})t_0 = (t_*^{\top}Gt_*)M_*F^{\top}\Sigma^{-1} \end{aligned} \tag{15.111}$$

and arbitrary $A \in \mathscr{M}_{r\times n}$. This is exactly the form of the maximin estimator indicated in (15.109). □

Remark 15.9. (i) An alternative derivation of the maximin estimator was given by KRAFFT (1986). He proposed to choose the matrix $\boldsymbol{A} \in \mathcal{M}_{r \times n}$ such that it minimizes the matrix valued risk $\mathsf{E}\{(\hat{\boldsymbol{\vartheta}}_A^* - \boldsymbol{\vartheta})(\hat{\boldsymbol{\vartheta}}_A^* - \boldsymbol{\vartheta})^\top | \boldsymbol{\vartheta} - \boldsymbol{\mu} = \boldsymbol{t}_0\}$ with respect to the usual (Löwner) semiordering in $\mathcal{M}_r^{\geqq}$. This amounts to choosing $\boldsymbol{A} = \boldsymbol{A}_* := (\boldsymbol{t}_*^\top \boldsymbol{G} \boldsymbol{t}_*) \boldsymbol{M}_* \boldsymbol{F}^\top \boldsymbol{\Sigma}^{-1}$, which then leads to the choice of $\hat{\boldsymbol{\vartheta}}_A^* = \boldsymbol{A}_*(\tilde{\boldsymbol{y}} - \boldsymbol{F}\boldsymbol{\mu}) + \boldsymbol{\mu}$.

(ii) Observing that $\boldsymbol{H}^{1/2}\boldsymbol{G}^{-1}\boldsymbol{H}^{1/2} = \boldsymbol{H} + \boldsymbol{V}^{-1}$, it holds $\lambda_* = \lambda_{\max}(\boldsymbol{U}, \boldsymbol{H} + \boldsymbol{V}^{-1})$, i.e., $\lambda_* = \max\{\lambda > 0 : \det[\boldsymbol{U} - \lambda(\boldsymbol{H} + \boldsymbol{V}^{-1})] = 0\}$.

From our characterization theorem 15.5 and the proof of Lemma 15.13 it is obvious that the maximin estimator $\hat{\boldsymbol{\vartheta}}_A^*$ is minimax in D_a if and only if $\boldsymbol{M}_* = \boldsymbol{t}_0 \boldsymbol{t}_0^\top$ is a least favourable moment matrix. Since $\boldsymbol{M}_* \in \mathcal{M}_r^{\geqq}$ and $\operatorname{tr} \boldsymbol{H}\boldsymbol{M}_* = \boldsymbol{t}_0^\top \boldsymbol{H} \boldsymbol{t}_0 = 1$, this is the case if and only if it holds

$$\begin{aligned} &\lambda_{\max}(\boldsymbol{H}^{-1/2}(\boldsymbol{M}_* + \boldsymbol{V})^{-1}\boldsymbol{V}\boldsymbol{U}\boldsymbol{V}(\boldsymbol{M}_* + \boldsymbol{V})^{-1}\boldsymbol{H}^{-1/2}) \\ &= \boldsymbol{t}_0^\top(\boldsymbol{M}_* + \boldsymbol{V})^{-1}\boldsymbol{V}\boldsymbol{U}\boldsymbol{V}(\boldsymbol{M}_* + \boldsymbol{V})^{-1}\boldsymbol{t}_0. \end{aligned} \tag{15.112}$$

Summarizing, we can state the following.

Corollary 15.10. *Let* $\Theta = \mathscr{E}(\boldsymbol{H}, \boldsymbol{\mu})$ *and assume that* $\operatorname{rank} \boldsymbol{F} = r$. *The maximin estimator* $\hat{\boldsymbol{\vartheta}}_A^*$ *given by* (15.109) *is minimax in* D_a^0 *if and only if* (15.112) *is fulfilled. The minimax risk then takes the value* $\lambda_* = \lambda_{\max}(\boldsymbol{U}, \boldsymbol{H} + \boldsymbol{V}^{-1})$.

(Cp. PILZ (1986 b), Section 3.)

A nontrivial example, for which the maximin estimator $\hat{\boldsymbol{\vartheta}}_A^*$, even in the full rank case $\operatorname{rank} \boldsymbol{U} = r$, comes out as minimax estimator may be found in KRAFFT (1986) and PILZ (1986 b).

In the special case of $\operatorname{rank} \boldsymbol{U} = 1$, i.e., $\boldsymbol{U} = \boldsymbol{c}\boldsymbol{c}^\top$ for some given $\boldsymbol{c} \in \mathbb{R}^r$, the convexity and symmetry of $\Theta = \mathscr{E}(\boldsymbol{H}, \boldsymbol{\mu})$ assures the existence of a least favourable moment matrix $\boldsymbol{M}_0$ having rank one (see Corollary 14.1 and Corollary 15.5).

Corollary 15.11. *Let the assumptions of Corollary* 15.10 *be satisfied and assume that* $\boldsymbol{U} = \boldsymbol{c}\boldsymbol{c}^\top$ *for some* $\boldsymbol{c} \in \mathbb{R}^r$. *Then any estimator of the form*

$$\hat{\boldsymbol{\vartheta}}_A^* = \left[\frac{\boldsymbol{c}\boldsymbol{c}^\top}{\boldsymbol{c}^\top\boldsymbol{c}}(\boldsymbol{H} + \boldsymbol{V}^{-1})^{-1}\boldsymbol{F}^{-1}\boldsymbol{\Sigma}^{-1} + \left(\boldsymbol{I}_r - \frac{\boldsymbol{c}\boldsymbol{c}^\top}{\boldsymbol{c}^\top\boldsymbol{c}}\right)\boldsymbol{A}\right](\tilde{\boldsymbol{y}} - \boldsymbol{F}\boldsymbol{\mu}) + \boldsymbol{\mu} \tag{15.113}$$

with arbitrary $\boldsymbol{A} \in \mathcal{M}_{r \times n}$ *is minimax in* D_a^0. *The minimax risk is given by*

$$\sup_{\boldsymbol{\vartheta} \in \Theta} R(\boldsymbol{\vartheta}, \hat{\boldsymbol{\vartheta}}_A^*) = \boldsymbol{c}^\top(\boldsymbol{H} + \boldsymbol{V}^{-1})^{-1}\boldsymbol{c}.$$

(Cp. LÄUTER (1975 a), KRAFFT (1986), PILZ (1986 a, b).)

Proof. For the eigenvalue and eigenvector defined in Lemma 15.13 we obtain $\lambda_* = \boldsymbol{c}^\top(\boldsymbol{H} + \boldsymbol{V}^{-1})^{-1}\boldsymbol{c}$ and $\boldsymbol{t}_* = \lambda_*^{-1/2}\boldsymbol{G}^{1/2}\boldsymbol{H}^{-1/2}\boldsymbol{c}$, respectively. Hence, according to the proof of Lemma 15.13, we have $\boldsymbol{t}_0 = (\boldsymbol{H} + \boldsymbol{V}^{-1})^{-1}\boldsymbol{c}/(\lambda_* \boldsymbol{t}_*^\top \boldsymbol{G} \boldsymbol{t}_*)^{1/2}$. Observing that $\boldsymbol{U}^+ = (\boldsymbol{c}^\top\boldsymbol{c})^{-2}\boldsymbol{c}\boldsymbol{c}^\top$ and

$$(t_*^\top G t_*) U^+ U t_0 t_0^\top = (c^\top c)^{-1} c c^\top (H + V^{-1})^{-1},$$

the assertion follows with $M_* = t_* t_*^\top$ (cp. (15.111)) from Corollary 15.10. □

In particular, choosing $A = (H + V^{-1})^{-1} F^\top \Sigma^{-1}$ we arrive at the Kuks-Ol'man estimator

$$\hat{\vartheta}_A^* = (F' \Sigma^{-1} F + H)^{-1} (F^\top \Sigma^{-1} \tilde{y} + H\mu) \tag{15.114}$$

which does not depend on U. This estimator can also be derived from (Tichonov's) regularization principle by minimizing

$$S_\lambda(\vartheta) := \|\tilde{y} - F\vartheta\|^2_{\Sigma^{-1}} + \lambda \|\vartheta - \mu\|^2_H, \quad \lambda > 0$$

with respect to ϑ and setting $\lambda = 1$ in the solution

$$\hat{\vartheta}_\lambda = (F^\top \Sigma^{-1} F + \lambda H)^{-1} (F^\top \Sigma^{-1} \tilde{y} + \lambda H \mu) \tag{15.115}$$

(simple regularization). From our preceding considerations it is clear that, in the sense of statistical decision theoretic notions such as Bayes optimality, minimaxity or admissibility, simple regularization is only optimal when estimating linear combinations $c^\top \vartheta$, $c \in \mathbb{R}^r$. If the full parameter vector ϑ is to be estimated then the simple regularization $\hat{\vartheta}_1 = (F^\top \Sigma^{-1} F + H)^{-1} (F^\top \Sigma^{-1} \tilde{y} + H\mu)$ does not result in an admissible estimator whenever $r \geqq 2$, admissibility can only be achieved when using $\hat{\vartheta}_\lambda$ with $\lambda \geqq r$ (see Section 16.1). For $\lambda \geqq r$ the estimator $\hat{\vartheta}_\lambda$ represents a Bayes affine estimator with respect to $P \in \Theta^*$ having (centered) second order moment matrix

$$M_P = \int_\Theta (\vartheta - \mu)(\vartheta - \mu)^\top P(\mathrm{d}\vartheta) = \frac{1}{\lambda} H^{-1}.$$

Since $\operatorname{tr} HM_P = r/\lambda$ it follows that an additional minimaxity of $\hat{\vartheta}_\lambda$ can at most be achieved for $\lambda = r$, provided U has a certain prescribed form.

The concept of quasi minimax estimation introduced by Trenkler/STAHLECKER (1987) also leads to the Kuks-Ol'man estimator $\hat{\vartheta}_\lambda$, $\lambda = 1$. These authors proposed to replace the maximum risk of $\hat{\vartheta} = Z(\tilde{y} - F\mu) + \mu \in D_a^0$ over $\Theta = \mathscr{E}(H, \mu)$,

$$\begin{aligned} \sup_{\vartheta \in \Theta} R(\vartheta, \hat{\vartheta}) &= \operatorname{tr} UZ\Sigma Z^\top + \lambda_{\max}(G(Z)), \\ G(Z) &= H^{-1/2} (ZF - I_r)^\top U (ZF - I_r) H^{-1/2}, \end{aligned} \tag{15.116}$$

by the upper bound $\operatorname{tr} UZ\Sigma Z^\top + \operatorname{tr} G(Z)$ and then to minimize this upper bound with respect to $Z \in \mathscr{M}_{r \times n}$. In the special case of $U = cc^\top$, $c \in \mathbb{R}^r$, we have identity $\lambda_{\max}(G(Z)) = \operatorname{tr} G(Z)$ and the quasi minimax estimator $\hat{\vartheta}_1$ is proper minimax. Flexible upper and lower bounds for the maximal eigenvalue of $G(Z)$ are given by

$$\left[\operatorname{tr}\left(\frac{1}{r} G(Z)^p\right)\right]^{1/p} \leqq \lambda_{\max}(G(Z)) \leqq [\operatorname{tr} G(Z)^p]^{1/p}, \tag{15.117}$$

$p > 0$, for $p \to \infty$ we have monotone convergence of the lower as well as the upper bounds to $\lambda_{\max}(\boldsymbol{G}(\boldsymbol{Z}))$. The minimization of (15.116) with $\lambda_{\max}(\boldsymbol{G}(\boldsymbol{Z}))$ being replaced by an appropriate upper or lower bound was termed by STAHLECKER (1987) a super- or subpessimistic approximation of the minimax estimator, respectively. Based on the monotone convergence of the bounds given in (15.117), he developed an iterative procedure for an approximate computation of the minimax estimator (see STAHLECKER (1987), Section 5.2.5, and also STAHLECKER/LAUTERBACH (1987), (1988) LAUTERBACH (1989)).

Alternatively, we can apply the iteration procedure described in Section 15.1.2, which can be easily implemented:According to (15.29), for the starting point we obtain

$$\boldsymbol{\vartheta}_0 = \boldsymbol{\mu} + \boldsymbol{H}^{-1/2}\boldsymbol{G}^{1/2}\boldsymbol{t}_*/(\boldsymbol{t}_*^\top \boldsymbol{G}\boldsymbol{t}_*)^{1/2},$$

where $\boldsymbol{t}_*$ and $\boldsymbol{G}$ are defined as in Lemma 15.13, and for the supporting points to be added at each stage of the iteration it holds $\boldsymbol{\vartheta}_s = \boldsymbol{\mu} + \boldsymbol{t}_s$, where $\boldsymbol{t}_s$ denotes the eigenvector corresponding to the maximum eigenvalue of

$$\boldsymbol{H}^{-1/2}(\boldsymbol{M}_s + \boldsymbol{V})^{-1}\boldsymbol{VUV}(\boldsymbol{M}_s + \boldsymbol{V})^{-1}\boldsymbol{H}^{-1/2}$$

with $$\boldsymbol{M}_s = \sum_{i=1}^{s} (\boldsymbol{\vartheta}_i - \boldsymbol{\mu})(\boldsymbol{\vartheta}_i - \boldsymbol{\mu})^\top, \qquad s = 1, 2, \ldots$$

We remark that the preceding results can easily be extended to the problem of determining minimax estimators in case of additional equality restrictions, where we have

$$\Theta = \{\boldsymbol{\vartheta} \in \mathbb{R}^r : (\boldsymbol{\vartheta} - \boldsymbol{\mu})^\top \boldsymbol{H}(\boldsymbol{\vartheta} - \boldsymbol{\mu}) \leqq 1, \quad \boldsymbol{K}(\boldsymbol{\vartheta} - \boldsymbol{\mu}) = \boldsymbol{0}_m\} \quad (15.118)$$

with some given matrix $\boldsymbol{K} \in \mathcal{M}_{m \times r}$ having rank $m < r$. With an appropriate partitioning of $\boldsymbol{\vartheta} = (\boldsymbol{\vartheta}_{(1)}^\top, \boldsymbol{\vartheta}_{(2)}^\top)^\top$; $\boldsymbol{\vartheta}_{(1)} = (\boldsymbol{\vartheta}_1, \ldots, \boldsymbol{\vartheta}_m)^\top$, $\boldsymbol{\vartheta}_{(2)} = (\boldsymbol{\vartheta}_{m+1}, \ldots, \boldsymbol{\vartheta}_r)^\top$, the subvector $\boldsymbol{\vartheta}_{(2)}$ may be written as a linear function $\boldsymbol{\vartheta}_{(2)} = \boldsymbol{K}_2\boldsymbol{\vartheta}_{(1)} + \boldsymbol{c}$ with some matrix $\boldsymbol{K}_2 \in \mathcal{M}_{(r-m)\times m}$ and some vector $\boldsymbol{c} \in \mathbb{R}^m$ of constants, and the problem reduces to finding a minimax estimator for $\boldsymbol{\vartheta}_{(1)}$ with pure ellipsoid restrictions. For the special case of rank $\boldsymbol{U} = 1$ STAHLECKER/TRENKLER (1987) derived minimax estimators for $\boldsymbol{\vartheta}$ subject to parameter restrictions (15.118) in a direct way.

We further remark that our approach can be extended to the problem of constructing minimax linear estimators which are robust against an incorrectly specified parameter region $\mathscr{E}(\boldsymbol{H}, \boldsymbol{\mu})$. This can be done within the framework of $\mathscr{P}$-minimax linear estimation, for example by assuming that $\boldsymbol{H} \leqq \boldsymbol{H}_0$ for some given $\boldsymbol{H}_0 \in \mathcal{M}_r^>$ and $\boldsymbol{\mu} \in \mathscr{K}$, where $\mathscr{K} \subset \mathbb{R}^r$ is compact and symmetric around some given center point, and then applying the results of Section 6.3.

15.4. Minimax estimation with linear inequality constraints

Let us now consider the case that prior knowledge suggests upper and lower bounds for the regression coefficients,

$$a_i \leqq \vartheta_i \leqq b_i, \qquad i = 1, \dots, r, \tag{15.119}$$

where a_i and b_i are real numbers such that $0 < b_i - a_i < \infty$, $i = 1, \dots, r$. Letting

$$\mu_i = \frac{1}{2}(a_i + b_i), \qquad m_i = \frac{1}{2}(b_i - a_i), \qquad i = 1, \dots, r \tag{15.120}$$

the restrictions (15.119) define a symmetric parameter region

$$\Theta = \{\boldsymbol{\vartheta} \in \mathbb{R}^r : |\vartheta_i - \mu_i| \leqq m_i; \quad i = 1, \dots, r\} \tag{15.121}$$

around the center point $\mu = (\mu_1, \dots, \mu_r)^\top$. This is a hyperrectangle the vertices of which have lengths $2m_i$.

Consider the translated region

$$\bar{\Theta} = \{\bar{\boldsymbol{\vartheta}} \in \mathbb{R}^r : |\bar{\vartheta}_i| \leqq m_i; \quad i = 1, \dots, r\}. \tag{15.122}$$

Unfortunately, for this region we cannot find a simple and tractable characterization of the set of second order moment matrices induced by the prior distributions from $\bar{\Theta}^*$ as it was the case with ellipsoidal restrictions (cp. (15.79)). Because of the convexity of $\bar{\Theta}$ the search for a least favourable prior distribution can be restricted to such distributions from $\bar{\Theta}^*$ whose supporting points are included in the set

$$T_m = \{\boldsymbol{t} \in \mathbb{R}^r : |t_i| = m_i; \quad i = 1, \dots, r\} \tag{15.123}$$

of all corner points of $\bar{\Theta}$, where $\boldsymbol{m} = (m_1, \dots, m_r)^\top$.

Lemma 15.14. *For the parameter region* (15.121) *there exists a least favourable moment matrix* $\boldsymbol{M}_0$ *such that*

$$\boldsymbol{M}_0 = \arg \inf_{\boldsymbol{M} \in \mathcal{M}(T_m)} \operatorname{tr} \boldsymbol{VUV}(\boldsymbol{M} + \boldsymbol{V})^{-1},$$

where

$$\mathcal{M}(T_m) = \left\{ \boldsymbol{M}_P = \int_{\bar{\Theta}} \bar{\boldsymbol{\vartheta}}\bar{\boldsymbol{\vartheta}}^\top P(\mathrm{d}\bar{\boldsymbol{\vartheta}}) : \operatorname{supp} P \subset T_m,\ |\operatorname{supp} P| \leqq \frac{s}{2}(2r - s + 1) \right\}$$

and $s = \operatorname{rank} \boldsymbol{VUV}$ (cp. Theorem 11.5).

Moreover, we do not need to consider all of the 2^r corner points of $\bar{\Theta}$, but at most 2^{r-1} points which are not symmetric to each other. This is due to the fact that for given $\boldsymbol{t} \in T_m \cap \operatorname{supp} P$ the symmetric counterpart $-\boldsymbol{t}$ yields the same contribution $\boldsymbol{tt}^\top = (-\boldsymbol{t})(-\boldsymbol{t})^\top$ to the information matrix $\boldsymbol{M}_P$. Without loss of generality, we can confine ourselves, for example, to those corner points whose first coordinate has the fixed value $+m_1$. In any case, a least favourable moment matrix $M_0 = (m_{ij}) \in \mathcal{M}(T_m)$ has diagonal elements $m_{ii} = m_i^2$, $i = 1, \ldots, r$, since supp $P \subset T_m$ implies that

$$\mathcal{M}(T_m) \subset \{M = (m_{ij}) \in \mathcal{M}_r^{\geqq} : m_{ii} = m_i^2; \quad i = 1, \ldots, r\}. \quad (15.124)$$

In the special case where U and V are diagonal matrices one easily verifies that M_0 must be diagonal, too, which leads us immediately to an explicit solution.

Theorem 15.10. *Assume Θ to be a hyperrectangle* (15.121) *and U and V to be diagonal matrices of full rank. Then the least favourable moment matrix is uniquely determined by*

$$M_0 = \operatorname{diag}(m_1^2, \ldots, m_r^2)$$

and

$$\begin{aligned}\hat{\vartheta}_M^a &= M_0(M_0 + V)^{-1}(\hat{\vartheta}_\Sigma - \mu) + \mu \\ &= (V^{-1} + M_0^{-1})^{-1}(F^\top \Sigma^{-1} \tilde{y} + M_0^{-1}\mu)\end{aligned}$$

is the unique minimax estimator for ϑ in D_a^0.

(For a proof see Pilz (1986 a), Theorem 6.)

Note that in the above theorem the solution M_0 does not depend on the magnitudes of the diagonal entries of U.

Setting $K = M_0^{-1} = \operatorname{diag}(m_1^{-2}, \ldots, m_r^{-2})$, the minimax estimator takes the form

$$\hat{\vartheta}_M^a = (F^\top \Sigma^{-1} F + K)^{-1}(F^\top \Sigma^{-1} \tilde{y} + K\mu) \quad (15.125)$$

of a generalized ridge estimator which shrinks $\hat{\vartheta}_\Sigma$ towards the center point μ, the ridge constants read $k_i = m_i^{-2} = 4(b_i - a_i)^{-2}$, $i = 1, \ldots, k$. For the prior restrictions (15.119) Toutenburg (1982) also proposed a generalized ridge estimator, however, with ridge constants $\bar{k}_i = k_i/r$ instead of our k_i's. This choice was motivated by an approximation using the minimum covering ellipsoid MCE (Θ) instead of the original region Θ and then forming the Kuks-Ol'man estimator based on this ellipsoid. Obviously, the minimum covering ellipsoid is given by MCE $(\Theta) = \mathcal{E}(K/r, \mu)$. The estimator $\hat{\vartheta}_M^a$ from (15.125) effects a stronger contraction towards μ than Toutenburg's ridge estimator with $\bar{K} = r^{-1}K$ instead of K, which is clear from his "relaxation" of the original prior bounds (15.119).

For arbitrary matrices $U \in \mathcal{M}_r^{\geqq}$ and $V \in \mathcal{M}_r^{>}$ the problem of determining a least favourable moment matrix amounts to finding optimal weights $p_1, \ldots, p_l$ to be attached to $l \leqq 2^{r-1}$ corner points $t_1, \ldots, t_l \in T_m$ of the hyperrectangle $\bar{\Theta}$ such that the trace functional $tr\, VUV(M + V)^{-1}$ is minimized over all matrices

$$M = \sum_{i=1}^{l} p_i t_i t_i^\top \quad \text{with} \quad \sum_{i=1}^{l} p_i = 1, \qquad p_i \in [0, 1], \qquad i = 1, \ldots, l.$$

Example 15.4. Let us consider a simple linear regression model

$$\mathbf{E}\,\tilde{y}(x_i) = \vartheta_1 + \vartheta_2 x_i, \quad i = 1, \ldots, n \quad \text{and} \quad \mathbf{Cov}\,\tilde{y} = \sigma^2 \boldsymbol{I}_n, \quad \sigma^2 > 0$$

with parameter restrictions $a_1 \leqq \vartheta_1 \leqq \beta_1$, $a_2 \leqq \vartheta_2 \leqq b_2$. The set of corner points of the corresponding rectangle $\bar{\Theta}$ is then given by

$$T_m = \{\boldsymbol{t}_1 = (m_1, -m_2)^\top, \quad \boldsymbol{t}_2 = (m_1, m_2)^\top, \quad \boldsymbol{t}_3 = -\boldsymbol{t}_1, \; \boldsymbol{t}_4 = -\boldsymbol{t}_2\},$$

where $\boldsymbol{m} = (m_1, m_2)^\top = \frac{1}{2}(b_1 - a_1, b_2 - a_2)^\top$. The search for a least favourable prior distribution within $\bar{\Theta}^*$ can be restricted to two-point measures

$$P = \{(\boldsymbol{t}_1, p), (\boldsymbol{t}_2, 1 - p)\}, \qquad p \in [0, 1].$$

Thus, the set $\mathcal{M}(T_m)$ defined in Lemma 15.14 consists of all moment matrices

$$\boldsymbol{M}_P = p\boldsymbol{t}_1\boldsymbol{t}_1^\top + (1 - p)\boldsymbol{t}_2\boldsymbol{t}_2^\top = \begin{bmatrix} m_1^2 & (1 - 2p)m_1 m_2 \\ (1 - 2p)m_1 m_2 & m_2^2 \end{bmatrix}, \tag{15.126}$$

which are completely determined through the weight p attached to $\boldsymbol{t}_1$. Abbreviating

$$f_l = \sum_{i=1}^{n} x_i^l, \qquad l = 1, 2, \qquad d = (nf_2 - f_1^2)/\sigma^2$$

and

$$s_0 = f_1/d - m_1 m_2, \qquad s_1 = n/d + m_2^2, \qquad s_2 = f_2/d + m_1^2,$$

the trace functional reads

$$\operatorname{tr} \boldsymbol{VUV}(\boldsymbol{M}_P + \boldsymbol{V})^{-1} = \frac{g_0 + 2g_{12}(s_0 + 2m_1 m_2 p)^2}{s_1 s_2 - (s_0 + 2m_1 m_2 p)^2} =: h(p),$$

where we have set $\boldsymbol{G} = (g_{ij})_{i,j=1,2} := \boldsymbol{VUV}$ and $g_0 = s_1 g_{11} + s_2 g_{22}$. Differentiating and equating $\mathrm{d}h(p)/\mathrm{d}p = 0$ we arrive at a quadratic equation

$$\begin{aligned} &q_2 p^2 + q_1 p + q_0 = 0 \quad \text{with} \\ &q_2 = 4g_{12} m_1^2 m_2^2, \quad q_1 = 2m_1 m_2 (g_0 + 2s_0 g_{12}), \\ &q_0 = s_0 g_0 + g_{12}(s_0^2 + s_1 s_2). \end{aligned} \tag{15.127}$$

If at least one of the zeroes of this equation falls in the interval [0, 1] then the corresponding values of the function $h(\cdot)$ have to be compared with those for the boundary points $p = 0$ and $p = 1$, respectively, in order to find the globally optimal weight p^*. If both zeroes fall outside the interval [0, 1] then we choose $p^* = 0$ in case that $h(0) < h(1)$, in the opposite case we choose $p^* = 1$.

In the special case where $g_{12} = 0$ (diagonality of $\boldsymbol{VUV}$) we have $q_2 = 0$, $q_1 = 2m_1 m_2 q_0 / s_0$ and the uniquely determined solution of (15.127) reads $p = -q_0/q_1 = 0.5 - f_1/2dm_1 m_2$. Hence, in this special case the optimal weight is given by

$$p^* = \begin{cases} 0 & \text{if } f_1 > dm_1 m_2, \\ 0.5 - f_1/2dm_1 m_2 & \text{if } |f_1| < dm_1 m_2, \\ 1 & \text{otherwise}. \end{cases}$$

When using an orthogonal design, i.e. $f_1 = 0$, then the optimal choice is $p^* = 0.5$ and, according to (15.126), the least favourable moment matrix is given by $\boldsymbol{M}_0 = \operatorname{diag}(m_1^2, m_2^2)$. This is in accordance with the result of Theorem 15.10, since the conditions $f_1 = 0$ and $g_{12} = 0$ imply that $\boldsymbol{V}$ and $\boldsymbol{U}$ are diagonal matrices. With $f_1 = 0$ and arbitrary diagonal weight matrix $\boldsymbol{U}$ we obtain the following minimax estimators for the regression coefficients:

$$\hat{\vartheta}_1 = \left(\sum_{i=1}^{n} \tilde{y}_i + \sigma^2 \mu_1 / m_1^2\right) \Big/ (n + \sigma^2/m_1^2),$$

$$\hat{\vartheta}_2 = \left(\sum_{i=1}^{n} x_i \tilde{y}_i + \sigma^2 \mu_2 / m_2^2\right) \Big/ (f_2 + \sigma^2/m_2^2),$$

where $\mu_i = (a_i + b_i)/2$, $i = 1, 2$, and $\tilde{y}_i = \tilde{y}(x_i)$ denotes the ith observation. In the limiting case $m_1 \to \infty$, $m_2 \to \infty$, which corresponds to noninformative prior knowledge, we have coincidence with the well-known least squares estimators $\hat{\vartheta}_1 = \sum_{i=1}^{n} \tilde{y}_i/n$, $\hat{\vartheta}_2 = \sum_{i=1}^{n} x_i \tilde{y}_i \Big/ \sum_{i=1}^{n} x_i^2$, formed with an orthogonal design.

Under the assumption that $f_1 = 0$ and $\boldsymbol{U} = \operatorname{diag}(u_1, u_2)$ for some given $u_1, u_2 \geqq 0$ the minimax risk takes the form

$$\sup_{\vartheta \in \Theta} R(\vartheta, \hat{\vartheta}_{\mathrm{M}}^{\mathrm{a}}) = \operatorname{tr} \boldsymbol{UV} - h(0.5) = \frac{u_1}{n\sigma^{-2} + m_1^{-2}} + \frac{u_2}{f_2\sigma^{-2} + m_2^{-2}}.$$

We see that for sufficiently small values of m_1, m_2 signifying fairly precise prior knowledge the improvements over the least squares risk $\operatorname{tr} \boldsymbol{UV} = \sigma^2(n^{-1}u_1 + f_2^{-1}u_2)$ will be appreciable, which is also documented by the numerical results in Pilz (1985).

Taking account of the inclusion (15.124) we can generalize the result of Theorem 15.10 dropping the restrictive assumption of diagonality of the matrices $\boldsymbol{U}$ and $\boldsymbol{V}$.

Theorem 15.11. *Let exist a matrix $\boldsymbol{M}^* = (m_{ij}^*) \in \mathcal{M}_r^{\geqq}$ with diagonal elements $m_{ii}^* = m_i^2$, $i = 1, \ldots, r$; and a diagonal matrix $\boldsymbol{A} \in \mathcal{M}_r^{\geqq}$ such that*

$$(\boldsymbol{M}^* + \boldsymbol{V})\boldsymbol{A}(\boldsymbol{M}^* + \boldsymbol{V}) = \boldsymbol{VUV}. \tag{15.128}$$

Then, provided that $\boldsymbol{M}^* \in \mathcal{M}(T_m)$, $\boldsymbol{M}^*$ is a least favourable moment matrix and $\hat{\vartheta}_{\mathrm{M}}^{\mathrm{a}} = \boldsymbol{M}^*(\boldsymbol{M}^* + \boldsymbol{V})^{-1}(\hat{\vartheta}_\Sigma - \boldsymbol{\mu}) + \boldsymbol{\mu}$ is a minimax-affine estimator for ϑ subject to the parameter region (15.121).

This result follows from Theorem 9 of CHALONER (1982), by differentiating $\operatorname{tr} \boldsymbol{VUV}(\boldsymbol{M}+\boldsymbol{V})^{-1}$ with respect to $\boldsymbol{M}$ chosen from the enlarged set of matrices $\{\boldsymbol{M}=(m_{ij}) \in \mathcal{M}_r^{\geqq} : m_{ii} = m_i^2\}$ using Lagrange multipliers. Note that, by virtue of (15.124), the solution to this problem does not necessarily represent a moment matrix of a probability measure from $\bar{\Theta}^*$, which necessitates the existence condition $\boldsymbol{M}^* \in \mathcal{M}(T_m)$ in the above theorem. In the special case of diagonality of $\boldsymbol{U}$ and $\boldsymbol{V}$ the result of Theorem 15.11 reduces to that given in Theorem 15.10, observing that $\boldsymbol{M}^* = \operatorname{diag}(m_1^2, \ldots, m_r^2) \in \mathcal{M}(T_m)$.

For regression models with $r \geqq 3$ unknown parameters and nondiagonal matrices $\boldsymbol{VUV}$ we recommend to employ an iteration procedure for the approximate computation of a least favourable moment matrix and a minimax estimator, respectively. A Fortran program for an iterative computation of the minimax estimator under the parameter restrictions (15.119), which is based on the iteration procedure described in Section 15.1.2, may be found in FRANKE (1987). There the hyperrectangle $\bar{\Theta}$ is transformed to the hypercube $[0, 1]^r$ so that every corner point can be identified with the corresponding dual number, and, at each stage of the iteration, the new supporting point $\boldsymbol{\vartheta}_s$ to be included in the measure P_s, $s = 1, 2, \ldots$, comes out as solution of an associated quadratic $0-1$-optimization problem. An improved version of this program with a modified rule for the choice of an appropriate stepsize α_s, which is included in a larger package of PC-routines for regression estimation using different kinds of prior knowledge, is available upon request.

15.5. Generalization to restricted minimax estimation

In this section we shortly want to point at a generalization of the preceding results to the problem of robust Bayes linear estimation which we have dealt with in Sections 6.2 and 6.3. To this consider the model

$$\begin{aligned} \tilde{\boldsymbol{y}} &= \boldsymbol{F}\tilde{\boldsymbol{\vartheta}} + \tilde{\boldsymbol{e}}, \qquad & \operatorname{Cov} \tilde{\boldsymbol{e}} &\leqq \boldsymbol{\Sigma}_0 \\ \mathbf{E}\tilde{\boldsymbol{\vartheta}} &\in \Theta \subset \mathbb{R}^r, \qquad & \operatorname{Cov} \tilde{\boldsymbol{\vartheta}} &\leqq \boldsymbol{\Phi}_0 \end{aligned} \tag{15.129}$$

with given matrices $\boldsymbol{\Sigma}_0 \in \mathcal{M}_n^{\geqq}$, $\boldsymbol{\Phi}_0 \in \mathcal{M}_r^{\geqq}$ and Θ being a compact region. Our objective is to find an estimator $\hat{\boldsymbol{\vartheta}}_{\mathcal{P}_1}$ such that

$$\sup_{\mathbf{E}\tilde{\boldsymbol{\vartheta}} \in \Theta} \mathbf{E}_{\tilde{\boldsymbol{\vartheta}}} R(\tilde{\boldsymbol{\vartheta}}, \hat{\boldsymbol{\vartheta}}_{\mathcal{P}_1}) = \inf_{\hat{\boldsymbol{\vartheta}} \in D} \sup_{\mathbf{E}\tilde{\boldsymbol{\vartheta}} \in \Theta} \mathbf{E}_{\tilde{\boldsymbol{\vartheta}}} R(\tilde{\boldsymbol{\vartheta}}, \hat{\boldsymbol{\vartheta}}), \tag{15.130}$$

$D \in \{D_1, D_a\}$, i.e., we are interested in finding a so-called $\mathcal{P}_1$-minimax estimator (cp. Definition 6.3) in the class of linear or affine-linear estimators, respectively, where

$$\mathcal{P}_1 = \{P_{\tilde{\boldsymbol{\vartheta}}, \tilde{\boldsymbol{e}}} : \mathbf{E}\tilde{\boldsymbol{\vartheta}} \in \Theta, \quad \operatorname{Cov} \tilde{\boldsymbol{\vartheta}} \leqq \boldsymbol{\Phi}_0, \quad \mathbf{E}\tilde{\boldsymbol{e}} = \boldsymbol{0}, \quad \operatorname{Cov} \tilde{\boldsymbol{e}} \leqq \boldsymbol{\Sigma}_0\}.$$

In particular, putting $\boldsymbol{\Phi}_0 = \boldsymbol{0}_{r,r}$, the problem reduces to the problem of finding a minimax estimator in D with $\boldsymbol{\vartheta}$ constrained to lie in the compact subset Θ of $\mathbb{R}^r$. Hence, together with the model (15.129), the above problem (15.130) represents a natural extension of the minimax estimation problem considered thus far.

In Section 6.3 we have shown that $\mathcal{P}_1$-minimax estimation in D_a is equivalent to finding a hierarchical Bayes affine-linear estimator with respect to some least favourable prior distribution for the hyperparameter $\mu = \mathbf{E}\tilde{\vartheta}$. Under the additional assumption of the symmetry of Θ around some center point $\mu_0 \in \mathbb{R}^r$ we know from Theorem 6.10 that the $\mathcal{P}_1$-minimax estimator is given by

$$\hat{\vartheta}_{\mathcal{P}_1} = (F^\top \Sigma^{-1} F + (\Phi_0 + M_0)^{-1})^{-1} (F^\top \Sigma_0^{-1} \tilde{y} + (\Phi_0 + M_0)^{-1} \mu_0),$$

provided that $\Phi_0 + M_0$ is regular, otherwise we have to put

$$\hat{\vartheta}_{\mathcal{P}_1} = (\Phi_0 + M_0)\,[F(\Phi_0 + M_0)F^\top + \Sigma_0]^{-1}(\tilde{y} - F\mu_0) + \mu_0\,. \qquad (15.131)$$

Here M_0 again plays the role of the (centered) second order moment matrix of a least favourable prior distribution over the space (of hyperparameters) Θ. This matrix can be obtained as solution of the convex optimization problem

$$\operatorname{tr} VUV(M_P + \Phi)^{-1} \to \inf_{M_P \in \mathcal{M}}\,, \qquad (15.132)$$

where $V = (F^\top \Sigma_0^{-1} F)^{-1}$, $\Phi = \Phi_0 + V$
and $\mathcal{M} = \left\{ M_P = \int\limits_{\Theta} (t - \mu_0)(t - \mu_0)^\top P(\mathrm{d}t) : P \in \Theta^* \right\}$ (see Lemma 6.6.).

Thus, M_0 is the information matrix of an L_B-optimal design measure with respect to the weight matrix VUV (cp. Definitions 10.2 and 10.4) and we can apply all the methods developed in Chapter 12 to find such a matrix.

For the special case where Θ is an ellipsoid or a hyperrectangle, respectively, we can easily modify the results presented in Sections 15.3 and 15.4 for finding solutions to the above problem. Likewise, for nonsymmetric regions Θ the approach set forth in Section 15.2 can be adapted to deal with $\mathcal{P}_1$-minimax estimation in D_a.

16. Admissible linear estimation under parameter restrictions

In Section 7.2 we have given characterizations of the class of (universally) admissible linear estimators in case of an unconstrained parameter region $\Theta = \mathbb{R}^r$. We have shown there that this class consists of exactly those estimators which are almost Bayes linear estimators, i.e., which are limit points of proper Bayes linear estimators. This class is very large, it includes for example the LSE, the principal component and Marquardt estimators $\hat{\vartheta}_{m,c}$ from (7.15), and similar estimators which, in their very nature, are non-Bayesian.

We will now deal with the characterization of universally admissible linear estimators when prior knowledge suggests a restricted compact parameter region $\Theta \subset \mathbb{R}^r$. In this case the aforementioned and some other well-known linear estimators are no longer admissible, not even within the class of linear estimators D_1. MARQUARDT (1970) showed that the (generalized) LSE is inadmissible, LA MOTTE (1978) showed that the Marquardt estimator itself is inadmissible for $m \geqq 1$ and can be dominated by a Bayes linear estimator. Moreover, in the sequel it will be seen that the linearly constrained LSE $\hat{\vartheta}_{K,q}$ (see Lemma 7.3) and also the ridge estimator (7.17) formed with a singular $\boldsymbol{H} \in \mathcal{M}_r^{\geqq}$ are not admissible in D_1.

In Section 16.1 we will show that in case of a compact region Θ the class of admissible (affine) linear estimators coincides with the class of proper Bayes (affine) linear estimators and every limit point of a sequence of such estimators again yields a proper Bayes (affine) linear estimator. The inadmissibility of the aforementioned estimators then results from the fact that these estimators do not belong to the class of proper Bayes linear estimators. In Section 16.2 we give conditions under which an admissible linear estimator dominates the generalized LSE over the whole compact parameter region.

Throughout this section, let the Assumptions 15.1 and 15.2 be satisfied (compactness of $\Theta \subset \mathbb{R}^r$ and rank $(\boldsymbol{F}^\top \boldsymbol{\Sigma}^{-1} \boldsymbol{F}) = r$).

16.1. Characterization of admissible linear estimators

We recall from Definitions 7.1 and 7.2 that an estimator $\hat{\vartheta} \in D_1$ is said to be universally Θ-admissible in D_1 if it is $(\boldsymbol{U}, \Theta)$-admissible in D_1 with respect to quadratic risk $R(\vartheta, \hat{\vartheta}) = \mathsf{E}_{\tilde{y}|\vartheta} \| \vartheta - \hat{\vartheta}(\tilde{y}) \|_U^2$ for any choice of $\boldsymbol{U} \in \mathcal{M}_r^{\geqq}$. We also recall that for universal Θ-admissibility it suffices to have $(\boldsymbol{U}, \Theta)$-admissibility for some full rank matrix $\boldsymbol{U}$ (see Lemma 7.1).

Under the assumption that rank $(\boldsymbol{F}^\top \boldsymbol{\Sigma}^{-1} \boldsymbol{F}) = \operatorname{rank} \boldsymbol{U} = r$, the class of Bayes linear estimators is given by

$$D_1^{\mathrm{B}} = \{\hat{\vartheta}_P = \boldsymbol{M}_P(\boldsymbol{M}_P + \boldsymbol{V})^{-1} \hat{\vartheta}_\Sigma : \boldsymbol{M}_P \in \mathcal{M}\}, \tag{16.1}$$

where

$$\mathcal{M} = \left\{ \boldsymbol{M}_P = \int_\Theta \vartheta \vartheta^\top \mathrm{P}(\mathrm{d}\vartheta) : P \in \Theta^* \right\} \subset \mathcal{M}_r^{\geqq} \tag{16.2}$$

denotes the class of second order moment matrices associated with the prior distributions over Θ. Obviously, there is a 1–1 correspondence between the estimators $\hat{\vartheta}_P$ from D_1^{B} and the underlying matrices $\boldsymbol{M}_P \in \mathcal{M}$. But, the basic difference to the preceding case of $\Theta = \mathbb{R}^r$ lies now in the fact that, under the Assumption 15.1, the above introduced set $\mathcal{M}$ is compact (with respect to the Frobenius norm). This implies that D_1^{B} contains all its boundary points, the almost Bayes linear estimators do not lead to an extension of D_1^{B} and represent proper Bayes linear estimators themselves. We will now give a precise formulation of this heuristic idea.

Lemma 16.1. (i) *Under the Assumption* 15.1 *the set* $\mathcal{M}$ *is convex and compact with respect to* $\| \boldsymbol{M}_P \| = (\operatorname{tr} \boldsymbol{M}_P^2)^{1/2}$.

(ii) *Under the Assumptions* 15.1 *and* 15.2 *the set* D_l^{B} *of Bayes linear estimators given by* (16.1) *is convex and compact with respect to the metric*

$$m(\hat{\vartheta}_{P_1}, \hat{\vartheta}_{P_2}) = \| \boldsymbol{M}_{P_1} - \boldsymbol{M}_{P_2} \|, \qquad P_i \in \Theta^*, \qquad \hat{\vartheta}_{P_i} \in D_l^{\mathrm{B}} \qquad (i = 1, 2). \tag{16.3}$$

Proof. Assertion (i) is clear from Lemma 15.1. Assertion (ii) then follows immediately from Assertion (i). □

Lemma 16.1 states that for any sequence $\{P_i\}_{i=1,2,\dots} \subset \Theta^*$, for which the sequence of associated moment matrices is convergent, the corresponding sequence of Bayes linear estimators converges, in turn, to a proper Bayes linear estimator, i.e.,

$$\hat{\vartheta}_{P_i} \xrightarrow[P_{\tilde{y}|\vartheta} - \text{a.e.}]{} \hat{\vartheta}_P = \boldsymbol{M}_P(\boldsymbol{M}_P + \boldsymbol{V})^{-1}\tilde{\boldsymbol{y}} \in D_l^{\mathrm{B}} \quad (\text{for all } \vartheta \in \Theta),$$

where

$$\boldsymbol{M}_P := \lim_{i \to \infty} \boldsymbol{M}_{P_i} \in \mathcal{M}.$$

We now prove a complete class theorem for the class of Bayes linear estimators by showing that any linear estimator which does not belong to this class D_l^{B} can be improved by some Bayes linear estimator. Moreover, we can show that there exists no complete class in D_l which is smaller than D_l^{B}.

Theorem 16.1. *Assume* $\boldsymbol{U}$ *to have full rank and let the Assumptions* 15.1 *and* 15.2 *be satisfied. Then the class of Bayes linear estimators* D_l^{B} *given by* (16.1) *forms a minimal complete class in* D_l.

Proof. First we show the completeness of D_l^{B}. To this we use a similar idea as in the proof of Theorem 3.17 of WALD (1950). Let $\hat{\vartheta}_0 \in D_l$ such that $\hat{\vartheta}_0 \notin D_l^{\mathrm{B}}$. Then consider the estimation problem on the basis of the modified quadratic risk

$$R_0(\vartheta, \hat{\vartheta}) := R(\vartheta, \hat{\vartheta}) - R(\vartheta, \hat{\vartheta}_0), \qquad \vartheta \in \Theta, \hat{\vartheta} \in D_l. \tag{16.4}$$

Clearly, all properties of linear estimators such as Bayes optimality, minimaxity a.s.o. and of the original risk function $R(\cdot, \hat{\vartheta})$ carry over to the modified estimation problem with the new risk function $R_0(\cdot, \hat{\vartheta})$, in particular, $R_0(\cdot, \hat{\vartheta})$ remains to be bounded over Θ, for any $\hat{\vartheta} \in D_l$. Then define $\hat{\vartheta}_1$ to be a minimax estimator in D_l with respect to the new risk R_0, in Section 15.1.1 we have shown that such an estimator exists and turns out to be a Bayes linear estimator with respect to a least favourable prior distribution, i.e. $\hat{\vartheta}_1 \in D_l^B$. Now, from the minimaxity of $\hat{\vartheta}_1$ it follows that

$$R_0(\vartheta, \hat{\vartheta}_1) = R(\vartheta, \hat{\vartheta}_1) - R(\vartheta, \hat{\vartheta}_0) \leqq 0 = R_0(\vartheta, \hat{\vartheta}_0)$$

for all $\vartheta \in \Theta$. But, $R(\cdot, \hat{\vartheta}_1) \neq R(\cdot, \hat{\vartheta}_0)$, since $\hat{\vartheta}_0 \notin D_l^{\mathrm{B}}$. Thus, $\hat{\vartheta}_1$ is a $(\boldsymbol{U}, \Theta)$-improvement over $\hat{\vartheta}_0$, which establishes the completeness of D_l^{B}. To show the minimum completeness observe that in case of rank $\boldsymbol{U} = r$ the estimator $\hat{\vartheta}_P \in D_1^B$ is the unique Bayes estimator in D_1 with respect to $P \in \Theta^*$ and is thus admissible, the assertion then follows from the fact that a minimal complete class (provided it exists) consists of exactly those estimators which are admissible (see, e.g. BERGER (1985), Section 8.1). □

Remark 16.1. (i) The first part of the preceding proof provides a method how to construct an estimator which improves a given inadmissible linear estimator.

(ii) The completeness of the class of (proper) Bayes linear estimators in case of a compact parameter region is essentially effected through the simultaneous existence of a minimax linear estimator and of a (proper) least favourable prior distribution. On the contrary, in case of $\Theta = \mathbb{R}^r$ there does not exist a proper prior distribution which is least favourable and the minimax estimators can only be shown to be almost Bayesian, thus all these estimators have to be taken into account, too, when constructing complete classes.

In case of a singular weight matrix $U \in \mathcal{M}_r^{\geqq}$ the class of all Bayes linear estimators is given by

$$D_1^B(U) = \{\hat{\vartheta}_P = [U^+ UM_P(M_P + V)^{-1} + (I_r - U^+ U)\, C]\, \hat{\vartheta}_\Sigma \colon M_P \in \mathcal{M},\ C \in \mathcal{M}_{r \times r}\}, \tag{16.5}$$

where U^+ denotes the Moore-Penrose inverse of U (cp. Theorem 5.3, Remark 5.3 and PILZ (1986a), Remark 2). Quite analogously to the proof of Theorem 16.1 we can then show the completeness of $D_1^B(U)$ in D_1. Obviously, it holds

$$D_1^B = \bigcap_{U \in \mathcal{M}_r^{\geqq}} D_1^B(U) = D_1^B(I_r)\,. \tag{16.6}$$

Corollary 16.1. *The class of universally Θ-admissible linear estimators coincides with the class of unique Bayes linear estimators, i.e.*

$$D_1^{ad} = D_1^B\,.$$

This follows with (16.6) immediately from Theorem 16.1. The result of Corollary 16.1 exhibits the central role of the Bayes linear estimators for the construction of "good" linear estimators for a parameter constrained to lie in a compact region and sharpens the characterization result given in Theorem 7.3 for admissible linear estimators in the unconstrained case $\Theta = \mathbb{R}^r$.

Corollary 16.2. *Under the Assumptions* 15.1 *and* 15.2 *the following assertions are equivalent:*

(i) $\hat{\vartheta}$ *is universally Θ-admissible in D_1.*
(ii) $\hat{\vartheta}$ *is Bayes linear with respect to mean square error risk.*
(iii) $\exists M \in \mathcal{M} \colon \hat{\vartheta} = M(M + V)^{-1} \hat{\vartheta}_\Sigma$.

Remark 16.2. If we drop Assumption 15.1 then the above characterizations continue to hold true when choosing the equivalent representation

$$D_1^{\mathrm{B}} = \{\hat{\boldsymbol{\vartheta}}_P = \boldsymbol{M}_P\boldsymbol{F}^\top(\boldsymbol{F}\boldsymbol{M}_P\boldsymbol{F}^\top + \boldsymbol{\Sigma})^{-1}\tilde{\boldsymbol{y}} \colon \boldsymbol{M}_P \in \mathcal{M}\} \tag{16.7}$$

and thus replacing $\hat{\boldsymbol{\vartheta}}$ in Corollary 16.2 (iii) by $\hat{\boldsymbol{\vartheta}} = \boldsymbol{M}\boldsymbol{F}^\top(\boldsymbol{F}\boldsymbol{M}\boldsymbol{F}^\top + \boldsymbol{\Sigma})^{-1}\tilde{\boldsymbol{y}}$. Correspondingly, we have to modify $D_1^{\mathrm{B}}(\boldsymbol{U})$ from (16.5) to reestablish the completeness of this class in D_1. The only difference is that in case of rank $(\boldsymbol{F}^\top\boldsymbol{\Sigma}^{-1}\boldsymbol{F}) < r$ (16.3) defines only a semimetric m.

In Section 15.1 we have characterized minimax linear estimators as Bayes linear estimators with respect to least favourable priors. The least favourable moment matrix $\boldsymbol{M}_0$ and thus the minimax linear estimator crucially depend on the choice of the weight matrix $\boldsymbol{U}$. However, any solution $\boldsymbol{M}_0$ associated with a full rank matrix $\boldsymbol{U}$ leads to a universally Θ-admissible minimax linear estimator.

Corollary 16.3. *Let* $\Theta \subset \Theta^r$ *compact*, rank $\boldsymbol{U} = r$ *and* $\boldsymbol{M}_0$ *a least favourable moment matrix, i.e.,* $\boldsymbol{M}_0$ *maximizes the Bayes risk* $\operatorname{tr}\boldsymbol{U}\boldsymbol{M} - \operatorname{tr}\boldsymbol{U}\boldsymbol{M}\boldsymbol{F}^\top(\boldsymbol{F}\boldsymbol{M}\boldsymbol{F}^\top + \boldsymbol{\Sigma})^{-1}\boldsymbol{F}\boldsymbol{M}$ *with respect to* $\boldsymbol{M} \in \mathcal{M} = \{\boldsymbol{M}_P \colon P \in \Theta^*\}$. *Then the corresponding minimax linear estimator* $\hat{\boldsymbol{\vartheta}}_{\mathrm{M}} = \boldsymbol{M}_0\boldsymbol{F}^\top(\boldsymbol{F}\boldsymbol{M}_0\boldsymbol{F}^\top + \boldsymbol{\Sigma})^{-1}\tilde{\boldsymbol{y}}$ *is universally* Θ*-admissible in* D_1.

This follows with Corollary 16.1 and Remark 16.2 immediately from the fact that $\hat{\boldsymbol{\vartheta}}_{\mathrm{M}} \in D_1^{\mathrm{B}}$.

Let us now consider in some more detail the special case of an ellipsoidal region

$$\Theta = \mathscr{E}(\boldsymbol{H}) := \{\boldsymbol{\vartheta} \in \mathbb{R}^r \colon \boldsymbol{\vartheta}'\boldsymbol{H}\boldsymbol{\vartheta} \leqq 1\}, \tag{16.8}$$

where $\boldsymbol{H}$ is a given positive definite matrix, and without loss of generality it can be assumed that the ellipsoid is centered at the origin. HOFFMANN (1980) gave a characterization of the universally admissible linear estimators for this region. He stated that $\hat{\boldsymbol{\vartheta}}$ is universally $\mathscr{E}(\boldsymbol{H})$-admissible in D_1 if and only if it can be written as $\hat{\boldsymbol{\vartheta}} = \boldsymbol{C}\boldsymbol{F}^\top\boldsymbol{\Sigma}^{-1}\tilde{\boldsymbol{y}}$ with a matrix $\boldsymbol{C} \in \mathcal{M}_{r\times r}$ satisfying

$$\boldsymbol{V}^{-1/2}\boldsymbol{C}\boldsymbol{V}^{-1/2} \in \mathscr{A}_{[0,1]} \quad \text{and} \quad \operatorname{tr}\boldsymbol{V}\boldsymbol{H}(\boldsymbol{I}_r - \boldsymbol{C}\boldsymbol{V}^{-1})^{-1} \leqq 1 + \operatorname{tr}\boldsymbol{V}\boldsymbol{H}. \tag{16.9}$$

We will give a simpler characterization based on the identity $D_1^{\mathrm{ad}} = D_1^{\mathrm{B}}$, and which makes essential use of the very simple structure of the set of moment matrices associated with $\mathscr{E}(\boldsymbol{H})$,

$$\mathcal{M} = \{\boldsymbol{M} \in \mathcal{M}_r^{\geqq} \colon \operatorname{tr}\boldsymbol{H}\boldsymbol{M} \leqq 1\}.$$

(See Lemma 12.2 and (15.79).)

From Corollary 12.6 we then get the following characterization.

Corollary 16.4. *Under the Assumption* 15.1, *an estimator* $\hat{\boldsymbol{\vartheta}} \in D_1$ *is universally* $\mathscr{E}(\boldsymbol{H})$*-admissible in* D_1 *if and only if it can be written in the form*

$\hat{\vartheta} = M(M+V)^{-1}\hat{\vartheta}_\Sigma$, *where* M *is such that*

$$M \in \mathcal{M}_r^{\geqq} \quad \text{and} \quad \operatorname{tr} HM \leqq 1 .$$

It is easily seen that the above conditions on M are equivalent to those stated in (16.9) for C using the representation $\hat{\vartheta} = CF^\top \Sigma^{-1}\tilde{y}$. In view of Remark 16.2, if we drop Assumption 15.1 then the assertion of Corollary 16.4 remains valid when choosing the more general representation $\hat{\vartheta} = MF^\top(FMF^\top + \Sigma)^{-1}\tilde{y}$.

From Corollary 16.2 it is fully clear that the LSE, the linearly constrained least squares estimator $\hat{\vartheta}_{K,q}$ from Lemma 7.3 and the Marquardt estimator (7.15) with $m \geqq 1$ are inadmissible in D_1 if we are faced with a compact parameter region, since they are not even Bayes linear estimators in the unconstrained case $\Theta = \mathbb{R}^r$. We shall now briefly investigate Bayes optimality and universal admissibility of the Marquardt estimator with $m = 0$ and the generalized ridge estimator in case of the constrained parameter region $\Theta = \mathscr{E}(H)$. To this we prove whether these estimators admit a representation as indicated in Corollary 16.4.

a) Consider the Marquardt estimator

$$\hat{\vartheta}_{0,c} = (c/d)\, ss^\top F^\top \Sigma^{-1}\tilde{y}, \qquad 0 < c < 1, \tag{16.10}$$

where d and s are the maximum eigenvalue and corresponding eigenvector of $V^{-1} = F^\top\Sigma^{-1}F$, respectively. Writing $\hat{\vartheta}_{0,c} = (c/d)\, ss^\top V^{-1}\hat{\vartheta}_\Sigma$ and equating $(c/d)\, ss^\top V^{-1} = M(M+V)^{-1}$ we obtain the solution

$$M = \left(\frac{d}{c} I_r - ss^\top V^{-1}\right)^{-1} ss^\top = cd(1-c)^{-1}\, ss^\top .$$

Obviously, $M \in \mathcal{M}_r^{\geqq}$ and the condition $\operatorname{tr} HM \leqq 1$ reads

$$s^\top Hs \leqq c^{-1}(1-c)\, d. \tag{16.11}$$

Thus, if (16.11) comes true then $\hat{\vartheta}_{0,c}$ is Bayes-linear and universally $\mathscr{E}(H)$-admissible in D_1.

b) The generalized ridge estimator $\hat{\vartheta}_{\mathrm{GR}} = (F^\top\Sigma^{-1}F + G)^{-1}F^\top\Sigma^{-1}\tilde{y}$ with a positive definite matrix G can be written as

$$\hat{\vartheta}_{\mathrm{GR}} = (V^{-1} + G)^{-1}V^{-1}\hat{\vartheta}_\Sigma = M(M+V)^{-1}\hat{\vartheta}_\Sigma \quad \text{with} \quad M = G^{-1}.$$

Hence, it is Bayes linear and universally $\mathscr{E}(H)$-admissible in D_1 if and only if $\operatorname{tr} HG^{-1} \leqq 1$.

In particular, choosing $G = H$ we obtain the Kuks-Ol'man estimator, for which we have $\operatorname{tr} HG^{-1} = r$. Thus, for $r \geqq 2$ this estimator is neither universally $\mathscr{E}(H)$-admissible in D_1 nor does there exist a prior distribution $P \in \mathscr{E}(H)^*$ with respect to which it comes out as a Bayes estimator. In fact, Kuks/Ol'man (1972) only aimed at estimating a linear combination of ϑ, which refers to the special case of rank $U = 1$. However, by a simple modification

$$\hat{\vartheta}_c = (F^\top \Sigma^{-1} F + cH)^{-1} F^\top \Sigma^{-1} \tilde{y}, \qquad c \geqq r \tag{16.12}$$

of the Kuks-Ol'man estimator we arrive at an estimator $\hat{\vartheta}_c$ which is universally $\mathscr{E}(H)$-admissible in D_1. Moreover, $\hat{\vartheta}_c$ is Bayes linear with respect to any $P \in \mathscr{E}(H)^*$ such that $M_P = (cH)^{-1}$.

Alternatively, choosing $G = c^{-1}(1-c)\, F^\top \Sigma^{-1} F$ with some $c \in (0,1)$ we arrive at the shrunken estimator $\hat{\vartheta}_c = c\hat{\vartheta}_\Sigma$, which is universally $\mathscr{E}(H)$-admissible in D_1 if and only if $\operatorname{tr} HV \leqq (1-c)/c$.

We will now indicate the corresponding characterization of the universsally Θ-admissible affine-linear estimators. The Bayes affine estimator makes explicit use of the first order moments of the prior distribution $P \in \Theta^*$ and takes the form

$$\hat{\vartheta}_P^{\mathrm{a}} = \Phi_P F^\top (F \Phi_P F^\top + \Sigma)^{-1} (\tilde{y} - F\mu_P) + \mu_P,$$

where (16.13)

$$\mu_P = \int_\Theta \vartheta P(\mathrm{d}\vartheta), \quad \Phi_P = \int_\Theta (\vartheta - \mu_P)(\vartheta - \mu_P)^\top P(\mathrm{d}\vartheta) = M_P - \mu_P \mu_P^\top.$$

(cp. (15.50) and Theorem 5.4). With the denotations

$$J^\top = \begin{bmatrix} 0_r^\top \\ I_r \end{bmatrix}, \quad F_0 = \begin{bmatrix} 1 & 0_r^\top \\ 0_n & F \end{bmatrix}, \quad \Sigma_0 = \begin{bmatrix} 0 & 0_n^\top \\ 0_n & \Sigma \end{bmatrix}, \quad \bar{M}_P = \begin{bmatrix} 1 & \mu_P^\top \\ \mu_P & M_P \end{bmatrix} \tag{16.14}$$

and $\tilde{y}_0^\top = (1, \tilde{y}^\top)$ this estimator can be written equivalently as

$$\hat{\vartheta}_P^{\mathrm{a}} = J \bar{M}_P F_0^\top (F_0 \bar{M}_P F_0^\top + \Sigma_0)^{-1} \tilde{y}_0 \tag{16.15}$$

(see the proof of Lemma 15.9). The Bayes estimator $\hat{\vartheta}_P^{\mathrm{a}}$ depends on the prior distribution only through the matrix $\bar{M}_P$ of first and second order moments. Under the assumption that F is of full rank we have a $1-1$ correspondence between $\hat{\vartheta}_P^{\mathrm{a}}$ and the underlying matrix $\bar{M}_P$. With the help of this matrix we introduce a metric (semimetric in case of rank $F < r$)

$$\bar{m}(\hat{\vartheta}_{P_1}^{\mathrm{a}}, \hat{\vartheta}_{P_2}^{\mathrm{a}}) := \| \bar{M}_{P_1} - \bar{M}_{P_2} \|; \qquad P_i \in \Theta^*, \qquad i = 1, 2 \tag{16.16}$$

in the class $D_{\mathrm{a}}^{\mathrm{B}} = \{\hat{\vartheta}_P^{\mathrm{a}}: P \in \Theta^*\}$ of Bayes affine estimators. From the proof of Lemma 15.9 we know that

$$\bar{\mathscr{M}} = \{\bar{M}_P \in \mathscr{M}_{r+1}^{\geqq}: P \in \Theta^*\}$$

is convex and compact with respect to $\| \bar{M}_P \| = (\operatorname{tr} \bar{M}_P^2)^{1/2}$. Therefore it holds

Lemma 16.2. *Under the Assumptions* 15.1 *and* 15.2 *the set* $D_{\mathrm{a}}^{\mathrm{B}}$ *of Bayes affine estimators is convex and compact with respect to* $\bar{m}$.

Thus, we do not need to consider almost Bayes affine estimators, since these are automatically included in $D_{\mathrm{a}}^{\mathrm{B}}$. Analogously to the proof of Theorem 16.1 we can show that, for a regular weight matrix U, the class D_a^B is (minimal) complete in D_a, for singular matrices U this class must be enlarged such that it includes any estimator of the form (15.50). The existence of a minimax estimator in D_a and of a least favourable prior distribution, which are essential prerequisites for the proof, are guaranteed by Lemma 15.9.

In complete analogy with Theorem 16.1 and Corollary 16.1 we can thus state the following result.

Theorem 16.2. *Assume $\Theta \subset \mathbb{R}^r$ to be compact. Then it holds:*

(i) *For any $U \in \mathcal{M}_r^>$, the set $D_a^B = \{\hat{\vartheta}_P^a : P \in \Theta^*\}$ forms a minimal complete class in D_a.*

(ii) *The class D_a^{ad} of admissible affine-linear estimators coincides with the class of Bayes affine estimators, i.e. $D_a^{ad} = D_a^B$.*

(iii) *Under the additional assumption that F has full rank an estimator $\hat{\vartheta} \in D_a$ is universally Θ-admissible in D_a if and only if there exists a prior distribution $P \in \Theta^*$ with covariance matrix Φ_P and expectation vector μ_P such that $\hat{\vartheta} = \hat{\vartheta}_P^a = \Phi_P(\Phi_P + V)^{-1}(\hat{\vartheta}_\Sigma - \mu_P) + \mu_P$.*

The third assertion follows from the second one observing Remark 5.3.

16.2. Universal improvements over the LSE

General results on improvements over the LSE for certain models with convex parameter region may be found in BUNKE/BUNKE (1986), explicit formulae for improvements were but only available for very few special cases. HOFFMANN (1980) gave a characterization of the class of all linear universal Θ-improvements over the LSE in case of ellipsoid restrictions. Below we give a general result on universal Θ-improvements in D_l in case of an arbitrary compact region Θ which specializes to Hoffmann's characterization when Θ happens to be an ellipsoid.

Clearly, we are only interested in improvements over the LSE which are (universally) admissible, so we start, again, from the class $D_l = \{\hat{\vartheta}_P = M_P(M_P + V)^{-1}\hat{\vartheta}_\Sigma : M_P \in \mathcal{M}\}$. Now, consider the minimum covering ellipsoid (MCE) of the compact region Θ

$$\mathscr{E}(C_0) = \{t \in \mathbb{R}^k : t^\top C_0 t \leqq 1\} \supseteqq \Theta, \tag{16.17}$$

i.e., the ellipsoid which has smallest volume among all ellipsoids $\mathscr{E}(C) = \{t \in \mathbb{R}^r : t^\top C t \leqq 1\}$, $C \in \mathcal{M}_r^>$, containing Θ as a subset. In other words, C_0 is the (uniquely determined) matrix minimizing $\det C^{-1}$ within the set of all positive definite matrices C such that

$$\vartheta^\top C \vartheta \leqq 1 \quad \text{for all} \quad \vartheta \in \Theta.$$

It is well-known that the MCE-problem is dual to the problem of finding a D-optimum design measure on Θ and both problems have a common solution, i.e.,

$$C_0 = \frac{1}{r} M_0^{-1}, \quad \text{where} \quad M_0 = \arg \sup_{M_P \in \mathcal{M}} \det M_P \tag{16.18}$$

(see SILVEY/TITTERINGTON (1973), TITTERINGTON (1980)). Moreover, it is an easy consequence of (16.18) and the classical KIEFER/WOLFOWITZ (1960) equivalence theorem that $\mathscr{E}(C_0)$ is the MCE of Θ if and only if

$$\sup_{\vartheta \in \Theta} \vartheta^\top C_0 \vartheta = 1, \tag{16.19}$$

and, moreover, any prior distribution P_0 assuring the optimum $C_0 = (1/r) M_{P_0}^{-1}$ only contains supporting points $\vartheta \in \Theta$ for which $\vartheta^\top C_0 \vartheta = 1$.

Theorem 16.3. *Let Assumptions* 15.1 *and* 15.2 *be satisfied and* $\mathscr{E}(C_0)$ *the MCE of* Θ. *Then the Bayes linear estimator* $\hat{\vartheta}_P = M_P(M_P + V)^{-1}\hat{\vartheta}_\Sigma$, $M_P \in \mathcal{M}$, *effects a universal* Θ*-improvement over* $\hat{\vartheta}_\Sigma$ *if it holds*

$$2M_P + V \geqq C_0^{-1},$$

where "$\geqq$*" refers to the usual semiordering in* $\mathcal{M}_r^{\geqq}$.

Proof. Denote $B_P = V(M_P + V)^{-1}$. Then the risk of $\hat{\vartheta}_P$ takes the form

$$R(\vartheta, \hat{\vartheta}_P) = \operatorname{tr} U(I_k - B_P) V (I_k - B_P)^\top + \vartheta^\top B_P^\top U B_P \vartheta .$$

Observing that $(M_P + V) B_P^\top = V$, we obtain:

$$\begin{aligned} R(\vartheta, \hat{\vartheta}_\Sigma) - R(\vartheta, \hat{\vartheta}_P) &= \operatorname{tr} U B_P [V + (M_P - \vartheta\vartheta^\top) B_P^\top] \\ &= \operatorname{tr} U B_P (2M_P + V - \vartheta\vartheta^\top) B_P^\top \\ &\geqq \operatorname{tr} U B_P (2M_P + V - C_0^{-1}) B_P^\top , \end{aligned}$$

where the latter inequality results from $\vartheta\vartheta^\top \leqq C_0^{-1}$ due to (16.19). Hence, assuming that $2M_P + V \geqq C_0^{-1}$, we have $R(\vartheta, \hat{\vartheta}_P) \leqq R(\vartheta, \hat{\vartheta}_\Sigma)$ for all $\vartheta \in \Theta$. Further, there exists at least one point $\vartheta_0 \in \Theta$ for which we have strict inequality, since, assuming for a moment that there exists $U \in \mathcal{M}_r^{\geqq}$, $U \neq 0_{r,r}$, such that $R(\cdot, \hat{\vartheta}_P) = R(\cdot, \hat{\vartheta}_\Sigma)$, this would imply $B_P^\top U B_P = 0_{r,r}$. But, by virtue of the regularity of B_P, this is only true if $U = 0_{r,r}$. □

Let us now have a look at the special case in which Θ happens to be an ellipsoid $\Theta = \mathscr{E}(H)$. Then, of course, $\mathscr{E}(H)$ and $\mathscr{E}(C_0)$ coincide, i.e., $C_0 = H$, and the sufficient condition for an improvement over $\hat{\vartheta}_\Sigma$ given in Theorem 16.3 proves to be necessary, too.

Theorem 16.4. *Let* $\Theta = \mathscr{E}(H)$ *and Assumption* 15.2 *be satisfied. Then the Bayes linear and universally* Θ*-admissible linear estimator* $\hat{\vartheta} = M(M + V)^{-1}\hat{\vartheta}_\Sigma$, *where* $M \in \mathcal{M}_r^{\geqq}$ *and* $\operatorname{tr} HM \leqq 1$, *effects a universal* Θ*-improvement over* $\hat{\vartheta}_\Sigma$ *if and only if holds* $2M + V \geqq H^{-1}$.

Proof. With account to the particular structure of the universally $\mathscr{E}(H)$-admissible linear estimators (cp. Corollary 16.4), one easily verifies that the condition $2\boldsymbol{M} + \boldsymbol{V} \geqq \boldsymbol{H}^{-1}$ is equivalent to the somewhat more complicated condition of Theorem 8 of HOFFMANN (1980), which he proved to be necessary and sufficient for attaining an improvement over $\hat{\boldsymbol{\vartheta}}_{\Sigma}$. □

Remark 16.3. Obviously, the conditions $\operatorname{tr} \boldsymbol{HM} \leqq 1$ and $\boldsymbol{M} \geqq (\boldsymbol{H}^{-1} + \boldsymbol{V})/2$ raised for universal $\mathscr{E}(\boldsymbol{H})$-admissibility and universal $\mathscr{E}(\boldsymbol{H})$-improvement over $\hat{\boldsymbol{\vartheta}}_{\Sigma}$, respectively, "pull" $\boldsymbol{M}$ into opposite directions. Thus, we have to make a compromise to guarantee both universal admissibility and universal improvement, and it depends on the size of $\mathscr{E}(\boldsymbol{H})$ and on the experimental conditions (through $\boldsymbol{V}$) whether there exists an estimator $\hat{\boldsymbol{\vartheta}} \in D_1^{\mathrm{B}}$ which enjoys both properties. According to HOFFMAN (1980, p. 383) this is the case if and only if

$$\sum_{j \in J} \lambda_j(\boldsymbol{H}^{1/2} \boldsymbol{V} \boldsymbol{H}^{1/2}) \geqq |J| - 2, \tag{16.20}$$

where J indexes all eigenvalues of $\boldsymbol{H}^{1/2}\boldsymbol{V}\boldsymbol{H}^{1/2}$ less than 1, a sufficient condition for (16.20) to come true is that there are at most two such eigenvalues.

As an example, consider the case where Θ is a sphere

$$\Theta = \{\boldsymbol{\vartheta} \in \mathbb{R}^r : \boldsymbol{\vartheta}^\top \boldsymbol{\vartheta} \leqq h^2\}, \quad h > 0$$

with radius h, i.e., $\boldsymbol{H} = h^{-2}\boldsymbol{I}_r$. If we look at the ordinary ridge estimator

$$\hat{\boldsymbol{\vartheta}}_{\mathrm{R}} = (\boldsymbol{V}^{-1} + c\boldsymbol{I}_r)^{-1}\boldsymbol{F}^\top \boldsymbol{\Sigma}^{-1}\tilde{\boldsymbol{y}} = \boldsymbol{M}(\boldsymbol{M} + \boldsymbol{V})^{-1}\hat{\boldsymbol{\vartheta}}_{\Sigma}, \qquad \boldsymbol{M} = c^{-1}\boldsymbol{I}_r$$

with some ridge constant $c > 0$ then $\hat{\boldsymbol{\vartheta}}_{\mathrm{R}}$ is universally Θ-admissible in D_1 iff $c \geqq r/h^2$. We remark that in this case the modified Kuks-Ol'man estimator defined by (16.12) also reduces to an ordinary ridge estimator. According to Theorem 16.4, $\hat{\boldsymbol{\vartheta}}_{\mathrm{R}}$ effects a universal Θ-improvement over $\hat{\boldsymbol{\vartheta}}_{\Sigma}$ iff

$$\boldsymbol{V} \geqq (h^2 - 2/c)\,\boldsymbol{I}_r \quad \text{iff} \quad c \leqq 2/(h^2 - \lambda_{\min}(\boldsymbol{V}))\,.$$

Necessary and sufficient conditions for a possible improvement of $\hat{\boldsymbol{\vartheta}}_{\Sigma}$ by a universally Θ-admissible ordinary ridge estimator are given by

$$\lambda_{\min}(\boldsymbol{V}) < h^2 \quad \text{and} \quad r \leqq 2h^2/(h^2 - \lambda_{\min}(\boldsymbol{V}))\,.$$

17. The minimax approach to fitting biased response surfaces

In this chapter we reconsider the regression estimation problem in inadequate models, i.e., we wish to approximate an unknown response surface known to belong to some function space $\mathscr{F}$ on the basis of a regression experiment in a given ideal family $\mathscr{G}$ of regression functions. A motivation for this sort of problem we have given in Section 9.2, where we have also derived a Bayes linear estimator for the response surface. Now we want to find a minimax linear estimator where, as in Section 9.2, we assume $\mathcal{F}$ and $\mathcal{G}$ to be finite-dimensional linear function spaces. The case of a nonparametric family $\mathcal{F}$ has been considered by MARCUS/SACKS (1977) and SACKS/YLVISAKER (1978), (1984) who derive minimax estimators and appropriate model robust designs. However, for families $\mathcal{G}$ with more than two parameters the resulting optimization problems are very complicated. Moreover, there are several papers dealing with optimum experimental design to lessen the sensitivity of standard estimators with respect to model inadequacy, a good literature survey is given in GAFFKE (1984) and SACKS/YLVISAKER (1984).

PETERSEN (1973) and LÄUTER (1975 b) discussed a minimax approach leading to a very intrinsic matrix optimization problem which is but not feasible for a solution. In BUNKE/BUNKE (1986), Section 2.7.2, this problem is traced back to the solution of a system of nonlinear matrix equations, which remains unfeasible, however. For the special case $\mathcal{G} \subset \mathcal{F}$ PILZ (1988 c) has given a minimax solution using prior knowledge about the parameters of the family $\mathcal{F}$ and about the goodness of approximation of the functions from $\mathcal{F}$ by those from $\mathcal{G}$. In the sequel we will generalize the results to the case where $\mathcal{F}$ and $\mathcal{G}$ are arbitrary finite dimensional spaces. As in Chapter 15 we will solve the problem by changing it to a Bayesian estimation problem with respect to a least favourable prior distribution. The solution coincides, as expected, with the minimax linear estimator developed in Section 15.1 when the model $\mathcal{G}$ is adequate, that is, when $\mathcal{G} = \mathcal{F}$. For the special case that $\mathcal{G} \subset \mathcal{F}$ we will see that, under an additional assumption on the regression experiment associated with $\mathcal{F}$ (block diagonality of the information matrix), an explicit form solution can be obtained on the basis of the minimax linear estimator for the parameter vector of the inadequate ("parsimonious") model $\mathcal{G}$.

17.1. The minimax approgression problem

We wish to approximate the (unknown) response surface $\eta(\cdot) = \mathbf{E}\tilde{y}(\cdot)$ known to belong to the family

$$\mathcal{F} = \{\eta(\cdot) : \eta(x) = \boldsymbol{\vartheta}^\top \boldsymbol{f}(\boldsymbol{x}); \quad \boldsymbol{x} \in X, \quad \boldsymbol{\vartheta} \in \Theta \subseteqq \mathbb{R}^r\} \tag{17.1}$$

by some function out of the family

$$\mathcal{G} = \{\gamma(\cdot) : \gamma(\boldsymbol{x}) = \boldsymbol{\beta}^\top \boldsymbol{g}(\boldsymbol{x}); \quad \boldsymbol{x} \in X, \quad \boldsymbol{\beta} \in B \subseteqq \mathbb{R}^s\}, \tag{17.2}$$

where $X \subseteqq \mathbb{R}^k$, $\boldsymbol{g}|X \to \mathbb{R}^s$ is a given function and $\boldsymbol{\beta} = (\beta_1, \ldots, \beta_s)^\top$ denotes an appropriate parameter vector. Again, attention is restricted to the class of linear approgression estimators of η,

$$D = \{\hat{\eta}(\cdot, \tilde{\boldsymbol{y}}) = \boldsymbol{g}(\cdot)^\top \boldsymbol{Z} \tilde{\boldsymbol{y}} : \boldsymbol{Z} \in \mathcal{M}_{s \times n}\}, \tag{17.3}$$

and the goodness of an estimator $\hat{\eta} \in D$ is evaluated on the basis of the risk function

$$R(\eta, \hat{\eta}) = \mathbf{E}_{\tilde{y}|\eta} \int_X (\eta(x) - \hat{\eta}(x, \tilde{y}))^2 \omega(\mathrm{d}x) \tag{17.4}$$

being induced by the $\mathbb{L}_2$-semimetric with respect to some given Borel measure ω on X (cp. Formulae (9.19) – (9.24)).

As before, we assume to be given prior knowledge such that Θ is a compact subset of $\mathbb{R}^r$. For example, when prior knowledge is given such that

$$\inf_{\gamma \in \mathcal{G}} \|\eta - \gamma\|^2 \leqq c \tag{17.5}$$

with some constant $c > 0$, where c gives an indication of how closely the members of $\mathcal{G}$ may come to approximate a certain function $\eta \in \mathcal{F}$, then, observing that $\|\eta - \gamma\|^2 = \boldsymbol{\vartheta}^\top S_{ff} \boldsymbol{\vartheta} - 2\boldsymbol{\vartheta}^\top S_{fg} \boldsymbol{\beta} + \boldsymbol{\beta}^\top S_{gg} \boldsymbol{\beta} = \boldsymbol{\vartheta}^\top \boldsymbol{S} \boldsymbol{\vartheta} + \|\boldsymbol{\beta} - \boldsymbol{S}_{gg}^{-1} \boldsymbol{S}_{fg}^\top \boldsymbol{\vartheta}\|_{S_{gg}}^2$ for any $\eta \in \mathcal{F}$ and $\gamma \in \mathcal{G}$, we have

$$\inf_{\gamma \in \mathcal{G}} \|\eta - \gamma\|^2 = \boldsymbol{\vartheta}^\top \boldsymbol{S} \boldsymbol{\vartheta}. \tag{17.6}$$

Thus, our prior knowledge describes an ellipsoidal region Θ provided that the matrix $\boldsymbol{S}$ is regular. If we have only vague prior knowledge on the parameter $\boldsymbol{\vartheta}$ then this can be modelled by assuming $\Theta = \{\boldsymbol{\vartheta} \in \mathbb{R}^r : \boldsymbol{\vartheta}^\top \boldsymbol{\vartheta} \leqq h^2\}$, $h > 0$, and considering the limiting situation as $h \to \infty$.

Now, our objective is to find a minimax linear approgression estimator $\hat{\eta}_{\mathrm{M}}$ satisfying

$$\sup_{\eta \in \mathcal{F}} R(\eta, \hat{\eta}_{\mathrm{M}}) = \inf_{\hat{\eta} \in D} \sup_{\eta \in \mathcal{F}} R(\eta, \hat{\eta}). \tag{17.7}$$

Using the denotations from (9.25) the risk can be written in the form

$$\begin{aligned} R(\eta, \hat{\eta}) &= \operatorname{tr} \boldsymbol{S}_{gg} \boldsymbol{Z} \boldsymbol{\Sigma} \boldsymbol{Z}^\top + \boldsymbol{\vartheta}^\top \boldsymbol{G}(\boldsymbol{Z}) \boldsymbol{\vartheta}, \quad \text{where} \\ \boldsymbol{G}(\boldsymbol{Z}) &= \boldsymbol{S} + (\boldsymbol{Z}\boldsymbol{F} - \boldsymbol{S}_{gg}^{-1} \boldsymbol{S}_{fg}^\top)^\top \boldsymbol{S}_{gg} (\boldsymbol{Z}\boldsymbol{F} - \boldsymbol{S}_{gg}^{-1} \boldsymbol{S}_{fg}^\top) \end{aligned} \tag{17.8}$$

and $\boldsymbol{S} = \boldsymbol{S}_{ff} - \boldsymbol{S}_{fg} \boldsymbol{S}_{gg}^{-1} \boldsymbol{S}_{fg}^\top \in \mathcal{M}_r^{\geqq}$ represents the minimum bias matrix (cp. (9.26) – (9.28)). Thus, our minimax approgression problem amounts to solving the problem

$$\sup_{\boldsymbol{\vartheta} \in \Theta} \{\operatorname{tr} \boldsymbol{S}_{gg} \boldsymbol{Z} \boldsymbol{\Sigma} \boldsymbol{Z}^\top + \boldsymbol{\vartheta}^\top \boldsymbol{G}(\boldsymbol{Z}) \boldsymbol{\vartheta}\} \to \inf_{\boldsymbol{Z} \in \mathcal{M}_{s \times n}}. \tag{17.9}$$

As in Chapter 15, the problem will be tackled by changing it to an equivalent Bayesian estimation problem with respect to a least favourable prior distribution.

From Theorem 9.1 we know that the Bayes linear approgression estimator with respect to a prior distribution $P_{\tilde{\vartheta}} \in \Theta^*$ is uniquely determined by

$$\begin{aligned} \hat{\eta}_P(x, \tilde{\boldsymbol{y}}) &= \boldsymbol{g}(x)^\top \boldsymbol{Z}_P \tilde{\boldsymbol{y}}, \quad \text{where} \\ \boldsymbol{Z}_P &= \boldsymbol{S}_{gg}^{-1} \boldsymbol{S}_{fg}^\top \boldsymbol{M}_P \boldsymbol{F}^\top (\boldsymbol{F} \boldsymbol{M}_P \boldsymbol{F}^\top + \boldsymbol{\Sigma})^{-1} \end{aligned} \tag{17.10}$$

and $\boldsymbol{M}_P = \mathbf{E}(\tilde{\boldsymbol{\vartheta}} \tilde{\boldsymbol{\vartheta}}^\top)$ denotes the second order moment matrix associated with $P = P_{\tilde{\vartheta}}$. This estimator minimizes the Bayes risk

$$\int_{\Theta} R(\eta, \hat{\eta}) P_{\tilde{\vartheta}}(\mathrm{d}\vartheta) = \operatorname{tr}[S_{gg} Z \Sigma Z^{\top} + G(Z) M_P]$$

$$= \varrho(M_P, Z), \tag{17.11}$$

say, among all estimators $\hat{\eta}(\cdot, \tilde{y}) = g(\cdot)^{\top} Z \tilde{y} \in D$ (cp. (9.30)). Now, since $\mathcal{M} = \{M_P : P \in \Theta^*\}$ is compact (with respect to the usual Frobenius norm) and $\varrho(M_P, Z)$ from (17.11) is convex with respect to $Z \in \mathcal{M}_{s \times n}$ and concave (actually linear) with respect to $M_P \in \mathcal{M}$, respectively, it follows along the same line of arguments used in the proof of Lemma 15.1 and Lemma 15.2 that

$$\inf_{\hat{\eta} \in D} \sup_{\eta \in \mathcal{F}} R(\eta, \hat{\eta}) = \sup_{M_P \in \mathcal{M}} \varrho(M_P, Z_P). \tag{17.12}$$

Hence, a minimax linear approgression estimator for η can be obtained as a Bayes linear approgression estimator with respect to a least favourable prior distribution leading to maximum Bayes risk among the prior distributions from Θ^*.

17.2. Computation of the least favourable prior

We are now going to determine a least favourable prior distribution, which, according to (17.12), is completely characterized through its second order moment matrix, and the corresponding minimax linear estimator for η in case that the design matrix underlying the observation model

$$\mathbf{E}\tilde{y} = F\vartheta, \qquad \mathbf{Cov}\,\tilde{y} = \Sigma \tag{17.13}$$

(cp. (9.20)) has full rank.

Assumption 17.1. rank $F = r \leqq n$.

Then, denoting $V = (F^{\top}\Sigma^{-1}F)^{-1}$ and applying one of the matrix inversion formulae given in A.33, the matrix Z_P from (17.10) can be written as

$$Z_P = S_{gg}^{-1} S_{fg}^{\top} M_P (M_P + V)^{-1} V F^{\top} \Sigma^{-1}, \tag{17.14}$$

implying that

$$\hat{\eta}_P(x, \tilde{y}) = g(x)^{\top} S_{gg}^{-1} S_{fg}^{\top} M_P (M_P + V)^{-1} \hat{\vartheta}_{\Sigma}, \tag{17.15}$$

where $\hat{\vartheta}_{\Sigma} = V F^{\top} \Sigma^{-1} \tilde{y}$ is the Gauss-Markov estimator (GLSE) of ϑ in the above observation model.

Theorem 17.1. *Assume Θ to be compact and let Assumption 17.1 be satisfied. Then $\hat{\eta}_{\mathrm{M}}$ is a minimax linear approgression estimator for η if and only if it is of the form*

$$\hat{\eta}_{\mathrm{M}}(x, \tilde{y}) = g(x)^{\top} S_{gg}^{-1} S_{fg}^{\top} M_0 (M_0 + V)^{-1} \hat{\vartheta}_{\Sigma},$$

where $M_0 = M_{P_0}$ represents the second order moment matrix of a least favourable prior distribution P_0 maximizing

$$\varrho(M_P, Z_P) = \operatorname{tr}[S_{ff}M_P - S_{gg}^{-1}S_{fg}^{\top}M_P(M_P+V)^{-1}M_PS_{fg}] \qquad (17.16)$$

among all priors $P \in \Theta^*$.

Proof. The necessity and sufficiency of the condition of Bayes optimality with respect to a least favourable prior $P_0 \in \Theta^*$ for establishing the minimaxity of $\hat{\eta}_M$ can be shown analogously to the proof of Lemma 15.3 observing (17.12) and A.11. Further, from Theorem 9.1 we know that

$$\varrho(M_P, Z_P) = \operatorname{tr}(S_{ff}M_P - Z_PFM_PS_{fg})$$

with Z_P being defined by (17.10). Inserting the equivalent representation (17.14) for Z_P we obtain the form of the Bayes risk indicated in (17.16). Finally, the existence of a least favourable moment matrix M_0 is guaranteed by the compactness of $\mathcal{M}$ and the fact that the Bayes risk (17.16) is continuous with respect to the Frobenius norm $\|M_P\| = (\operatorname{tr} M_P^2)^{1/2}$. □

Corollary 17.1. *The least favourable moment matrix* M_0 *needed for the computation of* $\hat{\eta}_M$ *minimizes the convex functional*

$$\Psi(M_P) = \operatorname{tr} VUV(M_P+V)^{-1} - \operatorname{tr} SM_P, \quad M_P \in \mathcal{M} \qquad (17.17)$$

with $U = S_{fg}S_{gg}^{-1}S_{fg}^{\top}$ *and* $S = S_{ff} - U$ *being the minimum bias matrix.*

Proof. Observing that $M_P(M_P+V)^{-1}M_P = M_P - V + V(M_P+V)^{-1}V$, the Bayes risk (17.16) may be rewritten as follows:

$$\begin{aligned}\varrho(M_P, Z_P) &= \operatorname{tr} SM_P - \operatorname{tr} U[V(M_P+V)^{-1}V - V]\\ &= -\Psi(M_P) + \operatorname{tr} UV.\end{aligned}$$

Thus, the search for a least favourable M_0 is equivalent to minimization of $\Psi(M_P)$ with respect to $M_P \in \mathcal{M}$. The convexity of the functional Ψ follows from the fact that both terms, $\Psi_1(M_P) = \operatorname{tr} VUV(M_P+V)^{-1}$ and $\Psi_2(M_P) = -\operatorname{tr} SM_P$, are convex functionals on $\mathcal{M}$. □

In the special case of model adequacy, i.e., $\mathcal{F} = \mathcal{G}$, we have $U = S_{ff}$ and the functional Ψ reduces to

$$\Psi_0(M_P) = \operatorname{tr} VS_{ff}V(M_P+V)^{-1}, \qquad M_P \in \mathcal{M}.$$

The corresponding minimax linear estimator for η takes the form $\hat{\eta}_M(x, \tilde{y}) = f(x)^{\top}\hat{\vartheta}_M(\tilde{y})$, where $\hat{\vartheta}_M(\tilde{y}) = M_0(M_0+V)^{-1}\hat{\vartheta}_\Sigma$ is the minimax linear estimator for ϑ in the model (17.13) with ϑ constrained to lie in the compact region $\Theta \subset \mathbb{R}^r$ and M_0 minimizes $\Psi_0(\cdot)$ over $\mathcal{M}$. This is exactly the estimation problem which we have dealt with in Sections 15.1.1 and 15.1.2, and for which we have also given explicit solutions for some special cases involving ellipsoid and linear inequality constraints (see Sections 15.3, 15.4).

Returning to the general case $\mathcal{F} \neq \mathcal{G}$, we note that the convexity of the functional Ψ defined in Corollary 17.1 allows at least the application of iteration procedures for finding a least favourable matrix M_0. For example, the steepeat descent algorithm formulated at the end of Section 15.1.2 can be easily adapted to the present case, the only modification to be made is to replace the directional derivate $\Delta_{L_B}(P, \delta_\vartheta)$ considered there by

$$\Delta_\Psi(P, \delta_\vartheta) = \lim_{\alpha \downarrow 0} \frac{\mathrm{d}}{\mathrm{d}\alpha} \Psi((1-\alpha)\boldsymbol{M}_P + \alpha\boldsymbol{\vartheta}\boldsymbol{\vartheta}^\top). \tag{17.18}$$

Lemma 17.1. *Let $P \in \Theta^*$ and $\vartheta \in \Theta$. Then it holds*

$$\Delta_\Psi(P, \delta_\vartheta) = \operatorname{tr} \boldsymbol{M}_P^S \boldsymbol{M}_P - \boldsymbol{\vartheta}^\top \boldsymbol{M}_P^S \boldsymbol{\vartheta}$$

with $\quad \boldsymbol{M}_P^S = \boldsymbol{S} + (\boldsymbol{M}_P + \boldsymbol{V})^{-1}\boldsymbol{VUV}(\boldsymbol{M}_P + \boldsymbol{V})^{-1}.$

Proof. From Lemma 11.7 (i) we know that $\Psi_1(\boldsymbol{M}_P) = \operatorname{tr} \boldsymbol{VUV}(\boldsymbol{M}_P + \boldsymbol{V})^{-1}$ has directional derivative

$$\Delta_{\Psi_1}(P, \delta_\vartheta) = \operatorname{tr}(\boldsymbol{M}_P^S - \boldsymbol{S})\boldsymbol{M}_P + \boldsymbol{\vartheta}^\top(\boldsymbol{M}_P^S - \boldsymbol{S})\boldsymbol{\vartheta}.$$

Further, the directional derivative of $\Psi_2(\boldsymbol{M}_P) = -\operatorname{tr} \boldsymbol{S}\boldsymbol{M}_P$ is easily seen to be

$$\begin{aligned}\Delta_{\Psi_2}(P, \delta_\vartheta) &= -\lim_{\alpha \downarrow 0} \frac{\mathrm{d}}{\mathrm{d}\alpha} \operatorname{tr} \boldsymbol{S}((1-\alpha)\boldsymbol{M}_P + \alpha\boldsymbol{\vartheta}\boldsymbol{\vartheta}^\top)\\ &= \operatorname{tr} \boldsymbol{S}\boldsymbol{M}_P - \boldsymbol{\vartheta}^\top \boldsymbol{S}\boldsymbol{\vartheta}.\end{aligned}$$

From this the result follows by observing that $\Delta_\Psi(P, \delta_\vartheta) = \Delta_{\Psi_1}(P, \delta_\vartheta) + \Delta_{\Psi_2}(P, \delta_\vartheta)$. □

Then, forming a sequence of measures $\{P_s\}_{s \subset \mathrm{N}}$ according to (15.26) and (15.27), with the supporting points chosen such that

$$\begin{aligned}\boldsymbol{\vartheta}_s &= \arg\inf_{\vartheta \in \Theta} \Delta_\Psi(P_{s-1}, \delta_\vartheta), \qquad && s = 1, 2, \ldots\\ \boldsymbol{\vartheta}_0 &= \arg\inf_{\vartheta \in \Theta} \Psi(\boldsymbol{\vartheta}\boldsymbol{\vartheta}^\top), && P_0 = \delta_{\vartheta_0}\end{aligned} \tag{17.19}$$

is follows from the corollary given in Böhning (1981) that

$$\lim_{s \to \infty} \Psi(\boldsymbol{M}_{P_s}) = \inf_{\boldsymbol{M}_P \in \mathcal{M}} \Psi(\boldsymbol{M}_P).$$

17.3. The special case of nested models

In this section we will deal with the special case $\mathcal{G} \subset \mathcal{F}$, i.e., we are interested in estimating $\eta(\boldsymbol{x}) = \boldsymbol{\vartheta}_1^\top \boldsymbol{f}_1(\boldsymbol{x}) + \boldsymbol{\vartheta}_2^\top \boldsymbol{f}_2(\boldsymbol{x})$ by some element out of the family

$$\mathcal{G} = \{\gamma(\cdot) : \gamma(\boldsymbol{x}) = \boldsymbol{\vartheta}_1^\top \boldsymbol{f}_1(\boldsymbol{x}); \quad \boldsymbol{x} \in X, \quad \boldsymbol{\vartheta}_1 \in \Theta_1\}, \tag{17.20}$$

where we have set $\boldsymbol{g}(\cdot) = \boldsymbol{f}_1(\cdot)$, $\boldsymbol{\beta} = \boldsymbol{\vartheta}_1$, with an appropriate partitioning of $\boldsymbol{f}(\boldsymbol{x})^\top = (\boldsymbol{f}_1(\boldsymbol{x})^\top, \boldsymbol{f}_2(\boldsymbol{x})^\top)$, $\boldsymbol{\vartheta}^\top = (\boldsymbol{\vartheta}_1^\top, \boldsymbol{\vartheta}_2^\top)$ and $\Theta = \Theta_1 \times \Theta_2$.

Let r_i denote the dimension of Θ_i, $i = 1, 2$, where $r = r_1 + r_2$. In the sequel we assume, without loss of generality, that the components of $\boldsymbol{f}(\cdot)$ form an orthonormal system with respect to the measure ω, i.e., let be satisfied

Assumption 17.2. $\int_X f(x)f(x)^\top \omega(\mathrm{d}x) = I_r$.

In particular, we then have $\int_X f_1(x)f_2(x)^\top \omega(\mathrm{d}x) = 0_{r_1, r_2}$.

For example, if ω is the Lebesgue measure on $X = [-1, 1]$ and η is assumed to be a third-order polynomial which we wish to approximate by some straight line then we put $\mathcal{F} = \{\eta(\cdot): \eta(x) = \vartheta_{11} + \vartheta_{12}x + 1/2\vartheta_{21}(3x^2 - 1) + 1/2\vartheta_{22}(5x^3 - 3x),\ x \in X\}$, i.e., a Legendre polynomial expansion with $\vartheta_1 = (\vartheta_{11}, \vartheta_{12})^\top$, $f_1(x) = (1, x)^\top$, $\vartheta_2 = (\vartheta_{21}, \vartheta_{22})^\top$, $f_2(x) = 1/2\,(3x^2 - 1, 5x^3 - 3x)^\top$ and $r_1 = r_2 = 2$.

In the special case to be considered here the minimax linear approgression estimator $\hat{\eta}_M$ for η is given by

$$\hat{\eta}_M(x, \tilde{y}) = f_1(x)^\top S_{ff_1}^\top M_0(M_0 + V)^{-1}\hat{\vartheta}_\Sigma, \tag{17.21}$$

where M_0 minimizes the functional Ψ defined by (17.17) and, by virtue of Assumption 17.2, the matrices S_{ff_1} and S occuring there take the forms

$$S_{ff_1} = \begin{pmatrix} I_{r_1} \\ 0_{r_2, r_1} \end{pmatrix}, \qquad S = \begin{pmatrix} 0_{r_1, r_1} & 0_{r_1, r_2} \\ 0_{r_2, r_1} & I_{r_2} \end{pmatrix}. \tag{17.22}$$

We will now consider a particular situation of practical importance for which the minimax linear approgression estimator can be obtained on the basis of the minimax linear estimator for ϑ_1 in the inadequate "parsimonious" model $\mathcal{G}$.

Assumption 17.3. (i) *$\Theta_1 \subset \mathbb{R}^{r_1}$ is compact.*

(ii) $\inf_{\gamma \in \mathcal{G}} \|\eta - \gamma\|^2 \leqq c$ *for some* $c > 0$.

(iii) $F_1^\top \Sigma^{-1} F_2 = 0_{r_1, r_2}$, *where* $F_j = (f_j(x_1), \ldots, f_j(x_n))^\top$, $j = 1, 2$.

The constant c in Assumption 17.3 (ii) gives an indication of the goodness of approximation of $\eta \in \mathcal{F}$ which is attainable by the members of $\mathcal{G}$; observing (17.5), (17.6) and (17.22), this assumption is equivalent to imposing a length constraint on ϑ_2,

$$\vartheta_2^\top \vartheta_2 \leqq c. \tag{17.23}$$

Similar restrictions may arise from some smoothness condition such as

$$\int_X [\eta^{(m)}(x)]^2\, \mathrm{d}x \leqq c, \tag{17.24}$$

where $\eta^{(m)}$ means the mth derivative of η, $1 \leqq m < r$.
For example, if η is assumed to be a third-order polynomial and we are interested in deviations from linearity (cp. the example following immediately after Assumption 17.2) then Condition (17.24) with $m = 2$ and $X = [-1, 1]$ leads to the ellipsoid constraint $3\vartheta_{21}^2 + 25\vartheta_{22}^2 \leqq c/6$ for $\vartheta_2 = (\vartheta_{21}, \vartheta_{22})^\top$.

Assumption 17.3 (iii) means that $V = (F^\top \Sigma^{-1} F)^{-1}$ has block diagonal structure, i.e.,

$$V = \begin{pmatrix} V_1 & \mathbf{0}_{r_1, r_2} \\ \mathbf{0}_{r_2, r_1} & V_2 \end{pmatrix}, \qquad V_j = (F_j^\top \Sigma^{-1} F_j)^{-1}, \qquad j = 1, 2 \quad (17.25)$$

which refers to a partially orthogonal experiment.

Accordingly, we subdivide the moment matrices M_P, $P \in \Theta^*$ such that

$$M_P = \begin{pmatrix} M_1(P_1) & M_{12}(P) \\ M_{12}^\top(P) & M_2(P_2) \end{pmatrix},$$

where $M_j(P_j) = \int_{\Theta_j} \vartheta_j \vartheta_j^\top P_j(\mathrm{d}\vartheta_j)$, $j = 1, 2$, $M_{12}(P) = \int_{\Theta} \vartheta_1 \vartheta_2^\top P(\mathrm{d}\vartheta)$. Here $M_1(P_1)$ and $M_2(P_2)$ are the moment matrices associated with the marginal prior distributions P_1 and P_2 of the subparameters ϑ_1 and ϑ_2 of ϑ, respectively.

Now, applying, the block inversion formula (A.37), it follows from Assumption 17.3 (iii) that

$$\begin{aligned} \Psi(M_P) &= \operatorname{tr} V_1^2 [V_1 + M_1 - M_{12} (V_2 + M_2)^{-1} M_{12}^\top]^{-1} - \operatorname{tr} M_2 \\ &\geqq \operatorname{tr} V_1^2 (V_1 + M_1)^{-1} - \operatorname{tr} M_2, \end{aligned}$$

since $M_{12}(V_2 + M_2)^{-1} M_{12}^\top$ is positive semidefinite whatever P is chosen. By virtue of the symmetry of $\Theta_2 = \{\vartheta_2 \in \mathbb{R}^{r_2} : \vartheta_2^\top \vartheta_2 \leqq c\}$ around zero we can restrict ourselves to measures $P \in \Theta^*$ with $M_{12}(P) = \mathbf{0}_{r_1, r_2}$ which implies that

$$\inf_{M_P \in \mathscr{M}} \Psi(M_P) = \inf_{M_1 \in \mathscr{M}_1} \{\operatorname{tr} V_1^2 (V_1 + M_1)^{-1}\} - c,$$

where

$$\mathscr{M}_1 = \{M_1(P_1) \in \mathscr{M}_{r_1}^{\leqq} : P_1 \in \Theta_1^*\}. \quad (17.26)$$

The problem of minimizing $\operatorname{tr} V_1^2 (V_1 + M_1)^{-1}$ with respect to $M_1 \in \mathscr{M}_1$ is equivalent to finding the second order moment matrix of a least favourable prior distribution for the parameter vector in the (inadequate) model

$$\mathsf{E}\,\tilde{y} = F_1 \vartheta_1, \qquad \mathsf{Cov}\,\tilde{y} = \Sigma \quad (17.27)$$

with ϑ_1 constrained to lie in the compact region Θ_1 and assuming (unweighted) quadratic loss. This is a problem which we have dealt with extensively in Sections 15.1, 15.3, 15.4, and for which we have also given explicit form solutions for the important special cases of ellipsoid and linear inequality constraints.

Theorem 17.2. *Under the Assumptions* 17.1 *through* 17.3 *the minimax linear approgression estimator for* η *is given by* $\hat{\eta}_{\mathrm{M}}(x, \tilde{y}) = f_1(x)^\top \hat{\vartheta}_1^{\mathrm{M}}(\tilde{y})$, *where*

$$\hat{\vartheta}_1^{\mathrm{M}}(\tilde{y}) = M_1^*(M_1^* + V_1)^{-1} V_1 F_1^\top \Sigma^{-1} \tilde{y}$$

is the minimax linear estimator for ϑ_1 *in the model* (17.27) *based on mean squared error risk and*

$$M_1^* = \arg \inf_{M_1 \in \mathcal{M}_1} \operatorname{tr} V_1^2 (M_1 + V_1)^{-1}$$

is the least favourable moment matrix associated with this model.

Thus, under the above assumptions, the minimax approgression problem reduces to the minimax estimation problem in the parsimonious model (17.27).

For example, in the special case where $\Theta_1 = \{\vartheta_1 \in \mathbb{R}^{r_1} : \vartheta_1^\top \vartheta_1 \leqq h\}$ we obtain $M_1^* = (h/\operatorname{tr} V_1) V_1$ (see Section 15.3, Example 15.1) and thus

$$\hat{\eta}_M(x, \tilde{y}) = \frac{h}{h + \operatorname{tr} V_1} f_1(x)^\top V_1 F_1^\top \Sigma^{-1} \tilde{y}.$$

This is in the form of a shrunken estimator $\hat{\eta}_M(x, \tilde{y}) = f_1(x)^\top \hat{\vartheta}_1^g(\tilde{y})$ where $\hat{\vartheta}_1^g$ is the shrunken GLSE of ϑ_1 in the model (17.27) with shrinkage factor $g = h/(h + \operatorname{tr} V_1)$, $0 < g < 1$. In the limiting case $h \to \infty$, which corresponds to increasing vagueness of prior knowledge concerning ϑ_1, the above estimator $\hat{\eta}_M$ coincides with the usual Gauss-Markov-estimator of η in the inadequate model (17.27) with $\Theta_1 = \mathbb{R}^{r_1}$.

Appendix

In this chapter we will recall, in a rough form, some basic facts and notions from probability theory, statistical decision theory, convex analysis and matrix algebra which we have used throughout the book. Detailed representations and proofs may be found in the literature, e.g. in BANDEMER et al. (1977), BECKENBACH/BELLMANN (1965), FERGUSON (1967), RAO (1973), STOER/WITZGALL (1970) and WALD (1950).

In the following denote $\mathcal{M}_{n\times r}$ the set of all real matrices of type $n \times r$, $\mathcal{M}_r^{>}$ and $\mathcal{M}_r^{\geqq}$ stand for the sets of positive definite and positive semidefinite matrices of type $r \times r$, respectively.

Probability Theory and Mathematical Statistics

Let $\tilde{x}$ be a random vector (or random variable $\tilde{x}$). Denote $P_{\tilde{x}}$ the probability distribution of $\tilde{x}$ and $p_{\tilde{x}}$ its probability density function (if it exists) with respect to Lebesgue measure.

A.1. The random variable $\tilde{x}$ is said to have a *gamma distribution* with parameters $\alpha \in \mathbb{R}^+$ and $\beta \in \mathbb{R}^+$ ($P_{\tilde{x}} = \mathrm{G}(\alpha, \beta)$) $\overset{\text{def}}{\Leftrightarrow}$

$$p_{\tilde{x}}(x) = \begin{cases} \dfrac{\beta^\alpha}{\Gamma(\alpha)} x^{\alpha-1} \mathrm{e}^{-\beta x} & \text{if} \quad x > 0, \\ 0 & \text{otherwise}. \end{cases}$$

Particularly, if it holds $\alpha = n/2$ and $\beta = 1/2$, where n is some positive integer, then $\tilde{x}$ is said to have a χ_n^2-distribution with n degrees of freedom.

If $P_{\tilde{x}} = \mathrm{G}(\alpha, \beta)$ then it holds $\mathbf{E}\tilde{x} = \alpha/\beta$, $\operatorname{Var}\tilde{x} = \alpha/\beta^2$.

A.2. The r-dimensional random vector $\tilde{x}$ is said to have a *normal distribution* with expectation $\boldsymbol{\mu} \in \mathbb{R}^r$ and covariance matrix $\boldsymbol{\Phi} \in \mathcal{M}_r^{>}$ ($P_{\tilde{x}} = \mathrm{N}(\boldsymbol{\mu}, \boldsymbol{\Phi})$) $\overset{\text{def}}{\Leftrightarrow}$

$$\forall \boldsymbol{x} \in \mathbb{R}^r: p_{\tilde{x}}(\boldsymbol{x}) = (2\pi)^{-r/2} (\det \boldsymbol{\Phi})^{-1/2} \exp\left(-\frac{1}{2} (\boldsymbol{x} - \boldsymbol{\mu})^\top \boldsymbol{\Phi}^{-1} (\boldsymbol{x} - \boldsymbol{\mu})\right).$$

Particularly, if $P_{\tilde{x}} = \mathrm{N}(\boldsymbol{0}, \boldsymbol{I}_r)$ and $\boldsymbol{A} \in \mathcal{M}_{r\times r}$ is some idempotent matrix (i.e.

$A^2 = A$) then it holds

$$X^\top A x \sim \chi^2_n \quad \text{where} \quad n = \operatorname{rank} A .$$

A.3. Let $\tilde{x} = (\tilde{x}_1, \ldots, \tilde{x}_r)^\top$ be a random vector and $\tilde{x}_0$ a random variable. The random vector $\tilde{z} = (\tilde{x}^\top, \tilde{x}_0)^\top$ is said to have a *normal-gamma-distribution* with parameters $\mu \in \mathbb{R}^r$, $\Phi \in \mathcal{M}_r^>$, $\alpha \in \mathbb{R}^+$ and $\beta \in \mathbb{R}^+$ ($P_{\tilde{z}} = \mathrm{NG}(\mu, \alpha, \Phi, \beta)$) $\overset{\text{def}}{=}$

$$P_{\tilde{x}|\tilde{x}_0 = x_0} = \mathrm{N}(\mu, x_0^{-1}\Phi) \quad \text{and} \quad P_{\tilde{x}_0} = \mathrm{G}\left(\frac{\alpha}{2} + 1, \frac{\beta}{2}\right).$$

A.4. The r-dimensional random vector $\tilde{x}$ is said to have a (Student) *t-distribution* with $n > 0$ degrees of freedom and parameters $\mu \in \mathbb{R}^r$, $\Phi \in \mathcal{M}_r^>$ ($P_{\tilde{x}} = t(n, \mu, \Phi)$) $\overset{\text{def}}{=}$

$$\forall x \in \mathbb{R}^r \colon p_{\tilde{x}}(x) = c(n + (x - \mu)^\top \Phi^{-1}(x - \mu))^{-(n+r)/2},$$

where

$$c = n^{(n+r)/2}\Gamma\left(\frac{n+r}{2}\right) \Big/ A\Gamma\left(\frac{n}{2}\right)(n\pi)^{r/2}(\det \Phi)^{1/2}.$$

If $P_{\tilde{x}} = t(n, \mu, \Phi)$ then it holds $\mathbf{E}\tilde{x} = \mu$, $\mathbf{Cov}\,\tilde{x} = \dfrac{n}{n-2}\Phi$ (provided that $n > 2$) and the marginal distribution of every subvector of $\tilde{x}$ is also a t-distribution with n degrees of freedom and its parameters are the corresponding subparameters of μ and Φ.

If $\tilde{z} = (\tilde{x}^\top, \tilde{x}_0)^\top$ has a normal-gamma-distribution $P_{\tilde{z}} = \mathrm{NG}(\mu, \alpha, \Phi, \beta)$ then the marginal distribution of $\tilde{x}$ is $P_{\tilde{x}} = t(\alpha, \mu, \beta\Phi)$.

A.5. The random $(r \times r)$-matrix $\tilde{X}$ is said to have a *Wishart distribution* with $n > r$ degrees of freedom and parameter $\Sigma \in \mathcal{M}_r^>$ ($P_{\tilde{X}} = \mathrm{W}_r(n, \Sigma)$) $\overset{\text{def}}{=}$

$$p_{\tilde{X}}(X) = \begin{cases} c(\det X)^{(n-r-1)/2} \exp\left(-\dfrac{1}{2}\operatorname{tr}\Sigma^{-1}X\right) & \text{if } X \in \mathcal{M}_r^>, \\ 0 & \text{otherwise}, \end{cases}$$

where

$$c^{-1} = 2^{nr/2}\pi^{r(r-1)/4}(\det \Sigma)^{n/2}\prod_{i=1}^{r}\Gamma\left(\frac{n+1-i}{2}\right).$$

If $P_{\tilde{X}} = \mathrm{W}_r(n, \Sigma)$ then it holds $\mathbf{E}\tilde{X} = n\Sigma$.

If $\tilde{x}_1, \ldots, \tilde{x}_n$ are r-dimensional random vectors each of which has a normal distribution $\mathrm{N}(\mathbf{0}, \Sigma)$ with some $\Sigma \in \mathcal{M}_r^>$ then it holds

$$\tilde{X} = \tilde{x}_1\tilde{x}_1^\top + \ldots + \tilde{x}_n\tilde{x}_n^\top \sim \mathrm{W}_r(n, \Sigma).$$

A.6. The complex-valued function φ is called *characteristic function* of the random vector $\tilde{x}$ $\overset{\text{def}}{=}$

$$\forall t \in \mathbb{R}^r \colon \varphi(t) = \mathbf{E}\{\exp(\mathrm{i}t^\top\tilde{x})\}.$$

Particularly, if $\tilde{x}$ has a normal distribution $P_{\tilde{x}} = \mathrm{N}(\boldsymbol{\mu}, \boldsymbol{\Phi})$ then it holds:

$$\forall \boldsymbol{t} \in \mathbb{R}^r: \varphi(\boldsymbol{t}) = \exp\left(\mathrm{i}\boldsymbol{t}^\top \boldsymbol{\mu} - \frac{1}{2}\boldsymbol{t}^\top \boldsymbol{\Phi}\boldsymbol{t}\right).$$

A.7. Let the distribution $P_{\tilde{x}}$ of the random vector $\tilde{x}$ be unimodal with mode $\boldsymbol{\mu} \in \mathbb{R}^r$ and density of the form

$$p_{\tilde{x}}(\boldsymbol{x}) = g(\|\boldsymbol{x} - \boldsymbol{\mu}\|_M^2) \quad \text{for all} \quad \boldsymbol{x} \in \mathbb{R}^r,$$

where $\boldsymbol{M}$ is some positive definite matrix and g is some real-valued nonnegative function. Then, for every continuous and monotonically increasing function $L|\mathbb{R}^+ \to \mathbb{R}^+$ and every matrix $\boldsymbol{U} \in \mathcal{M}_r^{\geqq}$ it holds:

$$\forall \boldsymbol{s} \in \mathbb{R}^r: \int_{\mathbb{R}^r} L(\|\boldsymbol{x} - \boldsymbol{s}\|_U^2)\, p_{\tilde{x}}(\boldsymbol{x})\,\mathrm{d}\boldsymbol{x} \geqq \int_{\mathbb{R}^r} L(\|\boldsymbol{x} - \boldsymbol{\mu}\|_U^2)\, p_{\tilde{x}}(\boldsymbol{x})\,\mathrm{d}\boldsymbol{x}.$$

A.8. Let $\mathcal{P}$ be a family of r-dimensional probability distributions all of which have the same expectation.

Then $P_{\tilde{x}}^1 \in \mathcal{P}$ is said to be *greater* than $P_{\tilde{x}}^2 \in \mathcal{P}(P_{\tilde{x}}^1 > P_{\tilde{x}}^2) \overset{\text{def}}{=}$

$$\forall \boldsymbol{U} \in \mathcal{M}_r^{\geqq}\ \forall \boldsymbol{s} \in \mathbb{R}^r: \int_{\mathbb{R}^r} \|\boldsymbol{x} - \boldsymbol{s}\|_U^2 P_{\tilde{x}}^1(\mathrm{d}\boldsymbol{x}) \geqq \int_{\mathbb{R}^r} \|\boldsymbol{x} - \boldsymbol{s}\|_U^2 P_{\tilde{x}}^2(\mathrm{d}\boldsymbol{x}).$$

The distribution $P_{\tilde{x}}^0 \in \mathcal{P}$ is called a *maximal element* of $\mathcal{P}$ with respect to the relation $> \overset{\text{def}}{=}$

$$\forall P_{\tilde{x}} \in \mathcal{P}: P_{\tilde{x}}^0 > P_{\tilde{x}}.$$

Let $\boldsymbol{C}_1$ and $\boldsymbol{C}_2$ be the covariance matrices corresponding to $P_{\tilde{x}}^1 \in \mathcal{P}$ and $P_{\tilde{x}}^2 \in \mathcal{P}$, respectively. Then it holds: $P_{\tilde{x}}^1 > P_{\tilde{x}}^2 \Leftrightarrow \boldsymbol{C}_1 - \boldsymbol{C}_2 \in \mathcal{M}_r^{\geqq}$. Let be $\boldsymbol{g}|\mathbb{R}^r \to \mathbb{R}^m$ some linear and Borel-measurable function, $P_{\tilde{x}}^1 \in \mathcal{P}$, $P_{\tilde{x}}^2 \in \mathcal{P}$ and $P^1_{\boldsymbol{g}(\tilde{x})}, P^2_{\boldsymbol{g}(\tilde{x})}$ the corresponding distributions induced by $\boldsymbol{g}$. Then it holds:

$$P_{\tilde{x}}^1 > P_{\tilde{x}}^2 \Rightarrow P^1_{\boldsymbol{g}(\tilde{x})} > P^2_{\boldsymbol{g}(\tilde{x})}.$$

A.9. Let $\tilde{y} = (\tilde{y}_1, \ldots, \tilde{y}_n)^\top$ be a random observation vector, $C_{\tilde{y}}$ the corresponding sample space and $K = \{P_{\tilde{y}}^{(\gamma)}: \boldsymbol{\gamma} \in \Gamma \subseteqq \mathbb{R}^m\}$ a parametric family of probability distributions. Denote $l(\cdot\,; \boldsymbol{y})$ the likelihood function of the realization $\boldsymbol{y} \in C_{\tilde{y}}$.

The measurable function $\boldsymbol{T}|C_{\tilde{y}} \to \mathbb{R}^m$ is called a *sufficient statistic* for K if there exist functions $k|\Gamma \times \mathbb{R}^m \to \mathbb{R}^+$ and $h|C_{\tilde{y}} \to \mathbb{R}^+$ such that

$$l(\boldsymbol{\gamma}; \boldsymbol{y}) = k(\boldsymbol{\gamma}, \boldsymbol{T}(\boldsymbol{y}))\, h(\boldsymbol{y}) \quad \text{for all} \quad \boldsymbol{\gamma} \in \Gamma, \quad \boldsymbol{y} \in C_{\tilde{y}}.$$

Statistical Decision Theory

Suppose the statistician to investigate some object called Nature, which can assume different states z from some set Z; the statistician is supposed to have available some set A of possible actions. Further, let $\tilde{y} = (\tilde{y}_1, \ldots, \tilde{y}_n)^\top$ be a random observation vector with sample space $C_{\tilde{y}} \subseteqq \mathbb{R}^n$ and distribution $P_{\tilde{y}|z}$ depending on the state of Nature.

A.10. Every measurable mapping $d \mid C_{\tilde{y}} \to A$ is called a *strategy* (or decision function); every measurable mapping $L \mid Z \times A \to \mathbb{R}^1$ is called a *loss function*.

The measurable mapping $R(\cdot\,; d) \mid Z \to \mathbb{R}^1$ is called *risk function* of the strategy $d \overset{\text{def}}{\equiv}$

$$\forall z \in Z \colon R(z; d) = \mathbf{E}_{\tilde{y}|z} L(z; d(\tilde{y})) = \int_{C_{\tilde{y}}} L(z; d(y))\, P_{\tilde{y}|z}(\mathrm{d}y).$$

If D is some nonempty set of strategies for which the risk function exists then the triplet $[Z, D, R]$ is called a *statistical decision problem.*

A.11. The strategy $d_1 \in D$ is called *better* than the strategy $d_2 \in D \overset{\text{def}}{\equiv}$

$$\forall z \in Z \colon R(z; d_1) \leqq R(z; d_2)$$
$$\wedge\ \exists z_0 \in Z \colon R(z_0; d_1) < R(z_0; d_2).$$

The strategy $d_0 \in D$ is called *admissible* in $[Z, D, R]$ if there is no strategy $d \in D$ which is better than d_0.

A.12. Let D_0 be some subset of strategies from D.

The class D_0 is called *complete* in $[Z, D, R]$ if for every $d \in D \setminus D_0$ there exists a strategy $d_0 \in D_0$ which is better than d.

The class D_0 is called *essentially complete* in $[Z, D, R] \overset{\text{def}}{\equiv}$

$$\forall d \in D \setminus D_0\, \exists d_0 \in D_0\, \forall z \in Z \colon R(z; d_0) \leqq R(z; d).$$

A.13. Let $\tilde{z}$ be a random variable taking values in Z according to some prior probability distribution $P_{\tilde{z}}$.

The function $\varrho(P_{\tilde{z}}; \cdot) \mid D \to \mathbb{R}^1$ defined by

$$\forall d \in D \colon \varrho(P_{\tilde{z}}; d) = \mathbf{E}_{\tilde{z}} R(\tilde{z}; d) = \int_Z R(z; d)\, P_{\tilde{z}}(\mathrm{d}z)$$

is called the *Bayesian risk function* in $[Z, D, R]$ with respect to $P_{\tilde{z}}$ and $\varrho(P_z; d)$ is called the *Bayes risk* of the strategy $d \in D$ with respect to $P_{\tilde{z}}$.

The strategy $d^* \in D$ is called a *Bayes strategy* in $[Z, D, R]$ with respect to $P_{\tilde{z}} \overset{\text{def}}{\equiv}$

$$\varrho(P_{\tilde{z}}; d^*) = \inf_{d \in D} \varrho(P_{\tilde{z}}; d).$$

Denote Z^* the linear space of all (prior) probability distributions over Z. Then $P_{\tilde{z}}^0 \in Z^*$ is called a *least favourable prior distribution* within Z^* if it holds

$$\inf_{d \in D} \varrho(P_{\tilde{z}}^0; d) = \sup_{P_{\tilde{z}} \in Z^*} \inf_{d \in D} \varrho(P_{\tilde{z}}; d).$$

A.14. The strategy $d_M \in \mathrm{D}$ is called a *minimax strategy* in $[Z, D, R] \overset{\text{def}}{\equiv}$

$$\sup_{z \in Z} R(z; d_M) = \inf_{d \in D} \sup_{z \in Z} R(z; d).$$

Assume that $d^* \in D$ is Bayesian in $[Z, D, R]$ with respect to $P_{\tilde{z}} \in Z^*$ and it holds $R(z; d^*) \leqq \varrho(P_{\tilde{z}}; d^*)$ for all $z \in Z$. Then d^* is a minimax strategy in $[Z, D, R]$ and $P_{\tilde{z}}$ is least favourable within Z^*.

If there exists a least favourable prior distribution $P^0_{\tilde{z}} \in Z^*$ and we have equality

$$\inf_{d \in D} \sup_{P_{\tilde{z}} \in Z^*} \varrho(P_{\tilde{z}}; d) = \sup_{P_{\tilde{z}} \in Z^*} \inf_{d \in D} \varrho(P_{\tilde{z}}; d)$$

then any minimax strategy in $[Z, D, R]$ is Bayesian with respect to $P^0_{\tilde{z}}$.

A.15. Let $P_{\tilde{z}}$ be some prior distribution over the state space and $\boldsymbol{y}$ some realization of the observation vector $\tilde{\boldsymbol{y}}$. Then the conditional distribution $P_{\tilde{z}|\boldsymbol{y}}$ is called the *posterior distribution* of $\tilde{z}$ for given observation $\boldsymbol{y}$.

If $P_{\tilde{z}}$ and the distributions $P_{\tilde{\boldsymbol{y}}|z}$, $z \in Z$, have probability densities $p_{\tilde{z}}$ and $p_{\tilde{\boldsymbol{y}}|z}$, respectively, with respect to some σ-finite measure ν then the posterior distribution of $\tilde{z}$ has density with respect to ν which is determined by

$$\forall z \in Z\colon p_{\tilde{z}|\boldsymbol{y}}(z) = p_{\tilde{\boldsymbol{y}}|z}(\boldsymbol{y})\, p_{\tilde{z}}(z) \Big/ \int_Z p_{\tilde{\boldsymbol{y}}|z'}(\boldsymbol{y})\, P_{\tilde{z}}(\mathrm{d}z').$$

A.16. If $\tilde{z}$ has posterior distribution $P_{\tilde{z}|\boldsymbol{y}}$ then

$$\mathbf{E}_{\tilde{z}|\boldsymbol{y}} L(\tilde{z}; d(\boldsymbol{y})) = \int_Z L(z; d(\boldsymbol{y}))\, P_{\tilde{z}|\boldsymbol{y}}(\mathrm{d}z)$$

is called the *posterior loss* of $d \in D$ for given observation $\boldsymbol{y} \in C_{\tilde{\boldsymbol{y}}}$. The strategy d^* is Bayesian in $[Z, D, R]$ with respect to $P_{\tilde{z}}$ if and only if

$$\forall \boldsymbol{y} \in C_{\tilde{\boldsymbol{y}}}\colon \mathbf{E}_{\tilde{z}|\boldsymbol{y}} L(\tilde{z}; d^*(\boldsymbol{y})) = \inf_{d \in D} \mathbf{E}_{\tilde{z}|\boldsymbol{y}} L(\tilde{z}; d(\boldsymbol{y})).$$

If $d^* \in D$ is the unique Bayes strategy (up to $P_{\tilde{\boldsymbol{y}}}$-equivalence) in $[Z, D, R]$ with respect to $P_{\tilde{z}}$ then d^* is admissible in $[Z, D, R]$.

Analysis

A.17. Let g_1 and g_2 be real-valued functions on some subset $B \subseteqq \mathbb{R}^r$, $r \geqq 1$. The function g_2 is called a *kernel* of g_1 (denotation: $g_1 \propto g_2$) if it holds $g_1(\cdot) = c g_2(\cdot)$ with some constant $c \in \mathbb{R}^1$. The relation $\propto$ thus defined is an equivalence relation within the class of all real-valued functions on B.

A.18. If g is a strictly convex function defined on some convex set $B \subseteqq \mathbb{R}^r$ then there exists a unique value $\boldsymbol{x}_0 \in B$ minimizing g over B.

If g is a concave, positive and real-valued function defined on some convex set $B \subseteqq \mathbb{R}^r$ then the function $g_0(\cdot) = 1/g(\cdot)$ is convex on B.

Every convex function defined on a convex subset B of $\mathbb{R}^r$ is continuous at all the interior points of B.

Let X be a nondegenerate subset of $\mathbb{R}^k$ and denote co(X) the convex hull of X. Then any point from co(X) can be written as a convex combination of at most $k+1$ different points of X (Carathéodory theorem).

A.19. Let $X \subset \mathbb{R}^k$ be compact and denote X^* the linear space of all probability distributions over X.

(i) Assume $\boldsymbol{f} = (f_1, \ldots, f_k)^\top$ is a vector of continuous functions $f_i \mid X \to \mathbb{R}^1$, $i = 1, \ldots, k$, then the set of matrices

$$\mathcal{M}(X^*) = \left\{ \boldsymbol{M}(\xi) = \int_X \boldsymbol{f}(\boldsymbol{x}) \boldsymbol{f}(\boldsymbol{x})^\top \xi(\mathrm{d}\boldsymbol{x}) : \xi \in X^* \right\}$$

is compact with respect to the sum of squares norm $\|\boldsymbol{M}(\xi)\| = (\operatorname{tr} \boldsymbol{M}(\xi)^2)^{1/2}$, $\xi \in X^*$ (see Bandemer et al. (1977), Lemma 5.1.8).

(ii) Assume $g \mid X \to \mathbb{R}^1$ to be a continuous function. Then it holds

$$\sup_{x \in X} g(\boldsymbol{x}) = \sup_{\xi \in X^*} \int_X g(\boldsymbol{x})\, \xi(\mathrm{d}\boldsymbol{x}).$$

A.20. Let B, C be linear spaces and $A \subset C$ be compact (with respect to some given metric m_C). Further, let $h \mid A \times B \to \mathbb{R}^1$ be a function such that $h(a, \cdot)$ is convex for all $a \in A$ and $h(\cdot, b)$ is concave and continuous (with respect to m_C) for all $b \in B$. Then it holds

$$\sup_{a \in A} \inf_{b \in B} h(a, b) = \inf_{b \in B} \sup_{a \in A} h(a, b)$$

(see Sion (1958), Theorem 4.2).

A.21. For all real numbers $a \in \mathbb{R}^+$, $b \in \mathbb{R}^+$ and $c \in (0, 1)$ it holds:

$$a^c b^{1-c} \geqq ac + b(1-c).$$

Matrix Algebra

For every $\boldsymbol{A} \in \mathcal{M}_r^{\geqq}$ let $\lambda_{\min}(\boldsymbol{A}) = \lambda_1(\boldsymbol{A}) \leqq \ldots \leqq \lambda_r(\boldsymbol{A}) = \lambda_{\max}(\boldsymbol{A})$ denote the ordered eigenvalues of $\boldsymbol{A}$. Further, let $\operatorname{tr} \boldsymbol{A}$ and $\det \boldsymbol{A}$ denote the trace and determinant of a matrix $\boldsymbol{A}$, respectively. For any two matrices $\boldsymbol{A}, \boldsymbol{B} \in \mathcal{M}_r^{\geqq}$, the notation $\boldsymbol{A} \leqq \boldsymbol{B}$ means that $\boldsymbol{B} - \boldsymbol{A} \in \mathcal{M}_r^{\geqq}$, which defines the usual (Loewner-)semiordering within $\mathcal{M}_r^{\geqq}$. Accordingly, if $\boldsymbol{A}$ and $\boldsymbol{B}$ are positive definite matrices then by $\boldsymbol{A} < \boldsymbol{B}$ we mean that $\boldsymbol{B} - \boldsymbol{A} \in \mathcal{M}_r^{>}$.

A.22. $\forall \boldsymbol{A} \in \mathcal{M}_r^{\geqq}: \ \operatorname{tr} \boldsymbol{A} = \sum_{i=1}^{r} \lambda_i(\boldsymbol{A}), \qquad \det \boldsymbol{A} = \prod_{i=1}^{r} \lambda_i(\boldsymbol{A}).$

A.23. $\forall \boldsymbol{A} \in \mathcal{M}_r^{>}: \ \det \boldsymbol{A}^{-1} = 1/\det \boldsymbol{A}, \qquad \lambda_{\min}(\boldsymbol{A}^{-1}) = 1/\lambda_{\max}(\boldsymbol{A}).$

A.24. $\forall \boldsymbol{A} \in \mathcal{M}_r^{\geqq}: \ \lambda_{\max}(\boldsymbol{A}) = \sup_{\boldsymbol{a} \in \mathbb{R}^r} \dfrac{\boldsymbol{a}^\top \boldsymbol{A} \boldsymbol{a}}{\boldsymbol{a}^\top \boldsymbol{a}} \geqq \dfrac{1}{r} \operatorname{tr} \boldsymbol{A},$

$$\lambda_{\min}(\boldsymbol{A}) = \inf_{\boldsymbol{a} \in \mathbb{R}^r} \frac{\boldsymbol{a}^\top \boldsymbol{A} \boldsymbol{a}}{\boldsymbol{a}^\top \boldsymbol{a}} \leqq \frac{1}{r} \operatorname{tr} \boldsymbol{A}.$$

A.25. $\forall (\boldsymbol{A}, \boldsymbol{B}) \in \mathcal{M}_r^{\geqq} \times \mathcal{M}_r^{\geqq}: \ \lambda_{\min}(\boldsymbol{A} + \boldsymbol{B}) \geqq \lambda_{\min}(\boldsymbol{A}) + \lambda_{\min}(\boldsymbol{B}).$

A.26. $\forall (A, B) \in \mathcal{M}_r^> \times \mathcal{M}_r^>$: $A \leqq B \Rightarrow \det A \leqq \det B$.

A.27. $\forall (A, B) \in \mathcal{M}_r^{\geqq} \times \mathcal{M}_r^{\geqq}$: $A \leqq B \Rightarrow \forall i \in \{1, \ldots, r\}$: $\lambda_i(A) \leqq \lambda_i(B)$.

A.28. $\forall (A, B, C) \in (\mathcal{M}_r^{\geqq})^3$: $A \leqq B \Rightarrow \operatorname{tr} AC \leqq \operatorname{tr} BC = \operatorname{tr} CB$.

A.29. $\forall (A, B) \in \mathcal{M}_r^> \times \mathcal{M}_r^>$: $A < B \Rightarrow B^{-1} < A^{-1}$.

A.30. $\forall (A, B) \in \mathcal{M}_r^> \times \mathcal{M}_r^> \ \forall \alpha \in (0, 1)$:

$$(\alpha A + (1 - \alpha) B)^{-1} < \alpha A^{-1} + (1 - \alpha) B^{-1}.$$

A.31. $\forall A \in \mathcal{M}_r^> \ \forall B \in \mathcal{M}_r^> \setminus \{A\} \ \forall \alpha \in (0, 1)$:

$$\det(\alpha A + (1 - \alpha) B) > (\det A)^{\alpha} (\det B)^{1-\alpha}.$$

A.32. Let $U \in \mathcal{M}_r^{\geqq}$ be such that $\operatorname{tr} UC > 0$ for all $C \in \mathcal{M}_r^{\geqq}$, $C \neq 0$. Then, for all $A \in \mathcal{M}_r^>$, $B \in \mathcal{M}_r^>$ and $\alpha \in (0, 1)$, it holds:

$$\operatorname{tr} U(\alpha A + (1 - \alpha) B)^{-1} < \alpha \operatorname{tr} UA^{-1} + (1 - \alpha) \operatorname{tr} UB^{-1}.$$

A.33. For all nonsingular $A, B \in \mathcal{M}_{r \times r}$ it holds:

$$(A + B)^{-1} = A^{-1} - A^{-1}(A^{-1} + B^{-1})^{-1} A^{-1}.$$

For all nonsingular matrices $A \in \mathcal{M}_{r \times r}$, $B \in \mathcal{M}_{n \times n}$ and arbitrary matrices $C_1, C_2 \in \mathcal{M}_{r \times n}$ it holds:

$$(A + C_1 B C_1^\top)^{-1} = A^{-1} - A^{-1} C_1 (C_1^\top A^{-1} C_1 + B^{-1})^{-1} C_1^\top A^{-1}.$$

$$(I_r + C_1 C_2^\top)^{-1} = I_r - C_1 (I_n + C_2^\top C_1)^{-1} C_2^\top.$$

A.34. Let $A \in \mathcal{M}_{r \times r}$ be nonsingular and $a, b \in \mathbb{R}^r$. Then it holds:

$$(A + ab^\top)^{-1} = A^{-1} - (1 + b^\top A^{-1} a)^{-1} A^{-1} a b^\top A^{-1}.$$

A.35. For every nonsingular $A \in \mathcal{M}_{r \times r}$ and arbitrary $B \in \mathcal{M}_{r \times n}$ it holds:

$$\det(A + BB^\top) = \det A \det(I_n + B^\top A^{-1} B).$$

A.36. Let $A \in \mathcal{M}_{r \times n}$. The matrix A^+ is called the *Moore-Penrose-inverse* (MP-inverse) of A if it satisfies the equations

$$AA^+A = A, \qquad A^+AA^+ = A^+, \qquad (A^+A)^\top = A^+A, \qquad (AA^+)^\top = AA^+.$$

A.37. Let $A \in \mathcal{M}_{r \times r}$ be nonsingular and subdivided into blocks according to

$$A = \begin{pmatrix} A_{11} & A_{12} \\ A_{21} & A_{22} \end{pmatrix}, \quad \text{where} \quad \begin{array}{ll} A_{11} \in \mathcal{M}_{s \times s}, & 0 \leqq s < r, \\ A_{22} \in \mathcal{M}_{(r-s) \times (r-s)}. & \end{array}$$

Then the inverse of A takes the form

$$A^{-1} = \begin{pmatrix} A^{11} & A^{12} \\ A^{21} & A^{22} \end{pmatrix},$$

where

$$A^{11} = (A_{11} - A_{12}A_{22}^{-1}A_{21})^{-1}, \qquad A^{22} = (A_{22} - A_{21}A_{11}^{-1}A_{12})^{-1},$$

and

$$A^{12} = -A_{11}^{-1}A_{12}A^{22}, \qquad A^{21} = -A^{22}A_{21}A_{11}^{-1}.$$

A.38. Let $A(\cdot) = (a_{ij}(\cdot)), i, j = 1, \ldots, r$, be a matrix whose elements are real-valued and differentiable functions. Define $\frac{\mathrm{d}}{\mathrm{d}t}A(\cdot) = \left(\frac{\mathrm{d}}{\mathrm{d}t}a_{ij}(\cdot)\right)$. If $A(\cdot)$ is nonsingular then it holds

$$\frac{\mathrm{d}}{\mathrm{d}t}A(\cdot)^{-1} = -A(\cdot)^{-1}\left\{\frac{\mathrm{d}}{\mathrm{d}t}A(\cdot)\right\}A(\cdot)^{-1},$$

$$\frac{\mathrm{d}}{\mathrm{d}t}\det A(\cdot) = \det A(\cdot)\,\mathrm{tr}\left\{A(\cdot)^{-1}\frac{\mathrm{d}}{\mathrm{d}t}A(\cdot)\right\}.$$

A.39. Let $A, B \in \mathcal{M}_r^>$. Further, let $\lambda_{\min}(A)$ have multiplicity $q \geqq 1$ and Q be the matrix of orthonormal eigenvectors of A corresponding to the minimum eigenvalue, so that $Q^\top AQ = \lambda_{\min}(A)\, I_q$. Then it holds

$$\lim_{\alpha \downarrow 0} \frac{\mathrm{d}}{\mathrm{d}\alpha}\lambda_{\min}(A + \alpha B) = \lambda_{\min}(Q^\top BQ)$$

(see KIEFER (1974), Section 4E).

A.40. For any two matrices $A \in \mathcal{M}_r^{\geqq}, B \in \mathcal{M}_r^>$ it holds

$$\sup_{x \in \mathbb{R}^r} \frac{x^\top Ax}{x^\top Bx} = \lambda_{\max}(A, B),$$

where $\lambda_{\max}(A, B)$ denotes the largest root of the generalized eigenvalue problem $\det(A - \lambda B) = 0$. Particularly, if $A = aa^\top$ for some $a \in \mathbb{R}^r$ then we have $\lambda_{\max}(A, B) = a^\top B^{-1}a$.

A.41. Let $s \mid \mathcal{M}_{r\times n} \to \mathbb{R}^1$ be a scalar matrix function. For any $Z = (z_{ij}) \in \mathcal{M}_{r\times n}$ define $\partial s/\partial Z = (\partial s/\partial z_{ij})$, $i = 1, \ldots, r$, $j = 1, \ldots, n$. Then for all matrices $A \in \mathcal{M}_{n\times r}$, $U \in \mathcal{M}_{r\times r}$, $B \in \mathcal{M}_{n\times n}$ it holds

$$\frac{\partial}{\partial Z}\mathrm{tr}\, AZ = A^\top, \qquad \frac{\partial}{\partial Z}\mathrm{tr}\, UZBZ^\top = UZB + U^\top ZB^\top.$$

References

AKAIKE, H. (1980): Ignorance prior distribution of a hyperparameter and Stein's estimator. Ann. Inst. Statist. Math. **32 (A)**, 171–179.

AKAIKE, H. (1983): On minimum information prior distributions. Ann. Inst. Statist. Math. **35 (A)**, 139–149.

ALAM, K. (1973): A family of admissible minimax estimators of the mean of a multivariate normal distribution. Ann. Statist. **1**, 517–525.

AL-BAYYATI, H. A./ARNOLD, J. C. (1972): On double-stage estimation in simple linear regression using prior knowledge. Technometrics **14**, 405–414.

ALI, M. A./SINGH, N. (1981): Optimal use of prior information in designing experiments. J. of Information & Optimization Sciences **2**, 64–72.

ALLEN, D. M. (1974): The relationship between variable selection and data augmentation and a method for prediction. Technometrics **16**, 125–127.

ALSON, P. (1988): Minimax properties for linear estimators of the location parameter of a linear model. Statistics **19**, 163–171.

ARNOLD, J. C./AL-BAYYATI, H. A. (1970): On double-stage estimation of the mean using prior knowledge. Biometrics **26**, 787–800.

ATKINSON, A. C. (1978): Posterior probabilities for choosing a regression model. Biometrika **65**, 39–48.

ATKINSON, A. C. (1982): Developments in the design of experiments. Int. Statist. Rev. **50**, 161–177.

ATKINSON, A. C. (1985): An introduction to the optimum design of experiments. In: A Celebration of Statistics (Eds.: A. C. ATKINSON and S. E. FIENBERG). Springer-Verlag, New York, 465–473.

ATKINSON, A. C./FEDOROV, V. V. (1975): Experiments for discriminating between several models. Biometrika **62**, 289–303.

ATWOOD, C. L. (1976): Convergent design sequences for sufficiently regular optimality criteria. Ann. Statist. **4**, 1124–1138.

BAKSALARY, J. K./MARKIEWICZ, A. (1986): Characterizations of admissible linear estimators in restricted linear models. J. Statist. Plann. Infer. **13**, 395–398.

BAKSALARY, J. K./LISKI, E. P./TRENKLER, G. (1989): Mean square error matrix improvements and admissibility of linear estimators. J. Statist. Plann. Infer. **23**, 313–325.

BANDEMER, H. et al. (1977): Theorie und Anwendung der optimalen Versuchsplanung, Vol. I, Akademie-Verlag, Berlin.

BANDEMER, H./NÄTHER, W. (1980): Theorie und Anwendung der optimalen Versuchsplanung, Vol. II, Akademie-Verlag, Berlin.

BANDEMER, H./NÄTHER, W./PILZ, J. (1987): Once more: Optimal experimental design for regression models (with discussion). Statistics **18**, 171–217.

BANDEMER, H./PILZ, J. (1978): Optimum experimental design for a Bayes estimator in linear regression. Transactions of the Eighth Prague Conference, Vol. A, 93–102.

BANDEMER, H./PILZ, J./FELLENBERG, B. (1986): Integral geometric prior distributions for linear Bayesian regression with bounded response. Statistics **17**, 323–335.

BANSAL, A. K. (1978): Robustness of a Bayes estimator for the mean of a normal population with nonnormal prior. Commun. Statist. A – Theory Methods **7**, 453–460.

BARANCHIK, A. J. (1970): A family of minimax estimators of the mean of a multivariate normal distribution. Ann. Math. Statist. **41**, 642–645.

BARLOW, R. E./BRUNK, H. D. (1972): The isotonic regression problem and its dual. J. Amer. Statist. Assoc. **67**, 140–147.

BAUR, F. (1984): Einige lineare und nicht-lineare Alternativen zum Kleinst-Quadrate-Schätzer im verallgemeinerten linearen Modell. Athenäum/Hain, Königstein/Taunus.

BAZARAA, M. S./SHETTY, C. M. (1979): Nonlinear Programming. Theory and Algorithms. J. Wiley & Sons, New York.

BARANCHIK, A. J. (1973): Inadmissibility of maximum likelihood estimators in some multiple regression problems with three or more independent variables. Ann. Statist. **1**, 312–321.

BECKENBACH, E. F./BELLMANN, R. (1965): Inequalities. Springer-Verlag, Berlin.

BEN MANSOUR, D. (1984): Présentation, dans les cadres classique et bayesien, des estimateurs de James-Stein généralises. Université de Rouen, Documents de travail, Nouvelle Série 01.

BERGER, J. O. (1976 a): Admissible minimax estimation of a multivariate normal mean with arbitrary quadratic loss. Ann. Statist. **4**, 223–226.

BERGER, J. O. (1976 b): Minimax estimation of a multivariate normal mean under arbitrary quadratic loss. J. Multivariate Anal. **6**, 256–264.

BERGER, J. O. (1980 a): Statistical Decision Theory: Foundations, Concepts, and Methods. Springer-Verlag, New York.

BERGER, J. O. (1980 b): A robust generalized Bayes estimator and confidence region for a multivariate normal mean. Ann. Statist. **8**, 716–761.

BERGER, J. O. (1982 a): Bayesian robustness and the Stein effect. J. Amer. Statist. Assoc. **77**, 358–368.

BERGER, J. O. (1982 b): Estimation in continuous exponential families: Bayesian estimation subject to risk restrictions and inadmissibility results. In: Statistical Decision Theory and Related Topics III (Eds.: S. S. GUPTA and J. O. BERGER). Academic Press, New York.

BERGER, J. O. (1984): The robust Bayesian viewpoint. In: Robustness of Bayesian analysis (Ed.: J. B. KADANE). Elsevier Science Publ. B. V., New York, 63–124.

BERGER, J. O. (1987): Robust Bayesian analysis: Sensitivity to the prior. Technical Report 87-10, Dept. of Statistics, Pardue Univ.

BERGER, J. O./BERLINER, L. M. (1984): Bayesian input in Stein estimation and a new minimax empirical Bayes estimator. J. Econometrics **25**, 87–108.

BERGER, J. O./BERLINER, L. M. (1986): Robust Bayes and empirical Bayes analysis with ε-contaminated priors. Ann. Statist. **14**, 461–486.

BERGER, J. O./BOCK, M. E. (1976): Combining independent normal mean estimation problems with unknown variances. Ann. Statist. **4**, 642–648.

BERGER, R. (1979): Gamma minimax robustness of Bayes rules. Commun. Statist. A – Theory Methods **8**, 543–560.

BERNARDO, J. M. (1979): Expected information as expected utility. Ann. Statist. **7**, 686–690.

BERNARDO, J. M./DE GROOT, M. H./LINDLEY, D. V./SMITH, A. F. M. (Eds.) (1980): Bayesian Statistics (Proc. of the 1st Int. Meeting, Valencia, 1979). Trab. Estadist. **31**, Valencia University Press.

BERNARDO, J. M./DE GROOT, M. H./LINDLEY, D. V./SMITH, A. F. M. (Eds.) (1985): Bayesian Statistics 2 (Proc. of the 2nd Int. Meeting, Valencia, 1983). North-Holland Publ. Co., Amsterdam.

BERNARDO, J. M./DE GROOT, M. H./LINDLEY, D. V./SMITH, A. F. M. (Eds.) (1989): Bayesian Statistics 3 (Proc. of the 3rd Int. Meeting, Valencia, 1987). Oxford Univ. Press.

BIBBY, J./TOUTENBURG, H. (1977): Prediction and Improved Estimation in Linear Models. J. Wiley & Sons, New York.

BICKEL, P. J. (1981): Minimax estimation of the mean of a normal distribution when the parameter space is restricted. Ann. Statist. **9**, 1301–1309.

BIRNBAUM, A./LASKA, E. (1967): Optimal robustness: a general method, with applications to linear estimators of the mean. J. Amer. Statist. Assoc. **62**, 1230–1240.

Blackwell, D. (1951): Comparison of experiments. Proc. Sec. Berkeley Symp., Univ. of Calif. Press, 93–102.

Blackwell, D. (1953): Equivalent comparisons of experiments. Ann. Math. Statist. **24**, 265–272.

Blight, B. J. N./Ott, L. (1975): A Bayesian approach to model inadequacy for polynomial regression. Biometrika **62**, 79–88.

Blum, J. R./Rosenblatt, J. (1967): On partial a priori information in statistical inference. Ann. Math. Statist. **38**, 1671–1678.

Bock, M. E. (1975): Minimax estimators of the mean of a multivariate normal distribution. Ann. Statist. **3**, 209–218.

Bock, M. E. (1982): Employing vague inequality information in the estimation of normal mean vectors. In: Statistical Decision Theory and Related Topics III, Vol. 1 (Eds.: S. S. Gupta and J. O. Berger). Academic Press, New York.

Bock, M. E. (1985): Minimax estimators that shift towards a hypersphere for location vectors of spherically symmetric distributions. J. Multivar. Anal. **17**, 127–147.

Böhning, D. (1981): On an assumption of Bandemer and Ketzel in optimal experimental design theory. Math. Operationsforsch. Statist., Ser. Statistics **12**, 497–502.

Böhning, D. (1982): Convergence of Simar's algorithm for finding the maximum likelihood estimate of a compound Poisson process. Ann. Statist. **10**, 1006–1008.

Böhning, D. (1985): Numerical estimation of a probability measure. J. Statist. Plann. Infer. **11**, 57–69.

Box, G. E. P. (1980): Sampling and Bayes' inference in scientific modelling and robustness (with discussion). J. Roy. Statist. Soc., Ser. A **143**, 383–430.

Box, G. E. P./Draper, N. R. (1959): A basis for the selection of a response surface design. J. Amer. Statist. Assoc. **54**, 622–654.

Box, G. E. P./Hill, W. J. (1967): Discrimination among mechanistic models. Technometrics **9**, 57–71.

Box, G. E. P./Lucas, H. L. (1959): Design of experiments in nonlinear situations. Biometrika **46**, 77–90.

Box, G. E. P./Tiao, G. C. (1962): A further look at robustness via Bayes's theorem. Biometrika **49**, 419–432.

Box, G. E. P./Tiao, G. C. (1965): Multiparameter problems from a Bayesian point of view. Ann. Math. Statist. **36**, 1468–1482.

Box, G. E. P./Tiao, G. C. (1968): A Bayesian approach to some outlier problems. Biometrika **55**, 119–129.

Box, G. E. P./Tiao, G. C. (1973): Bayesian Inference in Statistical Analysis. Reading Mass., Addison-Wesley Publishing Co., London.

Britney, R. R./Winkler, R. L. (1974): Bayesian point estimation and prediction. Ann. Inst. Statist. Math. **26**, 15–34.

Broemeling, L. D. (1985): Bayesian Analysis of Linear Models. Marcel Dekker, New York.

Broemeling, L. D./Chin Choy, J. H. (1981): Detecting structural change in linear models. Commun. Statist. A – Theory Methods **A 10**, 2551–2561.

Brooks, R. J. (1972): A decision theory approach to optimal regression designs. Biometrika **59**, 563–571.

Brooks, R. J. (1974): On the choice of an experiment for prediction in linear regression. Biometrika **61**, 303–311.

Brooks, R. J. (1976): Optimal regression designs for prediction when prior knowledge is available. Metrika **23**, 221–230.

Brooks, R. J. (1977): Optimal regression design for control in linear regression. Biometrika **64**, 319–325.

Brown, L. D. (1971): Admissible estimators, recurrent diffusions, and insoluble boundary value problems. Ann. Math. Statist. **42**, 855–904.

Bunke, H./Bunke, O. (1986): Statistical Methods of Model Building. J. Wiley & Sons, New York.

Bunke, H./Gladitz, J. (1974): Empirical linear Bayes decision rules for a sequence of linear models with different regressor matrices. Math. Operationsforsch. Stat. **5**, 235–244.

BUNKE, O. (1964): Bedingte Strategien in der Spieltheorie: Existenzsätze und Anwendung auf statistische Entscheidungsprobleme. Transactions Third Prague Conference, 35–43.

BUNKE, O. (1975a): Minimax linear, ridge and shrunken estimators for linear parameters. Math. Operationsforsch. Stat. **6**, 697–701.

BUNKE, O. (1975b): Improved inference in linear models with additional information. Math. Operationsforsch. Stat. **6**, 817–830.

BUNKE, O. (1977): Mixed models, empirical Bayes and Stein estimators. Math. Operationsforsch. Stat., Ser. Statistics **1**, 55–68.

BUNKE, O. (1987): Posterior distributions in semiparametric problems. Preprint No. 144, Humboldt-Univ., Sektion Mathematik.

BUONACCORSI, J. P./IYER, H. K. (1986): Optimal designs for ratios of linear combinations in the general linear model. J. Statist. Plann. Infer. **13**, 345–356.

CARROLL, R. J./RUPPERT, D. (1983): Robust estimators for random coefficient regression models. In: Contributions to Statistics, Essays in Honour of N. L. Johnson (Ed.: P. K. SEN), North-Holland Publ. Co., Amsterdam, 81–96.

CASELLA, G. (1980): Minimax ridge regression estimation. Ann. Statist. **8**, 1036–1056.

CASELLA, G./STRAWDERMAN, W. E. (1981): Estimating a bounded normal mean. Ann. Statist. **9**, 870–878.

CELLIER, D./FOURDRINIER, D./ROBERT, Ch. (1986): Controlled shrinkage estimators. Technical Report. Laboratoire de Calcul des Probabilités et Statistique, Université de Rouen.

CHALONER, K. (1982): Optimal Bayesian experimental design for linear models. Ph. D. Thesis, Carnegie-Mellon University, Technical Report No. 238.

CHALONER, K. (1984): Optimal Bayesian experimental design for linear models. Ann. Statist. **12**, 283–300. Corrigendum in: Ann. Statist. **13** (1985), 836.

CHALONER, K. (1986): Optimal Bayesian design for non-linear estimation. Univ. of Minnesota, School of Statistics, Technical Report No. 468.

CHALONER, K./LARNTZ, K. (1986): Optimal Bayesian design applied to logistic regression experiments. Univ. of Minnesota, School of Statistics, Technical Report No. 483.

CHAMBERLAIN, G./LEAMER, E. E. (1976): Matrix weighted averages and posterior bounds. J. Roy. Statist. Soc., Ser. B **38**, 73–84.

CHANG, D.-S. (1979): Design of optimal control for a regression problem. Ann. Statist. **7**, 1078–1085.

CHEN, L./EICHENAUER, J. (1987): Gamma minimax estimators for a multivariate normal mean. Preprint Nr. 1050, Technische Hochschule Darmstadt, Fachbereich Mathematik.

CHIN CHOY, J. H./BROEMELING, L. D. (1980): Some Bayesian inferences for a changing linear model. Technometrics **22**, No. 1, 71–78.

CHIPMAN, J. S./RAO, M. M. (1964): The treatment of linear restrictions in regression analysis. Econometrica **32**, 198–209.

CLEMMER, B. A./KRUTCHKOFF, R. G. (1968): The use of empirical Bayes estimators in a linear regression model. Biometrika **55**, 525–534.

COHEN, A. (1966): All admissible linear estimators of the mean vector. Ann. Math. Stat. **37**, 458–463.

COOK, R. D./NACHTSHEIM, C. J. (1982): Model robust, linear-optimal designs. Technometrics **24**, 49–54.

COOLEY, A. E./LIN, P. E. (1983): Bayes minimax estimators of a multivariate normal mean, with application to generalized ridge regression. Commun. Statist. A – Theory Methods **12** (24), 2861–2869.

COTE, R./MANSON, A. R./HADER, R. J. (1973): Minimum bias approximation of a general regression model. J. Amer. Statist. Assoc. **68**, 633–638.

COVEY-CRUMP, P. A. K./SILVEY, S. D. (1970): Optimal regression with previous observations. Biometrika **57**, 551–566.

DAGENAIS, M. (1974): Multiple regression analysis with incomplete observations, from a Bayesian viewpoint. In: Studies in Bayesian Econometrics and Statistics (Eds.: S. E. FIENBERG and A. ZELLNER). North-Holland Publ. Co., Amsterdam, 273–288.

DALAL, R. R./HALL, W. J. (1983): Approximating priors by mixtures of natural conjugate priors. J. Roy. Statist. Soc., Ser. B **45**, 278–286.

DAS GUPTA, A. (1985): Bayes minimax estimation in multiparameter families when the parameter space is restricted to a bounded convex set. Sankhyā A **47**, 326–332.

DAS GUPTA, A./STUDDEN, W. J. (1988 a): Robust Bayesian analysis and optimal experimental designs in normal linear models with many parameters-I. Technical Report 88–14, Purdue Univ., Dept. of Statistics.

DAS GUPTA, A./STUDDEN, W. J. (1988 b): Robust Bayesian analysis and optimal experimental designs in normal linear models with many parameters-II. Technical Report 88–34 C, Purdue Univ., Dept. of Statistics.

DAVIS, W. W. (1978): Bayesian analysis of the linear model subject to linear inequality constraints. J. Amer. Statist. Assoc. **73**, 573–579.

DAWID, A. P./STONE, M./ZIDEK, J. V. (1973): Marginalization paradoxes in Bayesian and structural inference (with discussion). J. Roy. Statist. Soc. **35** (2), 189–233.

DEMPSTER, A. P. (1973): Alternatives to least squares in multiple regression. In: Multivariate Statistical Inference, Proc. Dalhousie Univ., Halifax, 25–40.

DENISOV, V. I. (1977): Mathematical Implementation of the Computer-Experimenter-System. Regression and Variance Analysis (Russian). Nauka, Moscow.

DE FINETTI, B. (1964): Foresight: Its logical laws, its subjective sources. In: Studies in Subjective Probability (Eds.: H. E. KYBURG, JR. and H. E. SMOKLER). J. Wiley & Sons, New York, 93–158.

DE GROOT, M. H. (1970): Optimal Statistical Decisions. McGraw-Hill-Co., New York.

DE GROOT, M. H./RAO, M. M. (1963): Bayes estimation with convex loss. Ann. Math. Statist. **34**, 839–846.

DE GROOT, M. H./RAO, M. M. (1966): Multidimensional information inequalities and prediction. In: Multivariate Analysis (Ed.: P. R. KRISHNAIAH). Academic Press, New York, 287–313.

DE ROBERTIS, L./HARTIGAN, J. A. (1981): Bayesian inference using intervals of measures. Ann. Statist. **9**, 235–244.

DEUTSCH, R. (1965): Estimation Theory. Prentice Hall, Englewood Cliffs, N. J.

DEY, D./BERGER, J. O. (1983): Combining coordinates in simultaneous estimation of normal means. J. Statist. Plann. Infer. **8**, 143–160.

DIACONIS, P./FREEDMAN, D. (1983): Frequency properties of Bayes' rules. In: Scientific Inference, Data Analysis, and Robustness (Eds.: G. E. P. BOX, T. LEONARD, and C. F. WU). Academic Press, New York.

DIACONIS, P./YLVISAKER, D. (1979): Conjugate priors for exponential families. Ann. Statist. **7**, 269–281.

DIACONIS, P./YLVISAKER, D. (1985): Quantifying prior opinion. In: Bayesian Statistics 2 (Eds.: J. M. BERNARDO et al.). North-Holland Publ. Co., Amsterdam, 133–156.

DICKEY, J. (1975): Bayesian alternatives to the F-test and least-squares estimates in the normal linear model. In: Bayesian Studies in Econometrics and Statistics (Eds.: S. E. FIENBERG and A. ZELLNER). North-Holland Publ. Co., Amsterdam.

DICKEY, J. M./CHEN, C.-H. (1985): Direct subjective-probability modelling using ellipsoidal distributions. In: Bayesian Statistics 2 (Eds.: J. M. BERNARDO et al.). North-Holland Publ. Co., Amsterdam.

DICKEY, J. M./LINDLEY, D. V./PRESS, S. J. (1985): Bayesian estimation of the dispersion matrix of a multivariate normal distribution. Commun. Statist. A – Theory Methods **14** (5), 1019–1034.

DOOB, J. L. (1953): Stochastic Processes. J. Wiley & Sons, New York.

DRAPER, N. R./HUNTER, W. G. (1967): The use of prior distributions in the design of experiments for parameter estimation in nonlinear situations. Biometrika **54**, 147–153.

DRAPER, N. R./JOHN, J. A. (1981): Influential observations and outliers in regression. Technometrics **23**, 21–26.

DUBOV, E. L. (1981): Optimal Bayesian designs in nonlinear problems (Russian). In: Problems of Cybernetics (Eds.: V. V. NALIMOV and V. V. FEDOROV). Nauka, Moscow, 27–30.

DUNCAN, G./DE GROOT, M. H. (1976): A mean squared error approach to optimal design theory. Proc. of the 1976 conference on information: Sciences and Systems. The Johns Hopkins Univ., 217–221.

DUTTER, R./GUTTMAN, I. (1974): Procedures for investigating outliers when estimating in the general univariate linear situation – I. Full rank case. Technical Report, Univ. of Toronto.

DYKSTRA, O. JR. (1971): The augmentation of experimental data to maximize $|X^{\top}X|$. Technometrics **13**, 682–688.

EFRON, B./MORRIS, C. (1971): Limiting the risk of Bayes and empirical Bayes estimators – Part 1: The Bayes case. J. Amer. Statist. Assoc. **66**, 807–815.

EFRON, B./MORRIS, C. (1972): Limiting the risk of Bayes and empirical Bayes estimators – Part 2: The empirical Bayes case. J. Amer. Statist. Assoc. **67**, 130–139.

EFRON, B./MORRIS, C. (1973): Stein's estimation rule and its competitors – An empirical Bayes approach. J. Amer. Statist. Assoc. **68**, 117–130.

EFRON, B./MORRIS, C. (1976): Families of minimax estimators of the mean of a multivariate normal distribution. Ann. Statist. **4**, 11–21.

EHRENFELD, S. (1956): Complete class theorems in experimental designs. Proc. Third Berkeley Symp. **1**, 57–67, Berkeley – Los Angeles.

ELFVING, G. E. (1952): Optimum allocation in linear regression theory. Ann. Math. Statist. **23**, 255–262.

ESCOBAR, L. A./SKARPNESS, B. (1984): A closed form solution for the least squares regression problem with linear inequality constraints. Commun. Statist. A – Theory Methods **13**, 1127–1134.

EVANS, J. W. (1979): Computer augmentation of experimental designs to maximize $|X^{\top}X|$. Technometrics **21**, 321–330.

FAREBROTHER, R. W. (1975): The minimum mean square error linear estimator and ridge regression. Technometrics **17**, 127–128.

FAREBROTHER, R. W. (1986): Testing linear inequality constraints in the standard linear model. Commun. Statist. A – Theory Methods **15**, 7–32.

FEDOROV, V. V. (1972): Theory of optimal experiments. Transl. and ed. by W. J. STUDDEN and E. M. KLIMKO, Academic Press, New York, (Russian Original: Nauka, Moscow 1971).

FEDOROV, V. V./USPENSKI, A. (1975): Numerical aspects of the least quares method and the design of experiments. Moscow State Univ.

FELLMAN, J. (1974): On the allocation of linear observations. Commentationes Physico-Mathematicae **44**, No. 2–3, Ph. D. Thesis, Helsinki.

FEN LI, TZE (1982): A note on James-Stein and Bayes empirical Bayes estimators. Commun. Statist. A – Theory Methods **11** (9), 1029–1043.

FERGUSON, T. S. (1967): Mathematical Statistics – a Decision Theoretic Approach. Academic Press, New York.

FERGUSON, T. S. (1973): A Bayesian analysis of some nonparametric problems. Ann. Statist. **1**, 209–230.

FISHER, R. A./Yates, F. (1957): Statistical Tables for Biological, Agricultural and Medical Research. Fifth revised edition, Oliver and Boyd, Edinburgh.

FRANKE, T. (1987): Ein bayesscher Zugang zur minimax-linearen Regression und Approgression bei eingeschränktem Parameterraum. Diplomarbeit, Bergakademie Freiberg, Sektion Mathematik.

GAFFKE, N. (1984): Optimal designs for contaminated linear regression. In: Robustness of Statistical Methods and Nonparametric Statistics (Eds.: D. RASCH and M. L. TIKU), Deutscher Verlag der Wissenschaften, Berlin, 37–42.

GAFFKE, N. (1985): Directional derivatives of optimality criteria at singular matrices in convex design theory. Statistics **16**, 373–388.

GAFFKE, N./HEILIGERS, B. (1989): Bayes, admissible, and minimax linear estimators in linear models with restricted parameter space. Statistics **20**, 478–508.

GAFFKE, N./MATHAR, R. (1988): Linear minimax estimation under ellipsoidal parameter space and related Bayes *L*-optimal design. Technical Report No. 42, Universität Augsburg.

GEORGE, E. I. (1986 a): Minimax multiple shrinkage estimation. Ann. Statist. **14**, 188–205.

GEORGE, E. I. (1986 b): A formal Bayes multiple shrinkage estimator. Commun. Statist. A – Theory Methods **15**, 2099–2114.

GILES, D. E. A./RAYNER, A. C. (1979): The mean squared errors of the maximum likelihood and natural-conjugate Bayes regression estimators. J. Econom. **11**, 319–334.

GIOVAGNOLI, A./VERDINELLI, I. (1983): Bayes *D*-optimum and *E*-optimum block designs. Biometrika **70**, 695–706.

GLADITZ, J. (1981): Konstruktion optimaler Versuchspläne für lineare Regressionsmodelle mit zufälligen Koeffizienten. Thesis, Bergakademie Freiberg.

GLADITZ, J./PILZ, J. (1982 a): Construction of optimal designs in random coefficient regression models. Math. Operationsforsch. Statist., Ser. Statistics **13**, 371–385.

GLADITZ, J./PILZ, J. (1982 b): Bayes designs for multiple linear regression on the unit sphere. Math. Operationsforsch. Statist., Ser. Statistics **13**, 491–506.

GNOT, S. (1983): Bayes estimation in linear models: A coordinate-free approach. J. Multivar. Anal. **13**, 40–51.

GOEL, P. K. (1983): Information measures and Bayesian hierarchical models. J. Amer. Statist. Assoc. **78**, 408–410.

GOEL, P. K./DE GROOT, M. H. (1980): Only normal distributions have linear posterior expectations in linear regression. J. Amer. Statist. Assoc. **75 (372)**, 895–900.

GOEL, P. K./DE GROOT, M. H. (1981): Information about hyperparameters in hierarchical models. J. Amer. Statist. Assoc. **76 (373)**, 140–147.

GOLDBERGER, A. S. (1964): Econometric Theory. J. Wiley & Sons, New York.

GOLDBERGER, A. S./THEIL, H. (1961): On pure and mixed statistical estimation in economics. Internat. Economic Review **2**, 65–78.

GOLDSTEIN, M. (1974): Approximate Bayesian inference with incompletely specified prior distribution. Biometrika **61**, 619–621.

GOLDSTEIN, M. (1976): Bayesian analysis of regression problems. Biometrika **63**, 51–58.

GOLDSTEIN, M. (1980): The linear Bayes regression estimator under weak prior assumptions. Biometrika **67**, 621–628.

GOLDSTEIN, M. (1983): General variance modifications for linear Bayes estimators. J. Amer. Statist. Assoc. **78**, 616–618.

GOLDSTEIN, M./SMITH, A. F. M. (1974): Ridge-type estimators for regression analysis. J. Roy. Statist. Soc., Ser. B **36**, 284–291.

GRENANDER, U./SZEGÖ, G. (1958): Toeplitz Forms and their Applications. Univ. of Calif. Press, Berkeley.

GRIBIK, P. R./KORTANEK, K. O. (1977): Equivalence theorems and cutting plane algorithm for a class of experimental design problems. SIAM J. Appl. Math. **32**, 232–259.

GRUBER, M. H. J./RAO, P. S. R. S. (1982): Bayes estimators for linear models with less than full rank. Commun. Statist. A – Theory Methods **11**, 59–69.

GUTTMAN, I. (1971): A remark on the optimal regression designs with previous observations of Covey-Crump and Silvey. Biometrika **58**, 683–685.

GUTTMAN, I./DUTTER, R. (1974): A Bayesian approach to the detection of spuriousness and estimation in the general univariate linear model when outliers are present. Transactions Seventh Prague Conference, 175–186.

GUTTMAN, J./DUTTER, R./FREEMAN, P. R. (1978): Care and handling of outliers in the general linear model to detect spuriosity – a Bayesian approach. Technometrics **20**, 187–193.

HAHN, G. J. (1984): Experimental design in the complex world. Technometrics **26**, 19–31.

HAITOVSKY, Y./WAX, Y. (1980): Generalized ridge regression, least squares with stochastic prior information, and Bayesian estimators. Applied Math. and Comput. **7**, 125–154.

HALLUM, C. R./ODELL, P. L./BOULLION, T. L. (1977): An extension of the generalized Gauss-Markov Theorem to the mixed model. J. Industr. Math. Soc. **27** (2), 93–103.

HALPERN, E. F. (1973): Polynomial regression from a Bayesian approach. J. Amer. Statist. Assoc. **68**, 173–143.

HARRISON, P. J./STEVENS, C. F. (1976): Bayesian forecasting (with discussion). J. Roy Statist. Soc., Ser. B **38**, 205–247.

HARTIGAN, J. A. (1964): Invariant prior distributions. Ann. Math. Statist. **35**, 836–845.

HARTIGAN, J. A. (1969): Linear Bayesian methods. J. Roy. Statist. Soc., Ser. B **31**, 446–454.

HARTIGAN, J. A. (1983): Bayes Methods. Springer-Verlag, New York, Berlin.

HECKENDORFF, H. (1982): Grundlagen der sequentiellen Statistik. TEUBNER-TEXTE zur Mathematik **45**, Teubner-Verlag, Leipzig.

HEDAYAT, A./WALLIS, W. D. (1978): Hadamard matrices and their applications. Ann. Statist. **6**, 1184–1238.

HERING, F./TRENKLER, G./STAHLECKER, P. (1987): Partial minimax estimation in regression analysis. Statistica Neerlandica **41**, 111–128.

HOCKING, R. R. (1976): The analysis and selection of variables in linear regression. Biometrics **32**, 1–49.

HODGES, J. L./LEHMANN, E. L. (1952): The use of previous experience in reaching statistical decisions. Ann. Math. Statist. **23**, 396–407.

HOEL, P. G. (1965): Minimax designs in two dimensional regression. Ann. Math. Statist. **36**, 1097–1106.

HOEL, P. G./LEVINE, A. (1964): Optimal spacing and weighting in polynomial prediction. Ann. Math. Statist. **35**, 1553–1560.

HOERL, A. E./KENNARD, R. W. (1970): Ridge regression: Biased estimation for nonorthogonal problems. Technometrics **12**, 55–82.

HOFFMANN, K. (1977): Admissibility of linear estimators with respect to restricted parameter sets. Math. Operationsforsch. Statist., Ser. Statistics **8**, 425–438.

HOFFMANN, K. (1979): Characterization of minimax linear estimators in linear regression. Math. Operationsforsch. Statist., Ser. Statistics **10**, 19–26.

HOFFMANN, K. (1980): Admissible improvements of the least squares estimator. Math. Operationsforsch. Statist., Ser. Statistics **11**, 373–388.

HORN, A. (1954): Doubly stochastic matrices and the diagonal of a rotation matrix. Amer. J. Math. **76**, 620–630.

ITO, T. (1980): Methods of estimation for multi-market disequilibrium models. Econometrica **48**, 97–125.

JACKSON, D. A. et al. (1970): G_2-minimax estimators in the exponential family. Biometrika **57**, 439–443.

JAMES, W./STEIN, C. (1961): Estimation with quadratic loss. Proc. Fourth Berkeley Symp. **1**, Berkeley – Los Angeles, 361–379.

JEFFREYS, H. (1961): Theory of Probability. Clarendon Press, Oxford.

JENNRICH, R. I./OMAN, S. D. (1986): How much does Stein estimation help in multiple linear regression. Technometrics **28**, 113–122.

JUDGE, C./BOCK, M. (1978): The Statistical Implications of Pretest and Stein Rule Estimators in Econometrics. North-Holland Publ. Co., Amsterdam.

JUDGE, G. G./TAKAYAMA, T. (1966): Inequality restrictions in regression analysis. J. Amer. Statist. Assoc. **61**, 166–181.

JUDGE, G. G./YANCEY, T. A./BOCK, M. E./BOHRER, R. (1984): The non-optimality of the inequality restricted estimator under squared error loss. J. Econometrics **25**, 165–177.

KADANE, J. B. et al. (1980): Interactive elicitation of opinion for a normal linear model. J. Amer. Statist. Assoc. **75**, 845–854.

KAGAN, A. M./LINNIK, J. V./RAO, C. R. (1972): Characterization Problems in Mathematical Statistics (Russian). Nauka, Moscow.

KAHNEMAN, D./SLOVIK, P./TVERSKY, A. (1982): Judgement under Uncertainty: Heuristics and Biases. Cambridge Univ. Press.

KARLIN, S./STUDDEN, W. J. (1966): Optimal experimental designs. Ann. Math. Statist. **37**, 783–815.

KARSON, M. J./MANSON, A. R./HADER, R. J. (1969): Minimum bias estimation and experimental design for response surfaces. Technometrics **11**, 461–475.

KEMPTHORNE, P. J. (1986): Optimal minimax squared error risk estimation of the mean of a multivariate normal distribution. Commun. Statist. A – Theory Methods **15**, 2145–2158.

KHATRI, C. G. (1979): Characterizations of multivariate normality II. Through linear regressions. J. Multivariate Anal. **9**, 589–598.

KIEFER, J. (1959): Optimum experimental design. J. Roy. Statist. Soc., Ser. B **21**, 272–319.

KIEFER, J. (1961): Optimum designs in regression problems II. Ann. Math. Statist. **32**, 298–325.

KIEFER, J. (1973): Optimum designs for fitting biased multiresponse surfaces. In: Multivariate Analysis, Vol. III (Ed.: P. R. KRISHNAIAH). Academic Press, New York, 287–297.

KIEFER, J. (1974): General equivalence theory for optimum designs (approximate theory). Ann. Statist. **2**, 849–879.

KIEFER, J. (1975): Optimal design: variation in structure and performance under change of criterion. Biometrika **62**, 277–282.

KIEFER, J. (1980): Designs for extrapolation when bias is present. In: Multivariate Analysis, Vol. V (Ed.: P. R. KRISHNAIAH), North-Holland, Amsterdam, 79–93.

KIEFER, J./WOLFOWITZ, J. (1959): Optimum designs in regression problems. Ann. Math. Statist. **30**, 271–294.

KIEFER, J./WOLFOWITZ, J. (1960): The equivalence of two extremum problems. Canad. Journ. Math. **12**, 363–366.

KLEIN, R. W./BROWN, S. J. (1984): Model selection when there is "minimal" prior information. Econometrica **52**, 1291–1312.

KLOEK, T./VAN DIJK, H. K. (1978): Bayesian estimation of equation system parameters. An application of integration by Monte Carlo. Econometrica **46**, 1–19.

KOZÁK, J. (1985): Modified minimax estimation of regression coefficients. Statistics **16**, 363–371.

KRAFFT, O. (1978): Lineare statistische Modelle und optimale Versuchspläne. Verlag Vandenhoeck & Ruprecht, Göttingen.

KRAFFT, O. (1986): A maximin linear estimator for linear parameters under restrictions in form of inequalities. Statistics **17**, 3–8.

KUKS, J. (1972): Minimax estimation of regression coefficients (Russian). Izv. Akad. Nauk Est. SSR **21**, 73–78.

KUKS, J./OL'MAN, V. (1971): Minimax linear estimation of regression coefficients (Russian). Izv. Akad. Nauk Est. SSR **20**, 480–482.

KUKS, J./OL'MAN, V. (1972): Minimax linear estimation of regression coefficients II (Russian). Izv. Akad. Nauk Est. SSR **21**, 66–72.

KUPPER, L. L./MEYDRECH, E. F. (1973): A new approach to mean squared error estimation of response surfaces. Biometrika **60**, 573–579.

LA MOTTE, L. R. (1978): Bayes linear estimators. Technometrics **20**, 281–290.

LA MOTTE, L. R. (1982): Admissibility in linear estimation. Ann. Statist. **10**, 245–255.

LAUTERBACH, J. (1989): Zur Berechnung approximativer Minimax-Schätzer im linearen Regressionsmodell. Dissertationsschrift, Fachbereich Wirtschaftswissenschaften der Univ. Hannover.

LÄUTER, H. (1974): On the admissibility and nonadmissibility of the usual estimator for the mean of a multivariate normal population and conclusions to optimal design. Math. Operationsforsch. Stat. **5**, 591–598.

LÄUTER, H. (1975 a): A minimax linear estimator for linear parameters under restrictions in form of inequalities. Math. Operationsforsch. Stat. **6**, 689–695.

LÄUTER, H. (1975 b): Linear minimax estimation for inadequate models for the I. Petersen approach. Math. Operationsforsch. Statist. **6**, 769–774.

LAYCOCK, P. J. (1972): Convex loss applied to design in regression problems (with discussion). J. Roy. Statist. Soc., Ser. B **34**, 148–186.

LEAMER, E. E. (1973): Multicollinearity: A Bayesian interpretation. Rev. Econ. Statist. **55**, 371–381.

LEAMER, E. E. (1978): Specification Searches. J. Wiley & Sons, New York.

LEAMER, E. E. (1982): Sets of posterior means with bounded variance priors. Econometrica **50** (3), 725–736.

LEAMER, E. E./CHAMBERLAIN, G. (1976): A Bayesian interpretation of pretesting. J. Roy. Statist. Soc., Ser. B **38**, 85–94.

LE CAM, L. (1955): An extension of Wald's theory of statistical decision functions. Ann. Math. Stat. **26**, 69–81.

LEHMANN, E. L. (1983): Theory of Point Estimation. J. Wiley & Sons, New York.

LEMPERS, F. B. (1971): Posterior Probabilities of Alternative Linear Models – Some Theoretical Considerations and Empirical Experiments. University press, Rotterdam.

LEMPERS, F. B./LOUTER, A. S. (1971): A note on an extension of the table of the Student distribution. Report 7101 of the Econometric Institute, Netherlands School of Economics.

LEONARD, T. (1975): A Bayesian approach to the linear model with unequal variances. Technometrics **17**, 95–102.

LIEW, C. K. (1976): Inequality constrained least squares estimation. J. Amer. Statist. Assoc. **71**, 746–751.

LINDLEY, D. V. (1956): On a measure of the information provided by an experiment. Ann. Math. Statist. **27**, 986–1005.

LINDLEY, D. V. (1961): The use of prior probability distributions in statistical inference and decisions. Proc. Fourth Berkeley Symp. **1**, Berkeley – Los Angeles, 453–468.

LINDLEY, D. V. (1968): The choice of variables in multiple regression. J. Roy. Statist. Soc., Ser. B **30**, 31–66.

LINDLEY, D. V. (1971): The estimation of many parameters. In: Foundations of Statistical Inference (Eds.: V. P. GODAMBE and D. A. SPROTT). Toronto, 435–455.

LINDLEY, D. V./SMITH, A. F. M. (1972): Bayes estimates for the linear model. J. Roy. Statist. Soc. Ser. B **34**, 1–41.

LINDLEY, D. V./TVERSKY, A./BROWN, R. V. (1979): On reconciliation of probability assessments. J. Roy. Statist. Soc., Ser. A **142**, 146–180.

LOWERRE, J. M. (1974): On the mean square error of parameter estimates for some biased estimators. Technometrics **16**, 461–464.

MALLOWS, C. L. (1959): The information in an experiment. J. Roy. Statist. Soc., Ser. B **21**, 67–72.

MALLOWS, C. L. (1973): Some comments on C_p. Technometrics **15**, 661–675.

MARAZZI, A. (1980): Robust Bayesian estimation for the linear model. Research Report No. 27, Fachgruppe für Statistik, ETH Zürich.

MARAZZI, A. (1985): On constrained minimization of the Bayes risk for the linear model. Statistics & Decisions **3**, 277–296.

MARCUS, M. B./SACKS, J. (1977): Robust designs for regression problems. In: Statistical Decision Theory and Related Topics II (Eds.: S. S. GUPTA and D. S. MOORE). Academic Press, New York, 245–268.

MARITZ, J. S. (1970): Empirical Bayes Methods. Methuen and Co., London.

MARQUARDT, D. W. (1970): Generalized inverses, ridge regression, biased linear estimation and nonlinear estimation. Technometrics **12**, 591–612.

MARTZ, H. F./KRUTCHKOFF, R. G. (1969): Empirical Bayes estimators in a multiple regression model. Biometrika **56**, 367–374.

MASON, R. L. (1986): Latent root regression: a biased regression methodology for use with collinear predictor variables. Commun. Statist. A – Theory Methods **15**, 2651–2674.

MASRELIEZ, C. J./MARTIN, R. D. (1977): Robust Bayesian estimation for the linear model and robustifying the Kalman filter. IEEE Transact. Autom. Control, **AC-22**, 361–371.

MATHEW, T. (1985): Admissible linear estimation in singular linear models with respect to a restricted parameter set. Commun. Statist. A – Theory Methods **14**, 491–498.

MATHEW, T./RAO, C. R./SINHA, B. K. (1984): Admissible linear estimation in singular linear models. Commun. Statist. A – Theory Methods **13**, 3033–3045.

MAYER, L. S./HENDRICKSON, A. D. (1973): A method for constructing an optimal regression design after an initial set of input values has been selected. Commun. Statist. A – Theory Methods **2** (5), 465–477.

MAYER, L. S./WILLKE, T. A. (1973): On biased estimation in linear models. Technometrics **15**, 497–503.

MAYER, L. S./SINGH, J./WILLKE, T. A. (1974): Utilizing initial estimates in estimating the coefficients in a linear model. J. Amer. Statist. Assoc. **69**, 219–222.

MCELROY, F. W. (1976): Optimality of least squares in linear models with unknown error covariance matrix. J. Amer. Statist. Assoc. **71**, 374–377.

MEHTA, J. S./SWAMY, P. A. V. B. (1974): The exact finite sample distribution of Theil's compatibility test statistic and its application. J. Amer. Statist. Assoc. **69**, 154–158.

MELKMANN, A. A./MICHELLI, C. A. (1979): Optimal estimation of linear operators in Hilbert spaces from inaccurate data. SIAM J. Numer. Anal. **16**, 87–105.

MENGES, G. (1966): On the Bayesification of the minimax principle. Unternehmensforschung **10**, 81–91.

MILES, R. E. (1974): A synopsis of "Poisson flats in Euclidean spaces". In: Stochastic Geometry (Eds.: E. F. HARDING and D. G. KENDALL), J. Wiley & Sons, London.

MOCKUS, J. (1983): The Bayesian approach to global optimization. Mathematical Modelling in Immunology and Medicine, Proc. IFIP TC7 Working Conf., Moscow, 187–195.

MOCKUS, J. (1989): Bayesian Approach to Global Optimization. Theory and Applications. Kluwer Academic Publ., Dordrecht.

MOROZOV, V. A. (1984): Methods for Solving Incorrectly Posed Problems. Springer-Verlag, New York – Berlin.

MORRIS, C. (1983): Parametric empirical Bayes inference: Theory and Applications (with discussion). J. Amer. Statist. Assoc. **78**, 47–65.

MOUCHART, M./SIMAR, L. (1984): A note on least squares approximation in the Bayesian analysis of regression models. J. Roy. Statist. Soc. Ser. B **46**, 124–133.

NÄTHER, W. (1974): Optimale Versuchsplanung im linearen Regressionsmodell im Hinblick auf die Optimalität der Parameterschätzungen. Thesis, Bergakademie Freiberg.

NÄTHER, W. (1975): Semi-orderings between distribution functions and their application to robustness of parameter estimators. Math. Operationsforsch. Statist. **6**, 179–188.

NÄTHER, W. (1984): Bayes estimation of the trend parameter in random fields. Math. Operationsforsch. Stat., Ser. Statistics **15**, 553–558.

NÄTHER, W. (1985): Effective observation of random fields. TEUBNER-TEXTE zur Mathematik **72**. Teubner-Verlag, Leipzig.

NÄTHER, W./PILZ, J. (1980): Estimation and experimental design in a linear regression model using prior information. Zastosowania Matematyki **16**, 23–35.

NAYLOR, J. C./SMITH, A. F. M. (1982): Applications of a method for the efficient computation of posterior distributions. Applied Statistics **31**, 214–225.

NOVICK, M. R. (1969): Multiparameter Bayesian indifference procedures. J. Roy. Statist. Soc., Ser. B **31**, 29–51.

NUSSBAUM, M. (1985): Spline smoothing in regression models and asymptotic efficiency in L_2. Ann. Statist. **13**, 984–997.

OBENCHAIN, R. L. (1978): Good and optimal ridge estimators. Ann. Statist. **6**, 1111–1121.

O'HAGAN, A. (1973): Bayes estimation of a convex quadratic. Biometrika **60**, 565–571.

O'HAGAN, A. (1978): Curve fitting and optimal design for prediction (with discussion). J. Roy. Statist. Soc., Ser. B **40**, 1–42.

O'HAGAN, A. (1979): On outlier rejection phenomena in Bayes inference. J. Roy. Statist. Soc., Ser. B **41**, 358–367.

OL'MAN, V. (1983): Estimation of linear regression coefficients as an antagonistic game (Russian). Izv. Akad. Nauk Est. SSR **32**, 241–245.

OMAN, S. D. (1978): A Bayesian comparison of some estimators used in linear regression with multicollinear data. Commun. Statist. A – Theory Methods **7**, 517–534.

OMAN, S. D. (1982): Contracting towards subspaces when estimating the mean of a multivariate normal distribution. J. Multivar. Anal. **12**, 270–290.

OMAN, S. D. (1983): Regression estimation for a bounded response over a bounded region. Technometrics **25**, 251–261.

OMAN, S. D. (1984): A different empirical Bayes interpretation of ridge and Stein estimators. J. Roy. Statist. Soc., Ser. B **46**, 544–557

OMAN, S. D. (1985): Specifying a prior distribution in structured regression problems. J. Amer. Statist. Assoc. **80**, 190–195.

OWEN, R. J. (1970): The optimum design of a two-factor experiment using prior information. Ann. Math. Statist. **41**, 1917–1934.

PÁZMAN, A. (1974): The ordering of experimental designs–a Hilbert space approach. Kybernetika **10**, 373–388.

PÁZMAN, A. (1986): Foundations of Optimum Experimental Design. D. Reidel Publ. Co., Dordrecht.

PECK, J. E. L./DULMAGE, A. L. (1957): Games on a compact set. Canad. J. Math. **9**, 450–458.

PEELE, L./RYAN, T. P. (1982): Minimax linear regression estimators with application to ridge regression. Technometrics **24**, 157–159.

PETERSEN, I. (1973): Linear minimax estimation of inadequate models in L^2-metric. Math. Operationsforsch. Statist. **4**, 463–471.

PETTIT, L. I./SMITH, A. F. M. (1985): Outliers and influential observations in linear models. In: Bayesian Statistics 2 (Eds.: J. M. BERNARDO et al.). North-Holland Publ. Co., Amsterdam.

PILZ, J. (1977): Bayessche Schätzung und Versuchsplanung im linearen Regressionsmodell. Thesis, Bergakademie Freiberg.

PILZ, J. (1979 a): Entscheidungstheoretische Darstellung des Problems der bayesschen Schätzung und Versuchsplanung im linearen Regressionsmodell. Freiberger Forschungshefte D **117**, Deutscher Verlag für Grundstoffindustrie, Leipzig, 7–20.

PILZ, J. (1979 b): Das bayessche Schätzproblem im linearen Regressionsmodell. Freiberger Forschungshefte **D 117**, Deutscher Verlag für Grundstoffindustrie, Leipzig, 21–55.

PILZ, J. (1979 c): Optimalitätskriterien, Zulässigkeit und Vollständigkeit im Planungsproblem für eine bayessche Schätzung im linearen Regressionsmodell. Freiberger Forschungshefte **D 117**, Deutscher Verlag für Grundstoffindustrie, Leipzig, 67–94.

PILZ, J. (1979 d): Konstruktion von optimalen diskreten Versuchsplänen für eine Bayes-Schätzung im linearen Regressionsmodell. Freiberger Forschungshefte **D 117**, Deutscher Verlag für Grundstoffindustrie, Leipzig, 123–152.

PILZ, J. (1981 a): Robust Bayes and minimax-Bayes estimation and design in linear regression. Math. Operationsforsch. Statist., Ser. Statistics **12**, 163–177.

PILZ, J. (1981 b): Bayesian one-point designs in linear regression. Paper presented at the Fifth International Summer School on Problems of Model Choice and Parameter Estimation in Regression Analysis, March 81, Sellin.

PILZ, J. (1981 c): Bayessche Schätzung und Versuchsplanung für die multiple lineare Regression. IX. Internat. Kongreß über Anwendungen der Mathematik in den Ingenieurwissenschaften Weimar 1981, Vol. **4**, 54–57.

PILZ, J. (1983): Bayesian Estimation and Experimental Design in Linear Regression Models. TEUBNER-TEXTE zur Mathematik **55**, Teubner-Verlag, Leipzig.

PILZ, J. (1984): Robust Bayes regression estimation under weak prior knowledge. In: Robustness of Statistical Methods and Nonparametric Statistics (Eds.: D. RASCH and M. L. TIKU), Deutscher Verlag der Wissenschaften, Berlin, 85–89.

PILZ, J. (1985 a): Minimax straight line regression estimation using prior bounds for the parameters. Freiberger Forschungshefte **D 170**, Deutscher Verlag für Grundstoffindustrie, Leipzig, 33–48.

PILZ, J. (1986 a): Minimax linear regression estimation with symmetric parameter restrictions. J. Statist. Plann. Infer. **13**, 297–318.

PILZ, J. (1986 b): A note on Krafft's maximin linear estimator for linear regression parameters. Statistics **17**, 9–14.

PILZ, J. (1987): Beiträge zur Theorie der Bayes- und minimax-linearen Schätzungen in linearen Regressionsmodellen. Dissertation B (D. Sc. Dissertation), Bergakademie Freiberg.

PILZ, J. (1988 a): Minimax linear approgression estimation for finite dimensional classes of regression functions. In: Transactions 10th Prague Conf. on Inform. Theory, Statist. Dec. Functions, Random Proc. Academia, Prague, 237–246.

PILZ, J. (1988 b): Admissible and minimax linear estimation in linear models with compact parameter region. Schwerpunktprogramm der DFG, Univ. Augsburg, Report No. 100.

PILZ, J. (1988 c): Minimax estimation with ellipsoid and linear inequality constraints. Schwerpunktprogramm der DFG, Univ. Augsburg, Report No. 110.

PILZ, J. (1990): Minimax linear estimation for fitting biased response surfaces. Statistics & Decisions **8**, 47–60.

POLI, I. (1985): A Bayesian non-parametric estimate for multivariate regression. J. Econometrics **28**, 171–182.

PRESS, S. J./SCOTT, A. (1974): Missing variable in Bayesian regression. In: Studies in Bayesian Econometrics and Statistics (Eds.: S. E. FIENBERG and A. ZELLNER). North-Holland Publ. Co., Amsterdam, 259–272.

PUKELSHEIM, F. (1980): On linear regression designs which maximize information. J. Statist. Plann. Infer. **4**, 339–364.

PUKELSHEIM, F./TITTERINGTON, D. M. (1983): General differential and Lagrangian theory for optimal experimental design. Ann. Statist. **11** (1983), 1060–1068.

PURI, M. L./RALESCU, D. A./RALESCU, S. S. (1984): Linear minimax estimators for estimating a function of the parameter. Austr. J. Stat. **26**, 277–283.

RADNER, R. (1959): Minimax estimation for linear regression. Ann. Math. Statist. **30**, 1244–1260.

RAIFFA, H./SCHLAIFER, R. (1961): Applied Statistical Decision Theory. Harvard Univ., Boston.

RAIFFA, H. (1968): Decision Analysis. Addison-Wesley, Massachusetts.

RAO, C. R. (1965): The theory of least squares when the parameters are stochastic and its application to the analysis of growth curves. Biometrika **52**, 447–458.

RAO, C. R. (1973): Linear Statistical Inference and its Applications. Second ed., J. Wiley & Sons, New York.

RAO, C. R. (1975): Simultaneous estimation of parameters in different linear models and applications to biometric problems. Biometrics **31**, 545–554.

RAO, C. R. (1976 a): Characterization of prior distributions and solution to a compound decision problem. Ann. Statist. **4**, 823–835.

RAO, C. R. (1976 b): Estimation of parameters in a linear model. Ann. Statist. **4**, 1023–1037.

RAO, C. R./MITRA, S. K. (1971): Generalized Inverse of Matrices and its Applications. J. Wiley & Sons, New York.

RASCH, D./HERRENDÖRFER, G. (1982): Statistische Versuchsplanung. Deutscher Verlag der Wissenschaften, Berlin.

RIOS, S./GIRON, F. J. (1980): Quasi-Bayesian behaviour: a more realistic approach to decision making? In: Bayesian Statistics (Eds.: J. M. BERNARDO, M. H. DE GROOT, D. V. LINDLEY and A. F. M. SMITH). University Press, Valencia.

ROBBINS, H. (1964): The empirical Bayes approach to statistical decision problems. Ann. Math. Statist. **35**, 1–20.

ROTHENBERG, T. J. (1973): Efficient estimation with a priori information. Yale Univ., Monograph **23**, New Haven, Conn.-London.

SAVAGE, L. J. (1954): The Foundations of Statistics. J. Wiley & Sons, New York, London.

SACKS, J./YLVISAKER, D. (1978): Linear estimation for approximately linear models. Ann. Statist. **6**, 1122–1137.

SACKS, J./YLVISAKER, D. (1984): Some model robust designs in regression. Ann. Statist. **12**, 1324–1348.

SACKS, J./YLVISAKER, D. (1985): Model robust design in regression: Bayes theory. In: Proc. Berkeley Conf. in Honor of J. Neyman and J. Kiefer, Vol. II (Eds.: L. M. LE CAM and R. A. OLSHEN). Wadsworth, Inc., 667–679.

SACKS, J./STRAWDERMAN, W. (1982): Improvements on linear minimax estimates. In: Statistical Decision Theory and Related Topics III (Eds.: S. S. GUPTA and J. O. BERGER). Academic Press, New York, 287–304.

SAN MARTINI, A./SPEZZAFERRI, F. (1984): A predictive model selection criterion. J. Roy. Statist. Soc., Ser. B **46**, 296–303.

SANTALÓ, L. A. (1976): Integral geometry and geometric probability. Reading Mass., Addison Wesley, London.

SCHIPP, B./STAHLECKER, P./TRENKLER, G. (1985): Minimax estimation with additional linear restrictions: a simulation study. Forschungsbericht Nr. 85/1, Abteilg. Statistik, Univ. Dortmund.

SCHMIDT, P. (1981): Constraints on the parameters in simultaneous tobit and probit models. In: Structural Analysis of Discrete Data with Econometric Applications (Eds.: C. F. MANSKI and D. MCFADDEN). MIT Press, Cambridge.

SCLOVE, S. (1968): Improved estimators for coefficients in linear regression. J. Amer. Statist. Assoc. **63**, 596–606.

SCLOVE, S./MORRIS, C./RADHAKRISHNAN, R. (1972): Non-optimality of preliminary test estimators for the mean of a multivariate normal distribution. Ann. Math. Statist. **43**, 1481–1490.

SHINOZAKI, N. (1975): A study of generalized inverse of matrix and estimation with quadratic loss. Ph. D. Thesis, Keio University, Japan.

SILVEY, S. D. (1969): Multicollinearity and imprecise estimation. J. Roy. Statist. Soc., Ser. B **31**, 539–552.

SILVEY, S. D. (1980): Optimal Design. Chapman and Hall, London and New York.

SILVEY, S. D./TITTERINGTON, D. M. (1973): A geometric approach to optimal design theory. Biometrika **60**, 21–32.

SIMAR, L. (1984): A survey of Bayesian approaches to nonparametric statistics. Statistics **15**, 121–142.

SIMEONE, B./VERDINELLI, I. (1988): A feasible directions method for computing Bayes *E*-optimal block designs. Comput. Statistics & Data Analysis **7**, 23–38.

SINHA, B. K. (1970): A Bayesian approach to optimum allocation in regression problems. Calcutta Statistical Assoc. Bulletin **9**, 45–52.

SION, M. (1958): On general minimax theorems. Pacific J. Math. **8**, 171–176.

SMITH, A. F. M. (1973): A general Bayesian linear model. J. Roy. Statist. Soc., Ser. B **35**, 67–75.

SMITH, A. F. M. (1977): A Bayesian analysis of some time-varying models. In: Recent Develop. Statist., Amsterdam, 257–267.

SMITH, A. F. M. (1983): Bayesian approaches to outliers and robustness. In: Specifying Statistical Models (Eds.: FLORENS et al.). Lecture Notes in Statistics **16**, Springer-Verlag, New York.

SMITH, A. F. M./SPIEGELHALTER, D. J. (1980): Bayes factors and choice criteria for linear models. J. Roy. Statist. Soc., Ser. B **42**, 213–220.

SMITH, A. F. M./VERDINELLI, I. (1980): A note on Bayes designs for inference using a hierarchical linear model. Biometrika **67**, 613–619.

SMITH, G./CAMPBELL, F. (1980): A critique of some ridge regression methods (with discussion). J. Amer. Statist. Assoc. **75**, 74–103.

SMITH, J. Q. (1979): A generalization of the Bayesian steady forecasting model. J. Roy. Statist. Soc. Ser. B **41**, 375–387.

SOLOMON, D. L. (1972): Λ-minimax estimation of a multivariate location parameter. J. Amer. Statist. Assoc. **67**, 641–646.

SPENCE, E. (1967): A new class of Hadamard matrices. Glasgow Math. J. **8**, 59–62.

SPJØTVOLL, E. (1977): Random coefficient regression models. A review. Math. Operationsforsch. Statist., Ser. Statistics **8**, 69–93.

SPRUILL, C. (1984): Optimal designs for minimax extrapolation. J. Multivar. Anal. **15**, 52–62.

STAHLECKER, P. (1987): A priori Information und Minimax-Schätzung im linearen Regressionsmodell. Mathematical Systems in Economics **108**, Athenäum Verlag, Frankfurt am Main.

STAHLECKER, P./LAUTERBACH, J. (1987): Approximate minimax estimation in linear regression: Theoretical Results. Commun. Statist. A – Theory Methods **16**, 1101–1116.

STAHLECKER, P./TRENKLER, G. (1987): Quasi minimax estimation in the linear regression model. Statistics **18**, 219–226.

STEIN, C. (1956): Inadmissibility of the usual estimator for the mean of a multivariate normal distribution. Proc. Third Berkeley Symp. **1**, Berkeley – Los Angeles, 197–206.

STEIN, C. (1960): Multiple regression. In: Contributions to Probability and Statistics (Ed.: I. OLKIN). Univ. Press, Stanford.

STEIN, C. (1965): Approximation of improper prior measures by prior probability measures. In: Bernoulli, Bayes, Laplace Anniversary (Eds.: J. NEYMAN and L. LE CAM). Springer-Verlag, Berlin, 217–240.

STEIN, C. (1981): Estimation of the mean of a multivariate normal distribution. Ann. Statist. **9**, 1135–1151.

STEINBERG, D. M. (1985): Model robust response surface designs: Scaling two-level factorials. Biometrika **72**, 513–526.

STOER, J./WITZGALL, C. (1970): Convexity and Optimization in Finite Dimensions I. Springer-Verlag, Berlin.

STONE, M. (1959): Application of a measure of information to the design and comparison of regression experiments. Ann. Math. Statist. **30**, 55–70.

STONE, M. (1967): Generalized Bayes decision functions, admissibility and the exponential family. Ann. Math. Statist. **38**, 618–622.

STRAWDERMAN, W. E. (1971): Proper Bayes minimax estimators of the multivariate normal mean. Ann. Math. Statist. **42**, 385–388.

STRAWDERMAN, W. E. (1973): Proper Bayes minimax estimators of the multivariate normal mean vector for the case of common unknown variances. Ann. Statist. **1**, 1189–1194.

STRAWDERMAN, W. E. (1978): Minimax adaptive generalized ridge regression estimators. J. Amer. Statist. Assoc. **73**, 623–627.

SUZUKI, Y. (1964): On the use of some extraneous information in the estimation of the coefficients of regression. Ann. Inst. Statist. Math. **16**, 161–173.

SUZUKI, Y. (1980): A Bayesian approach to some empirical Bayes models. In: Recent Developments in Statistical Inference and Data Analysis (Proc. Internat. Conf., Inst. Statist. Math. Tokyo 1979). North-Holland Publ. Co. Amsterdam, 269–286.

SWAMY, P. A. V. B. (1971): Statistical inference in random coefficient regression models. Lecture Notes in Economics **55**, Springer-Verlag, Berlin.

SWAMY, P. A. V. B./MEHTA, J. S. (1975): Bayesian and non-Bayesian analysis of switching regressions and of random coefficient regression models. J. Amer. Statist. Assoc. **70**, 593–602.

Swamy, P. A. V. B./Mehta, J. S. (1976): Minimum average risk estimators for coefficients in linear models. Commun. Statist. A – Theory Methods **5**, 803–818.

Swindel, B. F. (1976): Good ridge estimators based on prior information. Commun. Statist. A – Theory Methods **5**, 1065–1075.

Szegö, G. (1959): Orthogonal polynomials. Amer. Math. Soc. Colloquium Publications, New York.

Teräsvirta, T. (1983): Restricted superiority of linear homogeneous estimators over ordinary least squares. Scand. J. Statist. **10**, 27–33.

Teräsvirta, T. (1986): Superiority comparisons of heterogeneous linear estimators. Commun. Statist. A – Theory Methods **15**, 1319–1336.

Teräsvirta, T. (1987): Incomplete ellipsoidal restrictions in linear models. Technical Report. Research Institute of the Finnish Economy.

Theil, H. (1963): On the use of incomplete prior information in regression analysis. J. Amer. Statist. Assoc. **58**, 401–414.

Theobald, C. M. (1974): Generalizations of mean square error applied to ridge regression. J. Roy. Statist. Soc., Ser. B **36**, 103–106.

Theobald, C. M. (1975): An inequality with application to multivariate analysis. Biometrika **62**, 461–466.

Tiao, G. C./Zellner, A. (1964): Bayes's theorem and the use of prior knowledge in regression analysis. Biometrika **51**, 219–230.

Tierney, L./Kadane, J. B. (1986): Accurate approximations for posterior moments and marginal densities. J. Amer. Statist. Assoc. **81**, 82–86.

Titterington, D. M. (1975): Optimal design: Some geometrical aspects of *D*-optimality. Biometrika **62**, 313–320.

Titterington, D. M. (1980 a): Aspects of optimal design in dynamic systems. Technometrics **22**, 287–299.

Titterington, D. M. (1980 b): Geometric approaches to design of experiment. Math. Operationsforsch. Statist., Ser. Statistics **11**, 151–163.

Titterington, D. M. (1985): Common structure of smoothing techniques in statistics. Internat. Statist. Review **53**, 141–170.

Toutenburg, H. (1975): Vorhersage in linearen Modellen. Akademie-Verlag, Berlin.

Toutenburg, H. (1976): Minimax-linear and MSE-estimators in generalized regression. Biometrical J. **18**, 91–100.

Toutenburg, H. (1980): On the combination of equality and inequality restrictions on regression coefficients. Biometrical J. **22**, 271–274.

Toutenburg, H. (1982): Prior Information in Linear Models. J. Wiley & Sons, Chichester.

Toutenburg, H. (1984): Minimax-linear estimation under incorrect prior information. In: Robustness of Statistical Methods and Nonparametric Statistics (Eds.: D. Rasch and M. L. Tiku). Deutscher Verlag der Wissenschaften, Berlin, 156–158.

Toutenburg, H./Roeder, B. (1978): Minimax-linear and Theil estimator for restrained regression coefficients. Math. Operationsforsch. Statist., Ser. Statist. **9**, 499–505.

Trader, R. L. (1983): A Bayesian predictive approach to the selection of variables in multiple regression. Commun. Statist. A – Theory Methods **12**, 1553–1567.

Trenkler, D. (1986): Verallgemeinerte Ridge Regression. Verlag Anton Hain Meisenheim, Frankfurt am Main.

Trenkler, G. (1981): Biased Estimators in the Linear Regression Model. Math. Systems in Economics, Vol. **58**, Oelgeschlager, Gunn & Hain, Cambridge, MA.

Trenkler, G./Trenkler, D. (1983): A note on superiority comparisons of homogeneous linear estimators. Commun. Statist. A – Theory Methods **12** (7), 799–808.

Ullah, A./Vinod, H. D. (1984): Improvement ranges for shrinkage estimators with stochastic target. Commun. Statist. A – Theory Methods **13**, 207–215.

Van Der Linde, A. (1985): Interpolation of regression functions in reproducing kernel Hilbert spaces. Statistics **16**, 351–363.

Viertl, R. (Ed.) (1987): Probability and Bayesian Statistics. Plenum, New York.

Villegas, C. (1971): On Haar priors. In: Foundations of Statistical Inference (Eds.: V. P. Godambe and D. A. Sprott). Holt, Rinehart and Winston, Toronto, Montreal, 409–414.

VINOD, H. D. (1978): A survey of ridge regression and related techniques for improvements over ordinary least squares. The Review of Economics and Statistics **60** (1), 121–131.

VUCHKOV, I. N. (1977): A ridge-type procedure for design of experiments. Biometrika **64**, 147–150.

WAHBA, G. (1977): A survey on some smoothing problems and the method of generalized cross-validation for solving them. In: Applications of Statistics (Ed.: P. R. KRISHNAIAH), North-Holland Publ. Co., Amsterdam.

WALD, A. (1950): Statistical Decision Functions. J. Wiley & Sons, New York.

WATSON, S. R. (1974): On Bayesian inference with incompletely specified prior distribution. Biometrika **61**, 193–196.

WEERHANDI, S./ZIDEK, J. V. (1981): Multi-Bayesian statistical decision theory. J. Roy. Statist. Soc., Ser. A **144**, 85–93.

WELCH, W. J. (1982): Branch-and-bound search for experimental designs based on D-optimality and other criteria. Technometrics **24**, 41–48.

WELCH, W. J. (1983): A mean squared error criterion for the design of experiments. Biometrika **70**, 205–213.

WEST, M. (1984): Bayesian aggregation. J. Roy. Statist. Soc., Ser. A **147**, 600–607.

WEST, M./HARRISON, J./MIGON, H. S. (1985): Dynamic generalized linear models and Bayesian forecasting. J. Amer. Statist. Assoc. **80**, 1–25.

WETHERILL, G. B. (1961): Bayesian sequential analysis. Biometrika **48**, 281–292.

WHITTLE, P. (1973): Some general points in the theory of optimal experimental design. J. Roy. Statist. Soc., Ser. B **35**, 123–130.

WIND, S. L. (1973): An empirical Bayes approach to multiple linear regression. Ann. Statist. **1**, 93–103.

WINKLER, R. L. (1967): The assessment of prior distributions in Bayesian analysis. J. Amer. Statist. Assoc. **62**, 776–800.

WINKLER, R. L. (1980): Prior information, predictive distribution and Bayesian model-building. In: Bayesian Analysis in Econometrics and Statistics (Ed.: A. ZELLNER), North-Holland Publ. Co., Amsterdam, 95–109.

WU, C. F. (1978): Some algorithmic aspects of the theory of optimal designs. Ann. Statist. **6**, 1286–1301.

WU, C. F./WYNN, H. P. (1978): The convergence of general step-length algorithms for regular optimum design criteria. Ann. Statist. **6**, 1273–1285.

WYNN, H. P. (1972): Results in the theory and construction of D-optimum experimental designs. J. Roy. Statist. Soc., Ser. B **34**, 133–147.

WYNN, H. P. (1975): Simple conditions for optimum design algorithms. In: A survey of statistical design and linear models (Ed.: J. N. SRIVASTAVA), North-Holland Publ. Co., Amsterdam.

YANCEY, T. A./JUDGE, G. G. (1976): A Monte Carlo comparison of traditional and Stein rule estimates under squared error loss. J. Econometrics **4**, 285–294.

YLVISAKER, D. (1987): Prediction and design. Ann. Statist. **15**, 1–15.

YU, P. M. V./BOULLION, T. L./WATKINS, T. A. (1978): The use of prior information in linear regression. Commun. Statist. A – Theory Methods **7** (1), 81–95.

ZACKS, S. (1977): Problems and approaches in design of experiments for estimation and testing in non-linear models. In: Multivariate Analysis, Vol. IV (Ed.: P. R. KRISHNAIAH). North-Holland Publ. Co., Amsterdam, 209–223.

ZARROP, M. B. (1979): Optimal Experimental Design for Dynamic System Identification. Lecture Notes in Control and Information Science **21**, Springer-Verlag, New York.

ZELLNER, A. (1971): An Introduction to Bayesian Inference in Econometrics. J. Wiley & Sons, New York.

ZELLNER, A./CHETTY, K. V. (1965): Prediction and decision problems in regression models from the Bayesian point of view. J. Amer. Statist. Assoc. **60**, 608–616.

ZELLNER, A./VANDAELE, W. (1975): Bayes-Stein estimators for k means, regression and simultaneous equation models. In: Studies in Bayesian Econometrics and Statistics (Eds.: S. E. FIENBERG and A. ZELLNER). North-Holland Publ. Co., Amsterdam.

ZELLNER, A. (1976): Bayesian and non-Bayesian analysis of the regression model with multivariate Student-t error terms. J. Amer. Statist. Assoc. **71**, 400–405.

Glossary

This glossary presents a description of the main notational conventions, abbreviations and symbols adopted throughout the text.

Notational conventions

$\mathbb{N}$, $\mathbb{R}^+$	Set of positive integers and of positive real numbers, respectively
$\mathbb{R}^r$	r-dimensional Euclidean space
$\mathcal{M}_{n \times r}$	Set of real matrices of type $n \times r$
$\mathcal{M}_r^>$, $\mathcal{M}_r^\geqq$	Set of positive definite and positive semidefinite matrices of order r, respectively
rank, det, tr	Rank, determinant and trace (of a matrix), respectively
$\boldsymbol{A}^\top$, $\boldsymbol{x}^\top$	Transpose of the matrix $\boldsymbol{A}$ or vector $\boldsymbol{x}$
$\boldsymbol{A}^+$	Moore-Penrose-inverse of the matrix $\boldsymbol{A}$
$\boldsymbol{A}^{1/2}$	Symmetric square root of $\boldsymbol{A} \in \mathcal{M}_r^\geqq$
$\boldsymbol{A} < \boldsymbol{B}$, $\boldsymbol{A} \leqq \boldsymbol{B}$	$\boldsymbol{B} - \boldsymbol{A}$ is positive definite and positive semidefinite, respectively
$\lVert\boldsymbol{x}\rVert$, $\lVert\boldsymbol{x}\rVert_A$	Norm and seminorm, respectively ($\lVert\boldsymbol{x}\rVert^2 = \boldsymbol{x}^\top\boldsymbol{x}$, $\lVert\boldsymbol{x}\rVert_A^2 = \boldsymbol{x}^\top\boldsymbol{A}\boldsymbol{x}$)
$\lambda_i(\boldsymbol{A})$	Eigenvalues of $\boldsymbol{A}$
$\lambda_{\min}(\boldsymbol{A})$, $\lambda_{\max}(\boldsymbol{A})$	Minimum and maximum eigenvalue of $\boldsymbol{A}$, respectively
card S	Cardinality of the set S
$\tilde{\boldsymbol{x}}$	Random vector $\tilde{x}$ (randomness is signified by a tilde)
$P_{\tilde{x}}$, $P_{\tilde{x}\mid\cdot}$	Unconditional and conditional probability distribution of $\tilde{x}$, respectively
$\mathbf{E}_{\tilde{x}}$, $\mathbf{E}_{\tilde{x}\mid\cdot}$	Expectation operator with respect to $P_{\tilde{x}}$ and $P_{\tilde{x}\mid\cdot}$, respectively
Cov $\tilde{x}$, **Cov** $(\tilde{x}\mid\cdot)$	Covariance matrix of $\tilde{x}$ with respect to $P_{\tilde{x}}$ and $P_{\tilde{x}\mid\cdot}$, respectively
Var $\tilde{x}$	Variance of the random variable $\tilde{x}$
supp $P_{\tilde{x}}$	Support of $P_{\tilde{x}}$
$\mathrm{G}(\alpha, \beta)$	Gamma distribution with parameters α, β (see A.1)
$\mathrm{N}(\boldsymbol{\mu}, \boldsymbol{\Phi})$	Normal distribution with expectation $\boldsymbol{\mu}$ and covariance matrix $\boldsymbol{\Phi}$ (see A.2)

$\mathrm{NG}(\boldsymbol{\mu}, \boldsymbol{\Phi}, \alpha, \beta)$	Normal-gamma distribution with parameters $\boldsymbol{\mu}, \boldsymbol{\Phi}, \alpha, \beta$ (see A.3)
$t(n, \boldsymbol{\mu}, \boldsymbol{\Phi})$	Multivariate Student distribution with parameters $n, \boldsymbol{\mu}, \boldsymbol{\Phi}$ (see A.4)
$[Z, D, R]$	Statistical decision problem (see A.10)
$f_1 \propto f_2$	Proportionality of the functions f_1 and f_2

Abbreviations

LSE	(Ordinary) least squares estimator
GLSE	Generalized least squares estimator
MSE	Mean square error

Symbols

Latin symbols

A_{B}	Bayes-A-optimality criterion
C_{B}	Bayes-C-optimality criterion
$C = C(v_n)$	Sample space of the observation vector
D	Class of regression estimators
D_1, D_{a}	Class of linear and affine-linear estimators, respectively
D_{B}	Bayes-D-optimality criterion
$\tilde{\boldsymbol{e}} = \tilde{\boldsymbol{e}}(v_n)$	Error vector
$e_Z(\xi)$	Efficiency of the design ξ with respect to the functional $Z(\cdot)$
E_{B}	Bayes-E-optimality criterion
$\boldsymbol{f} = (f_1, \ldots, f_r)^\top$	Vector of regression functions
$\boldsymbol{F} = \boldsymbol{F}(v_n)$	Design matrix of the exact design v_n
G	Statistical decision problem of regression estimation
$\boldsymbol{I}_r$	Identity matrix of order r
L	Loss function for regression estimation
L_{B}	Bayes-L-optimality criterion
$\boldsymbol{M}(\cdot), \boldsymbol{M}_{\mathrm{B}}(\cdot)$	Information matrix, Bayesian information matrix
$\boldsymbol{M}_P$	Moment matrix of the prior probability distribution P
n	Number of observations
$\boldsymbol{0}_r$	Null vector in $\mathbb{R}^r$
$\boldsymbol{0}_{n,r}$	Matrix of zeroes in $\mathcal{M}_{n \times r}$
R	Risk function for regression estimation
$\boldsymbol{U}$	Weight matrix associated with the (quadratic) loss function
v_n	Exact experimental design of size n
$\boldsymbol{V}$	Covariance matrix of the GLSE
w	Parameter index of the error distribution
W	Index set of parameters w
$\boldsymbol{x} = (x_1, \ldots, x_k)^\top$	Vector of regressor variables
$X_{\mathrm{E}}, X_{\mathrm{P}}$	Experimental region, prediction region
$\tilde{y}(\boldsymbol{x}), \tilde{\boldsymbol{y}}(v_n)$	Response variable, observation vector

$y(\boldsymbol{x}), \boldsymbol{y}$	Realizations of $\tilde{y}(\boldsymbol{x})$ and $\tilde{\boldsymbol{y}}(v_n)$, respectively
$Z(\cdot)$	General design functional (optimality criterion)

Script type and greek symbols

$\mathcal{M}$	Set of second-order moment matrices $\boldsymbol{M}_P$
$\mathcal{P}_{\tilde{e}}, \mathcal{P}_{\tilde{y}}, \mathcal{P}_{\tilde{\theta}}$	Family of probability distributions of $\tilde{\boldsymbol{e}}$, $\tilde{\boldsymbol{y}}$ and $\tilde{\boldsymbol{\vartheta}}$, respectively
δ_x	One-point measure giving weight one to the point $\boldsymbol{x}$
Δ_Z	Frechét directional derivative of $Z(\cdot)$
$\eta(\cdot)$	Response surface
$\hat{\eta}(\cdot, \tilde{\boldsymbol{y}})$	Estimator for $\eta(\cdot)$
$\boldsymbol{\vartheta}$	Regression parameter vector
Θ	Set of regression parameters
Θ^*	Set of all (prior) probability distributions over Θ
$\hat{\boldsymbol{\vartheta}}_{LS}, \hat{\boldsymbol{\vartheta}}_{\Sigma}$	LSE and GLSE for $\boldsymbol{\vartheta}$, respectively
$\hat{\boldsymbol{\vartheta}}_{B}, \hat{\boldsymbol{\vartheta}}_{M}$	Bayes and minimax estimator for $\boldsymbol{\vartheta}$, respectively
$\boldsymbol{\mu}$	Prior expectation (contraction point) of the regression parameter
ξ	Approximate (continuous) experimental design
Ξ	Set of all continuous designs (over X_E)
ϱ	Bayes risk
$\boldsymbol{\Sigma}$	Covariance matrix of the observation vector $\tilde{\boldsymbol{y}}$
$\boldsymbol{\Phi}$	Prior covariance matrix of the regression parameter

Subject Index